Lecture Notes in Computer Science 13881

Advanced Research in Computing and Software Science
Subline of Lecture Notes in Computer Science

More information about this series at https://link.springer.com/bookseries/558

Cezara Dragoi · Michael Emmi ·
Jingbo Wang (Eds.)

Verification, Model Checking, and Abstract Interpretation

24th International Conference, VMCAI 2023
Boston, MA, USA, January 16–17, 2023
Proceedings

 Springer

Editors
Cezara Dragoi
Inria, Amazon Web Services
Courbevoie, France

Michael Emmi
Amazon Web Services
Seattle, WA, USA

Jingbo Wang
University of Southern California
Los Angeles, CA, USA

ISSN 0302-9743 ISSN 1611-3349 (electronic)
Lecture Notes in Computer Science
ISBN 978-3-031-24949-5 ISBN 978-3-031-24950-1 (eBook)
https://doi.org/10.1007/978-3-031-24950-1

This Springer imprint is published by the registered company Springer Nature Switzerland AG
The registered company address is: Gewerbestrasse 11, 6330 Cham, Switzerland

Preface

This volume contains the proceedings of VMCAI 2023, the 24th International Conference on Verification, Model Checking, and Abstract Interpretation. VMCAI 2023 was part of the 50th ACM SIGPLAN Symposium on Principles of Programming Languages (POPL 2023), held at the Boston Park Plaza in Boston, USA, during January 16–17, 2023. VMCAI is a forum for researchers working in verification, model checking, and abstract interpretation. It attempts to facilitate interaction, cross-fertilization, and advancement of methods that combine these and related areas. The topics of the conference include program verification, model checking, abstract interpretation, program synthesis, static analysis, type systems, deductive methods, decision procedures, theorem proving, program certification, debugging techniques, program transformation, optimization, and hybrid and cyber-physical systems.

VMCAI 2023 received a total of 34 submissions. After a rigorous single blind review process, with each paper reviewed by at least three Program Committee (PC) members, followed by an online discussion, the PC accepted 17 papers for publication in the proceedings and presentation at the conference.

In addition to the contributed papers, the conference program included four keynotes: Aws Albarghouthi (University of Wisconson–Madison), Eric Koskinen (Stevens Institute of Technology), Sharon Shoham (Tel Aviv University), and Chao Wang (University of Southern California).

By now, artifact evaluation is a standard part of VMCAI. The artifact evaluation process complements the scientific impact of the conference by encouraging and rewarding the development of tools that allow for replication of scientific results as well as for shared infrastructure across the community. Authors of submitted papers were encouraged to submit an artifact to the VMCAI 2023 artifact evaluation committee (AEC). We also encouraged the authors to make their artifacts publicly and permanently available.

All submitted artifacts were evaluated in parallel with the papers. We assigned three members of the AEC to each artifact and assessed it in two phases. First, the reviewers tested whether the artifacts were working, e.g., there were no corrupted or missing files and the evaluation did not crash on simple examples. For those artifacts that did not work, we sent the issues to the authors, for clarifications. In the second phase, the assessment phase, the reviewers aimed at reproducing any experiments or activities and evaluated the artifact based on the following questions: 1. Is the artifact consistent with the paper and the claims made by the paper? 2. Are the results of the paper replicable through the artifact? 3. Is the artifact available? We awarded a badge for each of these question to each artifact that answered it in a positive way. Of the 14 accepted papers, there were five submitted artifacts with five that passed the second phase and were thus awarded one, two, or all three Artifact Evaluation Badges.

The VMCAI program would not have been possible without the efforts of many people. We thank the research community for submitting their results to VMCAI and for their participation in the conference. The members of the Program Committee, the artifact evaluation committee, and the external reviewers worked tirelessly to select a strong program, offering constructive and helpful feedback to the authors in their reviews. The VMCAI steering committee provided continued encouragement and advice. We warmly thank the keynote speakers for their participation and contributions. We also thank the general chair of POPL 2023, Andrew C. Myers, and the organization team for their support. We thank the publication team at Springer for their support, and EasyChair for providing an excellent conference management system.

December 2022

Cezara Dragoi
Michael Emmi
Jingbo Wang

Organization

Program Co-chairs

Cezara Dragoi   AWS, Inria Paris, and ENS, France
Michael Emmi   Amazon Web Services, USA

Artifact Evaluation Chair

Jingbo Wang   University of Southern California, USA

Program Committee

Mohamed Faouzi Atig   Uppsala University, Sweden
Ahmed Bouajjani   IRIF, Université Paris Cité, France
Bor-Yuh Evan Chang   University of Colorado Boulder and Amazon, USA
Jocelyn Chen   The University of Texas at Austin, USA
Yanju Chen   University of California, Santa Barbara, USA
Deepak D'Souza   Indian Institute of Science, Bangalore, India
Rayna Dimitrova   CISPA Helmholtz Center for Information Security,
                          Germany
Mihály Dobos-Kovács   Budapest University of Technology and Economics,
                          Hungary
Rui Dong   University of Michigan, USA
Cezara Dragoi   AWS, Inria Paris, and ENS, France
Michael Emmi   Amazon Web Services, USA
Constantin Enea   IRIF, Université de Paris, France
Ferhat Erata   Yale University, USA
Jerome Feret   Inria Paris, France
Jean-Christophe Filliatre CNRS, France
Bernd Finkbeiner   CISPA Helmholtz Center for Information Security,
                          Germany
Arie Gurfinkel   University of Waterloo, Canada
Liana Hadarean   Amazon Web Services, USA
Ákos Hajdu   Budapest University of Technology and Economics,
                          Hungary
Shaobo He   Amazon Web Services, USA
Zunchen Huang   University of Southern California, USA
Dejan Jovanović   Amazon Web Services, USA
Joomy Korkut   Princeton University, USA
Burcu Kulahcioglu Ozkan   Delft University of Technology, The Netherlands
Akash Lal   Microsoft, India
Yannan Li   University of Southern California, USA

Anthony Widjaja Lin	TU Kaiserslautern, Germany
Stephan Merz	Inria Nancy, France
Shouvick Mondal	Indian Institute of Technology, Madras, India
Suha Orhun Mutluergil	Koc University, Turkey
Kedar Namjoshi	Nokia Bell Labs, USA
Jorge A. Navas	SRI International, USA
Amirmohammad Nazari	University of Southern California, USA
Gennaro Parlato	University of Molise, Italy
Corina Pasareanu	CMU, NASA, and KBR, USA
Tatjana Petrov	University of Konstanz, Germany
Felipe R. Monteiro	Amazon Web Services, USA
Daniel Schwartz-Narbonne	Amazon Web Services, USA
Subodh Sharma	Indian Institute of Technology, Delhi, India
Mihaela Sighireanu	LMF, ENS Paris-Saclay, Université Paris-Saclay, and CNRS, France
Mandayam Srivas	Chennai Mathematical Institute, India
Abhishek Tiwari	University of Passau, Germany
Jingbo Wang	University of Southern California, USA
Yuhao Zhang	University of Wisconsin-Madison, USA
Zhen Zhang	Utah State University, USA
Yaoda Zhou	The University of Hong Kong, Hong Kong, China
Florian Zuleger	TU Wien, Austria

Additional Reviewers

Aktas, Ethem Utku	Magara, Seyma Selcan
Bajczi, Levente	Majumdar, Rupak
Barbot, Benoit	Mukhopadhyay, Diganta
Bilecen, Ali Enver	P., Habeeb
D'Souza, Meenakshi	Schmitt, Frederik
Duflot, Marie	Schoepe, Daniel
Garbi, Giulio	Stucki, Sandro
Habermehl, Peter	Szekeres, Dániel
La Torre, Salvatore	Vediramana Krishnan, Hari Govind

Contents

Distributing and Parallelizing Non-canonical Loops

Clément Aubert[1]([✉]) [iD], Thomas Rubiano[2], Neea Rusch[1] [iD],
and Thomas Seiller[2,3] [iD]

[1] School of Computer and Cyber Sciences, Augusta University, Augusta, USA
`caubert@augusta.edu`
[2] LIPN - UMR 7030 Université Sorbonne Paris Nord, Villetaneuse, France
[3] CNRS, Paris, France

Abstract. This work leverages an original dependency analysis to parallelize loops regardless of their form in imperative programs. Our algorithm distributes a loop into multiple parallelizable loops, resulting in gains in execution time comparable to state-of-the-art automatic source-to-source code transformers when both are applicable. Our graph-based algorithm is intuitive, language-agnostic, proven correct, and applicable to all types of loops. Importantly, it can be applied even if the loop iteration space is unknown statically or at compile time, or more generally if the loop is not in canonical form or contains loop-carried dependency. As contributions we deliver the computational technique, proof of its preservation of semantic correctness, and experimental results to quantify the expected performance gains. We also show that many comparable tools cannot distribute the loops we optimize, and that our technique can be seamlessly integrated into compiler passes or other automatic parallelization suites.

Keywords: Program transformation · Automatic parallelization · Loop optimization · Abstract interpretation · Program analysis · Dependency analysis

This research is supported by the Transatlantic Research Partnership of the Embassy of France in the United States and the FACE Foundation. Th. Rubiano and Th. Seiller are also supported by the Île-de-France region through the DIM RFSI project "CoHOp". N. Rusch is supported in part by the Augusta University Provost's office, and the Translational Research Program of the Department of Medicine, Medical College of Georgia at Augusta University.

1   Original Approaches to Automatic Parallelization

1.1   The Challenge of Unknown Iteration Space

Loop fission (a.k.a. loop distribution) is an optimization technique that breaks loops into multiple loops, with the same condition or index range, each taking only a part of the original loop's body. Such transformation creates opportunity for parallelization and reduces program's running time. For instance, the loop

```
while(t[i] != j){
    s1[i] = j*j;
    s2[i] = 1/j;
    i++;}
```
would become
```
while(t[i1] != j)
    {s1[i1] = j*j; i1++;}
while(t[i2] != j)
    {s2[i2] = 1/j; i2++;}
```
under

this transformation. In the transformed program, variable i is substituted with two copies, i1 and i2, and we obtain two while loops that can be executed in parallel[1] . The gain, in terms of time, results from the fact that the original loop could only be executed sequentially, while the transformed loops can each be assigned to one core. If we consider similarly structured loops that perform resource-intensive computation or that can be distributed in e.g., 8 loops running on 8 cores, it becomes intuitive how this technique can yield measurable performance gain.

This example straightforwardly captures the idea behind loop fission. Of course, as a loop with a short body, it misses the richness and complexities of realistic software. It is therefore very surprising that all the existing loop fission approaches fail at transforming such an elementary program! The challenge comes from the kind of loop presented. Applying loop fission to "canonical" (Definition 15) loops or loops whose number of iterations can be pre-determined is an established convention. But our example of a non-canonical loop with a (potentially) unknown iteration space cannot be handled by those approaches (Sect. 4).

In this paper we present a loop fission technique that can resolve this limitation, because it can be applied to all kinds of a loops[2]. The technique is applicable to any programming language in the imperative paradigm, lightweight and proven correct. The loop fission technique derives these capabilities from a graph-based dependency analysis, first introduced in our previous work [33]. Now we refine this dependency analysis and explain how it can be leveraged to obtain *loop-level parallelism*: a form of parallelism concerned with extracting parallel tasks from loops. We substantiate our claim of running time improvement by benchmarking our technique in Sect. 5. The results show, in cases where iteration space is unknown, that we obtain gain up to the number of parallelizable loops, and that in other cases the speedup is comparable to alternative techniques.

---

[1] In practice, private copies of i are automatically created by e.g., the standard parallel programming API for C, OpenMP. Its **pragma** directives are illustrated in Fig. 5.

[2] We focus on while loops, but other kinds of loops (for, do...while, foreach) can always be translated into while and general applicability follows.

1.2 Motivations for Correct, Universal and Automatic Parallelization

The increasing need to discover and introduce parallelization potential in programs fuels the demand for loop fission. To leverage the potential speedup available on modern multicore hardware, all programs—including legacy software—should instruct the hardware to take advantage of its available processors.

Existing parallel programming APIs, such as OpenMP [25], PPL [32], and oneTBB [22], facilitate this progression, but several issues remain. For example, classic algorithms are written sequentially without parallelization in mind and require reformatting to fit the parallel paradigm. Suitable sequential programs with opportunity for parallelization must be modified, often manually, by carefully inserting parallelization directives. The state explosion resulting from parallelization makes it impossible to exhaustively test the code running on parallel architectures [12]. These challenges create demand for *correct* automatic parallelization approaches, to transform large bodies of software to semantically equivalent parallel programs.

Compilers offer an ideal integration point for many program analyses and optimizations. Automatic parallelization is already a standard feature in developing industry compilers, optimizing compilers, and specialty source-to-source compilers. Tools that perform local transformations, generally on loops, are frequently conceived as compiler passes. How those passes are intertwined with sequential code optimizations can however be problematic [14]. As an example, OpenMP directives are by default applied early in the compilation and hence the parallelized source code cannot benefit from sequential optimizations such as unrolling. Furthermore, compilers tend to make conservative choices and miss opportunities to parallelize [14,21].

The loop fission technique presented in this paper offers an incremental improvement in this direction. It enables discovery of parallelization potential in previously uncovered cases. In addition, the flexibility of the system makes it suitable to integration and pipelining with existing parallelization tools at various stages of compilation, as discussed in Sect. 6.

1.3 Our Technique: Properties, Benefits and Limitations

Our technique possesses four notable properties, compared to existing techniques:

Suitable to loops with unknown iteration spaces—our method does not require knowing loop iteration space statically nor at compile time, making it applicable to loops which are often ignored.

Loop-agnostic—our method requires practically no structure from the loops: they can be while, do ... while or for loops, have arbitrarily complex update and termination conditions, loop-carried dependencies, and arbitrarily deep loop nests.

Language-agnostic—our method can be used on any imperative language, and without manual annotations, making it flexible and suitable for application and integration with tools and languages ranging from high-level to intermediate representations.

Correct—our method is easy to prove correct and intuitive, largely because it does not apply to loop bodies with pointers or complex function calls.

All the approaches we know of fail in at least one respect. For instance, polyhedral optimizations cannot transform loops with unknown iteration spaces, since they work on static control parts of programs, where all control flow and memory accesses are known at compile time [20, p. 36]. More importantly, all the "popular" [35] automatic tools fail to optimize do...while loops, and require for and while loops to have canonical forms, that generally require the trip count to be known at compilation time. We discuss these alternative approaches in detail in Sect. 4.

The main limitation of our approach is with function calls and memory accesses. Although we can treat loops with pure function calls, we exclude treatment of loops that contain explicit pointer manipulation, pointer arithmetic or certain function calls. We reserve the introduction of these enhancements as future extensions of our technique. In the meantime, and with these limitations in mind, we believe our approach to be a good complement to existing approaches. Polyhedral models [24]—that are also pushing to remove some restrictions [13]—, advanced dependency analyses, or tools developed for very precise cases (such as loop tiling [14]), should be used in conjunction with our technique, as their use cases diverge (Sect. 6).

1.4   Contributions: From Theory to Benchmarks

We deliver a complete perspective on the design and expected real-time efficiency of our loop fission technique, from its theoretical foundations to concrete measurements. We present three main contributions:

1. The loop fission transformation algorithm—Sect. 3.1—that analyzes dependencies of loop condition and body variables, establishes cliques between statements, and splits independent cliques into multiple loops.
2. The correctness proof—Sect. 3.2—that guarantees the semantic preservation of loop transformation.
3. Experimental results [8]—Sect. 5—that evaluate the potential gain of the proposed technique, including loops with unknown iteration spaces, and demonstrates its integrability with existing parallelization frameworks.

But first, we present and illustrate the dependency analysis that enables our loop fission technique.

2   Background: Language and Dependency Analysis

2.1   A Simple While Imperative Language with Parallel Capacities

We use a simple imperative while language, with semantics similar to C, extended with a parallel command, similar to e.g., OpenMP's directives [25], allowing to execute its arguments in parallel[3]. Our language supports arrays but not pointers, and we let for and do...while loops be represented using while loops. It is easy to map to fragments of C, Java, or any other imperative programming language with parallel support.

$$var ::= \texttt{i} \mid \texttt{j} \mid \ldots \mid \texttt{s} \mid \texttt{t} \mid \ldots \mid \texttt{x}_1 \mid \texttt{x}_2 \mid \ldots \mid \texttt{z}_n \mid var[exp] \qquad \text{(Variables)}$$
$$exp ::= var \mid val \mid op(exp, \ldots, exp) \qquad \text{(Expression)}$$
$$com ::= var = exp \mid \texttt{if } exp \texttt{ then } com \texttt{ else } com \mid$$
$$\texttt{while } exp \texttt{ do } com \mid \texttt{use}(var, \ldots, var) \mid \texttt{skip} \mid$$
$$com;com \mid \texttt{parallel}\{com\}\{com\} \cdots \{com\} \qquad \text{(Command)}$$

Fig. 1. A simple imperative while language

The grammar is given Fig. 1. A variable represents either an undetermined "primitive" datatype, e.g., not a reference variable, or an array, whose indices are given by an expression. We generally use s and t for arrays. An expression is either a variable, a value (e.g., integer literal) or the application to expressions of some operator op, which can be e.g., relational (==, <, etc.) or arithmetic (+, -, etc.). We let V (resp. e, C) ranges over variables (resp. expression, command) and W range over while loops. We also use combined assignment operators and write e.g., x++ for x += 1. We assume commands to be correct, e.g., with operators correctly applied to expressions, no out-of-bounds errors, etc.

A program is thus a sequence of statements, each statement being either an *assignment*, a *conditional*, a *while* loop, a *function call*[4] or a *skip*. *Statements* are abstracted into *commands*, which can be a statement, a sequence of commands, or multiple commands to be run in parallel. The semantics of parallel is the following: variables appearing in the arguments are considered local, and the value of a given variable x after execution of the parallel command is the value of the last modified local variable x. This implies possible race conditions, but our transformation (detailed in Sect. 3) is robust to those: it assumes given parallel-free programs, and introduces parallel commands that either uniformly update the (copy of the) variables across commands, or update them in only one command. The rest of this section assumes parallel-free programs, that will be given as input to our transformation explained in Sect. 3.1.

For convenience we define the following sets of variables.

---

[3] OpenMP's pragma omp parallel directive is illustrated in Sect. 5.

[4] The use command represents any command which does not modify its variables but use them and should not be moved around carelessly (e.g., a printf). In practice, we currently treat all function calls as use, even if the function is pure.

Definition 1. *Given an expression* e, *we define the variables occurring in* e *by:*

$$\mathrm{Occ}(x) = x \qquad\qquad \mathrm{Occ}(t\,[e]) = t \cup \mathrm{Occ}(e)$$
$$\mathrm{Occ}(val) = \emptyset \qquad \mathrm{Occ}(op(e_1, \ldots, e_n)) = \mathrm{Occ}(e_1) \cup \cdots \cup \mathrm{Occ}(e_n)$$

Definition 2. *Let* C *be a command, we let* $\mathrm{Out}(C)$ *(resp.* $\mathrm{In}(C)$, $\mathrm{Occ}(C)$*) be the set of variables* modified *by (resp.* used *by, occurring* in*)* C *as defined in Table 1. In the* use$(x_1, \ldots, x_n)$ *case,* f *is a fresh variable introduced for this command.*

Table 1. Definition of Out, In and Occ for commands

C	Out(C)	In(C)	$\mathrm{Occ}(C) = \mathrm{Out}(C) \cup \mathrm{In}(C)$
$x = e$	x	$\mathrm{Occ}(e)$	$x \cup \mathrm{Occ}(e)$
$t[e_1] = e_2$	t	$\mathrm{Occ}(e_1) \cup \mathrm{Occ}(e_2)$	$t \cup \mathrm{Occ}(e_1) \cup \mathrm{Occ}(e_2)$
if e then C_1 else C_2	$\mathrm{Out}(C_1) \cup \mathrm{Out}(C_2)$	$\mathrm{Occ}(e) \cup \mathrm{In}(C_1) \cup \mathrm{In}(C_2)$	$\mathrm{Occ}(e) \cup \mathrm{Occ}(C_1) \cup \mathrm{Occ}(C_2)$
while e do C	$\mathrm{Out}(C)$	$\mathrm{Occ}(e) \cup \mathrm{In}(C)$	$\mathrm{Occ}(e) \cup \mathrm{Occ}(C)$
use$(x_1, \ldots, x_n)$	f	$\{x_1, \ldots, x_n\}$	$\{x_1, \ldots, x_n, f\}$
skip	$\emptyset$	$\emptyset$	$\emptyset$
$C_1; C_2$	$\mathrm{Out}(C_1) \cup \mathrm{Out}(C_2)$	$\mathrm{In}(C_1) \cup \mathrm{In}(C_2)$	$\mathrm{Occ}(C_1) \cup \mathrm{Occ}(C_2)$

Our treatment of arrays is an over-approximation: we consider the array as a single entity, and that changing one value in it changes it completely. This is however satisfactory: since we do not split loop "vertically" (e.g., distributing the iteration space between threads) but "horizontally" (e.g., distributing the tasks between threads), we want each thread in the `parallel` command to have control of the array it modifies, and not to have to synchronize its writes with other commands.

2.2   Data-Flow Graphs for Loop Dependency Analysis

The loop transformation algorithm relies fundamentally on its ability to analyze data-flow dependencies between loop condition and variables in the loop body, to identify opportunities for loop fission. In this section we define the principles of this dependency analysis, founded on the theory of *data-flow graphs*, and how it maps to the presented `while` language. This dependency analysis was influenced by a large body of works related to static analysis [1,26,29], semantics [27,38] and optimization [33]; but is presented here in self-contained and compact manner.

We assume the reader is familiar with semi-rings, standard operations on matrices (multiplication and addition), and on graphs (union and inclusion).

Definition of Data-Flow Graphs. A data-flow graph for a given command C is a weighted relation on the set $\mathrm{Occ}(C)$. Formally, this is represented as a matrix over a semi-ring, with the implicit choice of a denumeration of $\mathrm{Occ}(C)$[5].

---

[5] We will use the order in which the variables occur in the program as their implicit order most of the time.

Definition 3 (DFG). *A data-flow graph (DFG) for a command* C *is a* | Occ(C)| × | Occ(C)| *matrix over a fixed semi-ring* $(\mathcal{S}, +, \times)$, *with* | Occ(C)| *the cardinal of* Occ(C). *We write* $\mathbb{M}(C)$ *the DFG of* C *and* $\mathbb{M}(C)(y, x)$ *for the coefficient in* $\mathbb{M}(C)$ *at the row corresponding to* x *and column corresponding to* y.

How a data-flow graph is constructed, by induction over the command, is explained in Sect. 2.3. To avoid resizing matrices whenever additional variables are considered, we identify $\mathbb{M}(C)$ with its embedding in a larger matrix, i.e., we abusively call the DFG of C any matrix containing $\mathbb{M}(C)$ and the multiplication identity element on the other diagonal coefficients, implicitly viewing the additional rows/columns as variables not in Occ(C).

2.3 Constructing Data-Flow Graphs

The data-flow graph (DFG) of a command is constructed by induction on the structure of the command. In the remainder of this paper, we use the semiring $(\{0, 1, \infty\}, \max, \times)$ to represent dependencies: ∞ represents *dependence*, 1 represents *propagation*, and 0 represents *reinitialization*.

Base Cases (assignment, Skip, Use). The DFG for an assignment C is computed using In(C) and Out(C):

Definition 4 (Assignment). *Given an assignment* C, *its DFG is given by:*

$$\mathbb{M}(C)(y, x) = \begin{cases} \infty & \text{if } x \in \text{Out}(C) \text{ and } y \in \text{In}(C) & \text{(Dependence)} \\ 1 & \text{if } x = y \text{ and } x \notin \text{Out}(C) & \text{(Propagation)} \\ 0 & \text{otherwise} & \text{(Reinitialization)} \end{cases}$$

We illustrate in Fig. 2 some basic cases and introduce the graphical conventions of using weighted relations, or weighted bi-partite graphs, to illustrate the matrices. Note that in the case of dependencies, In(C) is exactly the set of variables that are source of a dependence arrow, while Out(C) is the set of variables that either are targets of dependence arrows or were reinitialized.

Note that we over-approximate arrays in two ways: the dependencies of the value at one index are the dependencies of the whole array, and the index at which the value is assigned is a dependence of the whole array (cf. the solid arrow from i to t in the last example of Fig. 2). This is however enough for our purpose, and simplify our treatment of arrays.

The DFG for skip is simply the empty matrix, but the DFG of use function calls requires a fresh "effect" variable to anchor the dependencies.

C	Out(C), In(C)	M(C) (as a graph)	M(C)
$w = 3$	$\mathrm{Out}(C) = \{w\}$ $\mathrm{In}(C) = \emptyset$	reinitialization $w \qquad\qquad w$	$w \; (0)$
$x = y$	$\mathrm{Out}(C) = \{x\}$ $\mathrm{In}(C) = \{y\}$	$x \xrightarrow{\text{dependence}} x$ $y \dashrightarrow y$ propagation	$\begin{array}{c} x\;\;y \\ \begin{array}{c}x\\y\end{array}\begin{pmatrix} 0 & 0 \\ \infty & 1 \end{pmatrix} \end{array}$
$w = t[x+1]$	$\mathrm{Out}(C) = \{w\}$ $\mathrm{In}(C) = \{t,x\}$	$w \qquad\qquad w$ $t \dashrightarrow t$ $x \dashrightarrow x$	$\begin{array}{c} w\;\;t\;\;x \\ \begin{array}{c}w\\t\\x\end{array}\begin{pmatrix} 0 & 0 & 0 \\ \infty & 1 & 0 \\ \infty & 0 & 1 \end{pmatrix} \end{array}$
$t[i] = u + j$	$\mathrm{Out}(C) = \{t\}$ $\mathrm{In}(C) = \{i,u,j\}$	$t \qquad\qquad t$ $i \dashrightarrow i$ $u \dashrightarrow u$ $j \dashrightarrow j$	$\begin{array}{c} t\;\;i\;\;u\;\;j \\ \begin{array}{c}t\\i\\u\\j\end{array}\begin{pmatrix} 0 & 0 & 0 & 0 \\ \infty & 1 & 0 & 0 \\ \infty & 0 & 1 & 0 \\ \infty & 0 & 0 & 1 \end{pmatrix} \end{array}$

Fig. 2. Statement examples, sets, and representations of their dependences

Definition 5 (skip). *We let* $\mathbb{M}(\texttt{skip})$ *be the matrix with 0 rows and columns*[6].

Definition 6 (use). *We let* $\mathbb{M}(\texttt{use}(x_1, \ldots, x_n))$ *be the matrix with coefficients from each* x_i *to* f, *and from* f *to* f *equal to* ∞, *and 0 coefficients otherwise, for* f *a freshly introduced variable. Graphically, we get:*

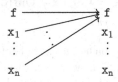

Composition and Multipaths. The definition of DFG for a (sequential) *composition* of commands is an abstraction that allows treating a block of statements as one command with its own DFG.

Definition 7 (Composition). *We let* $\mathbb{M}(C_1; \ldots; C_n)$ *be* $\mathbb{M}(C_1) \times \cdots \times \mathbb{M}(C_n)$.

For two graphs, the product of their matrices of weights is represented in a standard way, as a graph of length 2 paths; as illustrated in Fig. 3—where C_1 and C_2 are themselves already the result of compositions of assignments involving disjoint variables, and hence straightforward to compute.

---

[6] Identifying the DFG with its embeddings, it is hence the identity matrix of any size.

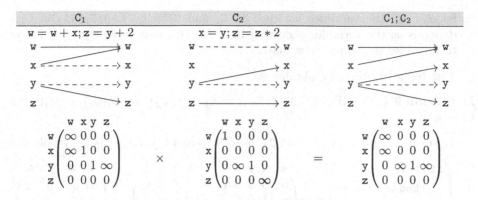

Fig. 3. Data-flow graph of composition.

Correction. Conditionals and loops both requires a *correction* to compute their DFGs. Indeed, the DFGs of if e then C_1 else C_2 and while e do C require more than the DFG of its body. The reason for this is that all the modified variables in C_1 and C_2 or C (e.g., $\mathrm{Out}(C_1) \cup \mathrm{Out}(C_2)$ or $\mathrm{Out}(C)$) depend on the variables occurring in e (e.g., in $\mathrm{Occ}(e)$). To reflect this, a *correction* is needed:

Definition 8 (Correction). *For* e *an expression and* C *a command, we define* e*'s correction for* C*,* $\mathrm{Corr}(e)_C$*, to be* $E^t \times O$*, for*

- E^t *the (column) vector with coefficient equal to* ∞ *for the variables in* $\mathrm{Occ}(e)$ *and 0 for all the other variables,*
- O *the (row) vector with coefficient equal to* ∞ *for the variables in* $\mathrm{Out}(C)$ *and 0 for all the other variables.*

As an example, let us re-use the programs C_1 and C_2 from Fig. 3, to construct w > x's correction for $C_1; C_2$, that we write $\mathrm{Corr}(w > x)_{C_1;C_2}$:

E^t		O		$E^t \times O$			
					w x y z		
w $\begin{pmatrix}\infty\end{pmatrix}$		w x y z		w $\begin{pmatrix}\infty & \infty & 0 & \infty\end{pmatrix}$			
x $\begin{pmatrix}\infty\end{pmatrix}$		$(\infty \ 0 \ 0 \ \infty)$	$(\mathrm{Out}(C_1))$	x $(\infty \ \infty \ 0 \ \infty)$			
y $\begin{pmatrix}0\end{pmatrix}$	+	$(0 \ \infty \ 0 \ \infty)$	$(\mathrm{Out}(C_2))$	y $(0 \ 0 \ 0 \ 0)$			
z $\begin{pmatrix}0\end{pmatrix}$	=	$(\infty \ \infty \ 0 \ \infty)$	$(\mathrm{Out}(C_1;C_2))$	z $(0 \ 0 \ 0 \ 0)$			

This last matrix represents the fact that w and x, through the expression w > x, control the values of w, x and z if C_1 and C_2's execution depend of it.

Conditionals. To construct the DFG of if e then C_1 else C_2, there are two aspects to consider:

1. First, our analysis does not seek to evaluate whether C_1 or C_2 will get executed. Instead, it will overapproximate and assume that both will get executed, hence using $\mathbb{M}(C_1) + \mathbb{M}(C_2)$.

2. Second, all the variables assigned in C_1 and C_2 (e.g., $\mathrm{Out}(C_1) \cup \mathrm{Out}(C_2)$) depends on the variables occurring in e. For this reason, $\mathrm{Corr}(e)_{C_1;C_2}$ needs to be added to the previous matrix.

Putting it together, we obtain:

Definition 9 (if). *We let* $\mathrm{M}(\texttt{if e then } C_1 \texttt{ else } C_2)$ *be* $\mathrm{M}(C_1) + \mathrm{M}(C_2) + \mathrm{Corr}(e)_{C_1;C_2}$.

Re-using the programs C_1 and C_2 from Fig. 3 and $\mathrm{Corr}(w > x)_{C_1;C_2}$, we obtain:

$$
\mathrm{M}\begin{pmatrix}\texttt{if(w>x)}\\ \quad\texttt{then w}=\texttt{w}+\texttt{x;}\\ \qquad\texttt{z}=\texttt{y}+2\\ \quad\texttt{else x}=\texttt{y;}\\ \qquad\texttt{z}=\texttt{z}*2\end{pmatrix} = \begin{array}{c} \\ w\\ x\\ y\\ z\end{array}\!\!\begin{array}{c}\begin{array}{cccc}w&x&y&z\end{array}\\ \left(\begin{array}{cccc}\infty&0&0&0\\ \infty&\boxed{1}&0&0\\ 0&0&1&\infty\\ 0&0&0&0\end{array}\right)\end{array} + \begin{array}{c} \\ w\\ x\\ y\\ z\end{array}\!\!\begin{array}{c}\begin{array}{cccc}w&x&y&z\end{array}\\ \left(\begin{array}{cccc}1&0&0&0\\ 0&\boxed{0}&0&0\\ 0&\infty&1&0\\ 0&0&0&\infty\end{array}\right)\end{array} + \begin{array}{c} \\ w\\ x\\ y\\ z\end{array}\!\!\begin{array}{c}\begin{array}{cccc}w&x&y&z\end{array}\\ \left(\begin{array}{cccc}\infty&\infty&\boxed{0}&\infty\\ \infty&\boxed{\infty}&0&\infty\\ 0&0&0&0\\ 0&0&0&0\end{array}\right)\end{array}
$$

The boxed value represents the impact of x on itself: C_1 has the value 1, since x is not assigned in it. On the other hand, C_2 has 0 for coefficient, since the value of x is reinitialized in it. The correction, however, has a ∞, to represent the fact that the value of x controls the values assigned in the body of C_1 and C_2—and x itself is one of them. As a result, we have again the value ∞ in the matrix summing them three, since x controls the value it gets assigned to itself—as it controls which branch ends up being executed. On the other hand, the circled value at (w, y) is a 0 since y's value is not controlled by w, since neither C_1 nor C_2 assign y: regardless of e's truth value, y's value will remain the same.

While Loops. To define the DFG of a command while e do C from $\mathrm{M}(C)$, we need, as for conditionals, the correction $\mathrm{Corr}(e)_C$, to account for the fact that all the modified variables in C depend on the variables used in e:

Definition 10 (while). *We let* $\mathrm{M}(\texttt{while e do } C)$ *be* $\mathrm{M}(C) + \mathrm{Corr}(e)_C$[7].

As an example, we let the reader convince themselves that the DFG of

```
while(t[i] != j){
    s1[i] = j*j;
    s2[i] = 1/j;
    i++
}
```

is
$$
\begin{array}{c} \\ t\\ i\\ j\\ s1\\ s2\end{array}\!\!\begin{array}{c}\begin{array}{ccccc}t&i&j&s1&s2\end{array}\\ \left(\begin{array}{ccccc}1&\infty&0&\infty&\infty\\ 0&\infty&0&\infty&\infty\\ 0&\infty&1&\infty&\infty\\ 0&0&0&0&0\\ 0&0&0&0&0\end{array}\right)\end{array}
$$
. Intuitively, one can note that

---

[7] This is different from our previous treatment of while loop [33, Definition 6], that required to compute the transitive closure of $\mathrm{M}(C)$: for the transformation we present in Sect. 3, this is not needed, as all the relevant dependencies are obtained immediately—this also guarantees that our analysis can distribute loop-carried dependencies.

the rows for s1 and s2 are filled with 0s, since those variables do not control any other variable and are assigned in the body of the loop. On the other hand, t, i and j all three control the values of i, s1 and s2, since they determine if the body of the loop will execute. The variables t and j are the only one whose value is propagated (e.g., with a 1 on their diagonal), since they are not assigned in this short example. The command i++ is the only command that has the potential to impact the loop's condition. We call it an update command:

Definition 11 (Update command). *Given a loop* W := while e do C, *the* update commands C_u *are the commands in* C *such that* $\mathbb{M}(W)(y,x) = \infty$ *for* $x \in \text{Out}(C_u)$ *and* $y \in \text{Occ}(e)$.

3   Loop Fission Algorithm

We now present our loop transformation technique and prove its correctness.

3.1   Algorithm, Presentation and Intuition

Our algorithm, presented in Algorithm 1, requires essentially to

1. Pick a loop at top level,
2. Compute its condensation graph (Definition 13)—this requires first the dependence graph (Definition 12), which itself uses the DFG,
3. Compute a covering (Definition 14) of the condensation graph,
4. Create a loop per element of the covering.

Even if our technique could distribute nested loops, it would require adjustments that we prefer to omit to simplify our presentation. None of our examples in this paper require to distribute nested loops. Note, however, that our algorithm handles loops containing themselves loops.

Definition 12 (Dependence graph). *The* dependence graph *of the loop* W := while e do $\{C_1; \cdots ; C_n\}$ *is the graph whose vertices is the set of commands* $\{C_1; \cdots ; C_n\}$, *and there exists a directed edge from* C_i *to* C_j *if and only if there exists variables* $x \in \text{Out}(C_j)$ *and* $y \in \text{In}(C_i)$ *such that* $\mathbb{M}(W)(y,x) = \infty$.

The last example of Sect. 2.3 gives s1[i] = j*j $\longrightarrow$ i++ $\longleftarrow$ s2[i] = 1/j . Note that all the commands in the body of the loop are the sources of dependence edges whose target is the update commands: for our example, this means that every command will be the source of an arrow whose target is i++. This comes from the correction, even if the condition does not explicitly appear in the dependence graph.

The remainder of the loop transforming principle is simple: once the graph representing the dependencies between commands is obtained, it remains to determine the cliques in the graph and forms *strongly connected components* (SCCs); and then to separate the SCCs into subgraphs to produce the final parallelizable loops that contain a copy of the loop header and update commands.

Definition 13 (Graph helpers). *Given the dependence graph of a loop* W,

- *its* strongly connected components *(SCCs) are its strongly connected sub-graphs,*
- *its* condensation graph G_W *is the graph whose vertices are SCCs and edges are the edges whose source and target belong to distinct SCCs.*

In our example, the SCCs are the nodes themselves, and the condensation graph is `s1[i] = j*j` $\longrightarrow$ `i++` $\longleftarrow$ `s2[i] = 1/j`. Excluding the update command `i++`, there are now two nodes in the condensation graph, and we can construct the parallel loops by 1. inserting a `parallel` command, 2. duplicating the loop header and update command, 3. inserting the command in the remaining nodes of the condensation graph in each loop. For our example, we obtain, as expected,

$$
\texttt{parallel} \left\{ \begin{array}{l} \texttt{while(t[i] != j)\{} \\ \quad \texttt{s1[i] = j*j;} \\ \quad \texttt{i++\}} \end{array} \right\} \left\{ \begin{array}{l} \texttt{while(t[i] != j)\{} \\ \quad \texttt{s2[i] = 1/j;} \\ \quad \texttt{i++\}} \end{array} \right\}.
$$

Formally, what we just did was to split the *saturated covering*.

Definition 14 (Coverings [16]). *A covering of a graph G is a collection of subgraphs $G_1, G_2, \ldots, G_j$ such that $G = \cup_{i=1}^{j} G_i$.*

A saturated covering of G is a covering $G_1, G_2, \ldots, G_k$ such that for all edge in G with source in G_i, its target belongs to G_i as well. It is proper if none of the subgraph is a subgraph of another.

The algorithm then simply consists in finding a proper saturated covering of the loop's condensation graph, and to split the loop accordingly. In our example, the only proper saturated covering is

$$
\{ \texttt{s1[i] = j*j} \longrightarrow \texttt{i++} , \texttt{i++} \longleftarrow \texttt{s2[i] = 1/j} \}.
$$

If the covering was not proper, then the `i++` node on its own would be in it, leading to create a useless loop that performs nothing but updating its own condition.

Sometimes, duplicating commands that are not update commands is needed to split the loop. We illustrate this principle with a more complex example that involve function call and multiple update commands in Fig. 4.

3.2 Correctness of the Algorithm

We now need to prove that the semantics of the initial loop W is equal to the semantics of $\widetilde{W}$ given by Algorithm 1. This is done by showing that for any variable x appearing in W, its final value after running W is equal to its final value after running $\widetilde{W}$. We first prove that the loops in $\widetilde{W}$ has the same iteration space as W:

Lemma 1. *The loops in $\widetilde{W}$ have the same number of iterations as* W.

Algorithm 1. Loop fission

Input: A loop $\mathtt{W} := \mathtt{while}\ \mathtt{e}\ \mathtt{do}\ \{\mathtt{C_1}; \cdots; \mathtt{C_n}\}$ ▷ Pick a loop W at top level

Compute the condensation graph $\mathtt{G_W}$ *of* W, ▷ cf. Def. 13

Compute the saturated covering $G_1, \ldots, G_j$ *of* $\mathtt{G_W}$*:* ▷ cf. Def. 14

while a node n in $\mathtt{G_W}$ is not part of a subgraph G_l **do**

  Create a new subgraph G_i containing n,

  Recursively add to G_i the nodes targeted by edges whose source is in G_i,

Compute the proper *saturated covering* $G_1, \ldots, G_k$ *of* $\mathtt{G_W}$*:*

for all G_i in the saturated covering **do**

  If $\exists G_l$ in the saturated covering s.t. G_i is a subgraph of G_l, then remove G_i

end for

Create one while *loop per subgraph in the proper saturated covering:*

for all G_i in the proper saturated covering **do**

  Let $\mathtt{W}_i := \mathtt{while}\ \mathtt{e}\ \mathtt{do}\ \{\mathtt{C}_{i_1}; \cdots; \mathtt{C}_{i_m}\}$ where $\{\mathtt{C}_{i_1}, \ldots, \mathtt{C}_{i_m}\}$ are the vertices of G_i, inserted in the same order as they are in W.

end for

Output: if $k > 1$, $\tilde{\mathtt{W}} := \mathtt{parallel\{W_1\}\{ \ldots \}\{W_k\}}$, else $\tilde{\mathtt{W}} := \mathtt{W}$.

$\mathbb{M}$

```
while(i != j){
  i++;
  j--;
  a = t[i];
  use(a);
  b = t[a]
}
```

=

	i	j	a	t	b	f
i	∞	∞	∞	0	∞	∞
j	∞	∞	∞	0	∞	∞
a	0	0	0	0	∞	∞
t	0	0	∞	1	∞	0
b	0	0	0	0	0	0
f	0	0	∞	0	0	∞

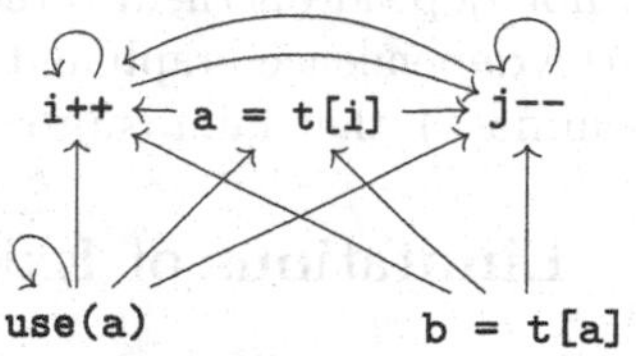

The proper saturated covering has two subgraphs: one contains everything but use(a) and the other contains everything but b = t[a]. Since both use(a) and b = t[a] depend on a = t[i], this latter command needs to be duplicated, even if it is not an update command:

```
parallel { while(i != j){      } { while(i != j){      }
             i++;                    i++;
             j--;                    j--;
             a = t[i];               a = t[i];
             use(a)}                 b = t[a]}
```

Fig. 4. Distributing a more complex while loop

Proof. Let $\mathtt{W}_i$ be a loop in $\tilde{\mathtt{W}}$. By property of the saturated covering, the update commands are in the body of $\mathtt{W}_i$: there is always an edge from any command to the update commands due to the loop correction, and hence the update commands are part of all the subgraphs in the saturated covering. Furthermore, if there exists a command C that is the target of an edge whose source is an update command $\mathtt{C_u}$, then C and $\mathtt{C_u}$ are always both present in any subgraph of the saturated covering. Indeed, since there are edges from $\mathtt{C_u}$ to C and from C to $\mathtt{C_u}$, they are part of the same node in the condensation graph.

Since the condition of W_i is the same as the condition of W, and since all the instructions that impact (directly or indirectly) the variables occurring in that condition are present in W_i, we conclude that the number of iterations of W_i and W are equal. □

Theorem 1. *The transformation* $W \rightsquigarrow \tilde{W}$ *given in Algorithm 1 preserves the semantic.*

Proof (sketch). We show that for every variable x, the value of x after the execution of W is equal to the value of x after the execution of $\tilde{W}$. Variables are considered local to each loop W_i in $\tilde{W}$, so we need to avoid race condition. To do so, we prove the following more precise result: for each variable x and each loop W_i in $\tilde{W}$ in which the value of x is modified, the value of x after executing W is equal to the value of x after executing W_i.

The previous claim is then straightforward to prove, based on the property of the covering. One shows by induction on the number of iterations k that for all the variables $x_1, \ldots, x_h$ appearing in W_i, the values of $x_1, \ldots, x_h$ after k loop iterations of W_i are equal to the values of $x_1, \ldots, x_h$ after k loop iterations of W. Note some other variables may be affected by the latter but the variables $x_1, \ldots, x_h$ do not depend on them (otherwise, they would also appear in W_i by definition of the dependence graph and the covering). Since the number of iteration match (Lemma 1), the claim is proven. □

4   Limitations of Existing Alternative Approaches

In the beginning of this paper, we made the bold claim that other loop fission approaches do not handle unknown iteration spaces, which makes our loop-agnostic technique interesting. In this section we discuss these alternative approaches, their capabilities, and provide evidence to support this claim. We also give justification for the need to introduce our loop analysis into this landscape.

4.1   Comparing Dependency Analyses

Since its first inception, loop fission [2] has been implemented using different techniques and dependency mechanisms. Program dependence graph (PDG) [18] can be used to identify when a loop can be distributed [3, p. 844], but other—sometimes simpler—mechanisms are often used in practice. For instance, a patch integrating loop fission into LLVM [28] tuned the simpler data dependence graph (DDG) to obtain a Loop Fission Interference Graph (FIG) [30]. GCC, on the other hand, build a partition dependence graph (PG) based on the data dependency given by a reduced dependence graph (RG) to perform the same task [19]. In this paper, we introduce another loop dependency analysis, not to further obfuscate the landscape, but because it allows us to express our algorithm simply and—more importantly—to verify it mathematically[8].

---

[8] This analysis also shares interesting links to a static analysis of values growth [9,10], as discussed more in-depth in a first draft [7].

We assume that the more complex mechanisms listed above (PDG, DDG or PG) could be leveraged to implement our transformation, but found it more natural to express ourselves in this language. We further believe that the way we compute the data dependencies is among the lightest, and with a very low memory footprint, as it requires only one pass on the source code to construct a matrix whose size is the number of variables in the program.

4.2 Assessment of Existing Automated Loop Transformation and Parallelization Tools

While we conjecture that other mechanisms *could*, in theory, treat loops of any kind like we do, we now substantiate our claim that none of them do: in short, any loop with non-basic condition or update statement is excluded from the optimizations we now discuss. We limit this consideration to tools that support C language transformations, because it is our choice implementation language for experimental evaluation in Sect. 5. We also focus on presenting the kinds of loops that other "popular" [35] automatic loop transformation frameworks *do not* distribute, but that our algorithm can distribute. In particular, we do not discuss loops containing control-flow modifiers (such as `break;` or `continue;`): neither our algorithm nor OpenMP nor the underlying dependency mechanisms of the discussed tools—to the best of our knowledge—can accommodate those.

Tools that fit the above specification include Cetus, a compiler infrastructure for the source-to-source transformation; Clava, a C/C++ source-to-source tool based on Clang; Par4All, an automatic parallelizing and optimizing compiler; Pluto, an automatic parallelizer and locality optimizer for affine loop nests; ROSE, a compiler-based infrastructure for building source-to-source program transformations and analysis tools; Intel's C++ compiler (icc), and TRACO, an automatic parallelizing and optimizing compiler, based on the transitive closure of dependence graphs. While these tools perform various automatic transformations and optimizations, only ROSE and icc perform loop fission [35, Section 3.1].

Based on our assessment, most of these tools process only *canonical loops*:

Definition 15 (Canonical Loop [25], **4.4.1 Canonical Loop Nest Form).** *A canonical loop is a loop of the form*
    `for (init-expr; test-expr; incr-expr) structured-block`
for `incr-expr` *a (single) increment or decrement by a constant or a variable, and* `test-expr` *a single comparison between a variable and a variable or a constant.*

Additional constraints on loop dependences are sometimes needed, e.g., the absence of loop-carried dependency for Cetus. It seems further that some tools cannot parallelize loops whose body contains e.g., `if` or `switch` statements [35, p. 18], but we have not investigated this claim further. However, our algorithm can handle `if`—and `switch` too, if it was part of our syntax—present in the body of the loop seamlessly.

It is always hard to infer the absence of support, but we evaluated the lack of formal discussion or example of e.g., `while` loop to be sufficient to determine that the tool cannot process while loops, unless of course they can trivially be transformed into for loops of the required form [39, p. 236]. We refer to a recent study [35, Section 2] for more detail on those notions and on the limitations of some of the tools discussed in Table 2.

Table 2. Feature support comparison of automated transformation and parallelization tools.

Name	Fission	for loop	while loop	do ...while loop	ref.
Cetus	−	In canonical form		−	[17, p. 39] , [11, p. 761]
Clava	−	In canonical form		−	[6]
icc	✓	Only if countable		−	[23, p. 2126]
Par4All	−	Unknown			[4,5]
Pluto	−	Only static control structures			[15]
ROSE	✓	In canonical form		−	[36, p. 124]
TRACO	−	In canonical form		−	[34]
OpenMP	−	In canonical form		−	[25]

5   Evaluation

We performed an experimental evaluation of our loop fission technique on a suite of parallel benchmarks. Taking the sequential baseline, we applied the loop fission transformation and parallelization. We compared the result of our technique to the baseline and to an alternative loop fission method implemented in ROSE.

We conducted this experiment in C programming language because it naturally maps to the syntax of the imperative while language presented in Sect. 2. We implement the **parallel** command as OpenMP directives. For instance, the sequential baseline program on the left of Fig. 5 becomes the parallel version on right[9], after applying our loop fission transformation and parallelization.

The evaluation experimentally substantiated two claims about our technique:

1. It can parallelize loops that are completely ignored by other automatic loop transformation tools, and results in appreciable gain, upper-bounded by the number of parallelizable loops produced by loop fission.
2. Concerning loops that other automatic loop transformation tools can distribute, it yields comparable results in speedup potential. We also demonstrate how insertion of parallel directives can be automated, which supports the practicality of our method.

These results combined confirm that our loop fission technique can easily be integrated into existing tools to improve the performances of the resulting code.

---

[9] This example is inspired by benchmark **bicg** from PolyBench/C and presented in our artifact.

```
                          #pragma omp parallel private(j)
                          { // Each "pragma" block below
                            // have its own copy of j.
                            #pragma omp single nowait
                            { // "nowait" lets the next
                              // block start in parallel.
                              j = 0;
   j = 0;                    while (j<M) {
   while (j<M)                 s[j] += r[j]*A[j];
   {                           j++;
     s[j] += r[j]*A[j];      }
     q[j] += A[j]*p[j];    }
     j++;                  #pragma omp single
   }                       {
                             j = 0;
                             while (j<M) {
                               q[j] += A[j]*p[j];
                               j++;
                             }
                           }
                          } // Both blocks must be terminated
                          // before passing this point.
```

Fig. 5. Code transformation example

5.1   Benchmarks

Special consideration was necessary to prepare an appropriate benchmark suite for evaluation. We wanted to test our technique on a range of standard problems, across different domains and data sizes, and to include problems containing while loops. Because our technique is specifically designed for loop fission, we also needed to identify problems that offered potential to apply this transformation. Finding a suite to fit these parameters is challenging, because standard parallel programming benchmark suites offer mixed opportunity for various program optimizations and focus on loops in canonical form.

We resolved this challenge by preparing a curated set, pooling from three standard parallel programming benchmark suites. PolyBench/C is a polyhedral benchmark suite, representing e.g., linear algebra, data mining and stencils; and commonly used for measuring various loop optimizations. NAS Parallel Benchmarks are designed for performance evaluation of parallel supercomputers, derived from computational fluid dynamics applications. MiBench is an embedded benchmark suite, with everyday programming applications e.g., image-processing libraries, telecommunication, security and office equipment routines. From these suites, we extracted problems that offered potential for loop fission, or already assumed expected form, resulting in 12 benchmarks. We detail these benchmarks in Table 4. Because these three suites are not mutually compatible, we leveraged the timing utilities from PolyBench/C to establish a common and comparable measurement strategy. To assess performance of other kinds of loops that our algorithm can distribute, but which do not occur prevalently in these benchmarks, we converted a portion of problems to use `while` loops.

Comparison Target. We compared our approach to ROSE Compiler. It is a rich compiler architecture that offers various program transformations and automatic parallelization, and supports multiple compilation targets. ROSE's built-in LoopProcessor tool supports loop fission for C-to-C programs. This input/output specification was necessary to allow observation of the transformation results and fit with the measurement strategy we defined previously. To our knowledge, ROSE is the only tool that satisfies these evaluation requirements.

Experimental Setup. We ran the benchmarks using a Linux 5.10.0-18-amd64 #1 SMP Debian 5.10.140-1 (2022-09-02) x86_64 GNU/Linux machine, with 4 Intel(R) Core(TM) i5-6300U CPU @ 2.40GHz processors, and `gcc` compiler version 7.5.0. The evaluation was performed in a containerized environment on Docker version 20.10.18, build b40c2f6. For each benchmark, we recorded the clock time 5 times, excluded min and max, and averaged the remaining 3 times to obtain the result. We constrained variance between recorded times not to exceed 5%. We ran experiments on 5 input data sizes, as defined in PolyBench/C: `MINI`, `SMALL`, `MEDIUM`, `LARGE` and `EXTRALARGE` (abbr. `XS`, `S`, `M`, `L`, `XL`). We also tested 4 `gcc` compiler optimization levels `-O0` through `-O3`. Speedup is the ratio of sequential and parallel executions, $S = T_{\text{Seq}}/T_{\text{Par}}$, where a value greater than 1 indicates parallel is outperforming the sequential execution. In presentation of these results, the sequential benchmarks are always considered the baseline, and speedup is reported in relation to the transformed versions. Our open source benchmarks, and instructions for reproducing the results, are available online [8]. It should be noted that some results may be sensitive to the particular setup on which those experiments are run.

5.2   Results

In analyzing the results, we distinguish two cases: distributing and parallelizing loops with potentially unknown iterations, and loops with pre-determined iterations (typically `while` and `for` loops, respectively). The difficulty of parallelizing the former arises from the need to synchronize evaluation of the loop recurrence and termination condition. Improper synchronization results in overshooting the iterations [37], rendering such loops effectively sequential.

Loop fission addresses this challenge by recognizing independence between statements and producing parallelizable loops. Special care is needed when inserting parallelization directives for such loops. This remains a limitation of automated tools and is not natively supported by OpenMP. We resolved this issue by using the OpenMP `single` directive, to prevent overshooting the loop termination condition and need for synchronization between threads, enabling parallel execution by multiple threads on individual loop statements. The strategy is simple, implementable, and we show it to be effective. However, it is also upper-bounded in speedup potential by the number of parallelizable loops produced by the transformation. This is a syntactic constraint, rather than one based on number of available cores.

The results, presented in Table 3, show that our approach, paired with the described parallelization strategy, yields a gain relative to the number of independent parallelizable loops in the transformed benchmark. We observe this e.g., for benchmarks `bicg`, `gesummv`, and `mvt`, as presented in Fig. 6. We also confirm that ROSE's approach did not transform these loops, and report no gain for the alternative approach.

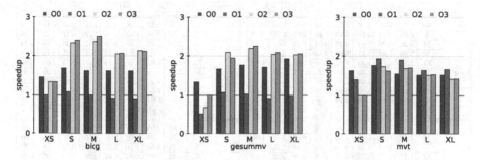

Fig. 6. Speedup of selected benchmarks implemented using `while` loops. Note the influence of various compiler optimization levels, -O0 to -O3 on each problem, and how parallelization overhead tends to decrease as input data size grows from MINI to EXTRALARGE. The gain is lower for `mvt` because it assumes fissioned form in the original benchmark. `bicg` and `gesummv` obtain higher gain from applied loop distribution.

Comparison with ROSE. The remaining benchmarks, with known iteration spaces, can be transformed by both evaluated loop fission techniques: ours and ROSE's LoopProcessor. In terms of transformation results, we observed relatively similar results for both techniques. We discovered one interesting transformation difference, with benchmark `gemm`, which ROSE handles differently from our technique.

After transformation, the program must be parallelized by inserting OpenMP directives. This parallelization step can be fully automatic and performed with e.g., ROSE or Clava, demonstrating that pipelining the transformed programs is feasible. For evaluations, we used manual parallelization for our technique and automatic approach for ROSE. However, we also noted that the automatic insertion of parallelization directives yielded, in some cases, suboptimal choices, such as parallelization of loop nests. This added unnecessary overhead to execution time, and negatively impacted the results obtained for ROSE, e.g., for benchmarks `fdtd-2d` and `gemm`, as observable in the results. It is possible this issue could be mitigated by providing annotations and more detailed instructions for applying the parallelization directives. In other experiments with alternative parallelization tools [7, Sect. 4.3], we have been successful at finding optimal parallelization directives automatically, and therefore conclude it is achievable. We again refer to Table 3 for a detailed presentation of the experimental evaluation results.

Table 3. Speedup comparison between original sequential and transformed parallel benchmarks, comparing our loop fission technique with ROSE Compiler, for various data sizes and compiler optimization levels. We note that the problems containing only `while` loop (in **bold**) are not transformed by ROSE and therefore report no gain. The other results vary depending on parallelization strategy, but as noted with e.g., problems `conjgrad` and `tblshft`, we obtain similar speedup for both fission strategies when automatic parallelization yields optimal OpenMP directives.

Benchmark Name	Size	-O0 ours	-O0 rose	-O1 ours	-O1 rose	-O2 ours	-O2 rose	-O3 ours	-O3 rose
3mm	XS	2.71	0.07	2.26	0.02	1.71	0.02	1.73	0.01
	S	2.80	0.22	3.78	0.09	3.49	0.05	3.35	0.05
	M	2.20	0.46	3.44	0.27	3.08	0.13	3.05	0.13
	L	2.85	1.92	3.11	1.16	2.89	0.66	2.97	0.66
	XL	2.16	2.31	3.13	1.83	2.24	1.05	2.25	1.04
bicg	XS	1.45	0.96	1.00	1.00	1.33	1.00	1.33	1.00
	S	1.68	0.98	1.08	1.00	2.33	1.01	2.39	1.02
	M	1.62	0.97	1.00	0.98	2.36	0.96	2.50	1.00
	L	1.61	0.96	0.90	0.94	2.05	0.95	2.06	0.95
	XL	1.62	0.96	0.89	0.95	2.13	0.93	2.11	0.94
colormap	XS	2.14	1.01	1.50	1.02	1.54	1.04	1.52	1.01
	S	2.08	0.97	1.57	1.00	1.54	1.02	1.43	0.99
	M	1.98	0.95	1.46	0.96	1.49	0.98	1.19	1.00
	L	1.93	1.03	1.42	0.98	1.44	0.98	1.20	1.01
	XL	1.82	1.00	1.53	0.97	1.55	0.99	1.16	1.00
conjgrad	XS	2.43	1.45	1.82	0.69	2.77	0.65	2.50	0.52
	S	2.50	2.39	1.91	2.03	2.84	1.88	2.96	1.65
	M	2.56	2.58	1.94	2.66	2.93	2.44	3.20	2.33
	L	2.38	2.62	1.73	2.96	2.92	2.92	3.24	2.91
	XL	2.29	2.61	1.59	2.55	2.72	2.57	2.99	2.39
cp50	XS	1.90	0.97	1.97	1.00	2.18	1.01	2.09	1.01
	S	1.94	0.95	2.00	1.02	2.08	1.00	2.07	1.00
	M	1.89	0.98	1.76	0.97	1.83	0.99	1.82	0.98
	L	1.74	0.98	1.49	0.96	1.51	0.96	1.50	0.96
	XL	1.63	0.99	1.16	0.96	1.07	0.98	1.11	0.96
deriche	XS	2.00	0.90	1.93	0.51	2.18	0.53	2.11	0.51
	S	2.30	1.49	2.16	1.05	2.17	1.04	2.14	1.03
	M	2.68	2.35	2.88	2.20	2.68	2.22	2.72	2.20
	L	1.79	1.75	2.08	2.03	2.05	2.05	2.07	2.04
	XL	1.12	1.12	1.65	1.61	1.67	1.67	1.60	1.64

Benchmark Name	Size	-O0 ours	-O0 rose	-O1 ours	-O1 rose	-O2 ours	-O2 rose	-O3 ours	-O3 rose
fdtd-2d	XS	2.34	0.27	1.48	0.05	1.81	0.06	1.15	0.03
	S	2.57	0.59	2.68	0.15	3.12	0.17	2.47	0.09
	M	2.23	0.82	2.01	0.29	2.47	0.30	2.60	0.24
	L	2.15	1.20	1.89	0.65	1.98	0.61	2.16	0.71
	XL	2.17	1.38	1.47	0.79	1.50	0.73	1.68	0.86
gemm	XS	2.73	0.09	2.33	0.02	2.43	0.02	1.20	0.01
	S	2.87	0.21	3.98	0.05	3.09	0.04	3.01	0.02
	M	2.57	0.56	3.42	0.12	3.40	0.12	2.73	0.05
	L	2.44	1.50	1.79	0.35	1.87	0.36	2.20	0.25
	XL	2.44	1.95	1.85	0.60	1.85	0.70	1.96	0.50
gesummv	XS	1.33	1.00	0.50	0.67	0.67	0.67	1.00	1.00
	S	1.67	0.95	1.08	1.03	2.09	1.03	1.94	1.01
	M	1.77	0.98	1.03	1.00	2.19	1.00	2.25	1.00
	L	1.71	0.94	0.90	0.93	2.04	0.93	2.08	0.97
	XL	1.92	0.98	0.96	0.98	2.03	0.99	2.05	0.98
mvt	XS	1.63	1.00	1.40	0.88	1.00	1.00	1.00	1.00
	S	1.76	1.01	1.93	1.01	1.73	1.02	1.62	1.00
	M	1.55	0.96	1.90	1.00	1.69	1.02	1.70	1.03
	L	1.52	0.98	1.64	0.97	1.51	0.98	1.53	1.00
	XL	1.52	0.98	1.66	0.99	1.42	1.00	1.42	1.00
remap	XS	1.43	0.97	0.54	1.00	0.54	1.00	0.64	1.00
	S	2.07	0.94	1.20	1.02	1.13	1.03	1.19	1.01
	M	2.43	0.99	3.13	0.96	3.36	0.98	2.89	0.97
	L	2.09	1.00	1.34	0.97	1.54	1.02	1.74	1.00
	XL	2.11	1.00	1.28	0.99	1.52	0.99	1.57	1.00
tblshft	XS	3.19	3.27	2.70	2.65	2.68	2.73	2.82	2.82
	S	3.37	3.45	2.82	2.84	2.89	2.86	3.05	3.08
	M	3.31	3.62	2.93	3.00	2.79	2.85	3.21	3.19
	L	3.05	3.40	2.17	2.32	2.38	2.32	2.40	2.39
	XL	3.08	3.48	1.91	1.85	1.64	1.69	1.96	1.96

Table 4. Descriptions of evaluated parallel benchmarks.

Benchmark	Description	for loop	while loop	Source
3mm	3D matrix multiplication	✓		PolyBench/C
bicg	BiCG sub kernel of BiCGStab linear solver		✓	PolyBench/C
colormap	TIFF image conversion of photometric palette		✓	MiBench
conjgrad	Conjugate gradient routine	✓		NAS-CG
cp50	Ghostscript/CP50 color print routine	✓	✓	MiBench
deriche	Edge detection filter	✓		PolyBench/C
fdtd-2d	2-D finite different time domain kernel	✓		PolyBench/C
gemm	Matrix-multiply C=alpha.A.B+beta.C	✓		PolyBench/C
gesummv	Scalar, vector and matrix multiplication		✓	PolyBench/C
mvt	Matrix vector product and transpose		✓	PolyBench/C
remap	4D matrix memory remapping	✓		NAS-UA
tblshift	TIFF PixarLog compression main table bit shift	✓	✓	MiBench

6   Conclusion

This work is only the first step in a very exciting direction. "Ordinary code", and not only code that was specifically written for e.g., scientific calculation or other resource-demanding operations, should be executed in parallel to leverage our modern architectures. As a consequence, the much larger codebase concerned with parallelization is much less predictable and offers more diverse loop structures. Focusing on resource-demanding programs led previous efforts not only to focus on predictable loop structures, but to completely ignore other non-canonical loops. Our effort, based on an original dependency analysis, leads to re-integrate such loops in the realm of parallel optimization. This alone, in our opinion, justifies further investigation in integrating our algorithm into specialized tools.

As presented in Fig. 6, our experimental results offer some variability, but they need to be put in context: loop distribution is often only *the first step* in the optimization pipeline. Loops that have been split can then be vectorized, blocked, unrolled, etc. , providing additional gain in terms of speed. Exactly as for loop fusion [31], a more global treatment of loops is needed to strike the right balance and find the optimum code transformation. Such a journey will be demanding and complex, but we believe this work enables it by reintegrating *all* loops in the realm of parallel optimization.

Acknowledgments. The authors wish to express their gratitude to João Bispo for explaining how to integrate AutoPar-Clava in the first version of their benchmark, to Assya Sellak for her contribution to the first steps of this work, and to the reviewers for their insightful comments.

References

1. Abel, A., Altenkirch, T.: A predicative analysis of structural recursion. J. Funct. Program. **12**(1), 1–41 (2002). https://doi.org/10.1017/S0956796801004191
2. Abu-Sufah, W., Kuck, D.J., Lawrie, D.H.: On the performance enhancement of paging systems through program analysis and transformations. IEE Trans. Comput. **30**(5), 341–356 (1981). https://doi.org/10.1109/TC.1981.1675792
3. Aho, A.V., Lam, M.S., Sethi, R., Ullman, J.D.: Compilers: Principles, Techniques, and Tools, 2nd edn. Addison Wesley, Boston (2006)
4. Amini, M.: Source-to-source automatic program transformations for GPU-like hardware accelerators. Theses, Ecole Nationale Supérieure des Mines de Paris, December 2012. https://pastel.archives-ouvertes.fr/pastel-00958033
5. Amini, M., et al.: Par4All: from convex array regions to heterogeneous computing. In: IMPACT 2012 : Second International Workshop on Polyhedral Compilation Techniques HiPEAC 2012. Paris, France, January 2012. https://hal-mines-paristech.archives-ouvertes.fr/hal-00744733
6. Arabnejad, H., Bispo, J., Cardoso, J.M.P., Barbosa, J.G.: Source-to-source compilation targeting OpenMP-based automatic parallelization of C applications. J. Supercomput. **76**(9), 6753–6785 (2019). https://doi.org/10.1007/s11227-019-03109-9

7. Aubert, C., Rubiano, T., Rusch, N., Seiller, T.: A novel loop fission technique inspired by implicit computational complexity, May 2022. https://hal.archives-ouvertes.fr/hal-03669387v1. draft
8. Aubert, C., Rubiano, T., Rusch, N., Seiller, T.: Loop fission benchmarks (2022). https://doi.org/10.5281/zenodo.7080145. https://github.com/statycc/loop-fission
9. Aubert, C., Rubiano, T., Rusch, N., Seiller, T.: MWP-analysis improvement and implementation: realizing implicit computational complexity. In: Felty, A.P. (ed.) 7th International Conference on Formal Structures for Computation and Deduction (FSCD 2022). Leibniz International Proceedings in Informatics, vol. 228, pp. 26:1–26:23. Schloss Dagstuhl-Leibniz-Zentrum für Informatik (2022). https://doi.org/10.4230/LIPIcs.FSCD.2022.26
10. Aubert, C., Rubiano, T., Rusch, N., Seiller, T.: pymwp: MWP analysis in Python, September 2022. https://github.com/statycc/pymwp/
11. Bae, H., et al.: The Cetus source-to-source compiler infrastructure: overview and evaluation. Int. J. Parallel Program. 41(6), 753–767 (2013). https://doi.org/10.1007/s10766-012-0211-z
12. Baier, C., Katoen, J., Larsen, K.: Principles of Model Checking. MIT Press, Cambridge (2008)
13. Benabderrahmane, M.-W., Pouchet, L.-N., Cohen, A., Bastoul, C.: The polyhedral model is more widely applicable than you think. In: Gupta, R. (ed.) CC 2010. LNCS, vol. 6011, pp. 283–303. Springer, Heidelberg (2010). https://doi.org/10.1007/978-3-642-11970-5_16
14. Bertolacci, I., Strout, M.M., de Supinski, B.R., Scogland, T.R.W., Davis, E.C., Olschanowsky, C.: Extending OpenMP to facilitate loop optimization. In: de Supinski, B.R., Valero-Lara, P., Martorell, X., Mateo Bellido, S., Labarta, J. (eds.) IWOMP 2018. LNCS, vol. 11128, pp. 53–65. Springer, Cham (2018). https://doi.org/10.1007/978-3-319-98521-3_4
15. Bondhugula, U., Hartono, A., Ramanujam, J., Sadayappan, P.: A practical automatic polyhedral Parallelizer and locality optimizer. SIGPLAN Not. 43(6), 101–113 (2008). https://doi.org/10.1145/1379022.1375595
16. Chung, F.R.K.: On the coverings of graphs. Discret. Math. 30(2), 89–93 (1980). https://doi.org/10.1016/0012-365X(80)90109-0
17. Dave, C., Bae, H., Min, S., Lee, S., Eigenmann, R., Midkiff, S.P.: Cetus: a source-to-source compiler infrastructure for multicores. Computer 42(11), 36–42 (2009). https://doi.org/10.1109/MC.2009.385
18. Ferrante, J., Ottenstein, K.J., Warren, J.D.: The program dependence graph and its use in optimization. ACM Trans. Programm. Lang. Syst. 9(3), 319–349 (1987). https://doi.org/10.1145/24039.24041
19. gcc.gnu.org git - gcc.git/blob - gcc/tree-loop-distribution.c. https://gcc.gnu.org/git/?p=gcc.git;a=blob;f=gcc/tree-loop-distribution.c;h=65aa1df4abae2c6acf40299f710bc62ee6bacc07;hb=HEAD#l39
20. Grosser, T.: Enabling Polyhedral Optimizations in LLVM. Master's thesis, Universität Passau, April 2011. https://polly.llvm.org/publications/grosser-diploma-thesis.pdf
21. Holewinski, J., et al.: Dynamic trace-based analysis of vectorization potential of applications. In: Proceedings of the 33rd ACM SIGPLAN Conference on Programming Language Design and Implementation, PLDI 2012, pp. 371–382. Association for Computing Machinery, New York (2012). https://doi.org/10.1145/2254064.2254108
22. Intel: oneTBB documentation (2022). https://oneapi-src.github.io/oneTBB/

23. Intel Corporation: Intel C++ Compiler Classic Developer Guide and Reference. https://www.intel.com/content/dam/develop/external/us/en/documents/cpp_compiler_classic.pdf
24. Karp, R.M., Miller, R.E., Winograd, S.: The organization of computations for uniform recurrence equations. J. ACM **14**(3), 563–590 (1967). https://doi.org/10.1145/321406.321418
25. Klemm, M., de Supinski, B.R. (eds.): OpenMP application programming interface specification version 5.2. OpenMP Architecture Review Board, November 2021. https://www.openmp.org/wp-content/uploads/OpenMP-API-Specification-5-2.pdf
26. Kristiansen, L., Jones, N.D.: The flow of data and the complexity of algorithms. In: Cooper, S.B., Löwe, B., Torenvliet, L. (eds.) CiE 2005. LNCS, vol. 3526, pp. 263–274. Springer, Heidelberg (2005). https://doi.org/10.1007/11494645_33
27. Laird, J., Manzonetto, G., McCusker, G., Pagani, M.: Weighted relational models of typed lambda-calculi. In: LICS, pp. 301–310. IEEE Computer Society (2013). https://doi.org/10.1109/LICS.2013.36
28. Lattner, C., Adve, V.S.: LLVM: a compilation framework for lifelong program analysis & transformation. In: 2nd IEEE / ACM International Symposium on Code Generation and Optimization (CGO 2004), 20–24 March 2004, San Jose, CA, USA, pp. 75–88. IEEE Computer Society (2004). https://doi.org/10.1109/CGO.2004.1281665, https://ieeexplore.ieee.org/xpl/conhome/9012/proceeding
29. Lee, C.S., Jones, N.D., Ben-Amram, A.M.: The size-change principle for program termination. In: Hankin, C., Schmidt, D. (eds.) Conference Record of POPL 2001: The 28th ACM SIGPLAN-SIGACT Symposium on Principles of Programming Languages, London, UK, 17–19 January 2001, pp. 81–92. ACM (2001). https://doi.org/10.1145/360204.360210
30. [loopfission]: Loop fission interference graph (fig). https://reviews.llvm.org/D73801
31. Mehta, S., Lin, P., Yew, P.: Revisiting loop fusion in the polyhedral framework. In: Moreira, J.E., Larus, J.R. (eds.) ACM SIGPLAN Symposium on Principles and Practice of Parallel Programming, PPoPP 2014, Orlando, FL, USA, 15–19 February 2014, pp. 233–246. ACM (2014). https://doi.org/10.1145/2555243.2555250
32. Microsoft: Parallel patterns library (PPL) (2021). https://docs.microsoft.com/en-us/cpp/parallel/concrt/parallel-patterns-library-ppl?view=msvc-170
33. Moyen, J.-Y., Rubiano, T., Seiller, T.: Loop quasi-invariant chunk detection. In: D'Souza, D., Narayan Kumar, K. (eds.) ATVA 2017. LNCS, vol. 10482, pp. 91–108. Springer, Cham (2017). https://doi.org/10.1007/978-3-319-68167-2_7
34. Palkowski, M., Klimek, T., Bielecki, W.: TRACO: an automatic loop nest parallelizer for numerical applications. In: Ganzha, M., Maciaszek, L.A., Paprzycki, M. (eds.) 2015 Federated Conference on Computer Science and Information Systems, FedCSIS 2015, Lódz, Poland, 13–16 September 2015. Annals of Computer Science and Information Systems, vol. 5, pp. 681–686. IEEE (2015). https://doi.org/10.15439/2015F34
35. Prema, S., Nasre, R., Jehadeesan, R., Panigrahi, B.: A study on popular auto-parallelization frameworks. Concurr. Comput. Pract. Exp. **31**(17), e5168 (2019). https://doi.org/10.1002/cpe.5168
36. Quinlan, D., et al.: Rose user manual: a tool for building source-to-source translators draft user manual (version 0.9.11.115). https://rosecompiler.org/uploads/ROSE-UserManual.pdf
37. Rauchwerger, L., Padua, D.A.: Parallelizing while loops for multiprocessor systems. In: Proceedings of the 9th International Symposium on Parallel Processing, IPPS 1995, pp. 347–356. IEEE Computer Society (1995)

38. Seiller, T.: Interaction graphs: full linear logic. In: Grohe, M., Koskinen, E., Shankar, N. (eds.) Proceedings of the 31st Annual ACM/IEEE Symposium on Logic in Computer Science, LICS 2016, New York, NY, USA, 5–8 July 2016, pp. 427–436. ACM (2016). https://doi.org/10.1145/2933575.2934568
39. Vitorović, A., Tomašević, M.V., Milutinović, V.M.: Manual parallelization versus state-of-the-art parallelization techniques. In: Hurson, A. (ed.) Advances in Computers, vol. 92, pp. 203–251. Elsevier (2014). https://doi.org/10.1016/B978-0-12-420232-0.00005-2

SMT-Based Modeling and Verification of Spiking Neural Networks: A Case Study

Soham Banerjee[1], Sumana Ghosh[1], Ansuman Banerjee[1(✉)], and Swarup K. Mohalik[2]

[1] Indian Statistical Institute, Kolkata, India
ansuman@isical.ac.in
[2] Ericsson Research, Bangalore, India

Abstract. In this paper, we present a case study on modeling and verification of Spiking Neural Networks (SNN) using Satisfiability Modulo Theory (SMT) solvers. SNN are special neural networks that have great similarity in their architecture and operation with the human brain. These networks have shown similar performance when compared to traditional networks with comparatively lesser energy requirement. We discuss different properties of SNNs and their functioning. We then use Z3, a popular SMT solver to encode the network and its properties. Specifically, we use the theory of Linear Real Arithmetic (LRA). Finally, we present a framework for verification and adversarial robustness analysis and demonstrate it on the Iris and MNIST benchmarks.

Keywords: Spiking neural networks · Satisfiability modulo theory · Verification · Adversarial robustness

1 Introduction

In recent times, Satisfiability Modulo Theory (SMT) solvers are being widely used for modeling and verification of both hardware and software designs. SMT solvers, being based on their SAT-counterparts, employ a wide arsenal of heuristics and proof techniques that enable them to scale to large and complex programs. This work is an exploration case study of the usage of SMT for modeling and verifying Spiking Neural Networks (SNNs). Specifically, we use it for property verification and adversarial robustness analysis, and examine the scalability issues.

Neural network based models, especially DNNs (Deep Neural Networks) are becoming mainstream due to the rapid progress of their performance in many application domains [22]. However, the large amount of energy consumed by these models during their training and inferencing phases remains a major area of concern, impacting their deployment on resource-constrained devices and on the OPEX (operating expense) budget. Neuromorphic computing [28] - mimicking the neuro-biological architecture and function in both hardware and software

- seems to be a promising approach to address the energy challenge. Therefore, recent times have seen tremendous growth in both neuromorphic hardware and software models such as Spiking Neural Networks [26], which are showing performance comparable to traditional DNNs with lesser energy requirements [21].

Due to comparable performance and better energy efficiency, it is envisaged that the application scope of neuromorphic models will extend, even to safety critical systems like self driving vehicles, automatic guidance and assistance systems like ADAS (Advanced Driver Assistance Systems) etc. Since these systems have extreme safety requirements, rigorous verification of these systems becomes indispensable to ensure that the models do not lead to critical failures or violation of their safety properties. Formal verification of neuromorphic models against specified properties such as safety and robustness is a promising approach to provide this level of rigor.

The main objective of the present work thus is to provide a framework [1] for formal verification of SNNs which is the most well-known neuromorphic software model. We show that a straightforward adaptation of Quantifier Free Linear Real Arithmetic constraints (QF_LRA) leads to a sound and complete model for a given SNN. Additionally, we show how a standard definition of adversarial robustness based on perturbations on pixel values in input images can be adapted to the concept of input spike trains used in the SNN context and encoded as a collection of SMT LRA constraints. Together, this gives us a novel application of SMT solvers for modeling, analysis and robustness checking for SNNs. Specifically, our framework has the following contributions:

- We present a method of encoding SNNs as Satisfiability Modulo Theory constraints that provides an expressive theory of the behaviour of SNNs.
- We use these encodings for verification and robustness analysis of SNNs.
- We finally demonstrate the verification framework on an SNN model trained for the Iris [12] dataset. We also demonstrate the adversarial robustness framework on the Iris and MNIST [7] datasets.

Though there exists a rich body of literature focusing on the modeling and verification of traditional neural networks [10,14,16,18,19], very few efforts are found focusing the same in the domain of SNNs [3,4]. To the best of our knowledge, this is the first SMT-based verification framework to capture the behavior and architecture of SNNs. An early work [4] has reported a Timed Automata [2] based model for an SNN, however, the scale and diversity of the SNNs they can handle is rather restricted. In contrast, we propose a generic framework that can handle any SNN variant and property type that is expressible in linear real arithmetic. We show experimental results on benchmark SNNs that demonstrates the working of our framework. This paper is an attempt to examine the power of SMT modeling on these new neural network variants.

The paper is organized as follows. Section 2 presents an overview of Spiking Neural Networks. Section 3 puts forward our proposal of an SMT encoding of SNNs, while Sect. 4 discusses the verification problem for these networks and our formulation for the same. Section 5 presents details of our implementation and experiments carried out on the Iris and MNIST benchmarks. Section 6 presents related work and Sect. 7 concludes this paper along with some future directions.

2 Spiking Neural Networks

A Spiking Neural Network tries to mimic both the architecture and the functioning of the human brain. It consists of a set of interconnected neurons. The functioning of these neurons is very different from those in traditional feed-forward neural networks where they act only as an instantaneous functions. Each neuron in SNN receives sequences of inputs through its input synapses, but only when they *spike*, they produce spikes on the output synapses depending upon the internal processing method. The inputs received by a neuron is modulated by the weight of the synapses. Therefore, given the weight of the synapses, a neuron can be modeled as a function which takes as input a sequence of spikes (*spike train*) per input synapse, and produces a spike train on its output.

In general, the spike trains, operations of the neurons and the SNN are defined in real time. However, for ease of presentation and for the feasibility of SMT based modeling presented in this paper, we present the discretized versions of these entities involved.

2.1 Discretized Simulation Time

Simulation time T and step size δ are hyperparameters of an SNN, which are used for defining spike trains towards training and prediction of the SNN. T specifies the length of the spike trains to be generated. Step size δ is chosen to be as large as possible so that the number of steps is minimum, while ensuring that no two spiking events occur within δ. For most systems, δ is considered as the length of the clock cycle of the underlying hardware on which the SNN is realized. For the rest of this discussion, we fix T and δ and define a discretized time domain $DI(\ell) = \{0, 1, \ldots, \ell\}$ where $\ell = \lfloor T/\delta \rfloor$, and $t_i - t_{i-1} = \delta \ \forall i \in \{1 \ldots \ell\}$.

2.2 Spike Trains and Encoding from Feature Inputs

Intuitively, a spike train is just a finite sequence of binary values (0 or 1) in the discretized time domain. Formally, we define a spike train as described below.

Definition 1 [Spike Train]. *Given simulation time T and step size δ, a spike train is defined as a mapping $DI \mapsto \{0, 1\}$ denoting spiking-time instances of a synapse. Given a spike train γ over DI, and $t \in \{0 \ldots \ell\}$, $\gamma[t] \in \{0, 1\}$ gives the spike value at time t.* ∎

Spike trains are either generated from external sources e.g. sensors or encoders, or as output from other neurons. For real life applications where an SNN based system consumes feature values that are either numeric or categorical, the values need to be encoded as spike trains of specified length of time. In the literature, a number of different encoding functions have been suggested [15] many of which are implemented in SNN platforms such as snnTorch [11]. For the experiments in this paper, we have used *rate encoding* which is discussed in Sect. 5.

2.3   Leaky Integrate and Fire (LIFR) Neuron

SNNs can greatly differ in their architecture and the types of constituent neurons. The most common type of neurons for SNNs is the Leaky-Integrate-and-Fire (LIFR) neurons. Formally, a LIFR neuron can be defined as in the following.

Definition 2 *(LIFR Neuron). A LIFR neuron is a 6-tuple $\langle \Psi, w, y, \theta, \lambda, \tau, p \rangle$: $\Psi = \{\psi_1, .., \psi_n\}$ is the set of input synapses, $w : \Psi \rightarrow \mathbb{R}$ gives the weights of the synapses, $y : DI(\ell) \rightarrow \{0, 1\}$ is the neuron output function, θ is the firing threshold of the neuron, τ is the refractory period, λ is the leak factor, and $p : DI(\ell) \rightarrow \mathbb{R}^+$ is the stored potential function initialized to 0. Given spike trains γ_i corresponding to each synapse ψ_i, the stored potential function p is evaluated at each $t \geq 1$ as:*

$$p(t) = \begin{cases} \lambda \cdot p(t-1) + \sum_{i=1}^{n} w(\psi_i) \cdot \gamma_i[t] & ; \ p(t-1) < \theta \\ \sum_{i=1}^{n} w(\psi_i) \cdot \gamma_i[t] & ; \ p(t-1) \geq \theta \end{cases}$$

Finally, the neuron output function y is given as:

$$y(t) = \begin{cases} 1; & p(t) \geq \theta \\ 0; & p(t) < \theta \end{cases}$$

Informally, in each timestep, the neuron collects the *instant potential* $\sum_{i=1}^{n} w(\psi_i) \cdot \gamma_i[t]$ which is the sum of the weights of the synapses ($w(\psi_i)$) that trigger at time t ($\gamma_i[t] = 1$). The instant potential is added to the stored potential (decayed by λ) and the total is checked if it crosses the firing threshold θ. If yes, then a spike is generated ($y(t) = 1$) and the stored potential is reset to 0, else, there is no spike and the stored potential is held. In the classical definition of an LIFR, after an output spike, the neuron waits for an amount of the refractory period τ before starting to update the stored potential. Most implementations of SNNs have a very small value of τ which is mostly ignored during computations. In this work, we assume $\tau = 0$.

We now define an SNN with LIFR neurons and define its behaviour based on the execution semantics of LIFR neurons as above.

Definition 3 *[SNN with LIFR neuron]. An SNN $\mathcal{N}$ with LIFR neurons is defined as a weighted directed acyclic graph $\langle N_{inp}, N, \psi, w \rangle$ where, N_{inp} refers to the set of input neurons and N refers to the set of LIFR neurons, $\psi \subseteq (N_{inp} \cup N) \times N$ refers to the set of synapses, and $w : \psi \rightarrow \mathbb{R}$ assigns a weight $w_{i,j}$ to each synapse (N_i, N_j).* ∎

The input neurons model the spike trains defined on $DI(\ell)$ as their output, and make it available for the LIFR neurons in N. Since the discretized domain is identical for all the spike trains, the input neurons provide a *complex spike train* from the space $\{0,1\}^{|N_{inp}| \times \ell}$. We denote this as γ. Individual spike train for each input neuron N_i is accessed as γ_i.

A subset of N is designated as the output neurons N_{out}. The output neurons do not have any outgoing synapses. For the LIFR neurons $N_i \in N$, we define $inSynapse(N_i) = \{(N_j, N_i) \mid (N_j, N_i) \in \Psi\}$, i.e., the incoming edges of N_i in the SNN graph. Each neuron $N_i \in N$ is then specifiable as a tuple $\langle inSynapse(N_i), w_i, y_i, \theta_i, \lambda_i, \tau_i, p_i \rangle$, where w_i is w restricted to $inSynapse(N_i)$.

The execution semantics of the SNN $\mathcal{N}$ is as follows. Let the complex spike train modeled by the input neurons be γ. Initially, the stored potential and output of each neuron is set to 0. At each timestep t, each input neuron N_i in N_{inp} produces $\gamma_i[t]$ as output. Thereafter, each neuron in N executes in a topological order and its output is propagated through the outgoing synapses. Thus, at each timestep, each output neuron produces an output (spike value 1 or 0). As a result, by the end of processing of γ, there is one output spike train from $\{0,1\}^\ell$ per output neuron. The complete output of the SNN is then a complex spike train in the space $\{0,1\}^{|N_{out}| \times \ell}$. We denote it as $\eta = \mathcal{N}(\gamma)$. Therefore, the SNN can be viewed as a function from $\{0,1\}^{|N_{inp}| \times \ell}$ to $\{0,1\}^{|N_{out}| \times \ell}$.

Example 1 Consider the SNN shown in Fig. 1 with the threshold of all neurons as 1 and initial potential as 0. We consider $T = 4\,\mathrm{s}$ and $\delta = 1\,\mathrm{s}$ and let the input spike trains for the two input neurons be $\gamma_0 = \{1, 0, 1, 1\}$ and $\gamma_1 = \{1, 1, 0, 1\}$. Each input spike train encodes the spiking behavior of an input neuron recorded over 4 timesteps. The first input neuron spikes at time steps 1, 3, and 4, while the second input spike at timesteps 1, 2, and 4 as shown in Columns 1 and 2 of Table 1. The spike execution behavior of the different neurons in the SNN in response to the input spike trains are presented in other columns of Table 1. At time 1, the stored potential of N_2 is updated by the instant potential of -0.7, which the sum of the weights of the synapses (N_0, N_2) and (N_1, N_2) (-0.3 and -0.4 resp.). Since it is less than the threshold, N_2 does not produce any spike at time 1. Similarly, N_3 and N_4 update their stored potential to 0.4 each. Since they also do not spike, the instant potential for N_5 at time 1 sums to 0. Therefore, the output neuron N_5 also does not spike at time 1. Going through the table, we find that only at time 4, N_3 and N_4 spike because their stored potentials cross the threshold of 1. Then, the stored potential of N_5 is updated to 2 and it produces a spike at time 4. Hence, the spike train of the output neuron N_5 is $\{0, 0, 0, 1\}$. ∎

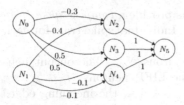

Fig. 1. SNN graph

Table 1. Execution of SNN

Time	Input	Stored potential				Spiking neurons
		N_2	N_3	N_4	N_5	
$t = 1$	N_0, N_1	-0.7	0.4	0.4	0	None
$t = 2$	N_1	-1.1	0.3	0.3	0	None
$t = 3$	N_0	-1.4	0.8	0.8	0	None
$t = 4$	N_0, N_1	-2.1	1.2	1.2	2	$\{N_3, N_4, N_5\}$

3 SMT Encoding of SNN

SMT encoding of a given SNN involves encoding the execution semantics of the internal and output neurons using the spike trains generated by the input neurons. Towards this, we first introduce the Boolean and real variables used in our SMT encoding. For all input neurons N_i, let $x_{i,t}$ be the binary variable denoting the value of the input spike train corresponding to N_i at time t. For all other neurons N_i, let the variable $x_{i,t}$ denote the binary output of N_i, $S_{i,t}$ the instant potential gained by N_i and $P_{i,t}$ the stored potential at time t.

LRA Constraints

Recall from the definition of an SNN that $\lambda \in \mathbb{R}$ is the leak factor of the neurons, $w_{i,j} \in \mathbb{R}$ gives the weight of the synapse between the adjacent neurons (N_i, N_j), and θ_i is the threshold of the neuron N_i.

ξ_0: *Initialization:* At the 0'th timestep, the potential variables are initialized to zero.

$$\xi_0(i, 0) \triangleq (P_{i,0} = 0)$$

ξ_1: *The instant potential of a neuron is computed from the weights of the incoming synapses that are triggered.* At every timestep, for a neuron N_i, if it receives a spike from another neuron N_j, the weight $w_{j,i}$ of the synapse between N_j and N_i is added to the potential of N_i. Formally, we encode the instant potentials as,

$$\xi_1(i, t) \triangleq (S_{i,t} = \sum_{j \in inSynapse(N_i)} x_{j,t} \cdot w_{j,i})$$

Note, the binary variable $x_{j,t}$ captures whether N_j has spiked or not at timestep t and thus affects the instant potential $S_{i,t}$ of neuron N_i.

ξ_2: *A neuron spikes if its stored potential crosses its threshold.* The stored potential of a neuron N_i at any timestep is the sum of the instant potential $S_{i,t}$ from the input synapses and the previously stored potential $P_{i,t-1}$. Therefore, we have the following LRA expression corresponding to ξ_2,

$$\xi_2(i, t) \triangleq ((S_{i,t} + \lambda \cdot P_{i,t-1}) \geq \theta_i \implies (x_{i,t} = 1)).$$

As time passes, the neuron potential starts to leak causing the stored potential to decay, hence, we use the leak factor λ in the formulation.

ξ_3: *A neuron does not spike if its potential does not cross its threshold.* Similar to ξ_2, we have the LRA expression for ξ_3 as,

$$\xi_3(i,t) \triangleq ((S_{i,t} + \lambda \cdot P_{i,t-1}) < \theta_i \implies (x_{i,t} = 0)).$$

ξ_4: *On crossing the threshold, the stored potential of a neuron is reset.* For a neuron N_i, once its threshold is crossed, the neuron's stored potential is set to 0. Hence, we have the LRA formula as,

$$\xi_4(i,t) \triangleq ((S_{i,t} + \lambda \cdot P_{i,t-1}) \geq \theta_i \implies (P_{i,t} = 0)).$$

ξ_5: *If the total potential (stored + instant) of a neuron does not cross the threshold, it is stored for the next timestep.* This can be obtained by updating the current (i.e., at the tth timestep) stored potential $P_{i,t}$ of the neuron N_i. Formally, the LRA expression is,

$$\xi_5(i,t) \triangleq (S_{i,t} + \lambda \cdot P_{i,t-1}) < \theta_{i,j} \implies (P_{i,t} = S_{i,t} + \lambda \cdot P_{i,t-1}).$$

Collecting the formulas $\xi_0, \xi_1, \xi_2, \xi_3, \xi_4$ and ξ_5, we define the SMT encoding $F_{\mathcal{N}}$ of the given SNN in the following.

$$F_{\mathcal{N}} \triangleq \left(\bigwedge_{i \in N} \xi_0(i,0) \right) \wedge$$

$$\left(\bigwedge_{t \in \{1...\ell\}} \bigwedge_{i \in N} \xi_1(i,t) \wedge \xi_2(i,t) \wedge \xi_3(i,t) \wedge \xi_4(i,t) \wedge \xi_5(i,t) \right) \tag{1}$$

As defined earlier, N denotes the set of LIFR neurons (ref. Definition 3). We note certain properties of the SMT encoding which allows us to verify an SNN through the SMT encoding.

Proposition 1. *Let $\mathcal{N}$ be an SNN, γ be a complex spike train and $F_{\mathcal{N}}$ be the SMT encoding of $\mathcal{N}$ as in Eq. 1. Let $\sigma_\gamma : x_{i,t} = \gamma_i[t]$ be the instantiation of the output variables for the input neurons $N_i \in N_{inp}$. Then, the formula $F_{\mathcal{N}}$ is satisfiable and there is a unique satisfying assignment for all the output variables.*

Since the output variables corresponding to N_{out} can be compiled into a spike train $\eta \in \{0,1\}^{|N_{out}| \times \ell}$, we can think of $F_{\mathcal{N}}$ as a function from $\{0,1\}^{|N_{inp}| \times \ell} \to \{0,1\}^{|N_{out}| \times \ell}$. In the following, we denote the output spike train $\eta = F_{\mathcal{N}}(\gamma)$. The one-to-one SMT-based encoding of the steps in the SNN execution semantics ensures the following proposition leading to the final corollary.

Proposition 2. *For all input spike trains* γ, $\mathcal{N}(\gamma) = F_{\mathcal{N}}(\gamma)$.

Corollary 1. *Given an SNN* $\mathcal{N}$, *its SMT encoding is sound and complete for all properties defined on its inputs and outputs.*

Thus, we are ensured of a sound and complete verification procedure for SNNs based on SMT encoding. Note that the number of variables used for the SMT encoding is $O(n \times \ell)$ where n refers to the number of neurons in the SNN, and $\ell = \lfloor T/\delta \rfloor$ refers to the number of timesteps. The number of LRA constraints is $O(8 \times n \times \ell)$ (3 constraints from ξ_0 and ξ_1–ξ_5).

4   The Proposed Framework

Our framework takes as input a given trained SNN and a property to be checked, and either provides a formal guarantee that the SNN satisfies the property for all inputs (either unconstrained or as constrained by the input constraints specified as part of the property), or a counterexample showing an input for which the property is violated. Through the SNN verification problem, one can check for various system properties, such as *adversarial robustness*. This property essentially says a minor perturbation to the input of an SNN does not affect its output, and thereby, the SNN is robust to adversarial perturbations (details in Sect. 4.3). Formally, the verification problem is defined as below.

Definition 4 *[SNN Verification].* *Given a 3-tuple* $\langle \mathcal{N}, P, Q \rangle$, *with the following* $\mathcal{N} : \{0,1\}^{|N_{inp}| \times \ell} \rightarrow \{0,1\}^{|N_{out}| \times \ell}$ *as an SNN,* $P : \{0,1\}^{|N_{inp}| \times \ell} \rightarrow \{0,1\}$ *and* $Q : \{0,1\}^{|N_{out}| \times \ell} \rightarrow \{0,1\}$ *respectively the input and output properties of N, the verification problem is to decide if there exists a* $\gamma \in \{0,1\}^{|N_{inp}| \times \ell}$ *for which* $P(\gamma) \wedge (\mathcal{N}(\gamma) = \eta) \wedge \neg Q(\eta)$ *is satisfiable.*   ∎

Note, $\eta \in \{0,1\}^{|N_{out}| \times \ell}$ is the output spike train obtained by running the SNN $\mathcal{N}$ for the input spike train γ. If there exists such an input spike train γ for which η does not satisfy the output property Q, γ then acts as a counter-example to the verification problem. Otherwise, we are ensured that for all γ satisfying the input property, the corresponding output $\eta = \mathcal{N}(\gamma)$ satisfies Q. The verification framework is enabled by the SMT encoding of SNNs. The SMT encoding of the input and output properties are as follows.

4.1   Encoding of Input and Output Properties

The input properties define a subset of complex input spike trains γ from the space $\{0,1\}^{|N_{inp}| \times \ell}$ for which the property is to be checked. Some of the common input properties and their respective encodings are given next. The SMT encoding F_P is over the set X_{inp} of input variables i.e. $\{x_{i,t} \mid N_i \in N_{inp}$ and $t \in \{1 \ldots \ell\}\}$. The SMT encoding, F_Q, for output properties is similar, except for the fact that F_Q uses the set $X_{out} = \{x_{i,t} \mid N_i \in N_{out}$ and $t \in \{1 \ldots \ell\}\}$.

IP1: All Input Spike Trains: This property holds for all input spike trains. It is most commonly used for property testing where independent of the input, the system should satisfy the safety properties. The respective encoding is: $F_P(X_{inp}) \triangleq \top$. Here, $\top$ refers to always true.

IP2: A Specific Input Spike Train: This property holds for a specific input spike train from the entire input space. It is commonly used for properties based on exceptions and corner cases. The encoding is done by assigning values to variables in X_{inp}. The SMT encoding is fairly direct:

$$F_P(X_{inp}) \triangleq \bigwedge_{N_i \in N_{inp}} \bigwedge_{t \in \{1...\ell\}} (x_{i,t} = \gamma_i[t])$$

For example, given the complex input spike train $\gamma = \{\{1,0,1\}, \{1,1,0\}\}$ for an SNN having two input neurons, the LRA expression is:

$$F_P(X_{inp}) \triangleq (x_{0,1} = 1) \wedge (x_{0,2} = 0) \wedge (x_{0,3} = 1) \wedge (x_{1,1} = 1) \wedge (x_{1,2} = 1) \wedge (x_{1,3} = 0).$$

IP3: Bounded Spike Counts: This property defines a condition on the number of spikes in the spike train for an input neuron N_i expressing the requirement that the number of spikes is bounded by a lower (c_1) and/or an upper (c_2) bound. The respective encoding is:

$$F_P(X_{inp}) \triangleq c_1 \leq \sum_{t=1}^{\ell} x_{i,t} \leq c_2$$

For example, given an arbitrary SNN with 3 timesteps, its first input neuron N_1 should have at least 2 and at most 5 spikes. The corresponding encoding is,

$$F_P(X_{inp}) \triangleq 2 \leq x_{1,1} + x_{1,2} + x_{1,3} \leq 5$$

IP4: Compound Properties: The conjunction of multiple input properties can be generically written as,

$$F_P(X_{inp}) \triangleq F_{P_1}(X_{inp}) \wedge F_{P_2}(X_{inp}) \wedge \cdots \wedge F_{P_k}(X_{inp})$$

4.2 Overall Verification Framework

Given a verification problem $\langle \mathcal{N}, P, Q \rangle$, the verification framework generates the SMT encodings $F_{\mathcal{N}}$, F_P and F_Q over the appropriate variables. It then calls Algorithm 1 which presents the formula $F_{\mathcal{N}} \wedge F_P \wedge \neg F_Q$ to an SMT-solver (e.g. Z3). If the solver returns SAT, the satisfying assignment is processed to extract the input and corresponding output spike trains as counterexample to the verification problem. If the solver return UNSAT, the algorithm reports that there are no violating instances and the verification problem is successful. Consider the following example.

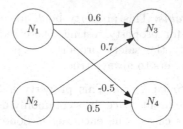

Fig. 2. A sample SNN

Example 2 Consider the SNN $\mathcal{N}$ shown in Fig. 2 and the input property P_1 and the output property Q_1 as stated below.

P_1: *the number of spikes generated by the neuron N_1 is more than the number of spikes generated by the neuron N_2.* Formally,

$$F_{P_1} \triangleq \sum_{t=1}^{\ell} x_{1,t} > \sum_{t=1}^{\ell} x_{2,t}.$$

Q_1: *the neuron N_4 never spikes.* Formally, $F_{Q_1} \triangleq \sum_{t=1}^{\ell} x_{4,t} = 0$.

We take the complement of Q_1, i.e., $\neg Q_1$: *neuron N_4 spikes at least once.* We then pass the SNN encoding $F_{\mathcal{N}}$ along with F_{P_1} and $\neg F_{Q_1}$ to the SMT-solver, which returns SAT along with the input spike train $\gamma_1 = \{0, 0, 1, 1, 1\}$ and $\gamma_2 = \{1, 1, 0, 0, 0\}$ for which the neuron N_4 does spike. Thus, $\mathcal{N}$ violates Q_1. Consider another input property P_2: *the neuron N_2 never spikes.* On invoking the SMT-solver for $F_{\mathcal{N}}$, F_{P_2}, and $\neg F_{Q_1}$, we get UNSAT. Hence, $\mathcal{N}$ satisfies Q_1. ■

The algorithm for SNN verification problem is outlined in Algorithm 1.

Algorithm 1. Algorithm for Verification of SNNs

Input: Verification problem $\langle \mathcal{N}, P, Q \rangle$
Output: $SUCCESS$ or $FAILURE(counterexample)$
    $F_{\mathcal{N}}$ = SMTEncode($\mathcal{N}$)
    F_P = SMTEncode(P)
    F_Q = SMTEncode(Q)
    $F = F_{\mathcal{N}} \wedge F_P \wedge \neg F_Q$
    if $Solver(F)$ **returns** SAT **then**
        $counterexample = getSolution(F)$
        return $FAILURE$
    else if $Solver(F)$ **returns** $UNSAT$ **then**
        return $SUCCESS$
    end if

In [16], it has been shown that the verification problem for feedforward neural networks is NP-hard. The proof contains a reduction of an instance of the 3-SAT problem to an instance of the neural network verification problem. Since SNN is a generalization of neural networks, the NP-hardness of the SNN verification problem follows. Moreover, there exists state-of-the-art solutions for verification of neural networks models (such as DNNs) using SMT [10,16], this motivated us to look at SNN encoding in SMT.

4.3 Verifying Adversarial Robustness of SNNs

For a classification problem with n-classes, one can design an SNN $\mathcal{N}$ with n output neurons. The SNN predicts a class y for a given complex input spike train γ when y is the predicted class, i.e. the index of the neuron with the maximum spikes. In other words,

$$\eta = \mathcal{N}(\gamma) \ \wedge \ y = argmaxspike(\eta)$$

where

$$argmaxspike(\eta) \triangleq \underset{i \in \{1...|N_{out}|\}}{argmax} \ (\sum_{j=1}^{\ell} \eta[i][j])$$

An SNN $\mathcal{N}$ trained for classification is adversarially robust when small perturbations to a given complex input spike train γ_0 does not lead to a change in the predicted class $y_0 = argmaxspike(\eta_0)$ where $\eta_0 = \mathcal{N}(\gamma_0)$. In this section, we describe how the proposed SMT based verification can be used to check if an SNN is adversarially robust with respect to a given input γ_0.

For the verification framework, we need to encode a formula $P(\gamma) \ \wedge \ \mathcal{N}(\gamma) = \eta \ \wedge \ \neg Q(\eta)$, where $P(\gamma)$ states that γ is a perturbation of a given γ_0, and $Q(\eta)$ states that $argmaxspike(\eta) = y_0$. Then, the corresponding SMT encoding and verification will ensure that if the SMT encoding is unsatisfiable, for all perturbations of γ_0, the predicted class is $argmaxspike(\eta_0)$ where $\eta_0 = \mathcal{N}(\gamma_0)$. On the other hand, when the SMT encoding is satisfiable, it produces an assignment to the variables which encodes a perturbation of γ_0 leading to a different prediction. We now formally define $P(\gamma)$ and $Q(\eta)$ and their SMT encoding.

Definition 5 [$\Delta-$Perturbation]. *For a given non-negative integer Δ, the $\Delta-$perturbation of an input spike train refers to changing its spike counts by at most Δ changes. If γ is obtained from the input spike train $\gamma_0 \in \{0,1\}^{|N_{inp}|\times \ell}$ after Δ-perturbation, we have*

$$\sum_{i=1}^{|N_{inp}|} \sum_{j=1}^{\ell} \mid \gamma[i][j] - \gamma_0[i][j] \mid \ \leq \Delta.$$

■

For most real-life applications, the spike trains generated by sensors are more susceptible to errors where there are missing intended spikes or more spikes than expected. Therefore, we prefer to use the notion of $\Delta-$perturbation in terms of spike counts for adversarial robustness. The difference, $\sum_{i=1}^{|N_{inp}|} \sum_{j=1}^{\ell} \mid \gamma[i][j] - \gamma_0[i][j] \mid$ is the Manhattan distance between the two vectors since these vectors consist of 0's and 1's only. Hence, we formally denote the Δ-perturbation as $\|\gamma - \gamma_0\|_1$, where $\|..\|_1$ refers to the first norm or Manhattan norm. So, $P(\gamma)$ can be concisely specified as $\|\gamma - \gamma_0\|_1 \leq \Delta$. The corresponding SMT encoding is:

$$F_P \triangleq \sum_{i=1}^{|N_{inp}|} \sum_{j=1}^{\ell} x_{i,j} \oplus \gamma_0[i][j] \leq \Delta.$$

Note, the symbol $\oplus$ indicates the XOR operation and $x_{i,j} \oplus \gamma_0[i][j]$ will be 1 only when $x_{i,j}$ and $\gamma_0[i][j]$ have opposite values. In order to formalize $Q(\eta)$, we note that the function $y = argmax(x_1, x_2, \cdots, x_n)$ can be encoded as below.

$$[\bigwedge_{j=1, j \neq i}^{n} (x_i > x_j) \iff (y = i)]$$

Therefore, the encoding of $Q(\eta)$ can be described as:

$$F_Q \triangleq \bigwedge_{j=1, j \neq y_0}^{|N_{out}|} (\sum_{t=1}^{\ell} x_{y_0,t} > \sum_{t=1}^{\ell} x_{j,t}),$$

which when satisfiable ensures that $y_0 = argmaxspike(\eta)$. Finally, the SMT encoding for verification of adversarial robustness of $\mathcal{N}$ for a given spike train γ_0 is produced as $F_P \wedge F_{\mathcal{N}} \wedge \neg F_Q$.

5   Implementation and Results

We implemented our approach as a tool [1] on top of Z3Py as the SMT solver. The Z3Py framework is a binding of the Z3 theorem prover [24] for the Python programming language. We carried out our work on some pre-trained SNN models on an Intel Core i7-10870H processor at 2.2 Ghz and 16 GB of RAM.

For our experiments, we considered the Iris dataset [12] and the MNIST handwritten digits dataset [7]. These datasets are image classification use cases of neural networks, used to classify flower types and handwritten digits respectively. The Iris dataset is composed of 150 data points (instances), each instance corresponding to the description of a type of Iris plant. Each row of the dataset contains four input features corresponding to the sepal length, sepal width, petal length and petal width of the instance. Each row also contains a classification label from the set {Iris-Setosa, Iris-Versicolor, Iris-Virginica}. The MNIST dataset contains 70,000 data points (image instances). Each row corresponds to a 28×28 gray-scale image with pixel values in the range 0–255 and a corresponding output label from the set of all single digit numbers (0 to 9). Since the real valued vectors cannot be fed as input to the SNN, we use *rate encoding* (discussed next) to convert them into spike trains.

Rate Encoding: Given a feature value v, normalized to $v \in [0, 1]$, the rate encoding of v for the discrete time domain $DI(\ell)$ generates a spike train $\gamma \in \{0, 1\}^{\ell}$ by running a Bernoulli trial $B(\ell, \bar{v})$. Thus, the probability of k number of spikes (spike value 1) in a generated spike train for v can be given by: $P(k) = \binom{\ell}{k} \bar{v}^k (1 - \bar{v})^{(\ell-k)}$, $0 \leq k \leq \ell$. Thus, an input value of 0 would produce a spike train with no spikes and a value of 1 would generate a spike train with spikes at each timestep. For the given datasets, we first normalize the values corresponding to each input feature to the range $[0, 1]$. For each input, we have a corresponding vector where all values are normalized. This input vector is then used to create an encoded spike train using the *spikegen* module of snnTorch.

Our models were trained using the snnTorch [11] with varying accuracy levels to see how networks with low/high accuracies perform when analyzed. An interesting observation is made with respect to the low accuracy obtained for the Iris models (Table 2 using snnTorch). This may be an artifact of the encoding used, however, our framework can take in any SNN model as input with any accuracy and carry out the analysis task.

Table 2. Verification experiments with Iris

Networks	Accuracy of model (%)	Properties							
		φ_1		φ_2		φ_3		φ_4	
		–	Time (s)	–	Time (s)	–	Time (s)	–	Time (s)
SNN-5	40	SAT	0.1	SAT	1.3	UNSAT	0.1	SAT	0.28
SNN-10	47	SAT	4.2	SAT	1.3	UNSAT	0.9	SAT	0.9
SNN-15	56	SAT	89.5	SAT	5.1	UNSAT	494.2	SAT	42.7
SNN-20	58	SAT	35.4	SAT	7.4	SAT	793.2	SAT	125.7
SNN-25	64	SAT	60.0	SAT	35.8	SAT	3752.2	SAT	4816.8

5.1   Verification Results on the Iris Dataset

We considered a number of SNNs (Column 1 of Table 2), each with 4 input neurons (corresponding to the input features), 1 hidden layer with 5 hidden neurons and 3 output neurons (corresponding to the 3 output classes). The networks differ in the number of time steps (i.e., $5, 10, 15, 20, 25$) for which they were trained and thus, in their accuracy of predictions as reported in the Column 2 of Table 2. These networks were then tested against 4 properties, φ_1 to φ_4. Table 2 lists the properties satisfied by each network along with the time taken in seconds for the verification task to be carried out by the SMT solver. We now explain the encoding of the properties. Let N_9 be the output neuron corresponding to the label 'Iris-Setosa', N_{10} be the output neuron corresponding to the label 'Iris-Versicolor', N_{11} the output neuron corresponding to the label 'Iris-Virginica' and ℓ the number of timesteps. We considered the following properties.

- φ_1: **There exists some input for which the network predicts the output class as 'Iris-Setosa'.**
  This property refers to the network's ability to label a given input as 'Iris-Setosa' for some input. If a network does not satisfy this property for any input, we can say that the network can never output the label corresponding to 'Iris-Setosa'. We need to check that the total number of spikes corresponding to the output Iris-Setosa is the highest. This property is a spike count based property encoded as:

$$F_{\varphi_1} \triangleq \left(\sum_{t \in 1 \cdots \ell} x_{9,t} > \sum_{t \in 1 \cdots \ell} x_{10,t} \right) \wedge \left(\sum_{t \in 1 \cdots \ell} x_{9,t} > \sum_{t \in 1 \cdots \ell} x_{11,t} \right).$$

- φ_2: **There exists some input for which the network predicts the output class as 'Iris-Versicolor'.**
- φ_3: **There exists some input for which the network predicts the output class as 'Iris-Virginica'.**
  φ_2 and φ_3 have encodings similar to F_{φ_1}. While the above are properties based on class labels, the following is an arbitrary property that formulates a requirement on the bound on the number of spikes. We generated multiple such properties as observations from the SNN traces and experimented with the same to show how our framework performs. In Table 2, we only report our findings on the above 4 properties. We could not find any standard property suite for the Iris or MNIST benchmarks, so these simple properties were created to check the sanity of the verification framework. The verification was carried out without any constraints on the input space, i.e. for all inputs.
- φ_4: **The total number of spikes at the output should not be more than 75% of the total number of spikes possible at the output.**
  This property refers to the network's ability to generate the output label without producing an excessive number of output spikes. The total number of spikes refers to the output spike train containing a maximum of (*number of timesteps* × *number of output labels*) spikes. This property is a spike count based property that can be encoded as follows.

$$F_{\varphi_4} \triangleq \left(\sum_{t \in 1 \cdots \ell} x_{9,t} + \sum_{t \in 1 \cdots \ell} x_{10,t} + \sum_{t \in 1 \cdots \ell} x_{11,t} \right) \leq 0.75 \times 3 \times \ell$$

We use our verification framework as outlined in Algorithm 1 for F_{φ_1} to F_{φ_4}. The satisfiability results along with the time (in seconds) taken by the SMT-solver are reported in Table 2. The FAILURE flag returned by Algorithm 1 indicates that the property is not satisfied by the model and is represented by 'UNSAT' in Table 2. Similarly, the SUCCESS flag returned by Algorithm 1 indicates that the property is satisfied and is represented by 'SAT' in Table 2. As we can see from the table, more the number of timesteps for which the network is trained, more is the accuracy and more is the time required for verification. This is due to the fact that an increasing number of timesteps also increases the number of variables in the network encoding.

5.2   Adversarial Robustness for Iris and MNIST

For carrying out experiments on the adversarial robustness of Iris, we took the model with the highest (64%) accuracy. We then randomly selected a small subset of instances from the Iris dataset. For each instance, we used our framework to verify the adversarial robustness of the trained network for different values of Δ, according to the notion of Δ-perturbation defined in Sect. 4.3. Table 3 presents the number of adversarially robust samples and the time taken for checking robustness. For the adversarial robustness experiment, we considered 72000 s as the timeout for the solver. We can observe that with increasing values of Δ (Column 1), the time taken for checking robustness also increases (Column 4). For larger values of Δ like 5 and 10, timeouts start to occur due to the large size of the adversarial input space. In cases where an adversarial input is found, the query finishes early (Column 6). For $\Delta = 1, 2$, all of the 5 instances were adversarially robust, whereas for $\Delta = 3$, 4 out of 5 samples were adversarially robust, and for one sample adversarial input was found in approximately 3.3 min. Similarly, for $\Delta = 5$, 2 out of 5 samples had adversarial inputs, one sample was found as adversarially robust, and for other 2 samples solver reported timeout. For $\Delta = 10$, 2 of the 5 samples had adversarial inputs which took time in the order of minutes to find.

Table 3. Adversarial robustness for SNN-Iris for different values of Δ

Δ	Total instances	#Robust instances		#Non-robust instances		Timeouts
		#Instances	Time (s)	#Instances	Time (s)	#Instances
1	5	5	22.63	0	–	0
2	5	5	288.28	0	–	0
3	5	4	6104.51	1	198.45	0
5	5	1	30425.76	2	341.34	2
10	5	0	–	2	259.26	3

For MNIST, we trained multiple SNNs each with 3 layers, an input layer with 784 neurons (28×28 is the MNIST input image dimension), an output layer with 10 neurons (corresponding to each single digit number) and 1 hidden layer with different numbers of neurons and trained for 5 timesteps. For these different networks (differing in the number of hidden layer neurons), we checked for adversarial robustness for $\Delta = 1$ to observe how an increase in network size can affect the time to check for robustness. Table 4 shows the time taken for the robustness check. We can observe that the time taken for robustness checking increases with increase in the number of neurons. However, increase in (a) the value of Δ, (b) the number of hidden layer neurons beyond 500, (c) timesteps for training – each led to scalability issues and timeouts in our experiments, for which we do not present the results here.

Table 4. Adversarial robustness for SNN-MNIST with $\Delta = 1$

# hidden layer neurons	Accuracy of model (%)	Timesteps	Adv. robust	Total time (s)
10	35.18	5	Yes	3109.05
50	68.07	5	Yes	16996.68
100	74.90	5	Yes	40397.20
500	91.31	5	Yes	419123.15

6 Related Work

In this section, we highlight some relevant existing works in the domain of modeling and verification of traditional neural networks as well as SNNs.

Modeling and Verification of Traditional Neural Networks (NNs): There exists a rich body of literature focusing on the modeling and verification of traditional NNs [10,14,16,18,19]. In [18], authors have presented a modeling and verification framework for DNNs using LRA. This paper also discusses the notion of scalable verification and how the different architectures can enable more verification friendly networks. An abstraction-refinement based verification for traditional NNs has been introduced in [10]. It has been shown that the proposed framework is a more efficient and scalable verification model for traditional NNs. Another popular framework for NN verification developed based on the simplex method is Reluplex [16]. Reluplex can only handle the rectified linear unit activation function, the most common activation functions used in traditional neural networks. A verification method developed based on mixed integer programming has been proposed in [27] to determine the exact adversarial accuracy of an MNIST classifier for some pre-specified perturbation measures.

Verification-Based Analysis of Traditional NNs: A significant amount of work exists in the literature that uses verification as an oracle to get some further improvised results on NNs. One such work [19] develops a sound framework that can prune and slice NNs by identifying redundant neurons and eliminating them. Another work [13] proposes a method for simplification of NNs by the elimination of dead neurons (that do not contribute to the output) without changing the behaviour of the original network. On the other hand, [14] presents a framework that repairs a given NNs by allowing a minimal modification with respect to a given counterexample of the input space. The verification framework is used to find the minimal amount of changes required to the network in order to account for the change in inputs and outputs.

Modeling and Verification of SNNs: There is a significant amount of literature focusing on various aspects of SNNs such as training methods of SNNs [8,9,26], variation of SNNs [6,17], hardware architectural development of SNNs [20,25]. However, very few efforts have been found in the formal modeling and verification of SNNs. The work [4] proposes a timed automata based encoding for modeling SNN functionalities and properties. Here, SNNs are formalised as sets of automata, one for each neuron, running in parallel and sharing channels according to the network structure. On the other hand, [3] translates a spiking neural system with weighted synapses into a class of timed safety automata. In both the papers, modeling and verification of system properties are done using tools and techniques developed for timed automata. A model checking approach to reduce a given SNN has been proposed in [23]. SNNs are modeled as discrete-time Markov chains and PRISM is used as the underlying verification tool. However, the effectiveness of both these frameworks is shown only on synthetic small examples. Formal encoding and verification of SNNs using Lustre programs (functional language operating on flows) has been proposed in [5]. Here, verification of temporal properties of neuronal archetypes are modeled as synchronous reactive systems, and thus represented as Lustre programs. In contrast to all these existing works, our work presents an application of SMT-based modeling and verification for SNNs.

7   Conclusion and Future Directions

In this paper, we propose an SMT-based encoding for Spiking Neural Networks with LIFR neurons. We address the property verification problem and the adversarial robustness problems for SNNs. We have implemented our methods into a tool that builds on the Z3 theorem prover. Experiments are presented on the Iris and MNIST benchmarks. As we can see from our experiments, the SMT-based encoding does not scale well to SNNs of large sizes. This has been reported in the context of traditional NNs as well. As one of the future directions, we plan to extend our framework with safe abstractions such that the analysis can be carried out in reasonable time. Another future direction to consider would be the comparison of the proposed model with other frameworks which use different mathematical formulations like Mixed Integer Linear Programming (MILP) for verification of SNNs. Additionally, it is worth investigating the effect of different encoding schemes (rate encoding versus other ones) and their effect on the scalability and verification of SNNs. Along with the improvements to the models, additional classes of properties can be studied further to improve the robustness of the framework. As part of this initial case study, we have adopted the rate encoding scheme with LIFR neurons, which has been reportedly one of the most popular and simpler variants of SNNs. We plan to extend the support to other variants going ahead. As future work, we also plan to extend the verification framework to wider range of SNNs (in terms of neurons and architectures) such as Convolutional SNNs and Recurrent SNNs.

References

1. Code and Benchmarks. https://github.com/Soham-Banerjee/SMT-Encoding-for-Spiking-Neural-Network

2. Alur, R.: Timed automata. In: Peled, D. (ed.) CAV 1999. LNCS, vol. 1633, pp. 8–22. Springer, Heidelberg (1999). https://doi.org/10.1007/3-540-48683-6_3

3. Aman, B., Ciobanu, G.: Modelling and verification of weighted spiking neural systems. Theoret. Comput. Sci. **623**, 92–102 (2016)

4. De Maria, E., Di Giusto, C., Laversa, L.: Spiking neural networks modelled as timed automata with parameter learning (2018)

5. De Maria, E., Muzy, A., Gaffé, D., Ressouche, A., Grammont, F.: Verification of temporal properties of neuronal archetypes modeled as synchronous reactive systems. In: Cinquemani, E., Donzé, A. (eds.) HSB 2016. LNCS, vol. 9957, pp. 97–112. Springer, Cham (2016). https://doi.org/10.1007/978-3-319-47151-8_7

6. Demin, V., Nekhaev, D.: Recurrent spiking neural network learning based on a competitive maximization of neuronal activity. Front. Neuroinf. **12**, 79 (2018)

7. Deng, L.: The MNIST database of handwritten digit images for machine learning research. IEEE Signal Process. Mag. **29**(6), 141–142 (2012)

8. Diehl, P.U., Cook, M.: Unsupervised learning of digit recognition using spike-timing-dependent plasticity. Front. Comput. Neurosci. **9**, 99 (2015)

9. Ding, J., Yu, Z., Tian, Y., Huang, T.: Optimal ANN-SNN conversion for fast and accurate inference in deep spiking neural networks (2021)

10. Elboher, Y.Y., Gottschlich, J., Katz, G.: An abstraction-based framework for neural network verification. In: Lahiri, S.K., Wang, C. (eds.) CAV 2020. LNCS, vol. 12224, pp. 43–65. Springer, Cham (2020). https://doi.org/10.1007/978-3-030-53288-8_3

11. Eshraghian, J.K., et al.: Training spiking neural networks using lessons from deep learning (2021)

12. Fisher, R.: Iris. UCI Machine Learning Repository (1988). https://archive.ics.uci.edu/ml/datasets/Iris

13. Gokulanathan, S., Feldsher, A., Malca, A., Barrett, C., Katz, G.: Simplifying neural networks using formal verification. In: Lee, R., Jha, S., Mavridou, A., Giannakopoulou, D. (eds.) NFM 2020. LNCS, vol. 12229, pp. 85–93. Springer, Cham (2020). https://doi.org/10.1007/978-3-030-55754-6_5

14. Goldberger, B., Katz, G., Adi, Y., Keshet, J.: Minimal modifications of deep neural networks using verification. In: LPAR23. LPAR-23: 23rd International Conference on Logic for Programming, Artificial Intelligence and Reasoning, vol. 73, pp. 260–278 (2020)

15. Guo, W., Fouda, M.E., Eltawil, A.M., Salama, K.N.: Neural coding in spiking neural networks: a comparative study for robust neuromorphic systems. Front. Neurosci. **15**, 638474 (2021)

16. Katz, G., Barrett, C., Dill, D.L., Julian, K., Kochenderfer, M.J.: Reluplex: a calculus for reasoning about deep neural networks. Formal Methods Syst. Design 1–30 (2021)

17. Kim, T., et al.: Spiking neural network (SNN) with memristor synapses having non-linear weight update. Front. Comput. Neurosci. **15**, 646125 (2021)

18. Kuper, L., Katz, G., Gottschlich, J., Julian, K., Barrett, C., Kochenderfer, M.: Toward scalable verification for safety-critical deep networks (2018)

19. Lahav, O., Katz, G.: Pruning and slicing neural networks using formal verification (2021)

20. Li, S., Zhang, Z., Mao, R., Xiao, J., Chang, L., Zhou, J.: A fast and energy-efficient SNN processor with adaptive clock/event-driven computation scheme and online learning. IEEE Trans. Circuits Syst. I Regul. Pap. **68**(4), 1543–1552 (2021)

21. Liu, T.Y., Mahjoubfar, A., Prusinski, D., Stevens, L.: Neuromorphic computing for content-based image retrieval. PLOS One **17**(4), 1–13 (2022). https://doi.org/10.1371/journal.pone.0264364
22. Malik, N.: Artificial neural networks and their applications (2005)
23. de Maria, E., Gaffé, D., Ressouche, A., Girard Riboulleau, C.: A model-checking approach to reduce spiking neural networks. In: BIOINFORMATICS 2018 - 9th International Conference on Bioinformatics Models, Methods and Algorithms, pp. 1–8 (2018)
24. de Moura, L., Bjørner, N.: Z3: an efficient SMT solver. In: Ramakrishnan, C.R., Rehof, J. (eds.) TACAS 2008. LNCS, vol. 4963, pp. 337–340. Springer, Heidelberg (2008). https://doi.org/10.1007/978-3-540-78800-3_24
25. Stimberg, M., Brette, R., Goodman, D.F.: Brian 2, an intuitive and efficient neural simulator. eLife **8**, e47314 (2019)
26. Tavanaei, A., Ghodrati, M., Kheradpisheh, S.R., Masquelier, T., Maida, A.S.: Deep learning in spiking neural networks (2018)
27. Tjeng, V., Xiao, K., Tedrake, R.: Evaluating robustness of neural networks with mixed integer programming (2017)
28. Yu, Z., Abdulghani, A.M., Zahid, A., Heidari, H., Imran, M.A., Abbasi, Q.H.: An overview of neuromorphic computing for artificial intelligence enabled hardware-based hopfield neural network. IEEE Access **8**, 67085–67099 (2020). https://doi.org/10.1109/ACCESS.2020.2985839

StaticPersist: Compiler Support for PMEM Programming

Sorav Bansal[✉]

Indian Institute of Technology Delhi, New Delhi, India
sbansal@iitd.ac.in

Abstract. Persistent Memory (PMEM) programs present unique programmability challenges. An important challenge involves ensuring that programs with mixed volatile-memory and persistent-memory ensure an important reachability invariant: at no point in the program execution should a persistent memory region contain a pointer to a volatile memory region. Such invariants are difficult to detect through testing methodologies, as the corresponding failures show up only in the presence of crashes. Prior work has leveraged runtime support in managed languages like Java (e.g., AutoPersist [31]) to check these invariants at runtime. However, such proposals incur a significant runtime cost. We propose a compile-time analysis that checks and maintains such reachability invariants statically with high precision. We implement this compile-time analysis in tool called *StaticPersist* which identifies such reachability-invariant violations and proposes fixes in C/C++ code.

Keywords: Persistent memory · Points-to analysis

1   Introduction

Fast byte-addressable persistent memory technologies, such as Intel 3D XPoint [20], Phase Change Memory (PCM) [36], and Resistive RAM [1], have the potential to re-define the programming interfaces and OS abstractions. Unlike traditional stable storage technologies like magnetic disks and Flash devices, these Persistent Memory (PMEM) technologies are nearly as fast as volatile DRAM and thus it is no longer tenable to pay the overheads of system calls and other traditional abstractions for stable storage. Instead, PMEM can now be managed

C. Dragoi et al. (Eds.): VMCAI 2023, LNCS 13881, pp. 44–65, 2023.
https://doi.org/10.1007/978-3-031-24950-1_3

simply using heap-allocated variables and existing load/store hardware instructions. This paradigm shift in the programming method opens a multitude of new problems and innovation opportunities. For example, programming language abstractions need to be adapted to allow easy and performant interfaces for mixing persistent and volatile memory objects, while minimizing the chances of programming mistakes. Some examples of PMEM programming frameworks are Intel PMDK [25], Mnemosyne [33], NVHeaps [7], Espresso [37], AutoPersist [31], Atlas [4], etc.

A major programmability issue associated with most of these programming models (except AutoPersist [31]) is the programmer burden associated with explicitly specifying PMEM objects and managing them separately from volatile memory objects, which we will also refer to as volatile objects. This manual disambiguation and management is both cumbersome and error-prone for the programmer. AutoPersist [31] proposed a new programming model in which the programmer only specifies the *durable roots*, i.e., the top-level named PMEM objects that need to PMEM-allocated. AutoPersist is then able to automatically mark all the other data structures that are *reachable* from the durable roots as persistent. The rationale of this logic is that it should not be acceptable for a program to have pointers to volatile memory within persistent objects—as such a situation would result in dangling pointers upon power failures.

AutoPersist implements the identification of these PMEM objects through a runtime reachability analysis implemented within a Java Virtual Machine. Because Java provides transparent support for object movement in memory, AutoPersist allows the JVM runtime to transparently and dynamically re-allocate volatile objects in PMEM, upon identification of a reachability path from a durable root to that object. AutoPersist implements optimizations that leverage profile-guided object placement to minimize re-allocation and copying at runtime. However, there are two primary drawbacks to the AutoPersist approach: (1) it is quite unclear how this approach can be adapted to C/C++ where transparent object movement may not be feasible due to support for raw pointers; (2) the runtime overheads of performing runtime reachability analysis, re-allocation and copying can be significant even after all the optimizations implemented in AutoPersist.

In our work, we are interested in designing compile-time analyses to automatically distinguish persistent memory from volatile memory. We borrow AutoPersist's programming model where the programmer is only required to specify the durable roots—the compiler then performs a static analysis to identify all *potentially* reachable objects from the durable roots, and marks them persistent. These persistent objects can then be allocated and accessed using PMEM-specific abstractions, e.g., to cater to failure-atomicity requirements. In theory, this problem of statically identifying all reachable memory objects from the durable roots is undecidable. Thus, we are interested in an over-approximate conservative solution. i.e., it is acceptable for our analysis to conservatively deem more objects as persistent. This can be seen more clearly with the C code example in Fig. 1. In this example, the `pmem_object` is allocated in persistent mem-

```
...
struct foo* pmem_object =
    pmem_malloc(M*sizeof(struct foo));
struct foo* volatile_object =
    malloc(N*sizeof(struct foo));
...
bool c = ...;
struct foo* p = c ? pmem_object
                  : volatile_object;
struct bar* reachable_object = ....;
p->child = reachable_object;
...
```

Fig. 1. Example showing the conservative nature of a static analysis.

ory (using `pmem_malloc()`) and the `volatile_object` is allocated in volatile memory. In this example, the `pmem_object` behaves like a durable root, as it is explicitly specified by the programmer to be allocated in persistent memory. The pointer `p` could potentially point to either the `persistent_object` or the `volatile_object`. Further, another object `reachable_object` could potentially be reached from the pointer `p`. Now it may not be feasible to statically determine if the pointer `p` points to the persistent object or the volatile object; and so the over-approximate analysis would conservatively mark `reachable_object` as persistent (because it could *potentially* be reached through a persistent object).

For correctness, the compile-time analysis needs to be *sound*, i.e., if a memory object O is potentially reachable at runtime from a persistent object, then O must be marked persistent. A *complete* compile-time analysis would ensure that *only* those objects are marked persistent that would actually be reachable from persistent objects at runtime; in other words, the objects that would not be reachable from persistent objects at runtime must be marked as volatile. Because the general problem of identifying reachability at compile-time is undecidable, it is not possible to develop an analysis that is both sound and complete. Instead, we are interested in developing a compile-time analysis that is sound and *precise*, i.e., the analysis must minimize the number of false-positives—here, a false-positive is a situation where an object is marked persistent even though it would not be reachable through a durable root. In our Fig. 1 example, if the Boolean condition `c` always evaluated to false at runtime, then the identification of `reachable_object` as persistent would be a false-positive. A false-positive does not compromise the correctness of the program (as it is always legal to allocate a volatile object in persistent memory) but it hurts the performance of the program (because the persistent memory object allocation and access would likely be slower than the allocation and access of a volatile memory object).

We present a compile-time sound reachability analysis algorithm that automatically identifies the persistent memory objects using the durable roots annotated by the programmer. We implement our analysis in a tool called *StaticPersist* that takes as input a C/C++ program annotated with durable roots, and identifies the objects that need to be allocated in persistent memory as they may potentially be reachable from the durable roots. A primary challenge in this line of investigation is the precision and scalability of such a compile-time analysis. We evaluate the precision and scalability of *StaticPersist* by testing it on programs implementing common data structures with up to 1000 Source Lines of Code (SLOC). In our current results, our algorithm and tool scales reasonably well and also produces usably precise results. Our current results indicate that it is possible to significantly reduce programmer burden associated with PMEM programming without incurring significant runtime overheads. Further, our algorithm is the first to support C/C++ programs. *StaticPersist* can be used either to automatically identify persistent objects, or to identify persistent-to-volatile-reachability bugs in a manually-written persistent memory program. Precision and scalability are paramount for both these applications. While we make significant advances on both these fronts in our work, we also identify several improvement opportunities in future work. Our tool is available for download at https://github.com/compilerai/counter/tree/static-persist.

2   Motivating Example

Most current PMEM programming models, that require the programmer to explicitly demarcate PMEM objects from volatile objects, are prone to both correctness and performance errors. An example of a correctness error is a situation where a volatile memory object is reachable through persistent memory object. Such an error would largely go undetected through traditional testing frameworks, but would stand exposed in the face of a crash failure. When a program would attempt to recover its state after a crash, it would observe a dangling pointer (to volatile memory) during recovery time; such a situation could potentially be catastrophic for security and safety critical programs.

Further, when the onus of carefully managing PMEM objects rests solely with the programmer, it is common for the programmer to choose code simplicity at the cost of potentially low-performance. This can be seen through the C++ code example in Fig. 2. In this example, the `callee()` function, that allocates and returns an object of type `bar`, is called twice by the `caller()` function: the result of the first call is used to store a pointer in a volatile memory object (heap-allocated variable referred-to by `foo *f`) while the result of the second call is used to store a pointer in a persistent memory object (global variable `pmem_f`). In this example, if the `callee()` returns a volatile pointer ((1)), then this program has a correctness bug, as the volatile object may be now reachable through a persistent object. On the other hand, if the `callee()` returns a persistent pointer, then this is a performance bug because we are unnecessarily storing a persistent object pointer in `f->child`, where a volatile pointer would have sufficed.

Our algorithm is capable of automatically transforming the program in Fig. 2 into the program shown in Fig. 3. In the transformed program shown in Fig. 3, a Boolean `persistent` flag is passed into the `callee()` program; depending on the

```
1 foo pmem_f;
2
3 bar* callee() {
4   //choose one
5   return new bar; //(1)
6   return pmem_new(bar);   //(2)
7 }
8 void caller() {
9   foo* f = new foo;
10   f->child = callee();
11   pmem_f.child = callee();
12   ...
13 }
```

Fig. 2. Motivating example.

```
foo pmem_f;

bar* callee(bool persistent) {
  if (persistent) {
    return pmem_new(bar);   //(2)
  } else {
    return new bar; //(1)
  }
}
void caller() {
  foo* f = new foo;
  f->child = callee(false);
  pmem_f.child = callee(true);
  ...
}
```

Fig. 3. Transformed version of the example in Fig. 2.

value of the flag, a persistent or a volatile object pointer is returned. Further, our analysis transforms all call-sites to the `callee()` function to additionally specify the value of the `persistent` flag. In this way, the transformed program is both correct and performant.

Because our compile-time analysis is capable of automatically determining such transformations, it relieves the programmer of complicating his/her code while still achieving the desired correctness and performance properties. Notice that because our analysis is compile-time, it incurs near-zero runtime overheads (unlike AutoPersist). Also, it may be applied to any programming language, irrespective of whether it is a managed language (like Java, Python, etc.) or an unmanaged language (like C/C++); this is again unlike AutoPersist which requires the language to support transparent object movement and thus limits the potential languages to managed languages only.

3 Algorithm

3.1 Points-To Algorithm Based on Allocation-Site

We now discuss our compile-time sound data-reachability analysis algorithm. The high-level algorithm resembles a *points-to analysis algorithm* [3] where, at each program PC (flow-sensitive), for each program variable, it maintains the set of memory locations that the variable could potentially point to. This points-to analysis is implemented as a DataFlow Analysis (DFA) [23] where the value is represented by a map from program variables to the set of memory locations. Because the set of memory locations may increase or decrease arbitrarily at runtime, it is not possible to precisely name all memory locations statically. Thus, we abstract the set of memory locations using the well-known *allocation-site* abstraction [2,5,21].

In the allocation-site abstraction, all memory objects allocated at a certain program PC value are coarsely referred-to by a single name. The nature of this coarsening can be seen through the example in Fig. 4. In this program, the allocation-site abstraction would determine that the variable pmem_p (a durable root intended to hold pointers to persistent memory objects) could potentially point to memory locations allocated at line 7. Similarly, the variable q may point to memory locations allocated at line 7. A more precise analysis would have been able to distinguish between objects allocated at line 7 in the first loop iteration (which may be reachable from pmem_p) from the objects allocated in subsequent loop iterations (which are not reachable from pmem_p but are reachable from q). However, an allocation-site based abstraction would coarsely determine that all objects allocated at line 7 are reachable from both pmem_p and q. As a result, the allocation-site based abstraction would conservatively determine that all objects allocated at line 7 must be allocated in persistent memory. Further, all

```
1  ...
2  bool first = true;
3  char* pmem_p = nullptr;
4  char* q = nullptr;
5  while (...) {
6    ...
7    x = malloc(...);
8    if (first) {
9      pmem_p = x;
10     first = false;
11   } else if (...) {
12     q = x;
13   }
14   q->child = r;
15 }
16 ...
```

Fig. 4. Allocation site abstraction.

objects reachable from the objects allocated at line 7 (e.g., r in line 14) are also determined to be persistent memory objects, as shown in the points-to graph for this example in Fig. 5. Even with these limitations, the allocation-site based abstraction (and its derivative based on bounded allocation-stack based abstraction) have been shown to be a practical abstraction as it often reflects the programmer's mental map for the program's logic.

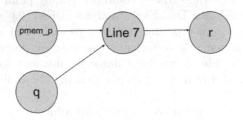

Fig. 5. Points-to graph for the C code example in Fig. 4.

3.2   Allocation-Stack with Bounded Depth

To make the allocation-site abstraction more precise, we use a bounded call-stack, instead of a single PC, to represent a single allocation site. To see the precision advantages of using a bounded call-stack abstraction, consider the motivating example in Fig. 2, previously discussed in Sect. 2. The points-to graph for this example, using the PC-based allocation-site abstraction is shown in Fig. 6. However, this points-to graph in Fig. 6 is too coarse-grained and would result in a conservative solution where the allocated in the `callee()` function is always identified as a persistent object. Recall however that we are instead interested in a more precise solution, as shown in Fig. 3.

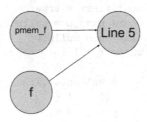

Fig. 6. Points-to graph for the C++ code example in Fig. 2.

In the bounded call-stack abstraction, we further categorize the memory objects allocated at line 5 into two mutually-exclusive sets:

1. Memory objects that were allocated at line 5 when the `callee()` function was called from the `caller()` function at line 10.

2. Memory objects that were allocated at line 5 when the `callee()` function was called from the `caller()` function at line 11.

Thus, in this abstraction, we are using a call-stack of depth-1 to further distinguish between memory objects by providing a finer-grained naming scheme. The resulting points-to graph is shown in Fig. 7. The points-to graph in Fig. 7 shows that while the global variable pmem_f may point to memory objects allocated when the program's callstack is represented by "Line 11; Line 5", the local variable f may point to memory objects allocated when the program's callstack is represented by "Line 10; Line 5". Because the two call-stacks are mutually exclusive, the corresponding sets of memory objects are disjoint, and thus we can conclude that only the memory objects corresponding to "Line 11; Line 5" need to be allocated on persistent storage (and it would be correct to allocate the memory objects corresponding to "Line 10; Line 5" in volatile memory). The corresponding code generation strategy involves passing a Boolean flag, `persistent` to the `callee()` function which will be initialized to `false` at Line 10 and `true` at Line 11 (as shown in Fig. 3).

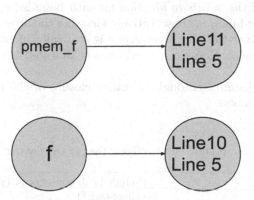

Fig. 7. Points-to graph for the C++ code example in Fig. 2 using the bounded call-stack abstraction.

Call-Stack Depth Bound. In the presence of recursion, there could be a loop in the function call-graph and so it becomes necessary to put a bound D on the *maximum depth of the call-stack context*. Higher values of D would necessarily yield more precise results at the cost of higher analysis time. In the worst-case, the analysis time could grow exponentially with an increase in D; fortunately, such worst-cases are relatively rare. We experiment with $D \leq 4$ in our evaluation.

Table 1. Inter-Procedural DFA for identifying the set of all bounded-depth allocation-stacks.

Domain	Bounded call-contexts × Sets of Bounded-depth Allocation-stacks
Direction	Forward
Intra-procedural transfer function	$f_a(cc, s) = (cc, s \cup \text{GEN}_a(cc))$
Caller-to-callee transfer function	$f_{\texttt{callee}}(cc,s) = (cc \odot \texttt{callee}, s)$
Callee-to-caller transfer function	$f_{\texttt{callerCtx}}(cc, s) = (\texttt{callerCtx}, s)$
Meet operator ∧	$(cc,s_1) \wedge (cc,s_2) = (cc, s_1 \cup s_2)$
Boundary condition	$\text{out}_{\texttt{main}}[n_{entry}] = (\texttt{main}, \{\})$

3.3   Inter-procedural DFA Specification

Our algorithm proceeds in three steps:

1. Identification of all bounded-depth allocation-stacks (Table 1).
2. Identification of the *points-to* function for each bounded call-context (Table 1). The points-to function conservatively identifies the objects that a program variable may point-to, where an object is statically identified through the bounded-depth allocation-stack at which it is allocated (as identified in Step 1).
3. Identifying reachability through transitive closure of the points-to function for the durable-roots.

Table 2. Inter-Procedural DFA for identifying the per call-context points-to function.

Domain	Bounded call-contexts × {v ↦ {allocStack}}
Direction	Forward
Intra-procedural transfer function	$f_a(cc, m) = (cc, \texttt{relax}(m, a))$
Caller-to-callee transfer function	$f_{\texttt{callee}}(cc,m) = (cc \odot \texttt{callee}, \texttt{retainMem}(m))$
Callee-to-caller transfer function	$f_{\texttt{callerCtx}}(cc, m) = (\texttt{callerCtx}, \texttt{retainMemAndRetVal}(m))$
Meet operator ∧	$(cc,m_1) \wedge (cc,m_2) = (cc, \texttt{meetMap}(m_1, m_2))$
Boundary condition	$\text{out}_{\texttt{main}}[n_{entry}] = (\texttt{main}, \phi)$

Identification of Bounded-Depth Allocation Stacks. The algorithm proceeds by first identifying the bounded-depth allocation-stacks that would be used as static proxies for the objects allocated in the calling contexts represented by those stacks. In Fig. 2, the allocation-stack (`caller.L10`; `callee.L5`) represents an allocation stack of depth 2 for example. To determine all such allocation stacks up to a depth D, we run the inter-procedural DFA shown in Table 1.

The forward DFA starts from the top-level `main` function with an empty set of allocation stacks (see boundary condition), and produces a set of bounded-depth allocation stacks for each call-context it encounters. In the example shown in Fig. 2, if we use the maximum call-context depth $D = 1$, then we obtain the following set of allocation stacks: {(`caller.L10`; `callee.L5`), (`caller.L10`; `callee.L6`), (`caller.L11`; `callee.L5`), (`caller.L11`; `callee.L6`), (`caller.L9`)}. On the other hand, if we use $D = 0$, then we obtain {(*; `callee.L5`), (*; `callee.L6`), (`caller.L9`)}. Here, the '*' represents an arbitrary string of call-sites.

At each step in the DFA algorithm, the value maintains both the current set of bounded-depth allocation stacks and the current call-context. On encountering an allocation statement (such as `new` or `malloc` or `pmem_malloc`), the DFA analysis creates and adds a new bounded-depth allocation-stack to the output value, as shown through the GEN_a function in the intra-procedural transfer function. Notice that the GEN_a function uses the current calling context to determine the bounded-depth allocation stack. The meet operator for this DFA is set-union.

When the DFA encounters a function call, the caller-to-callee transfer function is used to obtain the value at the callee's entry (using the value at the caller's function call site). The caller-to-callee transfer function changes the call-context by appending the callee's name to the current call-context using the $\odot$ operator. The $\odot$ operator is sensitive to the maximum call-depth parameter D, and if the call-context depth exceeds D after appending the callee's name, it truncates the call-context stack at the top and replaces it with the '*' symbol. The callee-to-caller transfer function is invoked upon function return; here, the current call-context is updated to the caller's context `callerCtx`.

Identification of per Call-Context Points-To Function. After we have identified all the bounded-depth allocation stacks, we run an inter-procedural points-to analysis to identify for each program variable, the subset of these allocation sets that the variable may point-to. This points-to analysis DFA algorithm's specification is shown in Table 2. The DFA maintains the current call-context along with the points-to function that maps a variable v to the potential allocation stacks it may point-to {allocStk}.

The set of variables v chosen for this analysis are based on the 3-Address-Code (3AC) Static Single Assignment (SSA) representation of the program—all intermediate SSA registers are used as the domain v of this points-to function[1]. Further, the domain of this points-to function includes a special variable V_a for each bounded-depth allocation-stack a that represents the memory region belonging to *all* objects allocated at context a. For example, the domain of the

---

[1] SSA Representation lends greater precision to the analysis as it avoids accumulating the points-to information in cases where the same variable is assigned twice with different values—the SSA representation uses two different variables in such cases and our analysis computes distinct points-to sets for each of these two variables.

Algorithm 1. Transitive closure of the durable roots in the points-to graph.

Input: droots, durable roots
Input: PointsToMaps, per call-context points-to maps
Output: PersistentAllocStacks representing the bounded-depth allocation-stacks
of objects that are determined to be allocated persistently.
1: PersistentAllocStacks ← droots
2: **while** (changed) **do**
3:     changed ← **false**
4:     **for each** (cc, m) ∈ PointsToMaps **do**
5:         NewPersistentAllocStacks ← PersistentAllocStacks
6:         **for each** pAllocStack ∈ PersistentAllocStacks **do**
7:             **if** any pAllocStack object points to some aStack not already present in
    PersistentAllocStacks **then**
8:                 NewPersistentAllocStacks ← NewPersistentAllocStacks ∪
    aStack
9:                 changed ← **true**
10:             **end if**
11:         **end for**
12:         PersistentAllocStacks ← NewPersistentAllocStacks
13:     **end for**
14: **end while**

---

points-to function for the example in Fig. 2 (using $D = 0$) is: {f, f->child, pmem_f, pmem_f.child, $V_{(*;callee.L5)}$, $V_{(*;callee.L6)}$}. These special $V_{...}$ variables are helpful in computing the transitive closure (step 3 of our overall algorithm).

The intra-procedural transfer function of the points-to DFA involves *relaxing* the points-to map based on the behaviour of the current statement a; for example, if the statement is of the form v = *w, then we update the points-to set of v to include the points-to set of *w. Similarly, if the statement is of the form *v = w, then the special $V_{...}$ variables' points-to information is updated to include the points-to values of the variable w. This part of the algorithm is similar to Andersen's algorithm [30,32]. If a statement of the form v = malloc(...) is encountered, the points-to set of v becomes the singleton set containing the bounded-depth allocation-stack a associated with the malloc() call. For other statements that do not de-reference memory, such as v = f(w,x), the points-to sets of w and x are union-ed into the points-to set of v.

The caller-to-callee and the callee-to-caller transfer functions of the points-to DFA update the current calling context. When going from the caller to the callee, we retain the mappings for all the special variables $V_{...}$ but drop the mappings for all other variables (which were local only to the caller). This is represented through the retainMem construct in Table 2. Similarly, when returning from the callee back to the caller, we retain the mappings for the special variables $V_{...}$ and also set the mapping for the returned variable in the caller (based on the corresponding mapping of the return value in the callee). This is represented through retainMemAndRetval.

The meet operator for this DFA involves iterating over all the variables of the points-to mapping function and taking a union of the sets of allocation-stacks that the variable may point-to (meetMap). The points-to analysis starts at the entry of the main function with the empty points-to function[2].

Transitive Closure of Durable Roots. After the points-to information is computed for each bounded-depth call-context, we finally compute the transitive closure of the durable roots to identify the objects that need to be (conservatively) allocated to persistent memory to avoid persistent-to-volatile dangling pointers. This transitive closure algorithm involves a fixed-point computation, as shown in Algorithm 1: at each step of the fixed-point procedure, we iterate over each bounded-depth call-context to accumulate the reachable allocation-stacks from the durable-roots. This fixed-point procedure is guaranteed to terminate and its fixed-point solution will represent the over-approximate set of memory locations that may be reachable from the durable roots.

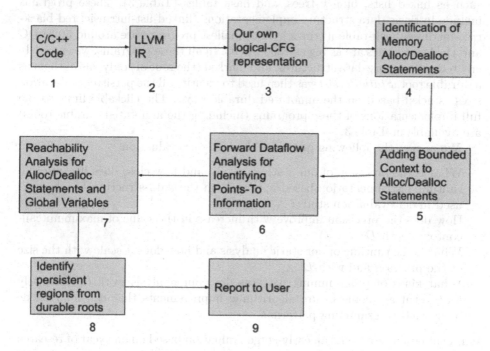

Fig. 8. Implementation flowchart.

----

[2] In the presence of global variables, we assume that the global variables may point to each other at the entry to main. We omit this discussion for brevity.

4   Evaluation

We have implemented our algorithm in a tool called *StaticPersist*, and we show the flowchart of the tool in Fig. 8. We model the intra-procedural transfer functions of almost all integer opcodes of LLVM that are generated from C/C++ code using `clang`. The identification of `malloc` and `new` functions is performed by looking for the function `call` opcode in LLVM and matching the callee-name with one of the library allocation routines. For C++, the `_Znwm` is used to match the `new` operator. Global variables are modeled both as variables (domain of the points-to mapping function) and as the pointee (range of the points-to mapping function). A global variable whose name begins with a special prefix `pmem_` is considered a durable root. Additionally, the user can use special `pmem_malloc` and `pmem_new` functions to specify allocation sites that allocate persistent objects only; such allocation sites also constitute the set of durable roots (see Fig. 2 for example).

We run *StaticPersist* on C/C++ programs involving common data-structures such as linked lists, binary trees, and hash tables (Table 3). These programs include intrinsic data structure implementations (linked-list-intrinsic, red-black-tree-intrinsic, hash-table-intrinsic). The smallest programs are around 50 SLOC and the largest ones are as big as 1500 SLOC. In all these programs, we explicitly created two or more data structures, and marked (the root of) only one of them as a durable root. *StaticPersist* was then used to identify all the persistent allocation sites (stacks) based on the annotated durable roots. The clickable links to the full implementations of these programs (including the annotated durable roots) are available in Table 3.

We answer the following questions through our evaluation:

- What is the precision of our static analysis, and how close does it come to a runtime instrumentation-based approach on the data-structure benchmarks used in our evaluation study?
- How does the precision improve with increase in the value of maximum call-context depth D?
- What is the runtime of our static analysis and how does it scale with the size of the program and with D?
- What kinds of programming patterns does our analysis work (im)precisely for? What are some future algorithmic improvements that can help tackle most such programming patterns?

Our evaluation represents an early-stage evaluation based on one year of research effort. A more complete evaluation should show extensive results on larger benchmarks. Our current evaluation points to future opportunities of improvement to our algorithm, and should be treated as such.

Figure 9 presents the results of applying *StaticPersist* on the benchmarks listed in Table 3. The X-axis represents the D value, ranging between 0 and 4, and the Y-axis represents the analysis time. Green bars are used to represent situations where the static analysis results are as precise as those possible with dynamic analysis (with runtime overheads); red bars represent situations where

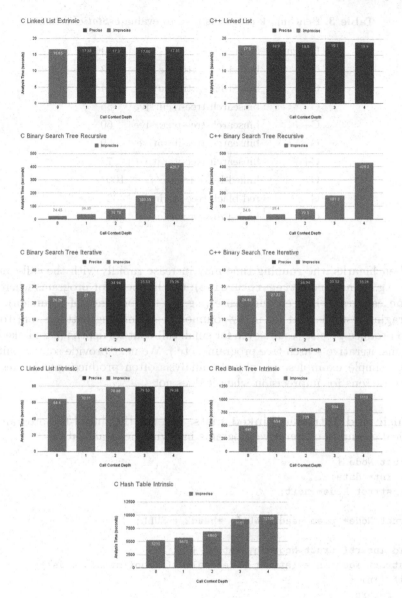

Fig. 9. Evaluation Results for the benchmarks listed in Table 3.

the static analysis results are imprecise compared to the results that could have been obtained using a dynamic runtime analysis. Even in cases where the static analysis produces imprecise results, the results are still often useful as they select only a subset of all allocation stacks as persistent.

To summarize the results: the analysis tool can scale up to a thousand SLOC with a call-context depth of 4, albeit it may have to run for hours. While for

Table 3. Benchmark programs used to evaluate *StaticPersist*.

Language	Name	SLOC
C	linked-list-extrinsic	43
C++	linked-list-extrinsic	54
C	binsearch-tree-recursive	55
C++	binsearch-tree-recursive	60
C	binsearch-tree-iterative	59
C++	binsearch-tree-iterative	73
C	linked-list-intrinsic	1007
C	red-black-tree-intrinsic	829
C	hash-table-intrinsic	1515

some benchmarks, the running time may increase rapidly with the call-context depth (e.g., `binsearch-tree-recursive`); this happens for programs where the function call graph is deep and dense (e.g., in the presence of recursion). The encouraging result is that for many common programming patterns, our tool is able to produce *precise* results even for small call-context depths (e.g., linked-list programs, iterative search tree programs, etc.). We next provide some intuition, through simple examples, on why the analysis often produces precise results, and the reasons for imprecision when it does not.

Extrinsic and Intrinsic Linked Lists. We use the linked-list programs as canonical examples of iterative programs involving heap allocations.

```
1 struct Node {
2     int* data;
3     struct Node* next;
4 };
5 struct Node* pmem_head = NULL, *head2 = NULL;
6
7 void insert(struct Node**h, int d) {
8   struct Node* n = (struct Node*)malloc(sizeof(struct Node));
9   if (!n) {
10     return;
11   }
12   n->data = malloc(sizeof(int));
13   *n->data = d;
14   n->next = *h;
15   *h = n;
16 }
17 int main() {
18   insert(&pmem_head, 1);
19   ...
20   insert(&head2, 2);
```

```
21    struct Node* n = (...) ? &pmem_head : &head2;
22    insert(n, 3);
23 }
```

Consider the linked-list program shown above, where the `insert()` function is used to insert into a linked-list. The `pmem_head` global variable is marked as a durable root, and so any node that is reachable from it should be allocated in PMEM; on the other hand, `head2` and nodes reachable from it should ideally be allocated in volatile memory. Our program makes three calls to the `insert()` funcction at lines 18, 20, and 22. The `insert()` function has two calls to the `malloc()` function at lines 8 and 12. Our allocation-stack abstraction thus creates six bounded-depth allocation stacks, namely 18-8, 18-12, 20-8, 20-12, 22-8, and 22-12 (a number represents a line number and the "-" represents a function call). Our points-to analysis identifies that the global variable `pmem_head` may point-to 18-8, 18-12, 22-8, and 22-12 (but not the remaining two). Thus, the analysis can be used to appropriately transform the `malloc` calls to `pmem_malloc` calls, and a Boolean `persistent` flag can be used to distinguish between different call contexts (just like it was done in Fig. 3).

As seen in our results, our analysis is also capable of correctly handling intrinsic data structures (and other programming patterns where the pointers may point to the middle of an object). We show an excerpt from the `linked-list-intrinsic` C program:

```
 1 /* List element. */
 2 struct list_elem {
 3   struct list_elem *prev;     /* Previous list element. */
 4   struct list_elem *next;     /* Next list element. */
 5 };
 6
 7 /* List. */
 8 struct list {
 9   struct list_elem head;      /* List head. */
11   struct list_elem tail;      /* List tail. */
12 };
13 #define list_entry(LIST_ELEM, STRUCT, MEMBER)          \
14         ((STRUCT *) ((uint8_t *) &(LIST_ELEM)->next    \
15                  - offsetof (STRUCT, MEMBER.next)))
16 struct node {
17   int data;
18   struct list_elem list_elem;
19 };
20
21 struct list ls;
22 struct list pmem_ls;
23
24 int
25 main(void)
26 {
27   list_init(&ls);
```

```
28   int i;
29   for (i = 0; i < 10; i++) {
30     struct node* n = malloc(sizeof(struct node));
31     list_push_front(&ls /* or pmem_ls*/, &n->list_elem);
32   }
33   ...
34 }
```

In the program excerpt above, the list node that is `malloc`-ed at line 30 is added to a list (using `list_push_front`) at line 31. Depending on whether the push happened to the volatile list (`ls`) or the persistent list (`pmem_ls`), the `malloc` should be converted to a volatile (or persistent) `malloc` respectively. Notice that the pointers in an intrinsic linked list are to the middle of allocated objects, and our static analysis is able to identify the precise solutions at call-context depths 3 or higher.

Recursion. Consider the standard binary-search tree algorithm shown in the code excerpt below:

```
1 struct Node {
2     int data;
3     struct Node* left, *right;
4 };
5 ...
6 void insert_helper(struct Node* h, struct Node* n)
7 {
8   if (h->data < n->data) {
9     if (!h->left) {
10       h->left = n;
11     } else {
12       insert_helper(h->left, n);
13     }
14   } else {
15     if (!h->right) {
16       h->right = n;
17     } else {
18       insert_helper(h->right, n);
19     }
20   }
21 }
```

The `insert_helper` function is called recursively at line 12 and line 18. Assuming that `insert_helper` represents the top-level entry function of the program, the corresponding call-contexts at depth-2 are 12–12, 12–18, 18–12, and 18–18. Similarly, the call-contexts at depth-3 are 12–12–12, 12–12–18, 12–18–12, 12–18–18, 18–12–12, 18–12–18, 18–18–12, and 18–18–18. Thus, in this example, the number of call-contexts grows exponentially with the number call-context depth. This exponential blow-up is also evident in the increase in running times of the recursive benchmarks with increasing call-context depth.

Further, because the recursion is potentially infinite, the bounded nature of the call-context depth forces the static analysis algorithm to eventually over-approximate the analysis results, causing imprecision.

Recursion in this form is a standard programming pattern, but our current algorithm is not equipped to handle it in an elegant manner – we simply use the standard bounded-depth call-context abstraction to deal with recursion currently which results in sound-but-imprecise results. In future work, we intend to design and implement more precise abstractions to handle such recursive programming patterns.

Intrinsic Hash Table. Our analysis currently produces imprecise results on the intrinsic hash table benchmark, which does not involve recursion. However, this program is a large program with deep calling depths and the use of function pointers. Our analysis currently treats function pointers as arbitrary pointers, and conservatively causes imprecise transfer function across calls to them. A more precise analysis should perform a value and type analysis on function pointers. We leave this for future work.

More Future Explorations. In future work, we intend to generalize this allocation stack approach to automatically identify *interesting PCs* which need not necessarily be limited to the calls to the allocation functions like malloc() and new. Based on our experience, we imagine that such an approach would drastically improve the precision generality of our approach.

5   Related Work

Programming models and frameworks for persistent memory has been a busy area of research, especially because of its promise to potentially re-define the hardware/software stacks vis-a-vis persistence.

Prior work on PMEM-based filesystems [11,22] discuss performant abstractions for filesystems based on direct-access persistent memory. The correctness implications of the cache-flush clflush and fence (sfence) instructions, and the resulting programmability and efficiency trade-offs, become apparent in these designs. Simple design decisions like the use of redo-logging (in favour of undo-logging) for PMEM have important performance implications (because it reduces the number of writes in the common case). The book on *programming persistent memory* [29] is an exhaustive resource for understanding the low-level PMEM programming models in C and C++. The basic ideas include (a) over-loading on the mmap system call interface; (b) having software facilities to convert from virtual addresses to physical addresses, e.g., through macros; and (c) using libraries to manage this space of persistent memory in a crash-consistent manner. These abstractions are complex and cumbersome, and that is perhaps a primary motivation for research in the space of compiler/runtime-based programmability support for PMEM. Log-free concurrent data structures for persistent memory [9] involve the careful use of lock-free abstractions and hardware-supported cache-flush and fence instructions to allow the implementation of failure-atomic data-structures without the need for redo/undo-logs. NVTraverse [16] proposes a framework to automatically and efficiently transform simple lock-free data-structures to a durably linearizable persistent data structure. Fragmentation in persistent memory has received much attention [6,28]: the use of segregated-fit algorithms for persistent memory could potentially benefit from more research as the lifetimes of persistent memory objects are usually very different from those of volatile memory objects.

Research work on higher-level programming models for persistent memory includes efforts like `go-pmem` [17] (garbage-collected persistent heap and transaction blocks for failure-atomicity). The use of a compiler to automatically do checkpointing and restore has also been much studied in the past [8,12]. Espresso [37] discusses PMEM abstractions in the context of the Java programming language: the traditional JPA (Java Persistence API) abstractions, based on the ORM model, is too expensive for PMEM devices. In constrast, PCJ (Persistent Collections for Java) store and manage persistent objects *off the heap*, i.e., the PMEM objects require their own synchronization and garbage collection. Espresso however proposes a programming framework (Persistent Java Heaps, or PJH) where the persistent objects can be managed on the heap, using the custom `pnew` (persistent new) operator. While the Espresso model is higher-level, it still has many complexities and leaves scope for several persistence bugs, including the possibility of dangling pointers from persistent memory to volatile memory.

AutoPersist [31] proposes a higher level PMEM programming abstraction for Java, compared to Espresso, where the programmer only needs to label the durable roots in the program (in contrast to having to identify all persistent objects manually). The runtime then uses a reachability analysis to automatically label all objects reachable from the durable roots, as durable (or persistent). This reachability analysis needs to be performed at runtime, and requires interposition on memory accesses. The paper develops efficient algorithms that do not cause significant overheads: they rely on the strong type system provided by Java (e.g., no pointer arithmetic, etc.) and on the fact that the JVM uses well-defined object headers that can be manipulated to efficiently implement incremental reachability analysis-based transformations. The paper shows that the programmer effort required to mark persistent regions is often reduced by over 10x. The execution and memory overheads of Autopersist are under 10%. In constrast to the AutoPersist work, our approach is static compile-time (does not incur any runtime overheads), can work for unmanaged languages like C/C++, and can handle low-level operations like pointer arithmetic.

There exists rich literature on context-sensitive alias analysis and points-to analysis [2,5,10,13–15,18,19,21,24,26,27,34,35]. Different approaches have explored different design points to carefully balance precision, generality, and scalability—for example, some of these approaches, such as DSA [24], exhibit higher scalability when type annotations are available. We are interested in a scalable analysis to conservatively disambiguate persistent memory pointers from volatile memory pointers for type-unsafe languages, and we present our initial algorithms and the associated results in this paper. Future work would involve improving these algorithms to make them more scalable for the common-case programming patterns.

6 Conclusions and Future Work

Our work suggests that a compile-time static reachability analysis to disambiguate persistent memory objects from volatile memory objects is a promising approach. The programmer is only required to specify the durable roots and the compiler automatically identifies all persistent objects in the program. Such an analysis tool needs to be evaluated on both its precision and its scalability. Our results demonstrate promise on both metrics, although there remains scope for future improvements. In future work, we are interested in better support for recursion, function pointers, and generalization of the allocation-stack abstraction to include non-allocation PCs as well.

Acknowledgements. This material is based upon work supported by Huawei Technologies Co. Ltd.

References

1. Akinaga, H., Shima, H.: Resistive random access memory (ReRAM) based on metal oxides. Proc. IEEE **98**(12), 2237–2251 (2010). https://doi.org/10.1109/JPROC.2010.2070830
2. Balakrishnan, G., Reps, T.: Recency-abstraction for heap-allocated storage. In: Yi, K. (ed.) SAS 2006. LNCS, vol. 4134, pp. 221–239. Springer, Heidelberg (2006). https://doi.org/10.1007/11823230_15
3. Berndl, M., Lhoták, O., Qian, F., Hendren, L., Umanee, N.: Points-to analysis using BDDs. ACM SIGPLAN Not. **38**(5), 103–114 (2003)
4. Chakrabarti, D.R., Boehm, H.J., Bhandari, K.: Atlas: leveraging locks for non-volatile memory consistency. ACM SIGPLAN Not. **49**(10), 433–452 (2014)
5. Chase, D.R., Wegman, M., Zadeck, F.K.: Analysis of pointers and structures. In: Proceedings of the ACM SIGPLAN 1990 Conference on Programming Language Design and Implementation, PLDI 1990, pp. 296–310. Association for Computing Machinery, New York (1990). https://doi.org/10.1145/93542.93585
6. Chen, F., Luo, T., Zhang, X.: CAFTL: a content-aware flash translation layer enhancing the lifespan of flash memory based solid state drives. In: Proceedings of the 9th USENIX Conference on File and Stroage Technologies, FAST 2011, p. 6. USENIX Association, USA (2011)
7. Coburn, J., et al.: NV-heaps: making persistent objects fast and safe with next-generation, non-volatile memories. ACM SIGARCH Comput. Archit. News **39**(1), 105–118 (2011)
8. Dahiya, M., Bansal, S.: Automatic verification of intermittent systems. In: VMCAI 2018. LNCS, vol. 10747, pp. 161–182. Springer, Cham (2018). https://doi.org/10.1007/978-3-319-73721-8_8

9. David, T., Dragojević, A., Guerraoui, R., Zablotchi, I.: Log-free concurrent data structures. In: Proceedings of the 2018 USENIX Conference on Usenix Annual Technical Conference, USENIX ATC 2018, pp. 373–385. USENIX Association, USA (2018)

10. Deutsch, A.: Interprocedural may-alias analysis for pointers: beyond k-limiting. In: Proceedings of the ACM SIGPLAN 1994 Conference on Programming Language Design and Implementation, PLDI 1994, pp. 230–241. Association for Computing Machinery, New York (1994). https://doi.org/10.1145/178243.178263

11. Dulloor, S.R., et al.: System software for persistent memory. In: Proceedings of the Ninth European Conference on Computer Systems, EuroSys 2014. Association for Computing Machinery, New York (2014). https://doi.org/10.1145/2592798.2592814

12. Elkhouly, R., Alshboul, M., Hayashi, A., Solihin, Y., Kimura, K.: Compiler-support for critical data persistence in NVM. ACM Trans. Archit. Code Optim. **16**(4), 1–25 (2019). https://doi.org/10.1145/3371236

13. Emami, M., Ghiya, R., Hendren, L.J.: Context-sensitive interprocedural points-to analysis in the presence of function pointers. SIGPLAN Not. **29**(6), 242–256 (1994). https://doi.org/10.1145/773473.178264

14. Fähndrich, M., Rehof, J., Das, M.: Scalable context-sensitive flow analysis using instantiation constraints. SIGPLAN Not. **35**(5), 253–263 (2000). https://doi.org/10.1145/358438.349332

15. Foster, J.S., Fähndrich, M., Aiken, A.: Polymorphic versus monomorphic flow-insensitive points-to analysis for C. In: Palsberg, J. (ed.) SAS 2000. LNCS, vol. 1824, pp. 175–198. Springer, Heidelberg (2000). https://doi.org/10.1007/978-3-540-45099-3_10

16. Friedman, M., Ben-David, N., Wei, Y., Blelloch, G.E., Petrank, E.: NVTraverse: in NVRAM data structures, the destination is more important than the journey. In: Proceedings of the 41st ACM SIGPLAN Conference on Programming Language Design and Implementation, PLDI 2020, pp. 377–392. Association for Computing Machinery, New York (2020). https://doi.org/10.1145/3385412.3386031

17. George, J.S., Verma, M., Venkatasubramanian, R., Subrahmanyam, P.: Go-pmem: native support for programming persistent memory in go. In: USENIX Annual Technical Conference (2020)

18. Ghiya, R., Hendren, L.J.: Connection analysis: a practical interprocedural heap analysis for C. Int. J. Parallel Program. **24**(6), 547–578 (1996)

19. Ghiya, R., Hendren, L.J.: Is it a tree, a DAG, or a cyclic graph? A shape analysis for heap-directed pointers in C. In: Proceedings of the 23rd ACM SIGPLAN-SIGACT Symposium on Principles of Programming Languages, POPL 1996, pp. 1–15. Association for Computing Machinery, New York (1996). https://doi.org/10.1145/237721.237724

20. Handy, J.: Understanding the intel/micron 3D XPoint memory. In: Proceedings of the SDC, vol. 68 (2015)

21. Jones, N.D., Muchnick, S.S.: A flexible approach to interprocedural data flow analysis and programs with recursive data structures. In: Proceedings of the 9th ACM SIGPLAN-SIGACT Symposium on Principles of Programming Languages, POPL 1982, pp. 66–74. Association for Computing Machinery, New York (1982). https://doi.org/10.1145/582153.582161

22. Kadekodi, R., Lee, S.K., Kashyap, S., Kim, T., Kolli, A., Chidambaram, V.: SplitFS: reducing software overhead in file systems for persistent memory. In: Proceedings of the 27th ACM Symposium on Operating Systems Principles, SOSP

2019, pp. 494–508. Association for Computing Machinery, New York (2019). https://doi.org/10.1145/3341301.3359631

23. Kildall, G.A.: Global expression optimization during compilation. Ph.D. thesis, USA (1972). aAI7228616

24. Lattner, C., Lenharth, A., Adve, V.: Making context-sensitive points-to analysis with heap cloning practical for the real world. SIGPLAN Not. **42**(6), 278–289 (2007). https://doi.org/10.1145/1273442.1250766

25. Lersch, L., Hao, X., Oukid, I., Wang, T., Willhalm, T.: Evaluating persistent memory range indexes. Proc. VLDB Endow. **13**(4), 574–587 (2019). https://doi.org/10.14778/3372716.3372728

26. Liang, D., Harrold, M.J.: Efficient points-to analysis for whole-program analysis. SIGSOFT Softw. Eng. Notes **24**(6), 199–215 (1999). https://doi.org/10.1145/318774.318943

27. Liang, D., Harrold, M.J.: Efficient computation of parameterized pointer information for interprocedural analyses. In: Cousot, P. (ed.) SAS 2001. LNCS, vol. 2126, pp. 279–298. Springer, Heidelberg (2001). https://doi.org/10.1007/3-540-47764-0_16

28. Persistent allocator design - fragmentation. https://pmem.io/2016/02/25/fragmentation.html

29. Scargall, S.: Programming Persistent Memory: A Comprehensive Guide for Developers. Springer, Heidelberg (2020). https://doi.org/10.1007/978-1-4842-4932-1, https://library.oapen.org/bitstream/id/e234e601-6128-4ee4-be45-32e8f2e417dd/1007325.pdf

30. Shapiro, M., Horwitz, S.: Fast and accurate flow-insensitive points-to analysis. In: Proceedings of the 24th ACM SIGPLAN-SIGACT Symposium on Principles of Programming Languages, pp. 1–14 (1997)

31. Shull, T., Huang, J., Torrellas, J.: Autopersist: an easy-to-use java nvm framework based on reachability. In: Proceedings of the 40th ACM SIGPLAN Conference on Programming Language Design and Implementation, PLDI 2019, pp. 316–332. Association for Computing Machinery, New York (2019). https://doi.org/10.1145/3314221.3314608

32. Steensgaard, B.: Points-to analysis in almost linear time. In: Proceedings of the 23rd ACM SIGPLAN-SIGACT Symposium on Principles of Programming Languages, pp. 32–41 (1996)

33. Volos, H., Tack, A.J., Swift, M.M.: Mnemosyne: lightweight persistent memory. ACM SIGARCH Comput. Archit. News **39**(1), 91–104 (2011)

34. Whaley, J., Lam, M.S.: Cloning-based context-sensitive pointer alias analysis using binary decision diagrams. In: Proceedings of the ACM SIGPLAN 2004 Conference on Programming Language Design and Implementation, PLDI 2004, pp. 131–144. Association for Computing Machinery, New York (2004). https://doi.org/10.1145/996841.996859

35. Wilson, R.P., Lam, M.S.: Efficient context-sensitive pointer analysis for C programs. SIGPLAN Not. **30**(6), 1–12 (1995). https://doi.org/10.1145/223428.207111

36. Wong, H.S.P., et al.: Phase change memory. Proc. IEEE **98**(12), 2201–2227 (2010). https://doi.org/10.1109/JPROC.2010.2070050

37. Wu, M., et al.: Espresso: Brewing java for more non-volatility with non-volatile memory. In: Proceedings of the Twenty-Third International Conference on Architectural Support for Programming Languages and Operating Systems, ASPLOS 2018, pp. 70–83. Association for Computing Machinery, New York (2018). https://doi.org/10.1145/3173162.3173201

Symbolic Abstract Heaps for Polymorphic Information-Flow Guard Inference

Nicolas Berthier[1,2]([✉]) and Narges Khakpour[3,4]([✉])

[1] OCamlPro, Paris, France
m@nberth.space
[2] University of Liverpool, Liverpool, UK
[3] Newcastle University, Newcastle upon Tyne, UK
narges.khakpour@ncl.ac.uk
[4] Linnæus University, Växjö, Sweden

Abstract. In the realm of sound object-oriented program analyses for information-flow control, very few approaches adopt flow-sensitive abstractions of the heap that enable a precise modeling of implicit flows. To tackle this challenge, we advance a new symbolic abstraction approach for modeling the heap in Java-like programs. We use a store-less representation that is parameterized with a family of relations among references to offer various levels of precision based on user preferences. This enables us to automatically infer polymorphic information-flow guards for methods via a co-reachability analysis of a symbolic finite-state system. We instantiate the heap abstraction with three different families of relations. We prove the soundness of our approach and compare the precision and scalability obtained with each instantiated heap domain by using the IFSPEC benchmarks and real-life applications.

1 Introduction

Information Flow Control (IFC) mechanisms offer an effective approach to prevent unwanted disclosure of confidential information, or illegal tampering of data. Their task is to ensure confidentiality and/or integrity, which are usually formalized as noninterference baseline properties [1]. Confidentiality demands that *high-sensitive* (secret) inputs do not influence *low-sensitive* (public) outputs. This means that any change in the value of a secret input must not induce a change in any public output. In other words, there must be no *information flow* from any secret to any public output. In software programs, information may flow *explicitly* via direct assignments, e.g., from y to x in x = y;, or *implicitly* when the execution of statements is guarded by a condition, e.g., from c to x in if (c > 0) x = 42;.

Many static analysis approaches to ensure noninterference have been advanced, that rely on type-systems [2–6], self-composition [7–9], theorem-proving [10], and abstract interpretation [11–14]. *Flow-insensitive* static analyses (with "flow" as in "control-flow") deal with a single set of facts that is valid for all possible executions of the whole program, whereas *flow-sensitive* analyses provide one set of facts for each statement. In general, flow-sensitivity increases precision, yet comes with an additional computational cost. Heap abstractions as computed by alias or points-to analyses obey the same principle [15]: a flow-insensitive heap analysis provides a single, finite representation of the conceptually infinite set of memory locations manipulated by all entire executions of the program, while a flow-sensitive variant gives an abstraction of the heap at each statement. Note that a flow-sensitive analysis for object-oriented programs may rely on a flow-insensitive heap abstraction; this means that the analysis must remain imprecise when dealing with heap-allocated structures.

We consider a Java-style low-level object-oriented language whose syntax is close to Jimple's [16], and design a static IFC analysis that *automatically* decides whether a program P implemented in this language *is secure*, i.e., P satisfies a desired noninterference property. Several approaches have been proposed to verify such properties for Java-style languages [4,5,10,14,17–21]. To the best of our knowledge, however, none of the sound and scalable solutions rely on a flow-sensitive heap abstraction. Our analysis is therefore *the first of its kind*, as it is *sound*, shows potential for *scalability* since it supports *modularity*, and both: (i) *captures implicit flows*; and (ii) relies on a *flow-sensitive heap abstraction*. Achieving these goals in combination is challenging because the analysis must track the information flows that result from manipulations of object fields and references performed in the program branches that are taken *as well as* in any program branch that is *not* taken: it must therefore reflect about the states of the heap in both taken and non-taken branches simultaneously.

To do so, we use a *security typing environment* that associates each memory location manipulated by the program P with a *security level*. In the case of confidentiality, such a level indicates whether the memory location may hold high-sensitive (secret) data, and P is secure if no value from a high-sensitive *source* flows to a *sink* statement. To deal with all information flows in the heap, we introduce the notion of *symbolic abstract heap domains*, that combine a flow-sensitive security typing environment for every object reachable from a given set of reference variables R, with a flow-sensitive representation of a set of *heap-related relations* pertaining to R (e.g., aliasing). The domains are parametric in a *family of heap-related relations*, which defines the relations that are captured flow-sensitively by the domain. This allows us to define multiple heap abstractions, each one with its level of precision. Abstract heaps in such a domain are *predicates* in a propositional logic, that provide a *store-less* model of the heap where irrelevant details related to the behaviors of the memory allocator and garbage collection are safely abstracted away. The semantics of reference and object mutations are specified using *predicate transformers*. These can be used to encode the security semantics of any method m of the program by means of a symbolic transition system S_m, where the desired noninterference property is reduced to a safety property φ_m [22].

Artifacts that we can infer for a method m include an *information-flow guard*, that is a predicate expressed on propositions about security levels and heap-related relations pertaining to m's formal arguments. This guard describes (sufficient) circumstances upon which m is secure, and it is *polymorphic* since it is valid in *any calling context*. More elaborate artifacts that additionally describe the *effects* of m on the heap enable *sound* inter-procedural analyses. In the present paper, however, we focus on our approach for abstracting the heap and concentrate our exposition on the inference of guards; we leave the inference of polymorphic effects for a future publication. We compute the guard for a method m via a co-reachability analysis of the system S_m w.r.t. its safety property φ_m. A co-reachability analysis finds all states from which a given set of states may be reached, and is typically solved using a fixed-point [23,24]. We have implemented the guard inference algorithm in a prototype tool called Guardies, available at http://nberth.space/symmaries, that is equipped with several instantiations of symbolic abstract heap domains using various families of heap-related relations.

Summary of Contributions

- We introduce a novel notion of symbolic abstract heap domains that uses a set of relations to represent the heap, and is the first flow-sensitive heap model used for information-flow control analysis. We define three different instances of this domain (deep, shal, dumb), each with a different set of relations (Sect. 3), and show that deep constitutes a secure heap abstraction (Sect. 4);
- We symbolically specify the security semantics of our input language to capture explicit and *implicit* flows via the heap, and infer polymorphic information-flow guards via a co-reachability analysis (Sect. 5). We prove that our analysis under a secure heap abstraction guarantees termination-insensitive noninterference [3];
- We empirically study the respective impacts of our three heap domains, in terms of precision on IFSPEC benchmarks [25], and in terms of scalability with 60 real-life ABM applications [26] (Sect. 6). Our experiments show that our approach offers the best precision, and the heap model precision has an inverse relationship with scalability. The heap domains dumb and deep, that are resp. the least- and most-precise heap model, improve the state-of-the-art precision by 2.4% and 4.2% respectively;
- While the existing approaches use an *ad hoc* (flow-insensitive) heap model, Guardies offers six different heap models, thereby allowing the user to choose a suitable heap model based on her preferences for precision and scalability.

2   Preliminaries

Input Programs. We consider a Java-style low-level language where the code of a method is a non-empty finite semicolon-separated sequence of statements built according to $\mathbb{S}$ in the grammar on the next page, where square brackets denote optional constructs. $\sqrt{}$ is an empty sequence of statements, and $lbl \in Labels$ is a label that uniquely identifies a statement. c is a class name, f_p (resp. f_r) is a primitive (resp. reference) field name, p is a scalar constant, and e is an expression. v (resp. r) depicts any local primitive (resp. reference) variable or formal argument used of the method in $\mathbb{S}$. $\text{output}_l(v)$ sends data v over a channel with the security label l. The label l belongs to a *two-level security domain* which is formalized as a lattice $\langle \mathbb{L}, \sqsubseteq, \sqcup \rangle$, where $\mathbb{L} = \{\bot, \top\}$ is the set of security levels, $\bot$ is the low-sensitive label, $\top$ is the high-sensitive label, $\sqsubseteq$ is a partial order defined over $\mathbb{L}$ with $\bot \sqsubseteq \top$, and $\sqcup$ gives the least-upper-bound[1]. Information may become public via *sink* statements, that we denote $\text{output}_\bot(x)$. Therefore, $\text{output}_\bot(v)$ is a sink for the value of v, and $\text{output}_\bot(r)$ is a sink for *every object that is reachable in the heap* via the reference r.

$$\mathbb{S} ::= [lbl :]\ a; \mathbb{S} \mid \sqrt{}$$
$$a ::= v = e \mid v = r.f_p \mid r.f_p = e$$
$$\mid r = r \mid r = r.f_r \mid r.f_r = r$$
$$\mid r = \text{new } c \mid r = \text{null} \mid r.m(w)$$
$$\mid \text{goto } lbl \mid \text{if } (e) \text{ goto } lbl$$
$$\mid \text{output}_l(v) \mid \text{output}_l(r)$$
$$e ::= p \mid v \mid \ominus e \mid e \oplus e \mid r == r$$

Symbolic Control-Flow Graphs—SCFGs. The transition systems that we use to encode the security semantics are traditional labeled transition systems augmented with sets of *state* and *input* variables, respectively denoted X and I. The values for input variables can be seen as coming from the environment of the system.

Definition 1 (Symbolic Control-flow Graph). *A symbolic control-flow graph is a tuple $S = \langle \Lambda, X, I, \Delta, \ell_0, X_0 \rangle$ where: Λ is a finite non-empty set of locations; X and I respectively denote state and input variables; Δ is a set of transitions labeled with a guard that is a predicate on state or input variables, and a possibly empty set of assignments to state variables, noted $[v_0 := e_0, \ldots, v_n := e_n]$, or $\varnothing$ if empty; $\ell_0 \in \Lambda$ is the initial location; and X_0 is a predicate that describes the entire set of possible initial valuations for all the state variables.*

Predicates and right-hand-side expressions in assignments are built using traditional logical connectives (i.e., $\neg$, $\vee$, $\wedge$ and $\Rightarrow$), along with a ternary conditional construct "if $\cdot$ then $\cdot$ else $\cdot$" with an obvious meaning. The symbolic variables we make use of typically take their value in the security domain $\mathbb{L}$,

---

[1] As is traditional, we will present our work by focusing on a standard two-level lattice $\mathbb{L} \overset{\text{def}}{=} \{\bot, \top\}$; minor adaptations would be necessary to support more complex lattices.

or the set of Booleans $\mathbb{B} \stackrel{\text{def}}{=} \{\text{ff}, \text{tt}\}$. We use a *merge operation* $\sqcup$ to merge variable assignments. This operation is obtained as a union where multiple expressions assigned to a variable v are combined using some connective $\sqcup_v$. The latter depends on the semantics of each variable: as we only use variables to hold over-approximations in our encoding, we use the disjunction $\vee$ for Booleans, and the least-upper-bound $\sqcup$ for security levels. For instance, $\{a := \text{tt}, b := \text{ff}\} \sqcup \{b := \text{tt}\} = \{a := \text{tt}, b := \text{ff} \vee \text{tt}\}$. S induces a model $[\![S]\!]$ that is a finite-state automaton whose *state-space* $\mathcal{Q}_S$ is the Cartesian product of the set of locations Λ and the set of all possible valuations for the state variables, i.e., $\text{Val}(X)$. $[\![S]\!]$ takes one transition whenever it receives a valuation for *all* the input variables, i.e., an element in $\text{Val}(I)$. In any location, there is always exactly one transition whose guard is satisfied by the valuations for all the variables. When this transition is taken, its assignments are applied to update the state variables. An *invariant* φ for the SCFG S is a mapping from locations to predicates on state variables. S *satisfies* φ iff every state q with location ℓ that is reachable by S is such that $q \models \varphi(\ell)$.

3 Symbolic Abstract Heap Domains

We first detail our design of heap abstractions that are suitable for the symbolic encoding of security semantics. In this approach, one predicate is used to model *a set of symbolic heaps*. Each symbolic heap represents a *parameterizable* set of *heap-related relations* between the portions of the heap that are reachable via a given set of references R, along with a *security typing environment* for every reachable portion of heap. Such predicates provide *storeless representations* since object locations are not explicitly represented. Predicate transformers describe the *effects of heap and reference variable mutations* on sets of symbolic heaps.

3.1 Families of Heap-Related Relations

Our definition of symbolic abstract heap domains is parameterized by a *family of heap-related relations*. A typical example of a heap-related relation is the aliasing relation, that we denote with the symbol $\sim$, and which is defined as an equivalence relation where $r \sim s$ holds iff r and s point to the same object. We define a family of heap-related relations as a pair $\text{hd} \stackrel{\text{def}}{=} \langle \mathcal{R}_{\text{Sen}}, \mathcal{R}_{\text{Insen}} \rangle$ where $\mathcal{R}_{\text{Sen}}$ is a set of *flow-sensitive relations*, and $\mathcal{R}_{\text{Insen}}$ are *constant flow-insensitive relations* (or facts). A relation is formally specified as a set of Boolean variables that each indicates whether two references taken from R are in the relation or not (i.e., we use predicate abstraction where a Boolean variable specifies whether a relation between two references holds). For instance, we need four propositions (therefore, as many Boolean variables) to represent the relation $\frown$ defined over $R = \{a, b\}$, i.e., $a \frown a$, $a \frown b$, $b \frown a$ and $b \frown b$. The proposition $x \frown y$ evaluates to true if x is in the relation $\frown$ with y. Further, a relation $\frown = \{(a, a), (b, a)\}$ is formalized as $a \frown a \wedge \neg a \frown b \wedge b \frown a \wedge \neg b \frown b$.

The propositions about flow-sensitive relations may be updated by the program statements, while the propositions about constant flow-insensitive relations are straightforwardly substituted with tt or ff, and serve the sole purpose of improving the precision of the predicate transformers that manipulate symbolic abstract heaps. For instance, in a heap domain that does not handle the aliasing relation flow-sensitively, two references of incompatible types can never alias each other. We formalize this with the pre-analysis function CanRelate, that returns three-valued *sound* facts about the relations in $\mathcal{R} = \mathcal{R}_{\mathsf{Insen}} \cup \mathcal{R}_{\mathsf{Sen}}$ w.r.t. the set of reference variables R. Given any relation $\frown \in \mathcal{R}$ and a pair of references $(r, s) \in R^2$, CanRelate$(r \frown s)$ returns yes if $r \frown s$ always holds, no if it cannot hold, or maybe otherwise. In its most trivial form, this pre-analysis function operates on a purely lexical level, e.g., by returning yes if queried for $r \frown r$ with $\frown$ a reflexive relation, maybe otherwise. It may additionally involve an analysis of the class hierarchy and take the declared type of the reference variables into account to give more precise facts. Note that CanRelate helps us simplify the heap formulae by reducing the number of propositions used to represent heaps. For instance, if $\frown$ is a reflexive relation in our previous example, we don't need to consider the propositions $a \frown a$ and $b \frown b$, and use the constant tt instead. We leave further specifications of the pre-analysis open for the sake of modularity.

We need to differentiate between "flow-sensitive heaps" and "flow-sensitive heap relations". The first case means that the heap changes during the execution while the latter states that the relation used to specify the heap structure changes during the execution (i.e., $\mathcal{R}_{\mathsf{Sen}}$). Therefore, a flow-sensitive symbolic heap abstraction models at least one heap relation flow-sensitively.

3.2 Symbolic Abstract Heap Domain

Formally, a *symbolic abstract heap domain* for the family of heap-related relations hd is defined as a pair HeapDom$_{\mathsf{hd}} \stackrel{\text{def}}{=} \langle \mathbb{H}_{\mathsf{hd}}, \mathbb{T}_{\mathsf{hd}} \rangle$ where $\mathbb{H}_{\mathsf{hd}}$ is the set of symbolic abstract heaps, and $\mathbb{T}_{\mathsf{hd}}$ is a set of predicate transformers to manipulate the abstract heaps. A *symbolic abstract heap* from this domain is a predicate $\mathfrak{h} \in \mathbb{H}_{\mathsf{hd}}$ defined on two sets of state variables $\mathbb{V}_{\overrightarrow{L}}$ and $\mathbb{V}_{\mathcal{R}}$. The set $\mathbb{V}_{\overrightarrow{L}}$ associates a *security level variable* $\overrightarrow{r}^{\mathfrak{h}}$ with each reference $r \in R$, that represents an *upper bound* on the security levels of any object that is reachable via r: these variables constitute the *security typing environment* for the abstract heap $\mathfrak{h}$. In turn, the set $\mathbb{V}_{\mathcal{R}}$ consists of Boolean variables that describe *over-approximations of flow-sensitive heap-related relations* between the references in R, i.e., a variable $r \frown^{\mathfrak{h}} s \in \mathbb{V}_{\mathcal{R}}$ holds whenever (r, s) *may* be in the heap-related relation $\frown$:

$$\mathbb{V}_{\mathcal{R}} \stackrel{\text{def}}{=} \left\{ r \frown^{\mathfrak{h}} s \mid (r, s) \in R^2, \frown \in \mathcal{R}_{\mathsf{Sen}}, \mathsf{CanRelate}(r \frown s) = \mathsf{maybe} \right\}.$$

Further, $\mathbb{V}_{\mathsf{ff}}$ and $\mathbb{V}_{\mathsf{tt}}$ are sets of constants that capture all relations that never hold and always hold according to the function CanRelate, respectively. (For the sake of readability, we will omit the exponent $\mathfrak{h}$ of security-level variables when a single abstract heap $\mathfrak{h}$ is involved, i.e., $\overrightarrow{r}^{\mathfrak{h}}$ will be denoted $\overrightarrow{r}$.)

Table 1. Denotations for symbolic abstract heaps

Two symbolic abstract heaps:	$\mathfrak{h}, \mathfrak{h}' \in \mathbb{H}_{\mathsf{hd}}$
Set of variables encoding a heap-related relation $\frown \in \mathcal{R}_{\mathsf{Sen}}$:	$V_{\mathcal{R}}$
Set of constants encoding non-membership facts, for any relation $\frown \in \mathcal{R}_{\mathsf{Sen}} \cup \mathcal{R}_{\mathsf{Insen}}$:	V_{ff}
Set of constants encoding membership facts, for any relation $\frown \in \mathcal{R}_{\mathsf{Sen}} \cup \mathcal{R}_{\mathsf{Insen}}$:	V_{tt}
Set of variables encoding the security levels:	$V_{\vec{L}}$

Variables & Predicate		Transformers ($\mathbb{T}_R$)	
Security level variable:	$\bar{r}^{\mathfrak{h}}$ (or $\bar{r}$)	Reference assignment:	$(\!as\!)$
Relation variable:	$r \frown^{\mathfrak{h}} s$	Mutation and allocation:	$(\!mu, \uparrow l\!)$
Initialization:	$(\!null\ R'\!)$	Bulk upgrade:	$\mathsf{BulkUpgr}_{\mathfrak{h} \leftarrow \mathfrak{h}'}$

with $as \in \{r = s, r = s.f_r, r = \mathbf{null}\}$, $mu \in \{r.f_p \leftsquigarrow, r.f_r = s, r = \mathbf{new}\}$, $(s, r) \in R^2$, $R' \subseteq R$, and any security level expression l.

1 **class** A { **int** fi; } **class** B { A fa; }	Statement Loc	Heap Transformer
2 **static void** m (A a, B b, **int** i) {	3	$(\!r = \mathbf{new}, \uparrow pc\!)$
3 B r = **new** B;	4	$(\!a.fi \leftsquigarrow, \uparrow \bar{i} \sqcup pc\!)$
4 a. fi = i;	5	$(\!r.fa = a, \uparrow \bar{a} \sqcup \bar{a} \sqcup pc\!)$
5 r.fa = a;	6	$\varnothing$
6 output$_\perp$(b);	$R = \{a, b, r\}$, $\mathcal{R} = \{\frown, \overset{*}{\hookrightarrow}\}$	
7 }	$V_{\vec{L}} = \{\bar{a}, \bar{b}, \bar{r}\}$, $V_{\mathcal{R}} = \{b \frown r, b \overset{*}{\hookrightarrow} a, r \overset{*}{\hookrightarrow} a\}$	

(a) Class definitions and method **m**.

(b) Heap Transformers and Variables

Fig. 1. Method that manipulates references, with a representation of heap transformers, heap-related relationships and variables.

$$V_{\mathsf{ff}} \overset{\text{def}}{=} \left\{ r \frown^{\mathfrak{h}} s \mid (r, s) \in R^2, \frown \in \mathcal{R}_{\mathsf{Sen}} \cup \mathcal{R}_{\mathsf{Insen}}, \mathsf{CanRelate}(r \frown s) = \mathsf{no} \right\}$$

$$V_{\mathsf{tt}} \overset{\text{def}}{=} \left\{ r \frown^{\mathfrak{h}} s \mid (r, s) \in R^2, \begin{matrix} (\frown \in \mathcal{R}_{\mathsf{Sen}}, \mathsf{CanRelate}(r \frown s) = \mathsf{yes}) \vee \\ (\frown \in \mathcal{R}_{\mathsf{Insen}}, \mathsf{CanRelate}(r \frown s) \neq \mathsf{no}) \end{matrix} \right\}.$$

We summarize in Table 1 the main denotations that we use to represent and manipulate symbolic abstract heaps. The leftmost column shows the variables that represent security levels and heap-related relations, along with the operator $(\!null\ R'\!)$. The right-hand side column lists the set of *predicate transformers* that can be applied on a symbolic abstract heap to alter its representation. The two first transformers in the column operate in accordance with a given reference assignment (as) or mutation (mu). The expression l given to the latter gives the security level of the information that flows to mutated objects. We give in Fig. 2 the definitions of all transformers. These definitions make use of functions specialized for each family of heap-related relations, detailed below. $(\!null\ R'\!)$ builds a predicate that constrains variables in $V_{\vec{L}}$ and $V_{\mathcal{R}}$ to account for the fact that a given set of references $R' \subseteq R$ is **null**—this notably entails that every object reachable via R' is low-sensitive. The *bulk upgrade* is a transformer used to capture implicit flows through the heap by joining two distinct heap abstractions $\mathfrak{h}$ and $\mathfrak{h}'$ that belong to the same domain. More specifically, this transformer assumes that $\bar{r}^{\mathfrak{h}'} \sqsubseteq \bar{r}^{\mathfrak{h}}$ for all $r \in R$, and: (i) copies the heap-

$$(r = \texttt{null}) \stackrel{def}{=} \textsf{UpdHpRel}(r = \_) \quad {}^{\sqcup}[\vec{r} := \bot]$$

$$(r = s) \stackrel{def}{=} \textsf{UpdHpRel}(r = s) \quad {}^{\sqcup}[\vec{r} := \vec{s}]$$

$$(r = s.f_r) \stackrel{def}{=} \textsf{UpdHpRel}(r = s.f_r) \quad {}^{\sqcup}[\vec{r} := \vec{s}]$$

$$(r = \texttt{new}, \uparrow l) \stackrel{def}{=} \textsf{UpdHpRel}(r = \_) \quad {}^{\sqcup}[\vec{r} := l]$$

$$(r.f_p \hookleftarrow, \uparrow l) \stackrel{def}{=} \qquad\qquad\qquad\qquad \textsf{UpdHpLev}(r, l)$$

$$(r.f_r = s, \uparrow l) \stackrel{def}{=} \textsf{UpdHpRel}(r.f_r = s) \quad {}^{\sqcup} \textsf{UpdHpLev}(r, l)$$

$$(\texttt{null } R') \stackrel{def}{=} \textsf{NullRefs}_{\mathfrak{h}}(R') \qquad \wedge \bigwedge_{r \in R'} \vec{r} = \bot$$

$$\textsf{BulkUpgr}_{\mathfrak{h} \leftarrow \mathfrak{h}'} \stackrel{def}{=} \textsf{CopyRels}_{\mathfrak{h} \leftarrow \mathfrak{h}'} \qquad {}^{\sqcup} \textsf{RstrLev}_{\mathfrak{h} \leftarrow \mathfrak{h}'}$$

with $\textsf{CopyRels}_{\mathfrak{h} \leftarrow \mathfrak{h}'} \stackrel{def}{=} \bigsqcup_{r \sim^{\mathfrak{h}'} s \in \mathbb{V}_{\mathcal{R}}}^{\sqcup} \left[r \sim^{\mathfrak{h}} s := r \sim^{\mathfrak{h}'} s \right]$ and $\textsf{NullRefs}_{\mathfrak{h}}(R') \stackrel{def}{=} \bigwedge_{r \sim^{\mathfrak{h}} s \in \mathbb{V}_{\mathcal{R}}, \{r,s\} \cap R' \neq \varnothing} \left(r \sim^{\mathfrak{h}} s = \texttt{ff} \right).$

Fig. 2. Definitions of generic transformers $\mathbb{T}_R$ for any symbolic abstract heap domain $\textsf{HeapDom}_{\textsf{hd}}$. Note $\mathfrak{h}$ and $\mathfrak{h}'$ belong to the same domain, i.e., $(\mathfrak{h}, \mathfrak{h}') \in \mathbb{H}_R{}^2$, and $R' \subseteq R$. See Fig. 3 for an example definition of $\textsf{UpdHpRel}(\cdot)$.

related relations from $\mathfrak{h}'$ to $\mathfrak{h}$; and (ii) upgrades the typing environment of $\mathfrak{h}$ via a pairwise join with the corresponding levels in $\mathfrak{h}'$.

Example 1. Consider method m given in Fig. 1(a). Figure 1(b) shows its references R, heap-related relations $\mathcal{R}$, references security levels $\mathbb{V}_{\vec{L}}$ and variables $\mathbb{V}_{\mathcal{R}}$ to specify the heap structure using relations $\mathcal{R}_{\textsf{Sen}}$. Further, the table in this Figure shows the heap transformers associated with each statement to update the heap relations and reference security levels. As an example transformer, consider $(\texttt{r} = \texttt{new}, \uparrow pc)$ that corresponds to the statement r = new B;. With a heap domain that captures the aliasing relation flow-sensitively, the resulting set of assignments includes (at least) $[b \sim^{\mathfrak{h}} r := \texttt{ff}, \vec{r} := l]$ where l is a security level expression s.t $l \sqsupseteq pc$.

3.3 Instances of Symbolic Abstract Heap Domains

We present three instances of the domain introduced above. We first define the *transitive "field-aliasing"* relation, denoted with the symbol $\stackrel{.*}{\hookrightarrow}$, which states for any given pair of references r

Table 2. Variables and constants involved in representing $\mathfrak{h}$ when analyzing m, for each domain. $\mathfrak{h}$ exponents have been omitted for readability.

	deep	shal	dumb
$\mathbb{V}_{\vec{L}}$		$\{\vec{a}, \vec{b}, \vec{r}\}$	
$\mathbb{V}_{\mathcal{R}}$	$\{b \sim r, b \stackrel{.*}{\hookrightarrow} a, r \stackrel{.*}{\hookrightarrow} a\}$	$\{b \sim r\}$	$\varnothing$
$\mathbb{V}_{\text{ff}}$	$\{a \sim b, a \sim r, a \stackrel{.*}{\hookrightarrow} a, a \stackrel{.*}{\hookrightarrow} b, a \stackrel{.*}{\hookrightarrow} r, b \stackrel{.*}{\hookrightarrow} b, b \stackrel{.*}{\hookrightarrow} r, r \stackrel{.*}{\hookrightarrow} b, r \stackrel{.*}{\hookrightarrow} r\}$		
$\mathbb{V}_{\text{tt}}$	$\{a \sim a, b \sim b, r \sim r\}$	$\{a \sim a, b \sim b, r \sim r, b \stackrel{.*}{\hookrightarrow} a, r \stackrel{.*}{\hookrightarrow} a\}$	$\{a \sim a, b \sim b, r \sim r, b \sim r, b \stackrel{.*}{\hookrightarrow} a, r \stackrel{.*}{\hookrightarrow} a\}$

and s, $r \stackrel{.*}{\hookrightarrow} s$ holds whenever a reference field of an object reachable via r is an alias of s. This relation allows heap domains to capture some useful facts about

$$\mathsf{UpdHpRel}(r = \_) \overset{\text{def}}{=} \bigsqcup_{d \in R}^{u} \left[d{\sim}r := \mathsf{ff}, r\overset{*}{\hookrightarrow}d := \mathsf{ff}, d\overset{*}{\hookrightarrow}r := \mathsf{ff} \right]$$

$$\mathsf{UpdHpRel}(r = s) \overset{\text{def}}{=} \bigsqcup_{d \in R}^{u} \left[d{\sim}r := d{\sim}s, r\overset{*}{\hookrightarrow}d := s\overset{*}{\hookrightarrow}d, d\overset{*}{\hookrightarrow}r := d\overset{*}{\hookrightarrow}s \right]$$

$$\mathsf{UpdHpRel}(r = s.f_r) \overset{\text{def}}{=} \bigsqcup_{d \in R}^{u} \left[d{\sim}r := s\overset{*}{\hookrightarrow}d, r\overset{*}{\hookrightarrow}d := s\overset{*}{\hookrightarrow}d, d\overset{*}{\hookrightarrow}r :=, d{\sim}s \vee d\overset{*}{\hookrightarrow}s \right]$$

$$\mathsf{UpdHpRel}(r.f_r = s) \overset{\text{def}}{=} \bigsqcup_{a\overset{*}{\hookrightarrow}b \in \mathbb{V}_{\mathcal{R}}}^{u} \left[a\overset{*}{\hookrightarrow}b := a\overset{*}{\hookrightarrow}b \vee \left((a{\sim}r \vee a\overset{*}{\hookrightarrow}r) \wedge (b{\sim}s \vee s\overset{*}{\hookrightarrow}b) \right) \right]$$

$$\mathsf{UpdHpLev}(r, l) \overset{\text{def}}{=} \bigsqcup_{s \in R}^{u} \left[\vec{\mathfrak{z}} := \vec{\mathfrak{z}} \sqcup \text{ if } s{\sim}r \vee s\overset{*}{\hookrightarrow}r \text{ then } l \text{ else } \bot \right]$$

$$\mathsf{RstrLev}_{\mathfrak{h} \leftarrow \mathfrak{h}'} \overset{\text{def}}{=} \bigsqcup_{(r,s) \in R^2}^{u} \left[\vec{\mathfrak{z}}^{\mathfrak{h}} := \vec{\mathfrak{z}}^{\mathfrak{h}} \sqcup \text{ if } s{\sim}^{\mathfrak{h}'}r \vee s\overset{*}{\hookrightarrow}^{\mathfrak{h}'} r \text{ then } \vec{r}^{\mathfrak{h}'} \text{ else } \bot \right]$$

Fig. 3. Specialized functions for updating security level and relation variables for each domain defined with $\mathsf{hd} \in \{\mathsf{deep}, \mathsf{shal}, \mathsf{dumb}\}$; $\mathfrak{h}$ exponents have been omitted when a single abstract heap is involved.

the structure of the graph of objects when it comes to maintaining object types such as security levels.

We now assume a sound pre-analysis function $\mathsf{CanRelate}$ over $\{\sim, \overset{*}{\hookrightarrow}\}$, and use the above relations to define the three families of heap-related relations based on which we shall instantiate our symbolic abstract heap domains:

- $\mathsf{HeapDom_{deep}}$, with $\mathsf{deep} \overset{\text{def}}{=} \langle \{\sim, \overset{*}{\hookrightarrow}\}, \varnothing \rangle$, uses symbolic variables to represent over-approximations of aliasing and field-aliasing relations ;
- $\mathsf{HeapDom_{shal}}$, with $\mathsf{shal} \overset{\text{def}}{=} \langle \{\sim\}, \{\overset{*}{\hookrightarrow}\} \rangle$, only maintains a flow-sensitive over-approximation of the aliasing relation, yet makes use of field-aliasing facts to improve the precision of transformers;
- $\mathsf{HeapDom_{dumb}}$, with $\mathsf{dumb} \overset{\text{def}}{=} \langle \varnothing, \{\sim, \overset{*}{\hookrightarrow}\} \rangle$, does not represent any flow-sensitive heap-related relation, yet makes use of flow-insensitive (field-) aliasing relations.

Example 2. Consider method m given in Fig. 1(a), and assume a class hierarchy pre-analysis. We instantiate the three symbolic abstract heap domains as $\mathsf{HeapDom_{hd}}$ and define an abstract heap $\mathfrak{h} \in \mathbb{H}_{\mathsf{hd}}$, for each $\mathsf{hd} \in \{\mathsf{deep}, \mathsf{shal}, \mathsf{dumb}\}$. Table 2 shows the sets of variables used by each one of these domains to represent $\mathfrak{h}$ ($\mathbb{V}_{\vec{L}}$ and $\mathbb{V}_{\mathcal{R}}$), along with the symbols that denote constants involved in capturing relations flow-insensitively ($\mathbb{V}_{\mathsf{ff}}$ and $\mathbb{V}_{\mathsf{tt}}$).

Regarding transformers, we give in Fig. 3 the specialized functions used by their definitions in Fig. 2. The domains defined with $\mathsf{hd} \in \{\mathsf{deep}, \mathsf{shal}, \mathsf{dumb}\}$ share these definitions. $\mathsf{UpdHpRel}(r = \_)$ updates relation variables in $\mathfrak{h}$ to reflect the erasing of a given reference r to either nil or a fresh reference by clearing variables from $\mathbb{V}_{\mathcal{R}}$. In turn, $\mathsf{UpdHpRel}(r = s)$ updates the variables in $\mathbb{V}_{\mathcal{R}}$ to encode the copy of a reference s to r. $\mathsf{UpdHpRel}(r = s.f_r)$ makes any reference d

that may be an alias of one of s's field a potential alias of r (when a corresponding variable $d\sim^{\flat}r$ belongs to $\mathbb{V}_{\mathcal{R}}$), and updates any variable that represents the $\overset{*}{\hookrightarrow}$ relation to reflect that s becomes a field-alias of r. $\mathsf{UpdHpRel}(r.f_r = s)$ makes s a field-alias of r while maintaining transitivity of $\overset{*}{\hookrightarrow}$. Storing a reference may only add elements in relation $\overset{*}{\hookrightarrow}$, hence the disjunction in every assignment defined by this operation. Observe that, as can be seen in the definition of $\mathsf{UpdHpRel}(r = s.f_r)$ for $\mathsf{hd} = \mathsf{shal}$, instead of simply setting $d\sim r := \mathsf{tt}$ for every potential alias d of r (i.e., blindly assuming that no information is known about the potential aliasing relation), a constant $s\overset{*}{\hookrightarrow}d \in \mathbb{V}_{\mathsf{tt}} \cup \mathbb{V}_{\mathsf{ff}}$ is used instead to further restrict the new potential aliases to the cases that have not been ruled out by the pre-analysis. The dumb domain involves some pre-established facts via a similar mechanism. $\mathsf{UpdHpLev}(r,l)$ takes a reference r and a security level expression l, and upgrades the security level associated with the objects reachable via r as well as that of every reference s that may transitively field-alias r (i.e., $s\overset{*}{\hookrightarrow}r$). $\mathsf{RstrLev}_{\mathfrak{h}\leftarrow\mathfrak{h}'}$ upgrades the typing environment of $\mathfrak{h}$ according to that of $\mathfrak{h}'$.

4 Secure Heap Abstraction

To specify the semantics of heap operations performed by a program, we define a *concrete heap domain* that maintains the value of *primitive fields* in addition to *the precise heap-related relations*. A concrete heap domain is defined similarly to that of abstract heap domains introduced in Sect. 3, *with the difference that the heap maintains the primitive fields instead of security levels*. The concrete heap domain is defined as $\mathsf{HeapDom}_{\mathsf{crt}} = \langle \mathbb{H}_{\mathsf{crt}}, \mathbb{T}_{\mathsf{crt}}\rangle$ where $\mathsf{crt} \overset{\mathrm{def}}{=} \langle\{\sim, \overset{f}{\hookrightarrow}\}, \varnothing\rangle$, the relation $\sim$ is an ordinary reference aliasing relation, and $\overset{f}{\hookrightarrow}$ is a field-aliasing relation, i.e., $r\overset{f}{\hookrightarrow}s$ holds iff the field f of the object referenced by r is an alias of s. See [27] for the details of the heap transformers for the concrete heap domain. The notation $\lfloor op\rfloor_{\hbar}$ shows the predicate transformer that corresponds to the operation op on a concrete heap $\hbar \in \mathbb{H}_{\mathsf{crt}}$.

Let h be a heap from an arbitrary heap domain and $R' \subseteq R$ be a set of references. The reference graph over R' induced by h is a labeled digraph $G^{\mathsf{h}}_{R'} \overset{\mathrm{def}}{=} (N_{\mathsf{h}}, E_{\mathsf{h}})$ where $N_{\mathsf{h}} = R'$ is the set of nodes, and the edges E_{h} show the heap-related relations between them, i.e., $E_{\mathsf{h}} = \{(r, \frown, r') \mid \mathsf{h} \models r \frown r', r \in R' \vee r' \in R'\}$. Let $\mathfrak{h}$ be an abstract heap and $G^{\mathfrak{h}}_{\perp}$ be a sub-graph of $G^{\mathfrak{h}}_{R}$ containing the low-sensitive references $R^{\mathfrak{h}}_{\perp} = \{r \mid \mathfrak{h} \models (\vec{r} = \perp)\}$. We say two concrete heaps are indistinguishable, if heap-related relations and primitive fields of their low-sensitive references are identical, i.e., (i) the reference graphs corresponding to their low-sensitive portions of the heaps are *isomorphic*, and (ii) the valuation of primitive fields of their low-sensitive references are identical.

Definition 2 (Indistinguishable Heaps). *We say two concrete heaps $\hbar$ and $\hbar'$ from $\mathbb{H}_{\mathsf{crt}}$, are indistinguishable w.r.t. an abstract heap $\mathfrak{h}$, noted by $\hbar =_{\mathfrak{h}} \hbar'$, iff (i) $G^{\hbar}_{\perp}$ and $G^{\hbar'}_{\perp}$ are isomorphic, denoted by $G^{\hbar}_{\perp} \cong G^{\hbar'}_{\perp}$, and (ii) $\forall x.\ \hbar \models r.f_p = x \Leftrightarrow \hbar' \models r.f_p = x$, for all $r \in R^{\mathfrak{h}}_{\perp}$ where f_p is a primitive field.*

Since we use an abstract heap domain to model and analyze information flow via heap, we should ensure that the analysis under abstract heap domains guarantees noninterference. To this end, we should show that the heap indistinguishability relation is preserved by the heap transformers. We define the concept of *secure heap abstraction*, which states that two indistinguishable heaps should remain indistinguishable after applying a heap operation and its corresponding transformer at the abstract heap domain level:

Definition 3 (Secure Heap Abstraction). *The concrete heap domain* $\mathsf{HeapDom}_{crt}$ *is secure w.r.t. an abstract heap domain* $\mathsf{HeapDom}_{hd}$, *if and only if it preserves the heap indistinguishability relation, i.e., given any concrete heaps* $(\hbar_1, \hbar_2) \in \mathbb{H}_{crt} \times \mathbb{H}_{crt}$, *and an abstract heap* $\mathfrak{h} \in \mathbb{H}_{hd}$ *s.t.* $\hbar_1 =_\mathfrak{h} \hbar_2$, *it holds that:*

(a) *for all pair* (as, as') *of reference assignment statements and their corresponding operations on abstract heaps where* $as \in \{\langle\!\langle r = s \rangle\!\rangle, \langle\!\langle r = s.f_r \rangle\!\rangle, \langle\!\langle r = \mathtt{null} \rangle\!\rangle\}$, $\hbar'_1 =_{\mathfrak{h}'} \hbar'_2$ *holds where* $\hbar'_i = \lfloor as \rfloor_{\hbar_i}$, $i \in \{1, 2\}$, *and* $\mathfrak{h}' = (\!|as'|\!)$;

(m) *for all pair* (mu, mu') *of mutation statements and their corresponding operations on heaps where* $mu \in \{\langle\!\langle r.f_p = e \rangle\!\rangle, \langle\!\langle r.f_r = s \rangle\!\rangle, \langle\!\langle r = \mathbf{new}\, c \rangle\!\rangle\}$, *for all* $l \in \mathbb{L}$ *where* $\vec{s} \sqsubseteq l$ *if* mu *is* $\langle\!\langle r.f_r = s \rangle\!\rangle$, *and* $\vec{e} \sqsubseteq l$ *if* mu *is* $\langle\!\langle r.f_p = e \rangle\!\rangle$, *it holds that* $\hbar'_1 =_{\mathfrak{h}'} \hbar'_2$ *where* $\hbar'_i = \lfloor mu \rfloor_{\hbar_i}$, $i \in \{1, 2\}$, *and* $\mathfrak{h}' = (\!|mu', \uparrow l|\!)$;

Theorem 1. *The concrete heap domain* $\mathsf{HeapDom}_{crt}$ *is secure w.r.t. the deep abstract heap domain* $\mathsf{HeapDom}_{deep}$ *according to Definition 3.*

Proof. See [27].

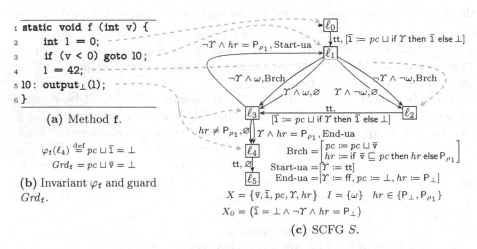

(a) Method f.

$\varphi_f(\ell_4) \overset{\text{def}}{=} pc \sqcup \bar{1} = \bot$
$Grd_f = pc \sqcup \bar{v} = \bot$

(b) Invariant φ_f and guard Grd_f.

(c) SCFG S.

Fig. 4. Example method with SCFG, invariant, and resulting guard.

5   Inferring Polymorphic Information-Flow Guards

Let us now put the heap abstraction aside, and focus on our approach for computing guards and capturing implicit flows. We start by considering the method f given in Fig. 4(a), that does not involve any reference variable. f implements a canonical pattern used to illustrate implicit flows: if its argument v is high-sensitive, then the information output on line 5 is also high-sensitive via an implicit flow induced by the assignment guarded by the condition on line 3. For this example, our requirement demands that executing the sink statement on line 5 does not leak confidential information. Therefore, the guard that we want to compute for f is a sufficient condition that allow us to decide whether any call to the method satisfies the confidentiality requirement based on a set of program facts available *whatever the calling context*. For f, the latter set of facts includes, for instance, the security level of the effective argument for v, or whether the call happens in a high context (i.e., if it is guarded by a condition on high-sensitive information). To achieve this, we build the SCFG that specifies the security semantics of any method in such a way that the set of all its potential initial states encodes all the possible calling contexts for the method. For instance, when encoding f the variable $\bar{v}$ assigned to the argument v is left uninitialized (contrary to the level $\bar{l}$ of local variable l). Another state variable that is left uninitialized is pc, as it denotes the security level of the calling context. We then associate a safety property φ that expresses constraints on security levels at states of the SCFG, and use a co-reachability analysis to find the set of *all initial states* from which no run ever leads to a violation of φ. We graphically represent in Fig. 4(c) the SCFG S obtained for f. The associated invariant is given in Fig. 4(b), along with the inferred guard.

5.1   Security Semantics

Our encoding of security semantics captures implicit flows (i.e., flows induced via the program control-flow structure) by constructing SCFGs that feature two *execution modes*, encoded with the help of a state variable Υ: (i) in *nominal* mode ($\Upsilon = \text{ff}$), updates to security levels reflect explicit information flows, and (ii) in *upgrade analysis* mode ($\Upsilon = \text{tt}$), the information flow from the *high* execution context pc to every variable updated in every possible execution path within the *Control Dependence Regions* (CDRs) of the current context are captured. A CDR ρ is a non-empty set of CFG (Control-Flow Graph) nodes that gathers every instruction that is control-dependent on a given branching statement. This use of CDRs is inspired by previous works [4,5,21,28][2]. The code in Fig. 4(a)

---

[2] The classical algorithm of Ball [29] for computing CDRs works by identifying as a *junction* each dominating node in the post-dominator tree of the CFG. Such a junction j is reached by every execution path that starts from any node in the set ρ of nodes that j post-dominates. Further, one can always find a *unique* branching node that precedes nodes in ρ and belongs to every path from the source of the CFG to any node in ρ, and ρ is therefore a CDR.

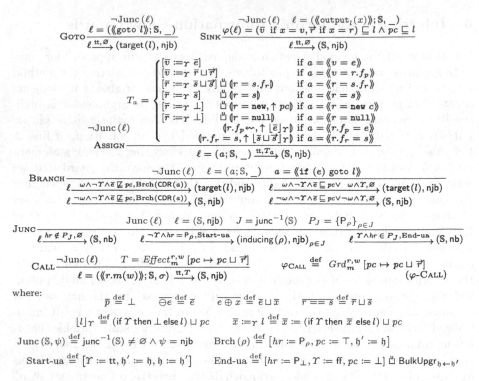

$$\text{GOTO} \frac{\neg\text{Junc}\,(\ell) \quad \ell = (\!\langle\text{goto } l\rangle\!\rangle; \mathbb{S}, \_)}{\ell \xrightarrow{\text{tt}, \varnothing} (\text{target}(l), \text{njb})} \quad \text{SINK} \frac{\neg\text{Junc}\,(\ell) \quad \ell = (\!\langle\text{output}_l(x)\rangle\!\rangle; \mathbb{S}, \_) \quad \varphi(\ell) = (\overline{v} \text{ if } x = v, \overline{r} \text{ if } x = r) \sqsubseteq l \wedge pc \sqsubseteq l}{\ell \xrightarrow{\text{tt}, \varnothing} (\mathbb{S}, \text{njb})}$$

$$\text{ASSIGN} \frac{\neg\text{Junc}\,(\ell) \quad T_a = \begin{cases} [\overline{v} := _{\gamma} \overline{e}] & \text{if } a = \langle\!\langle v = e\rangle\!\rangle \\ [\overline{v} := _{\gamma} \overline{r} \sqcup \overline{r}] & \text{if } a = \langle\!\langle v = r.f_p\rangle\!\rangle \\ [\overline{r} := _{\gamma} \overline{s} \sqcup \overline{s}] \; \overset{\smile}{\curlyvee}(r = s.f_r) & \text{if } a = \langle\!\langle r = s.f_r\rangle\!\rangle \\ [\overline{r} := _{\gamma} \overline{s}] \; \overset{\smile}{\curlyvee}(r = s) & \text{if } a = \langle\!\langle r = s\rangle\!\rangle \\ [\overline{r} := _{\gamma} \bot] \; \overset{\smile}{\curlyvee}(r = \text{new}, \uparrow pc) & \text{if } a = \langle\!\langle r = \text{new } c\rangle\!\rangle \\ [\overline{r} := _{\gamma} \bot] \; \overset{\smile}{\curlyvee}(r = \text{null}) & \text{if } a = \langle\!\langle r = \text{null}\rangle\!\rangle \\ (r.f_p \rightsquigarrow, \uparrow \lfloor\overline{e}\rfloor_{\gamma}) & \text{if } a = \langle\!\langle r.f_p = e\rangle\!\rangle \\ (r.f_r = s, \uparrow \lfloor\overline{s} \sqcup \overline{s}\rfloor_{\gamma}) & \text{if } a = \langle\!\langle r.f_r = s\rangle\!\rangle \end{cases}}{\ell = (a; \mathbb{S}, \_) \xrightarrow{\text{tt}, T_a} (\mathbb{S}, \text{njb})}$$

$$\text{BRANCH} \frac{\neg\text{Junc}\,(\ell) \quad \ell = (a; \mathbb{S}, \_) \quad a = \langle\!\langle\text{if } (e) \text{ goto } l\rangle\!\rangle}{\ell \xrightarrow{\omega \wedge \neg \Upsilon \wedge \overline{e} \not\sqsubseteq pc, \text{Brch}(\text{CDR}(a))} (\text{target}(l), \text{njb}) \qquad \ell \xrightarrow{\omega \wedge \neg \Upsilon \wedge \overline{e} \sqsubseteq pc \vee \omega \wedge \Upsilon, \varnothing} (\text{target}(l), \text{njb})}$$

$$\ell \xrightarrow{\neg\omega \wedge \neg \Upsilon \wedge \overline{e} \not\sqsubseteq pc, \text{Brch}(\text{CDR}(a))} (\mathbb{S}, \text{njb}) \qquad \ell \xrightarrow{\neg\omega \wedge \neg \Upsilon \wedge \overline{e} \sqsubseteq pc \vee \neg\omega \wedge \Upsilon, \varnothing} (\mathbb{S}, \text{njb})$$

$$\text{JUNC} \frac{\text{Junc}\,(\ell) \quad \ell = (\mathbb{S}, \text{njb}) \quad J = \text{junc}^{-1}(\mathbb{S}) \quad P_J = \{P_\rho\}_{\rho \in J}}{\ell \xrightarrow{hr \notin P_J, \varnothing} (\mathbb{S}, \text{nb}) \qquad \ell \xrightarrow{\neg \Upsilon \wedge hr = P_\rho, \text{Start-ua}} (\text{inducing}(\rho), \text{njb})_{\rho \in J} \qquad \ell \xrightarrow{\Upsilon \wedge hr \in P_J, \text{End-ua}} (\mathbb{S}, \text{nb})}$$

$$\text{CALL} \frac{\neg\text{Junc}\,(\ell) \quad T = \text{Effect}_m^{r,w} [pc \mapsto pc \sqcup \overline{r}]}{\ell = (\!\langle r.m(w)\rangle\!\rangle; \mathbb{S}, \sigma) \xrightarrow{\text{tt}, T} (\mathbb{S}, \text{njb})} \qquad \varphi_{\text{CALL}} \overset{\text{def}}{=} \text{Grd}_m^{r,w} [pc \mapsto pc \sqcup \overline{r}] \quad (\varphi\text{-CALL})$$

where:

$$\overline{p} \overset{\text{def}}{=} \bot \qquad \ominus e \overset{\text{def}}{=} \overline{e} \qquad e \oplus x \overset{\text{def}}{=} \overline{e} \sqcup \overline{x} \qquad r == s \overset{\text{def}}{=} \overline{r} \sqcup \overline{s}$$

$$\lfloor l\rfloor_{\gamma} \overset{\text{def}}{=} (\text{if } \Upsilon \text{ then } \bot \text{ else } l) \sqcup pc \qquad \overline{x} :=_{\gamma} l \overset{\text{def}}{=} \overline{x} := (\text{if } \Upsilon \text{ then } \overline{x} \text{ else } l) \sqcup pc$$

$$\text{Junc}\,(\mathbb{S}, \psi) \overset{\text{def}}{=} \text{junc}^{-1}(\mathbb{S}) \neq \varnothing \wedge \psi = \text{njb} \qquad \text{Brch}(\rho) \overset{\text{def}}{=} [hr := P_\rho, pc := \top, \mathfrak{h}' := \mathfrak{h}]$$

$$\text{Start-ua} \overset{\text{def}}{=} [\Upsilon := \text{tt}, \mathfrak{h}' := \mathfrak{h}, \mathfrak{h} := \mathfrak{h}'] \qquad \text{End-ua} \overset{\text{def}}{=} [hr := P_\bot, \Upsilon := \text{ff}, pc := \bot] \; \overset{\smile}{\curlyvee} \text{BulkUpgr}_{\mathfrak{h} \leftarrow \mathfrak{h}'}$$

Fig. 5. Translation rules and safety properties for encoding the security semantics.

features a single conditional branching statement on line 3, which induces the CDR ρ_1. The junction of ρ_1 is the statement $\text{output}_\bot$ (1). Two execution branches are possible within ρ_1: one branch executes no statement, whereas the other performs the assignment $1 = 42$ on line 4. This means that the execution of the latter is dependent on the condition on line 3. In the SCFG S, the upgrade analysis of ρ_1 starts whenever the model reaches location ℓ_3, which represents the *junction* of ρ_1, if the branching statement that induces ρ_1 (encoded by ℓ_1) was subject to a high condition. We use a state variable hr to record the CDR currently subject to a high-condition. We make use of the input variable ω to abstract away the actual branch condition in nominal mode (since our security semantics abstracts away the values of program variables). This is for instance the case on location ℓ_1 in S when $\Upsilon = \text{ff}$. The variable ω is also used to model upgrade analyses for multiple possible program paths which can be taken non-deterministically. In S, this is the case on location ℓ_1 as well, when $\Upsilon = \text{tt}$.

We give in Fig. 5 the set of translation rules that specify the security semantics of a program in terms of an SCFG. Each location of the resulting SCFG corresponds to a *semantic location*, that is defined as a pair $(\mathbb{S}, \psi)$ where $\mathbb{S}$ corresponds to a node in its CFG, and ψ is a *behavior mode* that belongs to $\{\text{njb}, \text{nb}\}$ (for nominal-or-junction and nominal behaviors, respectively). The junction step ψ is used in our encoding to distinguish the nominal mode from the upgrade analysis stage of junctions. Essentially, a semantic location that corresponds to a statement a that is the junction of a CDR behaves as a junction when $\psi = \mathsf{njb}$, and according to a when $\psi = \mathsf{nb}$. Thus, statements that are not junctions never give rise to semantic locations where $\psi = \mathsf{nb}$. To clarify the translation rules, we define the helper predicate $\mathrm{Junc}(\mathbb{S}, \psi)$ in Fig. 5 (where junc^{-1} is the retraction of junc: $\mathsf{junc}^{-1}(\mathbb{S})$ gives the set of CDRs of which $\mathbb{S}$ is the junction), that holds iff a semantic location $(\mathbb{S}, \psi)$ represents an actual junction. We use $\mathsf{target}(l)$ to denote the statement identified by a label l.

The ASSIGN rule encodes the security semantics of assignments. We use $\bar{e}$ to denote the security level of an expression e. from upgrade analyses in the rules. In nominal mode, $\lfloor l \rfloor_\Upsilon$ encodes the least upper-bound between l and the context level pc, and $\bar{x} :=_\Upsilon l$ models a *strong update* of the security level assigned to x with $\lfloor l \rfloor_\Upsilon$. In upgrade analysis mode, however, $\lfloor l \rfloor_\Upsilon$ is equal to the context level (i.e., $\top$), and $\bar{x} :=_\Upsilon l$ encodes a *weak update* of $\bar{x}$ with pc. Then, a statement $v = r.f_p$ that loads a primitive field translates into a transition that updates $\bar{v}$ with: the upper-bound between pc, $\bar{r}$, and the level of any object potentially pointed to by r as maintained by the heap abstraction $\mathfrak{h}$ (i.e., $\vec{r}$) when in nominal mode; the upper-bound between pc and $\bar{v}$ otherwise. BRANCH and JUNC encode the alternation of nominal and upgrade analyses, and do so with the help of a placeholder abstract heap $\mathfrak{h}'$ that belongs to the same abstract heap domain as $\mathfrak{h}$, and is also represented with state variables. According to BRANCH, when a high branch is reached, the transformer $\mathrm{Brch}(\rho)$: (i) sets the state variable hr used to record the CDR currently subject to a high-condition to P_ρ; (ii) updates pc; and (iii) stores the current heap abstraction to $\mathfrak{h}'$ by copying the values of all variables $\mathbb{V}_{\bar{\mathcal{L}}}$ (resp. $\mathbb{V}_{\mathcal{R}}$) to $\mathbb{V}_{\bar{\mathcal{L}}}$ (resp. $\mathbb{V}_{\mathcal{R}}$). The join of abstract heaps that ends upgrade analyses is performed using a bulk upgrade. In effect, $\mathsf{BulkUpgr}_{\mathfrak{h} \leftarrow \mathfrak{h}'}$: (i) upgrades the security typing environment for referenced portions of the heap according to the result of the upgrade analysis in $\mathfrak{h}$, by joining every security level from $\mathfrak{h}$ with the corresponding level in $\mathfrak{h}'$; and (ii) restores every heap-related relation as saved in $\mathfrak{h}'$ when entering the upgrade analysis mode. CALL encodes the security semantics of invocation of a method m based on its *polymorphic information-flow summary*, which is a *contract* that consists of:

- an information-flow *guard* Grd_m that specifies the invocation conditions under which the method call is secure, i.e., there is no illegal information flow in the method. This guard is described as constraints on the security types and heap structure of the method's formal arguments;
- an *effect* $Effect_m$ about its *worst potential* side-effects on security levels and heap structure, that is in principle a *transformer* describing how the heap structure and security labels *may be* updated by the method.

We use the guard to enforce the desired security properties upon an invocation of m: this boils down to ensure that Inv. (φ-CALL) holds for the location ℓ in which m is called. The effect is used in CALL to update the typing environment and the heap model. In detail, $Grd_m^{r,w}$ and $Effect_m^{r,w}$ correspond to the aforementioned guard and effect—or a combination of several summaries in case of virtual method dispatch, where guards are combined using a conjunction, and

---

Algorithm 1: SYNTHESIZEGUARD

Input: Method to analyze m

Result: Polymorphic information-flow guard Grd_m

{ Encode the security semantics of m as an SCFG S_m where every state variable related to the calling context is left uninitialized, and express the security requirement as a predicate $\varphi_m(\ell)$ on state variables for each location ℓ of S_m: }

1 $(S_m, \varphi_m) \leftarrow$ ENCODE(m)

2 $\mathcal{B}_0 \leftarrow \{\ell \mapsto \neg\varphi_m(\ell) \mid \ell \in \Lambda(S_m)\}$ *{ Define all known unsafe states }*

3 $\mathcal{B}_\infty \leftarrow$ COREACH$(S_m, \mathcal{B}_0)$ *{ All states that are co-reachable to $\mathcal{B}_0$ }*

    { Factor out the state variables that are not part of the calling context: *}*

4 $Grd_m \leftarrow$ cofactor $(\neg\mathcal{B}_\infty(\ell_0(S_m)), X_0(S_m))$

---

transformers are merged—, and after substitutions w.r.t. m's formal arguments. Furthermore, $e\,[v \mapsto l]$ denotes the substitution of security level expression l for variable v in e: this is required to upgrade the context pc w.r.t. the receiver object (the substitution in effects is performed in every expression on the right-hand side of assignments). For the sake of concision, we leave the computation of polymorphic effects out of the scope of this paper. In that respect, we want to mention that this computation is achievable, even for the cases of recursion, via an extension of our security semantics, accompanied by a dedicated processing of the co-reachability analysis results. Also note that a sound application of effects requires abstract heaps that capture object sharing relations, not just aliasing relations.

5.2 Guard Inference Procedure

We summarize the overall analysis procedure in Algorithm 1, where ENCODE(m) denotes the specification of the security semantics of a method m as an SCFG S_m and invariant φ_m as described above. We represent sets of states as mappings from locations to predicates on state variables, e.g., $\mathcal{B}_0$ is the set of all states that violate the invariant φ_m. $\mathcal{B}_0$ associates every location that corresponds to a sink statement with a predicate on security levels for program variables that violate security requirements encoded in φ_m. Then, the set of insecure states is back-propagated via a standard co-reachability analysis embodied by COREACH, i.e., finding all states from which a given set of states may be reached, and is typically solved using a fixed-point [23, 24]. On a symbolic finite-state system like S or S_m, this computation always terminates, and is traditionally performed using the least fixed-point (lfp)

$$\mathcal{B}_\infty \stackrel{\text{def}}{=} \text{lfp } \lambda\mathcal{B}_i.\mathcal{B}_0 \cup \text{pre}(\mathcal{B}_i), \tag{1}$$

where pre$(\mathcal{B})$ gives all predecessor states of $\mathcal{B}$. $\mathcal{B}_\infty$ associates each location with a predicate that must *not* hold for every subsequent path in S_m to represent secure executions. Therefore, the guard for m can be obtained by complementing $\mathcal{B}_\infty(\ell_0)$ and eliminating every state variable that does not represent a fact from m's calling context. This is done with the help of cofactor(f, g), which amounts to a partial evaluation of f w.r.t. all variables bound in g, i.e., this gives a predicate f' that does not involve any variable fully determined by g and $s.t$ $(g = \text{tt}) \Rightarrow (f = f')$.

Example 3 (Guard inference for m*).* The analysis first builds the SCFG given in Fig. 6, and associates the invariant $pc \sqcup \bar{\text{b}} \sqcup \vec{\text{b}} = \bot$ with location ℓ_3. This states that, for m to be secure, this statement must be executed in a low context and given a low-sensitive reference b (i.e., $\bar{\text{b}} = \bot$) that must only reach low-sensitive objects (i.e., $\vec{\text{b}} = \bot$). This gives the unsafe states shown in the first row of Table 3, where we report a trace of the co-reachability analysis and the resulting guard for each domain. The guard obtained with the dumb domain is the least precise of all three, as it basically describes m as insecure if it is called in high context, or whenever any of its effective arguments or objects they may reach in the heap is high-sensitive. With this domain, the statement r.fa = a (location ℓ_2) may raise the security level $\vec{\text{b}}$ since $\text{r} \sim \text{b} \in \mathbb{V}_{\text{tt}}$ (as $\mathsf{CanRelate}(\text{r} \sim \text{b}) = $ maybe). On the other hand, the inference with shal is able to distinguish whether b and r may alias on location ℓ_2, and then rules this case out thanks to the statement r = new B (location ℓ_0). However, the guard does not hold whenever $\bar{\text{i}} \neq \bot$, as the domain cannot distinguish whether b.fa aliases a or not: therefore, the statement a.fi = i (location ℓ_1) always raises the level $\vec{\text{b}}$ to that of $pc \sqcup \bar{\text{i}}$. At last, the domain deep distinguishes whether b.fa aliases a or not, and the guard indicates that m may not be secure if i is high-sensitive and a and b relate to each other via b.fa.

5.3   Soundness

We prove that any program guarded with a security guard inferred by our method guarantees termination-insensitive noninterference [3]. This notion states that, for any initial states q and q' whose secret parts may only differ, the observations sequence of the program running from the states q and q' will either be the same, or one is a prefix of the other. The reason for the latter case is that this notion is a termination-insensitive property. To prove soundness, we first define the full semantics of a program by an SCFG that extends the security semantics with its operational semantics [27].

Let $\mathcal{S}$ denote the (symbolic) full semantics of a program, and $[\![\mathcal{S}]\!] = \langle \mathcal{Q}, \mathcal{I}, \rightarrow, \mathcal{Q}_0 \rangle$ be an automaton that describes its concrete semantics, where $\mathcal{Q}$ is the set of states, $\mathcal{I}$ is the set of inputs, $\rightarrow \subseteq \mathcal{Q} \times \mathcal{Q}$ is the set of transitions and $\mathcal{Q}_0$ is the set of initial states. A program state q is defined as $\langle \ell, \mathcal{V}, \Omega, \mathcal{X}, \mathbf{h} \rangle$ where ℓ is the current location, $\mathcal{V}$ is the valuation for every one of the primitive program variables P, $\Omega : P \cup R \rightarrow \mathbb{L}$ is the *security typing environment* for the primitive variables P and references R, and $\mathcal{X}$ is the current valuation of the state variables in the security semantics except from the location and the typing environment for the references. We use the notation $\mathcal{X}_{\mathfrak{h}}$ to show $\mathcal{X}$'s heap abstraction $\mathfrak{h}$, and $\mathcal{X}_{\overline{L}}$ to denote its security typing environment. Furthermore, $\mathbf{h}$ is the valuation of the concrete heap's variables. Let $q \xrightarrow{\eta}_* q'$ be an execution of full semantics with a non-zero length (i.e., the reflexive and transitive closure of the concrete

Table 3. Polymorphic guard inference for method m, for different heap domains.

with $\mathcal{B}_0 = \{\ell_3 \mapsto pc \sqcup \bar{b} \sqcup \vec{b} \neq \bot\} \cup \{\ell_i \mapsto \text{ff}\}_{i \in \{0,1,2,4\}}$		
deep	$\mathcal{B}_1 = \mathcal{B}_0 \cup \{\ell_2 \mapsto pc \sqcup \bar{b} \sqcup \bar{b} \sqcup (\text{if } b \leadsto r \text{ then } \bar{a} \sqcup \vec{a} \text{ else } \bot) \quad \neq \bot\}$	
	$\mathcal{B}_2 = \mathcal{B}_1 \cup \{\ell_1 \mapsto pc \sqcup \bar{b} \sqcup \bar{b} \sqcup (\text{if } b \leadsto r \text{ then } \bar{a} \sqcup \vec{a} \sqcup \bar{i} \text{ else } \bot) \sqcup$	
	$\quad (\text{if } b \overset{*}{\hookrightarrow} a \text{ then } \bar{i} \text{ else } \bot) \quad \neq \bot\}$	
	$\mathcal{B}_\infty = \mathcal{B}_2 \cup \{\ell_0 \mapsto pc \sqcup \bar{b} \sqcup \bar{b} \sqcup (\text{if } b \overset{*}{\hookrightarrow} a \text{ then } \bar{i} \text{ else } \bot) \quad \neq \bot\}$	
	$Grd_m = \qquad pc \sqcup \bar{b} \sqcup \bar{b} \sqcup (\text{if } b \overset{*}{\hookrightarrow} a \text{ then } \bar{i} \text{ else } \bot) = \bot$	
shal	$\mathcal{B}_1 = \mathcal{B}_0 \cup \{\ell_2 \mapsto pc \sqcup \bar{b} \sqcup \quad \bar{b} \sqcup (\text{if } b \leadsto r \text{ then } \bar{a} \sqcup \vec{a} \text{ else } \bot) \quad \neq \bot\}$	
	$\mathcal{B}_2 = \mathcal{B}_1 \cup \{\ell_1 \mapsto pc \sqcup \bar{b} \sqcup \bar{i} \sqcup \bar{b} \sqcup (\text{if } b \leadsto r \text{ then } \bar{a} \sqcup \vec{a} \sqcup \bar{i} \text{ else } \bot) \neq \bot\}$	
	$\mathcal{B}_\infty = \mathcal{B}_2 \cup \{\ell_0 \mapsto pc \sqcup \bar{b} \sqcup \bar{i} \sqcup \bar{b} \quad \neq \bot\}$	
	$Grd_m = \qquad pc \sqcup \bar{b} \sqcup \bar{i} \sqcup \bar{b} = \bot$	
dumb	$\mathcal{B}_1 = \mathcal{B}_0 \cup \{\ell_2 \mapsto pc \sqcup \bar{b} \sqcup \bar{a} \sqcup \quad \bar{b} \sqcup \vec{a} \neq \bot\}$	
	$\mathcal{B}_2 = \mathcal{B}_1 \cup \{\ell_1 \mapsto pc \sqcup \bar{b} \sqcup \bar{a} \sqcup \bar{i} \sqcup \bar{b} \sqcup \vec{a} \neq \bot\}$	
	$\mathcal{B}_\infty = \mathcal{B}_2 \cup \{\ell_0 \mapsto pc \sqcup \bar{b} \sqcup \bar{a} \sqcup \bar{i} \sqcup \bar{b} \sqcup \vec{a} \neq \bot\}$	
	$Grd_m = \qquad pc \sqcup \bar{b} \sqcup \bar{a} \sqcup \bar{i} \sqcup \bar{b} \sqcup \vec{a} = \bot$	

Fig. 6. SCFG for m.

transition relation $\rightarrow$) from the state q to the state q', where $\eta \in \{o, \bot\}$. This execution either ends by executing a statement that outputs on a channel (i.e., $\eta = o$) or makes no observation (i.e., $\eta = \bot$). We denote an execution that never reaches an observation point by $q \overset{\bot}{\rightarrow}_*$. We define noninterference based on a *low-equivalence relation*, that states that the public parts of the two states q_1 and q_2 are indistinguishable.

Definition 4 (Indistinguishable Stores). *We say two valuations $\mathcal{V}_1 \in$ Val(P) and $\mathcal{V}_2 \in$ Val(P) are low-equivalent w.r.t. the typing environment $\Omega : P \cup R \rightarrow \mathbb{L}$, denoted by $\mathcal{V}_1 =_\Omega \mathcal{V}_2$, iff $\mathcal{V}_1(v) = \mathcal{V}_2(v)$ for all $v \in P$ where $\Omega(v) = \bot$.*

Definition 5 (Low-Bisimulation). *We say two states $q_i = \langle \ell, \mathcal{V}_i, \Omega_i, \mathcal{X}_i, \mathbf{h}_i \rangle$, $i \in \{1,2\}$ are compatible, denoted by $q_1 \approx q_2$, iff (i) $\Omega_1 = \Omega_2$, (ii) $\mathcal{X}_{1\overline{L}} = \mathcal{X}_{2\overline{L}}$, (iii) $\mathcal{V}_1 =_{\Omega_1} \mathcal{V}_2$, and (iv) $\mathbf{h}_1 =_{\mathcal{X}_{1\flat}} \mathbf{h}_2$. They are called low-bisimilar, denoted by $q_1 \sim_{low} q_2$, iff $q_1 \approx q_2$, and if $q_1 \overset{o}{\rightarrow}_* q_1'$, then either (a) there exists q_2' such that $q_2 \overset{o}{\rightarrow}_* q_2'$ and $q_1' \sim_{low} q_2'$, or (b) $q_2 \overset{\bot}{\rightarrow}_*$, and vice versa.*

Theorem 2 (Noninterference). *For any method m guarded by a security guard Grd_m, and any initial states q_1 and q_2 where $q_1 \approx q_2$ and $q_i \models Grd_m$, $i \in \{1,2\}$, it holds $q_1 \sim_{low} q_2$.*

Proof. To prove this theorem, we show that there exists a witnessing bisimulation relation for $q_1 \sim_{low} q_2$. See [27]. ∎

6   Implementation and Evaluation

To empirically validate our approach, we assess the respective performances of our three heap domains on actual code, both in terms of precision and scalability. We have first implemented a tool that relies on soot [16] to obtain the

Jimple code of a program, and translates it into our input language. Jimple is an intermediate language to represent Java byte-code at a higher level. The semantics of its instructions and reference manipulations correspond to that of the JVM. One Jimple statement roughly translates into one statement of our input language. We have then implemented the guard inference algorithm in a prototype tool called Guardies[3], that features multiple instantiations of our heap domains. Guardies's pre-analysis relies on a naive analysis of the class hierarchy to construct a graph that allows us to compute facts about heap-related relations (i.e., function CanRelate). This tool relies on ReaX [31] to solve the co-reachability problems. ReaX uses (Multi-terminal) Binary Decision Diagrams— (MT)BDDs—[32, 33] to represent symbolic expressions and compute the underlying fixed-points. To deal with guards and transformers that encode the semantics of library methods, we rely on *stubs*, given to Guardies, that describe the effects of these methods at a high level. We manually defined the security semantics of methods from the standard Java and Android libraries (about 1200 methods in total) in this way.

Table 4. IFSPEC precision results

Category	#Smpls	deep	shal	dumb	KEY	JOANA	CASSANDRA
explicit-flows	143	**80.4**	79.7	78.3	70.6	77.6	72.7
implicit-flows	21	**71.4**	**71.4**	**71.4**	57.1	57.1	61.7
simple	51	72.5	72.5	72.5	64.7	**76.4**	68.6
high-cond.	10	**80**	**80**	**80**	60	60	60
arrays	26	73	73	69.2	65.3	**76.9**	69.2
library	69	**88.4**	**88.4**	86.9	76.8	76.7	79.7
aliasing	7	**71.4**	**71.4**	57.1	57.1	42.8	42.8
average		**79.2**	78.6	77.4	68.9	75	71

The #Smpls column shows the number of included samples for each category; other figures are percentages.

Precision&Recall. We have employed the IFSPEC benchmark suite [25] to assess the *precision* of our different heap domains and compare our results to KEY [34], CASSANDRA [21], and JOANA [17]. The precision refers to a proportion of test cases that are correctly classified. The *recall* is the fraction of true positive and false negative test cases that are categorized correctly. IFSPEC provides 232 test cases that showcase various information-flow vulnerabilities in Java programs, with various forms of explicit and implicit information leaks. We report in Table 4 the precision results that we obtain for different abstract heap domains for the various categories of leaks and language features that IFSPEC covers. We have checked 164 out of 232 test cases supported by our sub-language: the

---

[3] Available as a software artifact [30], with user documentation and source code at http://nberth.space/symmaries.

excluded cases involve reflection, static class initializers, exceptions and method calls (11, 10, 9, and 39 samples respectively—we have excluded all cases in the latter category as they check the ability of the analysis to handle information-flows across method calls, while we left the problem of computing method effects aside). Note that a test case may belong to multiple categories.

Since our approach is sound, we obtain 100% *recall*, i.e., we correctly detect every insecure flow. Regarding precision, our experiments show that all the domains have close precision: the deep domain offers the highest precision of 79.2% and dumb offers the lowest precision of 77.4%. The false positives (i.e., the secure test cases that were restrictively classified as insecure) mainly occur because our domains are field-insensitive, value-insensitive, do not distinguish elements in some collections of data, or due to the over-approximations in heap-related relations. The results for the aliasing category are rather similar; 3 test cases in this category are insecure that are classified correctly by all three domains, as our analysis is sound. Two of the remaining 4 secure test cases are classified as insecure in all domains due to value- and field-insensitivity.

On average, the domain deep offers the best precision in five categories. It slightly underperforms the state-of-the-art for simple and arrays test cases only, notably due to value- and field-insensitivity. In some categories, the improvement is noticeable, i.e., it improves the best precision of the aliasing category offered by the existing tools by 14.3%, improves the library category by 8.7% and enhances the implicit-flows category by 9.5%. We attribute these substantial results in part to our precise handling of implicit flows across method calls (unlike CASSANDRA which forbids method calls in high-contexts for instance), and in part to our heap abstract domain, that is able to precisely track some intricate aliasing relations. That most of our domains obtain similar precision results on the aliasing category may indicate that these test cases are rather uniform in the facts about aliasing that need to be discovered to detect secure cases. Our findings show that while different domains had close precision results, they offer different computational complexity though. Further, IFSpec only partially covers the set of IFC problems one can encounter in practice; we therefore refrain from generalizing our results. Yet, IFSpec is the most extensive benchmark available for IFC that we know of.

Scalability Evaluation. We have conducted experiments on real-life web applications to compare different heap abstract domains in terms of scalability. To accommodate computationally intensive analyses, we interrupt any analysis after 5 min or if it uses more than 4 GB of memory. We have used applications from the ABM benchmark [26], a collection of 139 open-source projects that is dedicated to the evaluation of static analyzers for Java applications. Its content is deemed representative of real-world software, and has already been used for evaluating static taint analysis and dead code elimination approaches [26]. From this collection, we extracted the Java code from 60 applications with sizes ranging from 133 to 25K lines of Java. This provided us with a total of 22,512 analyzable methods. Overall, the deep domain led to 146 analyses being interrupted due to the time-outs or memory limitations (3 for shal, 0 for dumb). We plot in Fig. 7, for each

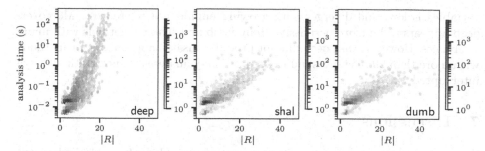

Fig. 7. Density plots showing the distributions of analyzed ABM methods w.r.t. both the number of reference variables (horizontal axes) and the analysis time (vertical axes). Note the shared log scales, including on the color-bars. (Color figure online)

domain, the distributions of successful analyses w.r.t. the number of reference variables and analysis time. As expected, analysis times grow with the amount of references, and by factors that depend on the heap-related relations captured flow-sensitively by the domains, e.g., deep is more expensive compared to dumb and shal. Further, many methods have fewer than 10 reference variables, and as a result most analysis times do not exceed 0.1s for every domain. Those figures empirically support the applicability of our approach on real-life applications.

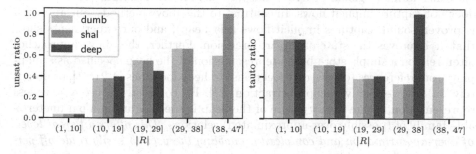

Fig. 8. Plots showing for each domain, the proportions of unsatisfiable (on the left) and tautological (on the right) guards *vs* number of reference variables.

Note that an ideal study on the scalability of different heap domains would compare the domains under different analysis techniques, provided by different tools. This, however, requires the support of the existing tools for modeling different heap domains. To the best of our knowledge, there was no such tool, as each implementation is typically tied with its own heap model, if any at all. Otherwise, extending the tools to support different heap domains is virtually infeasible since most existing tools rely on store-based models. To further compare the respective precision of each domain, we also report in Fig. 8 the ratios of unsatisfiable and tautological guards obtained for each domain. We observe that the precision of all domains seems similar when the number of reference variables is low, and diverges with growing numbers of references. Moreover, deep appears to be more permissive than dumb and shal for methods with many references. However, that deep did not produce unsatisfiable guards for methods with more than 29 references indicates that many analyses of such methods were interrupted.

7   Discussions

Static analysis approaches to ensure noninterference have been studied extensively in the community. The vast majority of suggested IFC solutions concentrates on type-systems [2–4], and various tools that target realistic programming languages have been developed for verifying such properties. Prominent examples include JFlow JIF [3], FlowCaml [2], CASSANDRA [21], and KEY [34]. Albeit sound, the aforementioned approaches often lack precision in practice (e.g., [21]), or require user intervention, such as the specification of loop invariants (e.g., [3]). Another line of research trades efficiency for soundness and/or precision, by exploiting more generic techniques like interprocedural dataflow analysis [35] or program slicing [36]. JOANA [17], DroidSafe [18] and FLOWDROID [19] are prominent frameworks in this category. Other solutions are dedicated to web applications [37] or Android apps [38], although most of them do not handle implicit flows or lack soundness (e.g., [17–19]). Tools that provide sound results via other forms of global program analyses include HORNDROID [39,40], which does not capture implicit flows. In contrast to the above methods, our approach is proven sound, captures implicit flows (via heap), and our experiments show that it improves the state-of-the-art precision. Further, the above approaches often rely on a simple store-based representation of the heap specified as a mapping from references (or abstract locations) to heap locations [4,21,41], or do not rely on a flow-sensitive heap abstraction [21]. By contrast, we use a store-less representation, where the structure of the heap is specified using a parameterizable family of (possibly over-approximated) relations. *This offers different levels of over-approximation and complexity, enabling the user to easily trade-off performance and scalability.*

Few works have addressed the problem of capturing implicit flows while exploiting flow-sensitive heap abstractions [14,42,43]. Khakpour [42] synthesizes sound security monitors that enforce IFC by using a symbolic discrete control algorithm. This work operates intraprocedurally on high-level programs and uses an *ad hoc* field-sensitive heap abstraction that does not scale well. Zanioli et al. [14] advance an abstract-interpretation-based analysis, where the construction of the heap abstraction is delegated to a separate analysis. Their analysis can operate on a flow-sensitive abstraction as produced by a TVLA-based shape analysis [44], yet it can only be applied to small, high-level programs. Other forms of symbolic heap abstractions have already been used in static program analysis. Separation logic [45] models a heap as a formula that comprises atomic predicates combined using the *separation* operator. While we use a store-less representation of the heap expressed using a proposition, symbolic heaps in separation logic are store-based, more expressive, and consequently are more complex for verification. Store-less heap abstractions are also polymorphic and enable us to operate on each method of the program in isolation. This is to be contrasted with traditional data-flow analysis [35], where flow functions must be distributive and expressed on finite domains (as typically provided by store-based abstractions).

We have introduced a generic abstract heap domain for modeling heaps and information flow via heap for low-level object-oriented programs, and instantiated it with different families of relations. Our experiments showed that our instantiated heap models improve the state-of-the-art precision, and that the precision has an inverse relationship with scalability. We are currently investigating the computation of method summaries in order to obtain a fully modular interprocedural IFC analysis. Guardies can be improved by implementing a more advanced analysis to reduce the amount of symbolic variables involved to represent the heap, thereby improving scalability. Further, the instantiated heap domains are field-insensitive, and a natural extension is introducing support for field-sensitive analyses.

Acknowledgement. The first author was supported by the UK Engineering and Physical Sciences Research Council (EPSRC) through grant EP/M027287/1, and the second author was supported by the Swedish Knowledge Foundation (KKs) via the grant No. 20160186.

References

1. Goguen, J.A., Meseguer, J.: Security policies and security models. In: 1982 IEEE Symposium on Security and Privacy, Los Alamitos, CA, USA, pp. 11–20. IEEE Computer Society (1982). https://doi.org/10.1109/SP.1982.10014
2. Pottier, F., Simonet, V.: Information flow inference for ML. ACM Trans. Program. Lang. Syst. **25**(1), 117–158 (2003). https://doi.org/10.1145/596980.596983. http://doi.acm.org/10.1145/596980.596983. ISSN 0164-0925
3. Sabelfeld, A., Myers, A.C.: Language-based information-flow security. IEEE J. Sel. Areas Commun. **21**(1), 5–19 (2003). https://doi.org/10.1109/JSAC.2002.806121. ISSN 0733-8716
4. Barthe, G., Pichardie, D., Rezk, T.: A certified lightweight non-interference Java bytecode verifier. In: De Nicola, R. (ed.) ESOP 2007. LNCS, vol. 4421, pp. 125–140. Springer, Heidelberg (2007). https://doi.org/10.1007/978-3-540-71316-6_10. http://dl.acm.org/citation.cfm?id=1762174.1762189. ISBN 978-3-540-71314-2
5. Liu, Y., Milanova, A.: Static information flow analysis with handling of implicit flows and a study on effects of implicit flows vs explicit flows. In: Proceedings of the 2010 14th European Conference on Software Maintenance and Reengineering, CSMR 2010, USA, pp. 146–155. IEEE Computer Society (2010). https://doi.org/10.1109/CSMR.2010.26. https://doi.org/10.1109/CSMR.2010.26. ISBN 9780769543215
6. Hedin, D., Sabelfeld, A.: Information-flow security for a core of JavaScript. In: Proceedings of the 2012 IEEE 25th Computer Security Foundations Symposium, CSF 2012, Washington, DC, USA, pp. 3–18. IEEE Computer Society (2012). https://doi.org/10.1109/CSF.2012.19. http://dx.doi.org/10.1109/CSF.2012.19. ISBN 978-0-7695-4718-3

7. Barthe, G., D'argenio, P.R., Rezk, T.: Secure information flow by self-composition. Math. Struct. Comput. Sci. **21**(6), 1207–1252 (2011)
8. Terauchi, T., Aiken, A.: Secure information flow as a safety problem. In: Hankin, C., Siveroni, I. (eds.) SAS 2005. LNCS, vol. 3672, pp. 352–367. Springer, Heidelberg (2005). https://doi.org/10.1007/11547662_24. ISBN 3-540-28584-9, 978-3-540-28584-7
9. Barthe, G., Crespo, J.M., Kunz, C.: Relational verification using product programs. In: Butler, M., Schulte, W. (eds.) FM 2011. LNCS, vol. 6664, pp. 200–214. Springer, Heidelberg (2011). https://doi.org/10.1007/978-3-642-21437-0_17. ISBN 9783642214363
10. Darvas, Á., Hähnle, R., Sands, D.: A theorem proving approach to analysis of secure information flow. In: Hutter, D., Ullmann, M. (eds.) SPC 2005. LNCS, vol. 3450, pp. 193–209. Springer, Heidelberg (2005). https://doi.org/10.1007/978-3-540-32004-3_20. ISBN 978-3-540-32004-3
11. Mizuno, M., Schmidt, D.: A security flow control algorithm and its denotational semantics correctness proof. Form. Asp. Comput. **4**(1), 754 (1992). https://doi.org/10.1007/BF03180570
12. Zanotti, M.: Security typings by abstract interpretation. In: Hermenegildo, M.V., Puebla, G. (eds.) SAS 2002. LNCS, vol. 2477, pp. 360–375. Springer, Heidelberg (2002). https://doi.org/10.1007/3-540-45789-5_26. ISBN 3540442359
13. Giacobazzi, R., Mastroeni, I.: Abstract non-interference: parameterizing non-interference by abstract interpretation. In: Proceedings of the 31st ACM SIGPLAN-SIGACT Symposium on Principles of Programming Languages, POPL 2004, pp. 186–197. Association for Computing Machinery, New York (2004). https://doi.org/10.1145/964001.964017. ISBN 158113729X
14. Zanioli, M., Ferrara, P., Cortesi, A.: Sails: static analysis of information leakage with sample. In: Proceedings of the 27th Annual ACM Symposium on Applied Computing, SAC 2012, pp. 1308–1313. Association for Computing Machinery, New York (2012). https://doi.org/10.1145/2245276.2231983. ISBN 9781450308571
15. Kanvar, V., Khedker, U.P.: Heap abstractions for static analysis. ACM Comput. Surv. **49**(2), 29:1–29:47 (2016). https://doi.org/10.1145/2931098. http://doi.acm.org/10.1145/2931098. ISSN 0360-0300
16. Vallée-Rai, R., Co, P., Gagnon, E., Hendren, L., Lam, P., Sundaresan, V.: Soot: a Java bytecode optimization framework. In: CASCON First Decade High Impact Papers, pp. 214–224 (2010)
17. Hammer, C., Snelting, G.: Flow-sensitive, context-sensitive, and object-sensitive information flow control based on program dependence graphs. Int. J. Inf. Secur. **8**(6), 399–422 (2009). https://doi.org/10.1007/s10207-009-0086-1. ISSN 1615-5262
18. Gordon, M.I., Kim, D., Perkins, J.H., Gilham, L., Nguyen, N., Rinard, M.C.: Information flow analysis of Android applications in DroidSafe. In: 22nd Annual Network and Distributed System Security Symposium, NDSS 2015, San Diego, California, USA, 8–11 February 2015. The Internet Society (2015). https://www.ndss-symposium.org/ndss2015/information-flow-analysis-android-applications-droidsafe
19. Arzt, S., et al.: FlowDroid: precise context, flow, field, object-sensitive and lifecycle-aware taint analysis for Android apps. In: Proceedings of the 35th ACM SIGPLAN Conference on Programming Language Design and Implementation, PLDI 2014, pp. 259–269. ACM, New York (2014). https://doi.org/10.1145/2594291.2594299. http://doi.acm.org/10.1145/2594291.2594299. ISBN 978-1-4503-2784-8

20. Johnson, A., Waye, L., Moore, S., Chong, S.: Exploring and enforcing security guarantees via program dependence graphs. In: Proceedings of the 36th ACM SIGPLAN Conference on Programming Language Design and Implementation, PLDI 2015, pp. 291–302. Association for Computing Machinery, New York (2015). https://doi.org/10.1145/2737924.2737957. ISBN 9781450334686

21. Lortz, S., Mantel, H., Starostin, A., Bähr, T., Schneider, D., Weber, A.: Cassandra: towards a certifying app store for Android. In: Proceedings of the 4th ACM Workshop on Security and Privacy in Smartphones & Mobile Devices, SPSM 2014, pp. 93–104. ACM, New York (2014). https://doi.org/10.1145/2666620.2666631. ISBN 9781450331555

22. Boudol, G.: Secure information flow as a safety property. In: Degano, P., Guttman, J., Martinelli, F. (eds.) FAST 2008. LNCS, vol. 5491, pp. 20–34. Springer, Heidelberg (2009). https://doi.org/10.1007/978-3-642-01465-9_2. ISBN 978-3-642-01465-9

23. Ramadge, P.J.G., Murray Wonham, W.: The control of discrete event systems. Proc. IEEE Spec. Issue Dyn. Discret. Event Syst. **77**(1), 81–98 (1989). https://doi.org/10.1109/5.21072

24. Pnueli, A., Rosner, R.: On the synthesis of a reactive module. In: Proceedings of the 16th ACM SIGPLAN-SIGACT Symposium on Principles of Programming Languages, POPL 1989, pp. 179–190. Association for Computing Machinery, New York (1989). https://doi.org/10.1145/75277.75293. ISBN 0897912942

25. Hamann, T., Herda, M., Mantel, H., Mohr, M., Schneider, D., Tasch, M.: A uniform information-flow security benchmark suite for source code and bytecode. In: Gruschka, N. (ed.) NordSec 2018. LNCS, vol. 11252, pp. 437–453. Springer, Cham (2018). https://doi.org/10.1007/978-3-030-03638-6_27. ISBN 978-3-030-03638-6

26. Do, L.N.Q., Eichberg, M., Bodden, E.: Toward an automated benchmark management system. In: Proceedings of the 5th ACM SIGPLAN International Workshop on State of the Art in Program Analysis, SOAP 2016, pp. 13–17. Association for Computing Machinery, New York (2016). https://doi.org/10.1145/2931021.2931023. ISBN 9781450343855

27. Berthier, N., Khakpour, N.: Symbolic abstract heaps for polymorphic information-flow guard inference (extended version). arXiv preprint arXiv:2211.03450 (2022)

28. Denning, D.E., Denning, P.J.: Certification of programs for secure information flow. Commun. ACM **20**(7), 504–513 (1977). https://doi.org/10.1145/359636.359712. ISSN 0001-0782

29. Ball, T.: What's in a region? Or computing control dependence regions in near-linear time for reducible control flow. ACM Lett. Program. Lang. Syst. **2**(1–4), 1–16 (1993). https://doi.org/10.1145/176454.176456. ISSN 1057-4514

30. Berthier, N., Khakpour, N.: Artifact for Paper (Symbolic Abstract Heaps for Polymorphic Information-flow Guard Inference) (2022). https://doi.org/10.5281/zenodo.7103855

31. Berthier, N., Marchand, H.: Discrete controller synthesis for infinite state systems with ReaX. In: 12th International Workshop on Discrete Event Systems, WODES 2014, pp. 46–53. IFAC (2014). https://doi.org/10.3182/20140514-3-FR-4046.00099. ISBN 978-3-902823-61-8

32. Bryant, R.E.: Graph-based algorithms for Boolean function manipulation. IEEE Trans. Comput. **35**(8), 677–691 (1986). https://doi.org/10.1109/TC.1986.1676819. ISSN 0018-9340

33. Billon, J.P.: Perfect normal forms for discrete programs. Technical report, Bull (1987)

34. Ahrendt, W., et al.: The KeY tool. Softw. Syst. Model. **4**(1), 32–54 (2004). https://doi.org/10.1007/s10270-004-0058-x

35. Reps, T., Horwitz, S., Sagiv, M.: Precise interprocedural dataflow analysis via graph reachability. In: Proceedings of the 22nd ACM SIGPLAN-SIGACT Symposium on Principles of Programming Languages, POPL 1995, pp. 49–61. ACM, New York (1995). https://doi.org/10.1145/199448.199462. http://doi.acm.org/10.1145/199448.199462. ISBN 0-89791-692-1

36. Kam, J.B., Ullman, J.D.: Monotone data flow analysis frameworks. Acta Inf. **7**(3), 305–317 (1977). https://doi.org/10.1007/BF00290339. ISSN 0001-5903

37. Hedin, D., Birgisson, A., Bello, L., Sabelfeld, A.: JSFlow: tracking information flow in JavaScript and its APIs. In: Proceedings of the 29th Annual ACM Symposium on Applied Computing, pp. 1663–1671. ACM (2014)

38. Li, L., et al.: Static analysis of Android apps: a systematic literature review. Inf. Softw. Technol. **88**, 67–95 (2017)

39. Calzavara, S., Grishchenko, I., Maffei, M.: HornDroid: practical and sound static analysis of Android applications by SMT solving. In: 2016 IEEE European Symposium on Security and Privacy (EuroS&P), pp. 47–62 (2016). https://doi.org/10.1109/EuroSP.2016.16

40. Calzavara, S., Grishchenko, I., Koutsos, A., Maffei, M.: A sound flow-sensitive heap abstraction for the static analysis of Android applications. In: 30th Computer Security Foundations Symposium, CSF 2017, pp. 22–36. IEEE (2017). https://doi.org/10.1109/CSF.2017.19

41. Amtoft, T., Bandhakavi, S., Banerjee, A.: A logic for information flow in object-oriented programs. In: Gregory Morrisett, J., Peyton Jones, S.L. (eds.) Proceedings of the 33rd ACM SIGPLAN-SIGACT Symposium on Principles of Programming Languages, POPL 2006, Charleston, South Carolina, USA, 11–13 January 2006, pp. 91–102. ACM (2006). https://doi.org/10.1145/1111037.1111046

42. Khakpour, N.: A field-sensitive security monitor for object-oriented programs. Comput. Secur. **108**, 102349 (2021). https://doi.org/10.1016/j.cose.2021.102349

43. Khakpour, N., Skandylas, C.: Synthesis of a permissive security monitor. In: Lopez, J., Zhou, J., Soriano, M. (eds.) ESORICS 2018. LNCS, vol. 11098, pp. 48–65. Springer, Cham (2018). https://doi.org/10.1007/978-3-319-99073-6_3

44. Sagiv, M., Reps, T., Wilhelm, R.: Parametric shape analysis via 3-valued logic. ACM Trans. Program. Lang. Syst. **24**(3), 217–298 (2002). https://doi.org/10.1145/514188.514190. ISSN 0164-0925

45. Reynolds, J.C.: Separation logic: a logic for shared mutable data structures. In: Proceedings of the 17th Annual IEEE Symposium on Logic in Computer Science, LICS 2002, Washington, DC, USA, pp. 55–74. IEEE Computer Society (2002). http://dl.acm.org/citation.cfm?id=645683.664578. ISBN 0-7695-1483-9

Satisfiability Modulo Custom Theories
in Z3

Nikolaj Bjørner[1], Clemens Eisenhofer[2(✉)], and Laura Kovács[2]

[1] Microsoft Research, Redmond, USA
nbjorner@microsoft.com
[2] TU Wien, Wien, Austria
{clemens.eisenhofer,laura.kovacs}@tuwien.ac.at

Abstract. We introduce *user-propagators* as a new feature of the Z3 SMT solver. User-propagation allows users to write custom theory extensions for Z3, by implementing callbacks via the Z3 API. These callbacks are invoked by Z3 and eliminate eager processing and instantiation of theory axioms with quantifiers. We report on application scenarios of user-propagation and describe further use-cases.

Keywords: SMT solving · SMT theories · Automated reasoning · Program verification

1 Introduction

Satisfiability Modulo Theories (SMT) solving [3] has become a backbone in formal verification, synthesis and optimization, see e.g. [12,18,23,34] One of the reasons for the success of SMT solvers is their ability to produce solutions, that is models or proofs, in fragments of first-order theories for data structures that are of relevance to software. As SMT-based reasoning is effective for first-order theories expressible in SMT-LIB logics [2], a challenging aspect in SMT solving is to come up with efficient extensions towards new theories, in particular in the presence of quantifiers.

A natural way to address this challenge is to extend SMT solvers with built-in decision procedures for relevant theories. This approach requires, however, expert knowledge about the functionalities and implementation choices of a respective state of the art SMT solver, such as Z3 [26] and CVC5 [1], limiting thus the general adaptation of this approach mainly to developers of the SMT solvers or heroic researchers.

In this paper, we advocate *Satisfiability Modulo Custom Theories* for a flexible approach towards improving SMT solving. We propose *user-propagators* to ease on-demand theory reasoning by implementing callback functions outside existing SMT solvers, and in particular in Z3. As such, user-propagators in Z3 allow users to write plugins through Z3's API in order to support custom theories. As the custom code is loaded dynamically, there is no need to recompile Z3's code base when having it extended with user-propagators. Moreover, the

C. Dragoi et al. (Eds.): VMCAI 2023, LNCS 13881, pp. 91–105, 2023.
https://doi.org/10.1007/978-3-031-24950-1_5

user is not confronted with the complexity of the whole system, but only has to implement a small number of functions (see Fig. 1).

The preliminary results of our extended abstract from [7] hint already that custom theory reasoning can strongly decrease Z3's computational resources (time and memory) on solving combinatorial problems. In this paper we go beyond [7] and bring the following contributions. We describe the overall framework of user-propagators in Z3, together with the supported callbacks to the core of Z3 (Sect. 3). We argue that user-propagators may increase performance by using a lazy problem encoding. We demonstrate how user-propagators can be used to speed up Boolean/bit-vector based program verification problems, in particular memory reasoning (Sect. 2 and Sect. 4), by lazily instantiating some of the required axioms of the encoding on-demand. We finally also discuss applications of user-propagators (Sect. 5), opening up many new research avenues when compared to [7].

2   Motivating Example

We motivate the benefits of user-propagators in Z3 by lazily instantiating Boolean based axioms, in particular in the setting of alive2 [22].

The alive2 framework [22] aims to find errors in transformations/optimizations applied to functions in LLVM intermediate representation [19]. Soundness of the applied LLVM optimizations is checked in alive2 by proving that the original (unoptimized) program and the optimized program are equivalent[1]. alive2 encodes the behaviour of finitely unrolled variants of the unoptimized and optimized LLVM intermediate representations into a set of SMT queries that are passed to Z3 for equivalence checking, via the Z3 API.

Generating the SMT encoding through the Z3 API, as well as proving in Z3, may take a significant amount of time for some inputs. One major source of such an inefficiency is that alive2 often has to compare all potential pairs of variables or functions, yielding a significant blow-up in the size of the SMT queries passed to Z3. For example, the *memory disjointness axiom* used by alive2 has a quadratic size with respect to the number of variables in the LLVM program; this axiom asserts that the physical memory addresses of two different allocations may not intersect. We note that alive2 uses a multi-memory encoding that assigns an array to each allocation, as well as a unique identifier [20]. In case the program compares addresses, alive2 associates an additional bit-vector with each memory block that represent their address. Thus, for encoding memory disjointness, alive2 may generate the formula

$$\bigwedge_{1 \leq i < j \leq n} (addr_i + size_i \leq addr_j \vee addr_j + size_j \leq addr_i), \tag{1}$$

for n globally allocated memory blocks given by their physical starting addresses ($addr$) and their sizes ($size$).

---

[1] Modulo undefined values, where undefined values in the unoptimized program may be replaced by any concrete value in the optimized program, but not vice-versa.

The encoding (1) of memory disjointness can be seen as a generalization of the following well-known problem in the constraint programming (CP) community: assume we have a set of n variables $x_1, \ldots, x_n$ that should be assigned to distinct values of a finite, but large, domain. Encoding this problem, and hence (1), eagerly requires an at least quadratically sized assertion of the form

$$\bigwedge_{1 \leq i < j \leq n} x_i \neq x_j. \tag{2}$$

Using (2) is however problematic: Assume, for example, that we have $1,000$ bit-vector variables x_i, each of them consisting of 32 bits. Encoding this eagerly[2] would require roughly half a million disequalities, although the chance of at least one conflicting assignment is below 1% in case of random variable assignments.

An alternative approach to eagerly encode and process the constraints (2) is offered through user-propagators, as follows: (i) observe the (bit-vector value) assignments to the variables x_i and (ii) add the constraint $x_i \neq x_j$ of (2) only in case x_i and x_j are assigned to the same value. We refer to such a conditional/on-demand addition of constraints through user-propagators as a *lazy approach*. A user-propagator encoding can naturally be adjusted to (1) to avoid generating unnecessary constraints: in case two intersecting intervals are derived during SMT solving, we propagate $addr_i + size_i \leq addr_j \vee addr_j + size_j \leq addr_i$ to alert the SMT solver about the violation of (1).

3   User-Propagators in Z3

A user-propagator can be seen as a set of functions that correspond to actions performed by the core of Z3. More precisely, each time a variable associated with the theory modelled by the user-propagator is involved in one of the actions covered by the user-propagator, the respective user-implemented function ("callback") is invoked.

3.1   Workflow for User-Propagators

The overall workflow of Z3, extended with a user-propagator, is presented in Fig. 1 and discussed in Sect. 3.2.

Most of the user-propagator callbacks listed in Fig. 1 already exist (directly or in a similar variant) internally in Z3 for those theories for which built-in theory reasoners are implemented. With user-propagator callbacks, such support can be extended to theories that are not natively supported in Z3. Each term in Z3 may be associated with a theory that is responsible for the semantics of the respective term and whose callback will be triggered. In most cases, the core Z3 code base does not distinguish whether the callback belongs to a built-in theory or is provided via the user-propagator, which is why user-propagators did not require fundamental changes in Z3's workflow.

---

[2] The special case of (2) has mostly efficient built-in implementations.

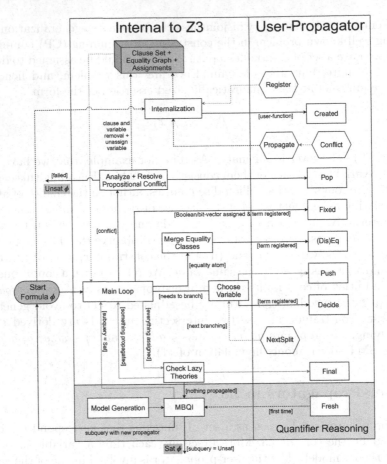

Fig. 1. Schematized workflow for user-propagators in Z3.

Upon using user-propagators in Z3, an initial (SMT) formula ϕ is passed to Z3 for satisfiability checking. As Z3 does not know which terms are relevant for the theory the user wants to model, all relevant (ground) terms in ϕ have to be registered by a Z3 API call. Registering a term in Z3 has two effects. (i) First, the user theory is associated with the registered terms, such that the callbacks of the user-propagator are called when one of the registered terms is processed. (ii) Second, Z3 will eagerly internalize the registered terms of ϕ; that is, the respective terms are translated into Z3 internal (normal form) representations so that Z3 can further reason with them. Notably, terms are associated with a union-find data-structure and for congruence closure reasoning, each term has pointers to parent terms. Mostly, step (ii) is done right before starting the actual reasoning process in Z3; for more details, see [25]. However, doing (ii) eagerly already during formula construction ensures that Z3 processes registered terms and does not eliminate it by some optimization. This way, even formulas that do not occur in the input formula ϕ can be registered, and hence also processed by the user-propagator. An example where such a case can be useful comes with registering a function application that extracts the kth bit of a bit-vector in ϕ: this way, user-propagators can also reason on the kth bit, even if this bit alone does not occur in the input formula ϕ.

3.2 Supported Callbacks

Currently, user-propagators in Z3 supports the following list of functions/callbacks, that can be implemented by the user.

Push and Pop Each decision procedure in Z3 provides a Push and a Pop function. In case Z3's SAT core branches on a Boolean variable, the Push function of each decision procedure is called. Theory solvers save their current state such that a subsequent call of Pop resets the reasoner in the same or an equisatisfiable state. As theory solvers in Z3 may backjump multiple decision levels, Pop may has to revert multiple pushes at once. The number of reverts is passed as an argument to the Pop callback.

Fixed The Fixed callback is invoked whenever a registered Boolean or bit-vector variable is fixed to a value. Boolean variables are considered as fixed as soon as their values are assigned by e.g., branching or Boolean constraint propagation. A bit-vector variable requires all its Boolean (bit) variables to be assigned before triggering Fixed. As theory solvers based on CDCL(T) fix the values of T-elements (e.g. integers, strings) mostly in a separate step after CDCL(T) reasoning (model generation) [3], the Fixed callback cannot be used to observe those values of T-elements.

Eq and Diseq In contrast to the Fixed callback, Eq and Diseq are not limited to Booleans/bit-vector terms. Eq is called when two equivalence classes containing registered terms are merged. The callback is guaranteed to report on all equalities between registered terms, but is only issued for enough terms to form a spanning tree of equal terms (for roots of an underlying union-find structure). Similar, Diseq is called whenever Z3 infers that two equivalence classes, both containing registered terms, are not equal. It does not report on all disequalities; the user-propagator can instead infer these from the complement of the reported equalities.

Decide Similar to the Push callback, Decide is called if Z3 branches on a Boolean variable. However, in contrast to Push, Decide is only called if Z3 decides to branch on a registered expression. The Decide callback is invoked with the variable Z3 will try to split on and the truth value that Z3 tried first as arguments. To implement custom heuristics, the user may change the variable to any other registered and unassigned Boolean term, as well as the truth value of the variable.

Final This callback is invoked in case all Boolean variables are assigned by the CDCL-based SAT solver [17,31]. The theory solvers are supposed to finally check if the current Boolean assignments and (dis)equalities are consistent with the theory. In case no decision procedure adds new variables, propagates new lemmas, or reports a conflict in this step, Z3 checks the relevant quantifiers and returns satisfiable in case they are. The order in which the Final functions of the decision-procedures are invoked is not fixed, so no assumptions can be made about being *the first/last* solver.

Fresh In case Z3 needs to spawn a new solver (e.g., for doing model based quantifier instantiation (MBQI) subqueries [13], or solving subgoals in parallel) the Fresh callback is called which returns a new user-propagator that receives the callbacks issued by the sub-solver. MBQI reuses the same sub-solver across different calls, so in typical use cases, Fresh is invoked at most once.

Created To track fresh terms created during search, such as by quantifier instantiation, user-propagators can define *user-functions*. All instances of user-functions are automatically registered and the Created callback is triggered the first time a particular instance of a user-function is encountered by Z3. This way, the user can additionally observe and register all (ground) arguments of the function application that are relevant for the theory. User-functions can be considered as an interpreted function/predicate symbols whose semantic is given by the user-propagator by adding conflicts or by propagating lemmas containing user-function instances.

We further note that a user can also call special functions within callbacks to participate in the solving process. We consider and support the following four functions[3] in this respect:

Propagate Recall that Eq yields equalities $t \simeq t'$ and Fixed yields assignments $t \simeq v$ that are true in the current branch, where t and t' are (registered) terms and v is a Boolean or bit-vector value. A *propagation* claim is

$$(E \wedge F) \Rightarrow G \qquad (3)$$

where E is a subset of the reported equalities of Eq, F a subset of Fixed, and G an arbitrary formula constructed by Z3's API. Z3 adds the propagation claim (3) as a lemma to the current scope by transforming the claim into an internal format, by resolving each of the premises in E and F with a set of internal decisions or propagations that *justify* the premises E and F. Thus, the equalities $t \simeq t'$ or $t \simeq v$ are not treated as potentially fresh atoms, but resolved with already existing literals in Z3.

Conflict A conflict is shorthand for a propagation claim $(E \wedge F) \Rightarrow \perp$.

NextSplit Controlling branching steps in Z3 with the Decide callback is not sufficient, as Decide is called only in case a registered term is chosen. NextSplit can mark yet unassigned Boolean/bit-vector terms to be chosen next for splitting. This way, Z3 uses these marked terms instead of choosing a potentially unregistered term next. The Decide callback is subsequently called for a marked term, as if the term was chosen by some internal Z3 heuristic.

Register This function registers a term to be processed by the user-propagator. Register can be both called during callbacks and (as already discussed) before starting the reasoning process.

---

[3] Referring to the C++ API.

Propagated lemmas and variables registered during reasoning may only be considered as asserted/registered as long as the scope in which they were propagated/registered has not been exited.

4 User-Propagators for Memory Reasoning in alive2

In this section, we reconsider the motivating example of Sect. 2 on proving memory disjointness in alive2. We detail our *lazy approach* based on user-propagators and show generalizations of this approach.

User-Propagators for Memory Disjointness. We successfully used user-propagators in Z3 to implement the lazy approach of Sect. 2 for proving (1). That is, (i) we first observe the values Z3 assigned to the memory addresses $addr_i$ and sizes $size_i$ by registering them. To this end, as alive2 uses bit-vectors, the Fixed callbacks will be called whenever all bits of the addresses $addr_i$ or the sizes and sizes $size_i$ are fixed. Further, (ii) in case both the address $addr_i$ and the size $size_i$ of a block i is fixed, we can check if there is a collision with another memory block j fixed so far. In case there is, we propagate a lemma to avoid this. As the lemma is not a logical consequence of any previous assignment, we choose $E = F = \emptyset$ in the propagation claim (3) and use the Propagate function.

Applying steps (i)-(ii) instead of an eager encoding has several advantages. We save computational time and memory, as we might not need to generate the whole disjointness constraint (1). Further, generating the constraints on-demand (lazily) prevents Z3 from handling all constraints and putting too much focus on unlikely relevant clauses. As the size of the memory blocks are mostly small constants, the constraints do not have to be eagerly enforced, as the chance of a collision is low. Using steps (i)-(ii) for the global case, our experiments show a significant performance improvement in alive2's performance. For benchmarking, we used the single-file versions of some well-known C programs[4]. For example, the gzip and bzip2 source codes contain each roughly 100 functions. However, only about 20 of them make relevant use of pointer comparisons/arithmetic after compilation to LLVM's intermediate representation. With the lazy implementation of the global disjointness changes in alive2, 6 additional function verification queries for the gzip source code and 3 for the bzip2 source code could be solved within a 60 second timeout by Z3. (A single function query generated by gzip could not be solved anymore with the changes.)

Further Generalizations. The lazy approach of steps (i)-(ii) can be further generalized to lazily instantiate arbitrary (quantified) parts of the formula at any positions in the formula and not necessarily constraints that can be added as conjunctions to the formula.

Consider for example the following (simplified) formula that is used as well by alive2:

$$\forall \bar{\mathbb{M}}(\bigwedge_{1 \leq i < j \leq n} (addr_i + size_i \leq addr_j \vee addr_j + size_j \leq addr_i) \Rightarrow G[\bar{\mathbb{M}}]) \qquad (4)$$

---

[4] http://people.csail.mit.edu/smcc/projects/single-file-programs/.

where $\bar{\mathbb{M}}$ is a sequence of n memory blocks (with addresses $addr_i$ and sizes $size_i$) and G is the encoding of the problem.

We can use user-propagation as well for lazy encoding the conjunction in the body of the quantifier in (4): we replace the premise of (4) by a user-function that is on-demand refined via user-propagator, similarly as in the case of the memory disjointness property of (1). That is, we replace (4) by

$$\forall \bar{\mathbb{M}}(disjoint(\bar{\mathbb{M}}) \Rightarrow G[\bar{\mathbb{M}}]), \tag{5}$$

where $disjoint(\bar{\mathbb{M}})$ denotes that all memory blocks $\bar{\mathbb{M}}$ are disjoint.

Upon quantifier instantiation in (5), we receive **Created** callbacks notifying the user-propagator about the new instances of (5). Note that (1) uses no quantifiers and the part that should be lazily instantiated occurs top-level. Therefore, unlike (1), the lazily instantiated part of (5) (i.e. the $disjoint$ instances) may be assigned both to true and to false. Therefore, we need to consider the following two cases in (5): (a) a $disjoint$ predicate instance is assigned to true but at least two memory blocks intersect, and (b) the $disjoint$ predicate instance is assigned to false, but there is no intersection between any memory blocks. For case (a), we propagate

$$disjoint(\bar{\mathbb{M}}) \Rightarrow disjoint^{\uparrow}(\bar{\mathbb{M}}) \tag{6}$$

whereas for (b) we propagate

$$disjoint^{\downarrow}(\bar{\mathbb{M}}) \Rightarrow disjoint(\bar{\mathbb{M}}), \tag{7}$$

where $disjoint^{\uparrow}$ and $disjoint^{\downarrow}$ are formulas (over- and under-approximations) that respectively witness the (non-)violation of (5). As usual in SMT solving, coming up with good witnesses/justifications is a crucial performance point. To this end, in case $\mathbb{M}_i$ and $\mathbb{M}_j$ intersect, $addr_i + size_i \leq addr_j \vee addr_j + size_j \leq addr_i$ can be used as $disjoint^{\uparrow}$ to witness a violation. A potential witness $disjoint^{\downarrow}$ for the non-violation is a total ordering of the memory blocks $addr_{\pi(1)} + size_{\pi(1)} \leq \ldots \leq addr_{\pi(n)}$, where π is the current ordering of the addresses reported by **Fixed**.

5   Using User-Propagators

In this section, we discuss existing and potential applications for user propagators and interface functionality of relevance.

Optimization Modulo Theories, Aggregates and Scheduling. It has long been recognized that branch and bound functionality can be added to SMT solvers as custom theories [27,29]. The custom theory maintains a running sum and bounds branches when the current aggregate exceeds a (best) bound. User propagators are used by [8] to implement multi-objective branch and bound optimization. Aggregates are also used to express packing constraints, namely to express that the sum of time spent on tasks at a given workstation does not exceed the maximal time assigned to it. The functionality relies on callbacks for Fixed and Conflict.

Partial Orders. Runtime Verification Inc.[5] encodes type constraints over an order-sorted type system as a reflexive-transitive extension of a relation on a finite (but large) set of sorts. Axiomatizations are possible, but grow quadratically in the size of the domain. User-propagators allow delaying the instantiation until they are really needed. The functionality relies on the callbacks from Fixed, Eq, Created, and the Conflict function.

CP Domains. Similarly, user-propagation can be used to model *combinatorial problems* mostly dealt with in the *constraint satisfaction problem (CSP)* community in a conflict-driven way. In our extended-abstract [7] we encode the n-queens problem lazily with the user-propagator such that only those parts of the problem encoding are kept in memory that are relevant for the currently considered part of the search space. It strongly reduces run-time and memory compared to full encodings in Z3, where all constraints are eagerly encoded either by bit-vector arithmetic or a direct propositional translation. We could also show that we can speed up enumerating propositional solutions by adding conflicts in the Final callback.

We hypothesize that user-propagators apply to other CSP domains as well, such as graph colouring problems. Additionally, well-known CSP heuristics, such as using least/most constrained values or minimum remaining values, and even very domain-specific branching heuristics can be encoded easily by the Decide and NextSplit. The core functionality relies on callbacks from Fixed and Conflict.

Quantifier Instantiation/Checking. Quantifier reasoning in Z3 is based on two built-in strategies: E-matching [24], using matching modulo congruence closures to find instantiations, and MBQI [13], that uses model checking of quantifiers. Controls for quantifier reasoning are provided as coarse grained knobs, such as programming patterns for E-matching [10] or assigning priorities to quantifiers. In one experiment, we used user-propagators to manually find instantiations using a cheap check: We considered the encoding of sorting-networks

$$WiringConstraints[a_1, \ldots, a_n] \wedge \qquad\qquad\qquad (8)$$
$$\forall x_1, \ldots, x_n, y_1, \ldots, y_n (ComparisonSwaps[a_1, \ldots, a_n, x_1, \ldots, x_n, y_1, \ldots, y_n]$$
$$\Rightarrow y_1 \leq \ldots \leq y_n)$$

that claims that there is some wiring (given by Booleans or bit-vectors $a_1, \ldots, a_n$), such that every input sequence of bits $(x_1, \ldots, x_n)$ results in a correctly sorted output sequence $y_1, \ldots, y_n$. Without user-propagation, Z3 will find a wiring that satisfies the wiring constraints and then check if this assignment can result in a non-sorted sequence in a subquery. However, doing this check manually already in the Final callback by a small amount of random inputs results in a counterexample in about 90% of all invalid cases. Thus, the MBQI subquery can be omitted and the instantiation (or a generalization of it) can be

---

[5] https://runtimeverification.com/.

propagated already by the user-propagator. Using this approach instead of Z3's default behaviour for finding a correct wiring with 5 input variables resulted in a 30% speed-up and with 6 variables in 65% on average (25.5 instead of 42 seconds). The described experiment requires the Fixed and Final callback as well as the Propagate function. We hypothesize that this use case is just a teaser for many creative ways to integrate domain specific heuristics for quantifier reasoning.

Algebraic Datatypes. Z3's decision procedure for Algebraic Datatypes does not build explicit values. For example, if t is of type (List (_ BitVec 8))[6], a list of 8-bit bit-vectors, the actual instantiation of t to (cons 0x00 (cons 0xF2 nil)) is only available implicitly when first t is constrained to be a cons-cell, then (tail t) is also constrained as a cons-cell, and finally (tail (tail t)) is nil. User-propagators can learn the shape of an algebraic datatype term t by registering the recognizer predicates ((_ is nil) t) and ((_ is cons) t). If the predicate ((_ is cons) t) is assigned to true, dually ((_ is nil) t) is assigned to false, the propagator can register the recognizers ((_ is nil) (tail t)) or ((_ is cons) (tail t)) until the shape of t is determined.

Strings. The solvers z3str [5] and S3 [33] were implemented on top of the old user-theory API [6]. To support string reasoning, it is relevant to integrate string with arithmetic reasoning. The z3str solver augmented Z3's API for pluggable theories with functions to query the *current value* (but not fixed) of integer variables and integer *bounds*. String solvers also benefit from a control loop around CDCL to explore finite bounded models by increasing string lengths incrementally. To our knowledge, there are currently five string solvers based on modified versions of Z3's source code. The prior experiences indicate that additional interfaces for *current values*, *bounds* and *iterative deepening* would be relevant for supporting external string solvers.

6   Related Work

A user-propagator can be seen as a revised version of the user-theory interface [6]. However, this interface was rather complicated: it was meant mainly for developers digging deep into the internals of SMT solving and does not fit smoothly in the overall Z3 reasoning process.

The OpenSMT solver [9] predates Z3's earlier effort and provides an interface with a similar custom theory reasoning motivation. The solver was mainly designed to allow users to build their own theory extensions by a minimalistic but easy to use interface. Users have to provide implementations similar to Z3's Push, Pop, Fixed, and Final to build their own decision procedures. While Z3's Conflict, Propagate, and NextSplit can arguably be realized by pushing the respective elements on a dedicated list, they are more restricted. For example, in contrast to OpenSMT, Z3 has also built-in support for more complex features

---

[6] Using SMT-LIB syntax.

in the user-propagator like quantifiers and bit-vectors without sacrificing Z3's overall good performance. As such and as argued in Sect. 5, we believe that our user-propagator within Z3 provides many new features and application domains compared to [9].

Injecting custom code into solvers is not limited to SMT solvers. The built-in SAT core clasp of the clingo answer set programming solver supports a so-called theory propagation [14] feature. Both clasp and Z3 allow watching assignments to Boolean variables and propagating clauses. In contrast to Z3's user-propagator, clasp's theory propagator works on a more basic level. Whereas Z3 users deal with terms of their formula that are internally mapped to the underlying variables, clasp users directly interact with the solver's clauses. Further, clasp's features are limited to those relevant for SAT solving, roughly corresponding to Z3's Fixed, Final, and Decide callbacks.

Within constraint programming/optimization, IBM's iLOG CPLEX optimization studio supports CP optimizer extensions [16]. This API allows the user to manually define the domain of variables in the problem encoding, by considering variable ranges depending on the current bounds of other variables during search and reporting conflicting assignments. The system also allows querying the bounds and values of fixed integer variables. The implementation inherits from the programmable propagator system in [28]. Callbacks are also available for MIP solvers, such as Gurobi [21].

The first-order prover SNARK [32] allows custom inference by providing procedural attachments. These are used to add new rewriting rules and define custom rules for instantiating/unifying selected function symbols. As SNARK operates on quantified terms directly by resolution and unification, their custom reasoning technique cannot be compared directly with the user-propagator and the other approaches, as all other solvers discussed here work (internally) on ground expressions and integrate with backtracking. However, unification on quantified clauses can be simulated by propagating quantified terms if needed.

7 Conclusions and Future Work

We introduced user-propagators to support satisfiability modulo custom theories in Z3. We argue that user-propagators open up new venues in SMT solving, as demonstrated by our discussed application domains. Custom theory reasoning is however not restricted to SMT solving, and as such improving custom extensions in other kind of reasoners, such as SAT solvers, constraint solvers or first-order reasoners, is an interesting venue for further work as well.

When to Use User-Propagators. Although we are convinced that custom reasoning via plugins can enrich the power of SMT solving, user-propagation should not be misunderstood as an answer to every kind of problem: A lot of problem classes can be already solved efficiently by the means of built-in Z3 features and do not require any special customization. We encountered three main reasons for using user-propagation: Defining custom theories that would require using (higher-order) quantification (e.g., strings and graph reachability), utilizing additional knowledge (e.g., heuristics and quantifier testing/instantiation), and lazily instantiating lengthy constraints that are unlikely to be used by the solver in their full form (e.g., disjointness axioms and conflict driven encoding of combinatorial problems).

While we believe it is impossible to give a general recipe when user propagator should (not) be used, we list a few issues that can favorably solved based on user propagators. As most callbacks (Fixed, Decide, Eq, Push, Pop) are called quite frequently during search, the overhead of each of these calls should be calibrated against the overhead of Boolean propagation by the CDCL solver. This can be achieved e.g., by postponing checks and propagations to Final. As Z3 ignores all conflicts/lemmas reported as soon as the solver's state becomes inconsistent, the user-propagator should try to report conflicts as soon as possible and abort the current callback early. Although custom heuristics/quantifier instantiations can speed-up reasoning considerably, Z3 has built-in support for a lot of additional options (e.g., alternative variable selection/assignment heuristics, multiple MBQI instantiations, pattern for E-matching) that should be considered before trying to come up with custom extensions. Furthermore, lemmas with many literals are prone to be ineffective for propagation and new expressions introduced in lemmas may amplify the search space of case splits.

Enabling Scenarios. User-propagators were primarily designed for encoding simple Boolean-based constraints, as those presented in [8], and extended to allow building general decision procedures for potentially complex, not necessarily Boolean-based, theories by observing equalities and declaring user-functions. However, the user-propagation may be diverted to extend Z3 in other ways:

Many software analysis tools use SMT solvers as logical cores for their own special-purpose reasoners. The number of potential extensions for classical SMT queries is huge (new reasoning problems, different logical semantics, . . .). User-propagation could be used to help people to integrate Z3 much tighter in their own solver without having to change the Z3 source. Recently developed solvers that could potentially benefit from the user-propagator include MonoSAT [4] for monotonic theories (utilizing the MiniSAT [11] SAT solver that is well known for its simplicity and adaptability) and Zord [15] for partial order constraints for multi-threaded program verification (using a fork of the Z3 source code).

Another potential use-case of the user-propagator might be monitoring the solver. In most cases in which the solver does not terminate or timeout, the user does not get any relevant information why Z3 could not prove a claim or find a model. The reasons for Z3 to fail can be numerous, including that the solver focused on a rather uninteresting part of the formula or made too many unrewarding quantifier instantiations. User-propagation can be used here as well to trace the solver in a rather understandable way.

Future Interfaces. There are many possible extensions to the user-propagator API. The case of string reasoning suggested there is functionality that can be exposed in addition to the feature set presented in this paper. Notably, as arithmetic reasoning based on Simplex does not fix values to arithmetic terms until all Boolean decisions have been made, the Fixed method does not apply for integer and real valued terms. Instead, it is possible to query a *current viable* value and *lower and upper bounds*. To support iterative deepening requires giving a solver access to initializing search state with assumptions that can be retracted and refined. The internal interface for decision procedures in Z3 can register a callback that is invoked whenever a new term is created for a given sort. It allows solvers, such as the bit-vector solver, to track all bit-vector sub-terms. We have not yet encountered a scenario where surfacing a *sort*-Created callback has been necessary, but it will likely be. A flexible approach for Propagate to produce certificates that can be checked by a trusted core is a significant topic in itself. The functions Decide and NextSplit provide one half of the functionality required to support branch prediction using machine learning [30], the other half of the functionality relies on inspecting the global search state (the current set of clauses). Access to the current search state and low level clause information like in [14] would also be required for user defined in-processing.

Acknowledgements. We thank Nuno Lopes for his support on the alive2 use-cases. The work described in this paper was supported by the ERC Consolidator Grant ARTIST 101002685, the TU Wien Doctoral College SecInt, and the FWF project SpyCoDe SFB-F85.

References

1. Barbosa, H., et al.: cvc5: a versatile and industrial-strength SMT solver. In: Fisman, D., Rosu, G. (eds.) TACAS 2022. LNCS, vol. 13243, pp. 415–442. Springer, Cham (2022). https://doi.org/10.1007/978-3-030-99524-9_24
2. Barrett, C., Fontaine, P., Tinelli, C.: The Satisfiability Modulo Theories Library (SMT-LIB) (2016). https://www.SMT-LIB.org
3. Barrett, C.W., Sebastiani, R., Seshia, S.A., Tinelli, C.: Satisfiability modulo theories. In: Handbook of Satisfiability. Frontiers in Artificial Intelligence and Applications, 2nd edn., vol. 336, pp. 1267–1329. IOS Press (2021)
4. Bayless, S., Bayless, N., Hoos, H.H., Hu, A.J.: SAT modulo monotonic theories. In: AAAI, pp. 3702–3709. AAAI Press (2015)
5. Berzish, M., Ganesh, V., Zheng, Y.: Z3str3: a string solver with theory-aware heuristics. In: FMCAD, pp. 55–59. IEEE (2017)
6. Bjørner, N.: Engineering theories with Z3. In: Yang, H. (ed.) APLAS 2011. LNCS, vol. 7078, pp. 4–16. Springer, Heidelberg (2011). https://doi.org/10.1007/978-3-642-25318-8_3
7. Bjørner, N.S., Eisenhofer, C., Kovács, L.: User-propagators for custom theories in SMT solving. In: SMT. CEUR Workshop Proceedings, vol. 3185, pp. 71–79. CEUR-WS.org (2022)
8. Bjørner, N., Nachmanson, L.: Navigating the universe of Z3 theory solvers. In: Carvalho, G., Stolz, V. (eds.) SBMF 2020. LNCS, vol. 12475, pp. 8–24. Springer, Cham (2020). https://doi.org/10.1007/978-3-030-63882-5_2
9. Bruttomesso, R., Pek, E., Sharygina, N., Tsitovich, A.: The OpenSMT solver. In: Esparza, J., Majumdar, R. (eds.) TACAS 2010. LNCS, vol. 6015, pp. 150–153. Springer, Heidelberg (2010). https://doi.org/10.1007/978-3-642-12002-2_12

10. Dross, C.: Generic decision procedures for axiomatic first-order theories. Ph.D. thesis, University of Paris-Sud, Orsay, France (2014). (in French)
11. Eén, N., Sörensson, N.: An extensible SAT-solver. In: Giunchiglia, E., Tacchella, A. (eds.) SAT 2003. LNCS, vol. 2919, pp. 502–518. Springer, Heidelberg (2004). https://doi.org/10.1007/978-3-540-24605-3_37
12. Feser, J.K., Madden, S., Tang, N., Solar-Lezama, A.: Deductive optimization of relational data storage. Proc. ACM Program. Lang. **4**(OOPSLA), 170:1–170:30 (2020)
13. Ge, Y., de Moura, L.: Complete instantiation for quantified formulas in satisfiabiliby modulo theories. In: Bouajjani, A., Maler, O. (eds.) CAV 2009. LNCS, vol. 5643, pp. 306–320. Springer, Heidelberg (2009). https://doi.org/10.1007/978-3-642-02658-4_25
14. Gebser, M., Kaminski, R., Kaufmann, B., Ostrowski, M., Schaub, T., Wanko, P.: Theory solving made easy with clingo 5. In: ICLP. OASIcs, vol. 52, pp. 2:1–2:15. Schloss Dagstuhl - Leibniz-Zentrum für Informatik (2016)
15. He, F., Sun, Z., Fan, H.: Satisfiability modulo ordering consistency theory for multi-threaded program verification. In: PLDI, pp. 1264–1279. ACM (2021)
16. IBM-Corporation: IBM ILOG CPLEX Optimization Studio CP Optimizer Extensions User's Manual (2017). Version 12.8
17. Bayardo, R.J., Jr., Schrag, R.: Using CSP look-back techniques to solve real-world SAT instances. In: AAAI/IAAI, pp. 203–208 (1997)
18. Hari Govind, V.K., Shoham, S., Gurfinkel, A.: Solving constrained horn clauses modulo algebraic data types and recursive functions. Proc. ACM Program. Lang. **6**(POPL), 1–29 (2022)
19. Lattner, C., Adve, V.S.: LLVM: a compilation framework for lifelong program analysis & transformation. In: CGO, pp. 75–88. IEEE Computer Society (2004)
20. Lee, J., Kim, D., Hur, C.-K., Lopes, N.P.: An SMT encoding of LLVM's memory model for bounded translation validation. In: Silva, A., Leino, K.R.M. (eds.) CAV 2021. LNCS, vol. 12760, pp. 752–776. Springer, Cham (2021). https://doi.org/10.1007/978-3-030-81688-9_35
21. LLC, GO: Gurobi Optimizer Reference Manual (2022). Version 9.5
22. Lopes, N.P., Lee, J., Hur, C., Liu, Z., Regehr, J.: Alive2: bounded translation validation for LLVM. In: PLDI, pp. 65–79. ACM (2021)
23. Mariano, B., Chen, Y., Feng, Y., Lahiri, S.K., Dillig, I.: Demystifying loops in smart contracts. In: ASE, pp. 262–274. IEEE (2020)
24. de Moura, L., Bjørner, N.: Efficient e-matching for SMT solvers. In: Pfenning, F. (ed.) CADE 2007. LNCS (LNAI), vol. 4603, pp. 183–198. Springer, Heidelberg (2007). https://doi.org/10.1007/978-3-540-73595-3_13
25. de Moura, L.M., Bjørner, N.S.: Proofs and refutations, and Z3. In: LPAR Workshops. CEUR Workshop Proceedings, vol. 418. CEUR-WS.org (2008)
26. de Moura, L., Bjørner, N.: Z3: an efficient SMT solver. In: Ramakrishnan, C.R., Rehof, J. (eds.) TACAS 2008. LNCS, vol. 4963, pp. 337–340. Springer, Heidelberg (2008). https://doi.org/10.1007/978-3-540-78800-3_24
27. Nieuwenhuis, R., Oliveras, A.: On SAT modulo theories and optimization problems. In: Biere, A., Gomes, C.P. (eds.) SAT 2006. LNCS, vol. 4121, pp. 156–169. Springer, Heidelberg (2006). https://doi.org/10.1007/11814948_18
28. Puget, J., Leconte, M.: Beyond the glass box: constraints as objects. In: ILCP, pp. 513–527. MIT Press (1995)

29. Sebastiani, R., Tomasi, S.: Optimization in SMT with $\mathcal{LA}(\mathbb{Q})$ cost functions. In: Gramlich, B., Miller, D., Sattler, U. (eds.) IJCAR 2012. LNCS (LNAI), vol. 7364, pp. 484–498. Springer, Heidelberg (2012). https://doi.org/10.1007/978-3-642-31365-3_38
30. Selsam, D., Bjørner, N.: Guiding high-performance SAT solvers with unsat-core predictions. In: Janota, M., Lynce, I. (eds.) SAT 2019. LNCS, vol. 11628, pp. 336–353. Springer, Cham (2019). https://doi.org/10.1007/978-3-030-24258-9_24
31. Silva, J.P.M., Sakallah, K.A.: GRASP - a new search algorithm for satisfiability. In: ICCAD, pp. 220–227. IEEE Computer Society/ACM (1996)
32. Stickel, M.E., Waldinger, R.J., Chaudhri, V.K.: A guide to SNARK. Technical report, SRI International Menlo Park United States (2000)
33. Trinh, M., Chu, D., Jaffar, J.: S3: a symbolic string solver for vulnerability detection in web applications. In: SIGSAC, pp. 1232–1243. ACM (2014) .
34. Zhang, H., Gupta, A., Malik, S.: Syntax-guided synthesis for lemma generation in hardware model checking. In: Henglein, F., Shoham, S., Vizel, Y. (eds.) VMCAI 2021. LNCS, vol. 12597, pp. 325–349. Springer, Cham (2021). https://doi.org/10.1007/978-3-030-67067-2_15

Bayesian Parameter Estimation with Guarantees via Interval Analysis and Simulation

Michele Boreale[(✉)] and Luisa Collodi

Dipartimento di Statistica, Informatica, Applicazioni "G. Parenti",
Università di Firenze, Florence, Italy
{michele.boreale,luisa.collodi}@unifi.it

Abstract. We give a method to compute guaranteed estimates of Bayesian a posteriori distributions in a model where the relation between the observation y and the parameters θ is a function, possibly involving additive noise parameters ψ, say $y = f(\theta) + h(\psi)$. This model covers the case of (noisy) ODE parameters estimation and the case when f is computed by a neural network. Applying a combination of methods based on uncertain probability (P-boxes), Interval Arithmetic (IA) and Monte Carlo (MC) simulation, we design an efficient randomized algorithm that returns guaranteed estimates of the posterior CDF of the parameters θ, and moments thereof, given that the observation y lies in a (small) rectangle. Guarantees come in the form of confidence intervals for the CDF values and its moments. Comparison with state-of-the-art approaches on ODEs benchmarks shows significant improvement in terms of efficiency and accuracy.

Keywords: Bayesian parameter estimation · Interval arithmetic · Monte Carlo · P-box · Ordinary differential equations · Neural networks

1 Introduction

We investigate the problem of estimating posterior parameter distributions given an observation. The proposed framework encompasses a wide variety of systems, including ordinary differential equations (ODEs) with noisy state observations, and neural networks. We take a Bayesian standpoint, that is, we assume a known prior distribution on the unknown parameter values. Computationally, the most widespread approach to Bayesian posterior estimation relies on Monte Carlo (MC) simulation, and specifically on Markov Chain Monte Carlo (MCMC) [18, 27,29] and on the particles-based Sequential Monte Carlo (SMC) [9]. The use of these techniques is justified by asymptotic results, saying that in the limit of an infinite number of simulation steps or particles, the samples produced by these methods converge in distribution to the exact posterior [29]. If the simulation is performed with only a finite number of steps or particles, which of course is always the case in practice, formal guarantees of correctness for the obtained samples are extremely hard to achieve.

In this paper, we explore a hybrid approach, which combines imprecise probability in the form of *P-boxes* [11] (see below) with MC simulation. The goal is to obtain sharp estimates of posterior quantities, such as Cumulative Distribution Functions (CDFs), and their expectations and higher moments, equipped with formal guarantees of correctness. At the same time, we aim at reducing the computational effort required by the simulation phase, in comparison to the above mentioned classical MC methods. Instrumental in achieving these goals are the following three elements: (1) leveraging the power of *Interval Arithmetic* (IA) [28] in order to drastically reduce the parameter search space; (2) accepting a level of *controlled* uncertainty on the computed estimates, introduced by the MC simulation phase of our method; (3) switching from conditioning on an individual observation y^* to conditioning on a (small) *set* of potential observations S^*. We give a more detailed account of our approach below.

The proposed method consists of two phases. The first phase is entirely deterministic. We assume a functional relation among observations y, parameters θ and noise ψ, say $y = g(\theta, \psi) := f(\theta) + h(\psi)$, for known, real vector valued functions f, h. With this functional model, the problem of estimating the probability of θ given that $y \in S^*$ can be recast as a problem of volume—or better, probability measure—estimation (Sect. 2). Rather then going for a direct MC estimation of the involved measures, though, we first reduce the search space: we compute a tight overapproximation of the set of *feasible* parameters θ, those that can actually be mapped into S^* by g for some instance of the noise vector ψ. Such an overapproximation can be effectively computed by relying on an *interval extension* of the function f: via IA, one can often determine at once if a whole rectangular region of the parameter space is unfeasible and discard it right away. This process can be repeated in a branch-and-bound fashion, and gives rise to a well-known refinement algorithm [22] (Sect. 3). The reduced parameter space obtained in this way can be significantly smaller than the original one. In any case, as discussed below, even a moderate volume reduction brings significant benefits in the subsequent phase of MC estimation. Moreover, the reduced space comes partitioned into axis-aligned hyper-rectangles, from which it is easy to draw samples.

In the second phase, we use a randomized algorithm $\mathcal{A}$ to actually compute the confidence intervals of the wanted posterior quantities (Sect. 4). We start by building a pair of lower and upper approximations of the true posterior CDF: this pair is commonly referred to as a P-box [11], as the graphs of the approximate CDFs form an envelope for that of the exact CDF. Being the outcome of a randomized algorithm, unlike classical ones our P-boxes have a confidence level attached: such objects are known in the literature as *confidence bands* [23]. From confidence bands, confidence intervals for the posterior expectation and other moments can be easily built (Sect. 5). The core algorithm $\mathcal{A}$ is very simple, and involves the extraction of a number of independent samples (θ, ψ), with θ drawn from the reduced parameter space, and considering the fraction of these pairs that are mapped into S^* by g. Confidence levels for the resulting estimates can be established relying on an exponential tail inequality for the sum of independent random variables, Hoeffding's bound [19]. We show that, by sampling from the reduced parameter space, the number of samples necessary to guarantee a given confidence level drops to a fraction μ^2 of the number necessary with the original space, where $\mu \in [0, 1]$ is the measure of the reduced space (Sect. 6).

We have put our algorithm at work on a few problems of parameter estimation from the literature (Sect. 7). Specifically, we have considered: the set of benchmarks for noisy ODE parameter estimation proposed in [5], where DSA, a method based on imprecise probability, is put forward; and a problem of feature relevance estimation for neural network classifiers proposed in [1]. For ODEs, we also offer a comparison with the results obtained with the state-of-the-art approaches (DSA, MCMC, SMC) from [5]. This comparison shows clearly the benefits of our method. A few concluding remarks are contained in the final section of the paper (Sect. 8).

Related Work. We shall limit our discussion on related work to Bayesian inference, a framework where a prior distribution on parameters is presupposed. Bayesian parameter inference has found application in a variety of fields, ranging from biological models [7,15,31] to linear hybrid dynamical systems [14] and more recently probabilistic programming [17]; cf. the extensive literature review in [5, Sect. 1]. As argued above, a problem of MCMC/SMC Bayesian inference methods is the difficulty of establishing formal guarantees for the obtained estimates. Moreover, these methods are computationally demanding (cf. our Table 1 in Sect. 7) and require an explicit expression of the likelihood, the function mapping θ to the probability of obtaining a certain observation given θ; this expression is often not available. An *Approximate Bayes Computation* (ABC) approach [25] has been proposed in recent years that also works in the absence of an explicit likelihood, and is more similar in spirit to ours. However, ABC shares the same difficulties as MCMC and SMC about formal guarantees, and is even more demanding from a computational point of view.

Closely related to ours is a method recently proposed by Chou and Sankaranarayanan [5] for ODE parameter inference. This method too is based on imprecise probability [8,11,13,30], which is a way of dealing with uncertainty. Like in our case, the parameter space is divided into disjoint cells. Differently from our approach, first likelihoods bounds for each cells are computed analytically; then these bounds are normalized to obtain bounds on posterior probabilities. The method of [5] avoids MC estimation, hence the computed bounds on probabilities are certain. On the contrary, in our case the MC phase introduces a level of controlled aleatoric uncertainty, hence our bounds come equipped with confidence levels. As clearly shown by the comparison in Sect. 7, the MC phase allows us to trade off a small level of (un)certainty for greater efficiency and accuracy. Additional differences between [5] and our approach are discussed in Sect. 7.

An important computational ingredient of our approach is the SIVIA refinement algorithm [21,22], which has been used in several works on parameter estimation. Notably, Jaulin in [20] proposes the use of IA for Bayesian estimation, in the following sense: given $\alpha \in [0,1]$ and a posterior probability density function, compute a minimal volume region whose probability w.r.t. the density equals α. Note that this is very different from the problem considered here: we apply SIVIA to the *model* function f, not to the posterior density function. In fact, we do not even presuppose an explicit knowledge of the posterior density. Another recent proposal is the application of SIVIA to feature relevance in neural networks [1]; this is further discussed in Sect. 7.

A proposed method for obtaining confidence bands from empirical CDF functions relies on the Dvoretzky-Kiefer-Wolfowitz (DKW) inequality [10,23,26]. This is an exponential tail inequality that bounds the probability that the empirical CDF deviates from the exact CDF by more than a given ϵ. When compared to our measure-based approach, a serious drawback of empirical CDFs in the present setting is that they require *exact* sampling from the posterior, a nontrivial problem in itself. General exact sampling schemes, like rejection sampling [29], might turn out to be very expensive. We leave for future work an experimental comparison with the DKW approach.

2 Problem Statement

Let $f : \mathbb{R}^n \longrightarrow \mathbb{R}^m$ and $h : \mathbb{R}^\ell \longrightarrow \mathbb{R}^m$ be functions, with f continuous. Here, $y = g(\theta, \psi) := f(\theta) + h(\psi)$ will be interpreted as a functional relation among the *observations* y, the *parameters* θ and the (independent) *nuisance parameters* ψ. For instance, the parameters ψ might represent additive noise: $\ell = m$, $h =$ identity, hence $y = f(\theta) + \psi$. A probability measure[1] μ on the space of all parameters $\mathbb{R}^n \times \mathbb{R}^\ell$ is given that factorizes as $\mu(A \times B) = \mu_0(A) \cdot \mu_1(B)$, where μ_0 and μ_1 are two probability measures over $\mathbb{R}^n$ and $\mathbb{R}^\ell$, respectively. The functions f, h are assumed to be measurable under μ_0, μ_1, respectively. This induces a triple of random variables (Θ, Ψ, Y) where $Y := g(\Theta, \Psi) = f(\Theta) + h(\Psi)$, and Θ and Ψ are independent. Moreover, in all applications, we shall consider a μ_0 with a finite-diameter support[2]. Given a (typically, small) measurable set of observations $S^* \subseteq \mathbb{R}^m$, for any $t \in (\mathbb{R} \cup \{+\infty\})^n$ one is interested in computing the quantity $F(t|S^*)$ defined below, which is the a posteriori CDF of Θ given $Y \in S^*$. Here, $\leq$ on vectors is taken componentwise, $\mathbb{R}^n_{\leq t} := \{\theta \in \mathbb{R}^n : \theta \leq t\}$ and $P_t := \mathbb{R}^n_{\leq t} \times \mathbb{R}^\ell$. Provided $\Pr(Y \in S^*) > 0$, we define:

$$F(t|S^*) := \Pr(\Theta \leq t \,|\, Y \in S^*)$$
$$= \frac{\mu(P_t \cap g^{-1}(S^*))}{\mu(g^{-1}(S^*))}. \tag{1}$$

We introduce a notion of correctness for randomized algorithms that approximate $F(t|S^*)$.

---

[1] All probability measures considered here are assumed to be absolutely continuous w.r.t. the Lebesgue measure, hence admit a density.

[2] supp(μ_0) is the smallest closed measurable set $T \subseteq \mathbb{R}^n$ s.t. $\mu_0(T) = 1$.

Definition 1 (algorithms for confidence intervals). *Let f, h, g, μ and S^* be fixed as specified above. Consider a randomized algorithm $\mathcal{A}$ that, taken as input a tuple $t \in (\mathbb{R} \cup \{+\infty\})^n$, returns as output a pair a real valued random variables, written $\mathcal{A}(t) = [\underline{\mathcal{A}}(t), \overline{\mathcal{A}}(t)]$. For any $\delta \geq 0$, we say that $\mathcal{A}$ approximates $F(\cdot|S^*)$ with confidence $1 - \delta$ if for each $t \in (\mathbb{R} \cup \{+\infty\})^n$*

$$\Pr\left(\underline{\mathcal{A}}(t) \leq F(t|S^*) \leq \overline{\mathcal{A}}(t) \right) \geq 1 - \delta.$$

For each t, the probability $\Pr(\cdot)$ is taken only on the internal random choices in the execution of $\mathcal{A}(t)$.

In other words, for each t, $[\underline{\mathcal{A}}(t), \overline{\mathcal{A}}(t)]$ is a confidence interval for $F(t|S^*)$. In the above definition, we do not impose requirements on the accuracy of the approximation, that is on the width of the interval $[\underline{\mathcal{A}}(t), \overline{\mathcal{A}}(t)]$: this can only be judged a posteriori.

Based on an algorithm $\mathcal{A}$ for $F(t|S^*)$, one can easily build P-boxes. We shall limit our discussion to the important special case of marginal CDFs. For $\lambda \in \mathbb{R}$, let $F(\lambda|S^*)$ abbreviate $F(t|S^*)$ with $t = (\lambda, +\infty, ..., +\infty)$; similarly, let $\mathcal{A}(\lambda) := \mathcal{A}(t)$. Note that $F(\lambda|S^*) = \Pr(\Theta_1 \leq \lambda|Y \in S^*)$ is the first marginal posterior CDF of Θ (the same reasoning applies to the other marginals). Now choose $k + 1 \geq 2$ node points on the real line, say $\lambda_0 < \lambda_1 < \cdots < \lambda_k$, such that $F(\lambda_0|S^*) = 0$ and $F(\lambda_k|S^*) = 1$. Based on $\mathcal{A}(\lambda_1), ..., \mathcal{A}(\lambda_{k-1})$, we define F^- and F^+, stepwise lower and upper approximations of F, as follows. Below, for the sake of uniform notation, we convene that $\underline{\mathcal{A}}(\lambda_0)$ denotes 0, and that $\overline{\mathcal{A}}(\lambda_k)$ denotes 1. Moreover $j = 0, ..., k - 1$.

$$F^-(\lambda|S^*) := \begin{cases} 0 & \text{if } \lambda < \lambda_0 \\ \underline{\mathcal{A}}(\lambda_j) & \text{if } \lambda \in [\lambda_j, \lambda_{j+1}) \\ 1 & \text{if } \lambda \geq \lambda_k \end{cases} \quad F^+(\lambda|S^*) := \begin{cases} 0 & \text{if } \lambda < \lambda_0 \\ \overline{\mathcal{A}}(\lambda_{j+1}) & \text{if } \lambda \in [\lambda_j, \lambda_{j+1}) \\ 1 & \text{if } \lambda \geq \lambda_k. \end{cases}$$

$$(2)$$

Note that F^-, F^+ are in turn random variables, depending on the random variables $\mathcal{A}(\lambda_1), ..., \mathcal{A}(\lambda_{k-1})$. Importantly from the computational point of view, these $k - 1$ calls to $\mathcal{A}$ are not required to be independent: as we shall see, there is a way of computing them at essentially the same cost of a single call (see Sect. 4, Remark 2). The next result says that, with high probability, the pair (F^-, F^+) is a P-box for the exact marginal posterior CDF $F(\lambda|S^*)$. Otherwise said, the pair (F^-, F^+) is a *confidence band* [23] for $F(\lambda|S^*)$. The proof of the proposition is an immediate consequence of the previous definition of $\mathcal{A}$ and of a union bound on probabilities.

Proposition 1 (confidence bands). *Suppose $\mathcal{A}$ approximates $F(\cdot|S^*)$ with confidence $1 - \delta$. Then, with probability at least $1 - (k - 1)\delta$, we have that for all $\lambda \in \mathbb{R}$:*

$$F^-(\lambda|S^*) \leq F(\lambda|S^*) \leq F^+(\lambda|S^*). \tag{3}$$

From the confidence band (F^-, F^+), confidence intervals for a variety of statistics, including moments of the true posterior, can be easily computed. We will detail this point in Sect. 5. We will design a correct core algorithm $\mathcal{A}$ under certain mathematical and computational assumptions, listed below. Mathematically, we assume the following.

1. $S^* = I_1 \times \cdots \times I_m$ is an axis-aligned hyper-rectangle (from now on, *rectangle* for short), where each $I_j = [a_j, b_j]$ $(a_j < b_j)$ is a closed interval of $\mathbb{R}$. In typical use cases, S^* might be a small rectangle centered at a given observation $y^* \in \mathbb{R}^m$;
2. there exists an *interval extension* $\overline{f}$ of the function f, see the next section for the precise definition.

Computationally, we assume we have efficient algorithms to:

(a) compute $f, \overline{f}$, and h;
(b) compute $\mu_0(R)$ for any rectangle $R \subseteq \mathbb{R}^n$;
(c) for any non-zero measure rectangle $R \subseteq \mathbb{R}^n$, sample from the the random variable $\Theta_{|R}$ obtained by conditioning[3] Θ on the event $\Theta \in R$;
(d) sample from Ψ.

In the next section, we will review in detail these prerequisites and explore some instances of the model where they are fulfilled.

3   Interval Arithmetic, Discretized ODEs, Neural Networks

We review the mathematical and computational prerequisites we will rely upon and explore possible instances of the model.

3.1   Interval Arithmetic, Coverings, Set Inversion

Interval Arithmetic (IA) [28] offers a framework to compute rigorously with abstract versions of functions, where individual points are replaced by intervals. The abstract functions conservatively extend their concrete counterparts (see below). IA can be used, for instance, to compute certified bounds on the error of numerical operations. In what follows, we quickly introduce the terminology of IA we need.

Formally, an *interval* I is a finite, closed interval $[a, b] \subseteq \mathbb{R}$. A *rectangle* is a cartesian product of intervals, $R = I_1 \times \cdots I_k$. We let $\mathbb{IR}$ denote the set of all intervals included in $\mathbb{R}$. For $J = (I_1, ..., I_k) \in \mathbb{IR}^k$ a tuple of intervals, we define the rectangle $[J] := I_1 \times \cdots \times I_k \subseteq \mathbb{R}^k$. An *interval extension* of $f : \mathbb{R}^n \to \mathbb{R}^m$ is a function $\overline{f} = (\overline{f}_1, ..., \overline{f}_n) : \mathbb{IR}^n \longrightarrow \mathbb{IR}^m$ that is compatible with f, that is:

1. whenever $x \in [I_1, \cdots, I_n]$ then $f(x) \in [\overline{f}(I_1, ..., I_n)]$; and
2. whenever $I_i \subseteq I_i'$ for $i = 1, ..., n$ then $[\overline{f}(I_1, ..., I_n)] \subseteq [\overline{f}(I_1', ..., I_n')]$.

---

[3] Explicitly, $\Theta_{|R}$ is the random variable induced by the measure on $\mathbb{R}^n$ defined by $\mu_R(R') := \mu_0(R \cap R')/\mu_0(R)$.

A basic fact about IA is that the set of interval extensions is closed under composition: if $\overline{f}$ and $\overline{g}$ are interval extensions of f, g respectively, where $f : \mathbb{R}^n \to \mathbb{R}^m$ and $g : \mathbb{R}^m \to \mathbb{R}^p$, then $\overline{g} \circ \overline{f}$ is an interval extension of $g \circ f$. By slight abuse of notation, for a rectangle $R = I_1 \times \cdots \times I_n$, we let $\overline{f}(R) := [\overline{f}(I_1, ..., I_n)]$. The volume of the difference $\overline{f}(R) \setminus f(R)$ is a measure of how accurate the interval extension $\overline{f}$ is with respect to the concrete function f. Functions most commonly found in applications, including all polynomials, exponentials and trigonometric functions, do possess accurate interval extensions [28]. Moreover, every monotonic function has an interval extension.

Given a set $A \subseteq \mathbb{R}^n$, a *covering* of A is a finite set of rectangles $\mathcal{C} = \{R_1, ..., R_K\}$ such that: (a) $A \subseteq \bigcup_{i=1}^{K} R_i$, and (b) the rectangles $R_1, ..., R_K$ are *almost-disjoint* according to a fixed measure $\mu_0(\cdot)$ on $\mathbb{R}^n$, that is for each $1 \le i < j \le K$, $\mu_0(R_j \cap R_i) = 0$ [4]. Given a function $f : \mathbb{R}^n \to \mathbb{R}^m$ and rectangles $S \subseteq \mathbb{R}^m$ and $R_0 \subseteq \mathbb{R}^n$, we will be interested in computing a covering of the set

$$f^{-1}(S) \cap R_0 = \{x \in \mathbb{R}^n \, : \, f(x) \in S\} \cap R_0.$$

This problem is referred to as *set inversion* in [21,22], where a practical branch-and-bound algorithm based on IA is offered: SIVIA, standing for *Set Inversion Via Interval Analysis*. In our application of SIVIA, S will be a suitable superset of S^*, and R_0 a superset of the support of μ_0. We give a pseudocode description of SIVIA as Algorithm 1. SIVIA maintains a set L of rectangles, each represented as a n-tuple of intervals, initially containing only R_0. At each iteration, a rectangle R is extracted from L, and IA is used to check if the $\overline{f}$-image of R is: (a) entirely inside S (feasible), or (b) entirely outside S (unfeasible), or (c) indeterminate. In case (a), R is inserted into a set of *inner rectangles*, $\mathcal{C}_{\text{in}}$, and will not be reconsidered; in case (b), R is simply discarded; in case (c), R is bisected[5] and the resulting halves R_1, R_2 are inserted into L for later consideration. An exception to the last rule is when the width of R is less than a given resolution threshold, ρ: in this case, R is inserted into a set of *outer rectangles*, $\mathcal{C}_{\text{out}}$ and will not be reconsidered. Informally, outer rectangles are those found at the border of the covering. The output of the algorithm is the pair $(\mathcal{C}_{\text{in}}, \mathcal{C}_{\text{out}})$. Note that the set theoretic operations involving $\overline{f}(R)$ and S in lines 5 and 7 can be efficiently implemented, as both sets are rectangles.

Lemma 1. *Algorithm 1 always terminates returning a pair $(\mathcal{C}_{\text{in}}, \mathcal{C}_{\text{out}})$ of sets of rectangles. Moreover, (1) $\mathcal{C}_{\text{in}} \cup \mathcal{C}_{\text{out}}$ is a covering of $f^{-1}(S) \cap R_0$; (2) $\bigcup \mathcal{C}_{\text{in}} \subseteq f^{-1}(S) \cap R_0$.*

Remark 1 (complexity). The worst case time complexity of SIVIA is exponential in n [22], not surprisingly given its branch-and-bound structure. This theoretical complexity is less of a concern for our purposes than it may seem at first glance,

---

[4] E.g. R_i and R_j might share part of a face.

[5] Explicitly, if $R = I_1 \times \cdots \times I_n$ and $I_j = [a, b]$ is the largest among the intervals involved in the product, we have: $width(R) := b - a$, $R_1 := I_1 \times \cdots \times [a, c] \times \cdots \times I_n$ and $R_2 = I_1 \times \cdots \times [c, b] \times \cdots \times I_n$, where $c = \frac{a+b}{2}$.

Algorithm 1. SIVIA [22]

> **Input:** $S \subseteq \mathbb{R}^m$, a rectangle to be inverted; $R_0 \in \mathbb{IR}^n$, a rectangle; $\overline{f} : \mathbb{IR}^n \longrightarrow \mathbb{IR}^m$, an interval extension of function f; $\rho > 0$, a resolution threshold.
> **Output:** $(\mathcal{C}_{\text{in}}, \mathcal{C}_{\text{out}})$, sets of rectangles such that $\mathcal{C}_{\text{in}} \cup \mathcal{C}_{\text{out}}$ covers $f^{-1}(S) \cap R_0$.

```
 1: Cin, Cout ← ∅
 2: L ← {R0}
 3: while L ≠ ∅ do
 4:     R ← remove(L)                                    ▷ extract a rectangle from L
 5:     if f̄(R) ⊆ S then                                ▷ if the rectangle is feasible
 6:         Cin ← Cin ∪ {R}                              ▷ then insert it into Cin
 7:     else if f̄(R) ∩ S ≠ ∅ then        ▷ otherwise if the rectangle is indeterminate
 8:         if width(R) < ρ then                 ▷ if its width less than resolution
 9:             Cout ← Cout ∪ {R}                        ▷ then insert it into Cout
10:         else
11:             R1, R2 ← Bisect(R)                  ▷ otherwise bisect the rectangle
12:             L ← L ∪ {R1, R2}               ▷ insert the resulting halves into L
13:         end if
14:     end if
15: end while
16: return (Cin, Cout)
```

---

for two reasons. First, we can set a relatively large resolution threshold ρ, as the subsequent Monte Carlo estimation of $\mu_0(f^{-1}(S))$ can greatly benefit even from a conservative covering; this point will be made precise in Sect. 6. Second, in our application of SIVIA, the set S will be typically quite small: as a consequence, one may expect that in roughly half of the iterations the rectangle R will be unfeasible hence will not lead to further bisections.

3.2 Discretized ODEs and Neural Networks

Ordinary differential equations (ODEs) and neural networks are models that naturally fit in the framework introduced in Sect. 2.

Let us consider ODEs first. Let $z = (z_1, ..., z_p)$ be a vector of distinct variables. Consider an initial value problem defined by: a system of p (nonlinear) ODEs that also depend on a tuple of n parameters θ, written $\dot{z}(t) = \phi(z(t), \theta)$, and a fixed initial condition $z(0) = z_0 \in \mathbb{R}^p$. Under suitable regularity assumptions on ϕ, for any $\theta^* \in \mathbb{R}^n$ a unique solution $z(t; z_0; \theta^*)$ to this problem exists in a time interval containing 0. Assuming ϕ, z_0 and a prior probability distribution $\mu_0(\theta)$ are known, the goal is to estimate the posterior distribution of θ given a vector y^* of k observations, obtained by a measurement of the solution $z(t; z_0; \theta^*)$ at fixed time points $t_1 < ... < t_k$, say $y^* = (y_1^*, ..., y_k^*) \in \mathbb{R}^m$, with $y_i^* \in \mathbb{R}^p$ and $m = k \cdot p$. Such measurements will be assumed to be affected by an additive noise $\psi \in \mathbb{R}^m$, generated by an independent random variable Ψ, induced by a known probability measure μ_1 over $\mathbb{R}^m$. To recast this in the setting of Sect. 2, let us fix a tolerance threshold $\gamma > 0$, let $\theta \in \mathbb{R}^n$, $\psi \in \mathbb{R}^m$ and define:

$$S^* = \Pi_{i=1}^k \Pi_{j=1}^p [y_{i,j}^* - \gamma, \, y_{i,j}^* + \gamma]$$
$$f(\theta) = (z(t_1; z_0; \theta), ..., z(t_k; z_0; \theta))$$
$$g(\theta, \psi) = f(\theta) + \psi.$$

An interval version of $f(\theta)$ can be computed via set reachability techniques for ODEs, see e.g. [2–4]; but this is quite expensive. In our experiments, from the outset we will replace the original model with an accurate discretized version of the ODE, obtained by applying Euler's scheme, which we now quickly introduce. For a fixed z_0 and time step $\tau > 0$, consider the recurrence relation ($s \geq 0$): $\tilde{z}_0 := z_0$ and $\tilde{z}_{s+1} = \tilde{z}_s + \tau \cdot \phi(\tilde{z}_s, \theta)$. It can be seen that $\tilde{z}_s \approx z(s \cdot \tau; z_0; \theta)$, and this approximation can be made arbitrarily accurate by choosing τ sufficiently small. Making the dependence on θ explicit in the notation, let us denote by $\tilde{z}_s(\theta)$ the elements of this sequence. Assuming that each $s_i = t_i/\tau$ is an integer for $i = 1, ..., k$, we will replace the above $f(\theta)$ with the following

$$f(\theta) := (\tilde{z}_{s_1}(\theta), ..., \tilde{z}_{s_k}(\theta)).$$

If ϕ has an interval extension $\overline{\phi}$, it is easy to compute $\overline{f}$, an interval extension of f.

Example 1 (simple ball/1). We use a toy model also considered in [5] as a running example. The vertical motion of a ball obeys the following ODE, where in $z = (z_1, z_2)$, z_1 is the position, z_2 is the velocity, and the parameter θ, on which we want to make inference, is gravity acceleration: $\dot{z} = (\dot{z}_1, \dot{z}_2) = \phi(z, \theta) := (z_2, -\theta)$. We take $z_0 := (0, -4)$ as the initial condition. Although this ODE is trivial to solve analytically, for the purpose of illustration we will consider its Euler's discretization. We choose $\tau = 0.1$ and consider the recurrence relation for $\tilde{z}_s = (\tilde{z}_{1,s}, \tilde{z}_{2,s})$ given by: $\tilde{z}_0 := z_0$ and $\tilde{z}_{s+1} := \tilde{z}_s + \tau \cdot (\tilde{z}_{2,s}, -\theta)$. Making the dependence on the parameter θ explicit, let us write this as $\tilde{z}_s(\theta)$. We choose to observe the system once at time $t = 1$, hence set $f(\theta) := \tilde{z}_{10}(\theta)$. For $\theta^* = 9.8$, one has $f(\theta^*) = (-8.4, -13.8)$: we choose this as our observation y^* and fix the tolerance $\gamma = 0.5$, hence $S^* = [-8.9, -7.9] \times [-14.3, -13.3]$. We assume a noise vector $\psi \in [-1, 1]^2$ and let h be the identity: consequently, $g(\theta, \psi) := f(\theta) + \psi$ is the functional description of our model. To complete the description, we have to specify the probability measures μ_0 and μ_1: we choose μ_0 to be the uniform distribution on the finite support $R_0 := [7, 12]$, and μ_1 to consist of a pair of independent truncated gaussian distributions on $[-1, 1]$ of standard deviation 0.1. Application of SIVIA to this example is postponed to Example 2.

Let us now consider neural networks. Generally speaking, a trained feedforward neural network with k hidden layers [16] implements a function $f : \mathbb{R}^n \longrightarrow \mathbb{R}^m$ defined as

$$f(\theta) = (f_{k+1} \circ f_k \circ \cdots \circ f_0)(\theta)$$

where each layer f_j has the structure $f_j(\xi_j) = \alpha_j(W_j \cdot \xi_j + b_j)$: here, ξ_j is a column vector, W_j, b_j are fixed and known weight matrix and bias vector of appropriate dimensions, respectively, while α_j is an activation function, applied componentwise to $W_j \cdot \xi_j + b_j$; here $\xi_0 = \theta$ is seen as a column vector. Provided each of the activation functions α_j possesses an interval extension $\overline{\alpha}_j$, an interval extension $\overline{f}$ exists and can be easily computed. Commonly encountered activation functions, such as various versions of Linear Unit (LU), hyperbolic tangent, and in general monotonic activation functions, do possess interval extensions. Recasting this in the framework of Sect. 2, one can be interested in inferring a posterior distribution of θ given the result of an application of the function, $y^* = f(\theta^*)$. Noise is not considered in this setting. Moreover, one is often interested in the restriction of f to few selected components of θ. This is the case in the application to feature relevance, that will be discussed in detail in Sect. 7.

4   The Core Algorithm $\mathcal{A}$

Consider equality (1). We will first discuss how to estimate the denominator $\mu(g^{-1}(S^*))$. Let $R_0 \subseteq \mathbb{R}^n$ be a rectangle, $R_0 \supseteq \mathrm{supp}(\mu_0)$, the support of μ_0: in the discussion in Sect. 1, R_0 corresponds to the *original space*, the one from which one would sample θ in the absence of further information. We take the *feasible space* to be

$$\mathcal{F} := \mathrm{pr}_{1..n}\left(g^{-1}(S^*)\right) \cap R_0.$$

This set[6] contains all θ's that can be sampled and mapped into S^* for some choice of ψ. Now assume we have a covering $\mathcal{C} = \{R_1, ..., R_K\}$ of $\mathcal{F}$: the union of the rectangles in $\mathcal{C}$ forms the *reduced space*. We will discuss later in the section how to compute $\mathcal{C}$.

Lemma 2. $\mu(g^{-1}(S^*)) = \mu\left(g^{-1}(S^*) \cap (R_1 \times \mathbb{R}^\ell)\right) + \cdots + \mu\left(g^{-1}(S^*) \cap (R_K \times \mathbb{R}^\ell)\right)$.

Proof. Let $A = R_1 \cup \cdots \cup R_K$. We have $g^{-1}(S^*) \cap (R_0 \times \mathbb{R}^\ell) = g^{-1}(S^*) \cap (A \times \mathbb{R}^\ell)$, because for every $\theta \in R_0 \setminus A \subseteq \mathcal{F}^c$ we have $(\theta, \psi) \notin g^{-1}(S^*)$ for any $\psi \in \mathbb{R}^\ell$. Then by elementary set-theoretic reasoning

$$g^{-1}(S^*) \cap (R_0 \times \mathbb{R}^\ell) = \left(g^{-1}(S^*) \cap (R_1 \times \mathbb{R}^\ell)\right) \cup \cdots \cup \left(g^{-1}(S^*) \cap (R_K \times \mathbb{R}^\ell)\right).$$

By assumption, the rectangles $R_1, ..., R_K$ are almost disjoint w.r.t. μ_0, which implies the above union is almost disjoint w.r.t. μ. Moreover, $R_0 \times \mathbb{R}^\ell \supseteq \mathrm{supp}(\mu)$. Consequently:

$$\mu(g^{-1}(S^*)) = \mu\left(g^{-1}(S^*) \cap (R_0 \times \mathbb{R}^\ell)\right)$$
$$= \mu\left(g^{-1}(S^*) \cap (R_1 \times \mathbb{R}^\ell)\right) + \cdots + \mu\left(g^{-1}(S^*) \cap (R_K \times \mathbb{R}^\ell)\right).$$

---

[6] $\mathrm{pr}_{1..n}$ is projection on the first n coordinates. Elements of $\mathcal{F}$ outside the support of μ_0 play no role.

Now each of the summands in Lemma 2, $\mu\left(g^{-1}(S^*) \cap (R_i \times \mathbb{R}^\ell)\right)$ for $i = 1, ..., K$, can be estimated via a MC simulation: informally speaking, one draws a number N_i of samples (θ, ψ) from $R_i \times \mathbb{R}^\ell$ and computes the fraction r_i of them such that $g(\theta, \psi) = f(\theta) + h(\psi) \in S^*$. Then

$$\mu\left(g^{-1}(S^*) \cap (R_i \times \mathbb{R}^\ell)\right) \approx r_i \cdot \mu(R_i \times \mathbb{R}^\ell)$$
$$= r_i \cdot \mu_0(R_i) \tag{4}$$

(the approximate equality above will rendered rigorously below). Note that, by our assumption of independence, (θ, ψ) can be sampled by separately drawing θ from $\mathbb{R}^n$ via $\mu_{0|R_i}$, and ψ from $\mathbb{R}^\ell$ via μ_1, which we assume we know how to do. Overall, the more tightly the union of the R_i's overapproximates $\mathcal{F}$, the more efficient this process is: this will made precise in Sect. 6.

Concerning the actual computation of a covering $\mathcal{C}$ of the feasible space, we proceed as follows. Let $\underline{h}, \overline{h}$ be the vectors of inf's and sup's values of h over $\mathbb{R}^\ell$, taken componentwise, possibly equal to $\pm\infty$. That is, for $i = 1, ..., m$

$$\underline{h}_i := \inf_{\psi \in \mathbb{R}^\ell} h_i(\psi) \qquad \overline{h}_i := \sup_{\psi \in \mathbb{R}^\ell} h_i(\psi).$$

The following lemma is an easy consequence of the definition of $\mathcal{F}$. We let $A + B$ denote the Minkowski sum of two subsets of $\mathbb{R}^m$, $A + B := \{a + b : a \in A \text{ and } b \in B\}$, with the sum taken componentwise, and $[-\overline{h}, -\underline{h}] := \Pi_{i=1}^m [-\overline{h}_i, -\underline{h}_i]$.

Lemma 3. $\mathcal{F} \subseteq \mathcal{X}_0 := f^{-1}(S^* + [-\overline{h}, -\underline{h}]) \cap R_0.$

Therefore a covering $\mathcal{C}$ of $\mathcal{X}_0$ is also a covering of $\mathcal{F}$, cf. the figure on the right. Accordingly, we will consider coverings of $\mathcal{X}_0$ from now on. Moreover, will always consider cases where both $\underline{h}$ and $\overline{h}$ are finite. Note that too large values of $|\underline{h}|, |\overline{h}|$ will tend to make $\mathcal{X}_0$ coincide with R_0, trivializing the proposed method. A covering $\mathcal{C}$ of $\mathcal{X}_0$ can be computed via the 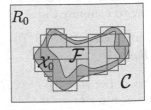 SIVIA algorithm presented in Sect. 3: $\mathcal{C} := \mathcal{C}_{\text{in}} \cup \mathcal{C}_{\text{out}}$, where $(\mathcal{C}_{\text{in}}, \mathcal{C}_{\text{out}}) = \text{SIVIA}(S, R_0, \overline{f}, \rho)$, with $S := S^* + [-\overline{h}, -\underline{h}]$ and $\rho > 0$ a chosen resolution threshold. Note that S too is a rectangle in $\mathbb{R}^m$, since S^* is by assumption.

The estimation of the numerator $\mu(g^{-1}(S^*) \cap P_t)$ in (1) proceeds similarly, but $g^{-1}(S^*)$ must be replaced with $g^{-1}(S^*) \cap P_t$, where $P_t = \mathbb{R}^n_{\leq t} \times \mathbb{R}^\ell$. Also, one must ensure that $\mathcal{C}$ *refines* $\mathbb{R}^n_{\leq t}$ (see details in Algorithm 2, step 1). The summation in Lemma 2 is replaced by one that involves only the rectangles contained in $\mathbb{R}^n_{\leq t}$.

Having identified the basic ingredients, we proceed now to a formal presentation of the core algorithm $\mathcal{A}$: see Algorithm 2. The algorithm consists of three steps. Step 1 is entirely deterministic, and just consists in the refinement of $\mathcal{C}$, if required. Step 2 introduces the basic random variables. Step 3 introduces the random variables that correspond to the actual simulation part, consisting in an

Algorithm 2. core algorithm $\mathcal{A}$

Input: $t \in (\mathbb{R} \cup \{+\infty\})^n$, a n-tuple of real numbers or $+\infty$.

Output: $\mathcal{A}(t) = [\underline{\mathcal{A}}(t), \overline{\mathcal{A}}(t)]$, a pair of random variables defining a confidence interval for $F(t|S^*)$.

Fixed parameters: S^*, f, h, μ_0, μ_1, as in Section 2; $\mathcal{C} = \{R_1, ..., R_K\}$, a covering of $\mathcal{X}_0$ s.t. $\mu_0(R_i) > 0$ for each $i = 1, ..., K$; $N = N_1 + \cdots + N_K$ ($N_i \geq 1$), an integer number of samples to draw in the simulation step (budget); $\epsilon > 0$, an error threshold.

1: If necessary, split the rectangles of $\mathcal{C}$ to make it a *refinement* of $\mathbb{R}^n_{\leq t}$, that is: for each $R \in \mathcal{C}$, either $R \subseteq \mathbb{R}^n_{\leq t}$ or $\mu_0(R \cap \mathbb{R}^n_{\leq t}) = 0$. Let $H_t := \{j : 1 \leq j \leq K \text{ and } R_j \subseteq \mathbb{R}^n_{\leq t}\}$.

2: For each $i = 1, ..., K$, recalling that $\Theta_{|R_i}$ is drawn from R_i according to $\mu_{|R_i}(\cdot)$, define the random variable
$$X_i := \mu_0(R_i) \cdot \mathbb{1}_{\{g(\Theta_{|R_i}, \Psi) \in S^*\}}.$$

3: For each $i = 1, ..., K$, let $X_{i1}, ..., X_{iN_i}$ be N_i i.i.d. copies of X_i. Let $X := \sum_{i=1}^K \frac{1}{N_i} \sum_{j=1}^{N_i} X_{ij}$ and $X_t := \sum_{h \in H_t} \frac{1}{N_h} \sum_{j=1}^{N_h} X_{hj}$. Return the following, where the right endpoint by convention is 1 if $X - \epsilon < 0$.
$$\mathcal{A}(t) = [\underline{\mathcal{A}}(t), \overline{\mathcal{A}}(t)] := \left[\frac{X_t - \epsilon}{X + \epsilon}, \frac{X_t + \epsilon}{X - \epsilon} \right].$$

---

overall N independent samplings of the random variables defined in the previous step, and in the construction of the actual confidence interval. Here $\epsilon > 0$ represents an error threshold, which has an impact on the width of the returned confidence interval. Note that the quantity $r_i \cdot \mu_0(R_i)$ in (4) of the informal derivation above, corresponds in step 3 of the algorithm to a realization of the random variable $\sum_{j=1}^{N_i} \frac{1}{N_i} X_{ij} = (\frac{1}{N_i} \sum_{j=1}^{N_i} \mathbb{1}_{\{g(\Theta_{|R_i}^{(ij)}, \Psi^{(ij)}) \in S^*\}}) \cdot \mu_0(R_i)$, with the superscript (ij) used here to denote different i.i.d. copies of a random variable. As part of the parameters, we presuppose a partition of the sampling budget over the K rectangles of the covering $\mathcal{C}$, $N = \sum_{i=1}^K N_i$: an optimal way of determining this partition will be discussed in Sect. 6.

We proceed to prove the correctness of $\mathcal{A}$, which is based on the following well-known result. Note that the random variables considered in the statement are required to be independent, but need not be identically distributed.

Lemma 4 (Hoeffding's bound [19]). *Let $Z_1, ..., Z_k$ be independent random variables such that $a_i \leq Z_i \leq b_i$ for $i = 1, ..., k$. Let $Z := \sum_{i=1}^k Z_i$ and $\epsilon > 0$. Then $\Pr(|Z - E[Z]| > \epsilon) \leq 2 \exp(-\frac{2\epsilon^2}{\sum_{i=1}^k (b_i - a_i)^2})$.*

Theorem 1 (correctness of $\mathcal{A}$). *For any $t \in (\mathbb{R} \cup \{+\infty\})^n$, let $\mathcal{C}, N, N_i, X, X_t, \epsilon$ and $\mathcal{A}(t)$ be as defined in Algorithm 2.*

1. $E[X] = \mu(g^{-1}(S^*))$ *and* $\Pr(|X - E[X]| > \epsilon) \leq \delta_0 := 2 \exp\left(-\frac{2\epsilon^2}{\sum_{i=1}^K \frac{\mu_0(R_i)^2}{N_i}} \right)$.

2. $E[X_t] = \mu(P_t \cap g^{-1}(S^*))$ *and* $\Pr(|X_t - E[X_t]| > \epsilon) \leq \delta_1 := 2 \exp\left(-\frac{2\epsilon^2}{\sum_{h \in H_t} \frac{\mu_0(R_h)^2}{N_h}} \right)$.

3. $F(t|S^*) = \frac{E[X_t]}{E[X]} \in [\underline{A}(t), \overline{A}(t)]$ *with probability at least* $1 - \delta$, *where* $\delta = \delta_0 + \delta_1 \leq 2\delta_0$. *In other words,* $\mathcal{A}$ *approximates* $F(\cdot|S^*)$ *with confidence* $1 - \delta$.

Proof. We consider the three parts separately.

1. Let $i \in \{1, ..., K\}$ and consider the definition of X_i in step 2, $X_i = \mu_0(R_i) \cdot \mathbb{1}_{\{g(\Theta_{|R_i}, \Psi) \in S^*\}}$. Now $Z := \mathbb{1}_{\{g(\Theta_{|R_i}, \Psi) \in S^*\}}$ is a Bernoulli random variable with success (1) probability p equal to

$$p = E[Z] = \Pr(g(\Theta_{|R_i}, \Psi) \in S^*) = \frac{\mu(g^{-1}(S^*) \cap R_i \times \mathbb{R}^\ell)}{\mu_0(R_i)}.$$

Hence, for each i, j: $E[X_{ij}] = E[X_i] = \mu_0(R_i)E[Z] = \mu(g^{-1}(S^*) \cap R_i \times \mathbb{R}^\ell)$. Applying the linearity of expectation, we have:

$$E[X] = \sum_{i=1}^{K} \frac{1}{N_i} \sum_{j=1}^{N_i} E[X_{ij}] = \sum_{i=1}^{K} \frac{1}{N_i} \cdot N_i \cdot \mu(g^{-1}(S^*) \cap R_i \times \mathbb{R}^\ell) \quad (5)$$

$$= \sum_{i=1}^{K} \mu(g^{-1}(S^*) \cap R_i \times \mathbb{R}^\ell) = \mu(g^{-1}(S^*)) \quad (6)$$

where the last step stems from Lemma 2. The upper bound on $\Pr(|X - E[X]| > \epsilon)$ is obtained by applying Hoeffding's bound (Lemma 4) to X, seen as the sum of the N independent random variables $\frac{X_{ij}}{N_i}$, for $i = 1, ..., K$ and $j = 1, ..., N_i$. Here we take into account the fact that, for each such i, j we have $0 \leq \frac{X_{ij}}{N_i} \leq \frac{\mu_0(R_i)}{N_i}$.

2. The derivation for $E[X_t]$ is similar to the previous case, but only the variables X_{hj} for $h \in H_t$, corresponding to rectangles contained in $\mathbb{R}^n_{\leq t}$, contribute to the summation. Therefore in place of (5)–(6), we have

$$E[X_t] = \sum_{h \in H_t} \frac{1}{N_h} \sum_{j=1}^{N_h} E[X_{hj}] = \sum_{h \in H_t} \frac{1}{N_h} \cdot N_h \cdot \mu(g^{-1}(S^*) \cap R_h \times \mathbb{R}^\ell)$$

$$= \sum_{h \in H_t} \mu(g^{-1}(S^*) \cap R_h \times \mathbb{R}^\ell) = \mu(P_t \cap g^{-1}(S^*)).$$

3. The previous two parts imply that $F(t|S^*) = \frac{E[X_t]}{E[X]}$ (note that by assumption $E[X] = \mu(g^{-1}(S^*)) > 0$). The event $\frac{E[X_t]}{E[X]} \notin [\underline{A}(t), \overline{A}(t)]$ can be decomposed as: $(\frac{E[X_t]}{E[X]} < \underline{A}(t))$ or $(\frac{E[X_t]}{E[X]} > \overline{A}(t))$. We analyse these two events separately.

   (a) $\frac{E[X_t]}{E[X]} < \underline{A}(t) = \frac{X_t - \epsilon}{X + \epsilon}$ implies $E[X_t] < X_t - \epsilon$ or $E[X] > X + \epsilon$, given the positivity of $X + \epsilon$. In turn, this implies $|E[X_t] - X_t| > \epsilon$ or $|E[X] - X| > \epsilon$.

   (b) $\frac{E[X_t]}{E[X]} > \overline{A}(t)$ implies $X - \epsilon > 0$ and $\frac{E[X_t]}{E[X]} > \overline{A}(t) = \frac{X_t + \epsilon}{X - \epsilon}$ (note that $X - \epsilon \leq 0$ would imply $\overline{A}(t) = 1$ by definition of $\mathcal{A}$, but $\frac{E[X_t]}{E[X]} \leq 1$). Given the positivity of $X - \epsilon$, this in turn implies $E[X_t] > X_t + \epsilon$ or $E[X] < X - \epsilon$. In turn, this implies $|E[X_t] - X_t| > \epsilon$ or $|E[X] - X| > \epsilon$.

We have therefore proved that

$$F(t|S^*) \notin [\underline{A}(t), \overline{A}(t)] \text{ implies } (|E[X_t] - X_t| > \epsilon \text{ or } |E[X] - X| > \epsilon).$$

In terms of probability, by applying the bounds obtained in part 1 and 2 and a union bound, we obtain:

$$\Pr \left(F(t|S^*) \notin [\underline{A}(t), \overline{A}(t)] \right) \leq \Pr \left(|E[X_t] - X_t| > \epsilon \right) + \Pr \left(|E[X] - X| > \epsilon \right)$$
$$\leq \delta_1 + \delta_0 = \delta.$$

From this inequality, the thesis for this part immediately follows.

Example 2 (simple ball/2). Consider the model defined in Example 1. For the additive noise Ψ, we have the range $\underline{h} = (-1, -1)$ and $\overline{h} = (1, 1)$, hence, in the notation discussed in this section, we can set $S = [-9.9, -6.9] \times [-15.3, -12.3]$ and $\mathcal{X}_0 = f^{-1}(S) \cap R_0$. A covering $\mathcal{C}$ of $\mathcal{X}_0$ can be computed calling SIVIA($S, R_0, \overline{f}, \rho$), where we set the resolution to $\rho = 0.2 \times 5$, that is the 20% of the width of R_0. After five bisections, we obtain a covering of $\mathcal{X}_0$ composed of four rectangles, specifically: $\mathcal{C} = \{[8.25, 8.875], [8.875, 9.5], [9.5, 10.75], [10.75, 11.375]\}$; we have $\mu_0(\cup \mathcal{C}) = 0.625$. Now, suppose we want to estimate $F(t|S^*)$ for $t = 8.8$. We set $\epsilon = 0.01$, $N = 15000$, and the N_i's proportional to $\mu_0(R_i)$, getting $\delta_0 < 0.001$ hence $\delta < 0.002$. We run $\mathcal{A}(t)$. In step 1, we refine $\mathcal{C}$ by splitting $[8.25, 8.875]$ into $[8.25, 8.8], [8.8, 8.875]$, thus obtaining five rectangles $R_1, ..., R_5$. In step 2, we define the basic r.v.'s X_i. In step 3, we run the actual simulation, in which a value for X and one for X_t are computed. In detail, X will take on the value $\sum_{i=1}^5 r_i \cdot \mu_0(R_i)$, where r_i the fraction of the N_i i.i.d. samples (θ, ψ) with $\theta \in R_i$ that are mapped into S^*; and X_t will take on the value $r_1 \cdot \mu_0(R_1)$ where $R_1 = [8.25, 8.8]$. In a specific simulation, we have found $X = 0.592$ and $X_t = 0.095$, so that, taking into account $\epsilon = 0.01$, the r.v. $\mathcal{A}(t) = [\underline{A}(t), \overline{A}(t)]$ takes on confidence interval $[0.141, 0.181]$.

Remark 2 (enhancements of $\mathcal{A}$). We outline two straightforward enhancements of the core algorithm $\mathcal{A}$.

1. It is sometimes possible to identify rectangles $R \in \mathcal{C}$ such that $g(R \times [\underline{h}, \overline{h}]) \subseteq S^*$. In case $\mathcal{C} = \mathcal{C}_{in} \cup \mathcal{C}_{out}$ is a set of rectangles obtained with SIVIA, one can check the rectangles $R \in \mathcal{C}_{in}$ using an interval version of g, $\overline{g} := \overline{f} + [\underline{h}, \overline{h}]$. In any case, let $\mathcal{C}_0$ be the set of identified rectangles that satisfy this property. Letting $v_0 := \mu_0(\cup \mathcal{C}_0)$ and $v_{0,t} := \mu_0(\cup \{R \in \mathcal{C}_0 : R \subseteq \mathbb{R}^n_{\leq t}\})$, one defines $\tilde{X} := v_0 + \sum_{R_i \in \mathcal{C} \setminus \mathcal{C}_0} \sum_{j=1}^N X_{ij}$ and $\tilde{X}_t := v_{0,t} + \sum_{i \in H_t \text{ s.t. } R_i \in \mathcal{C} \setminus \mathcal{C}_0} \sum_{j=1}^N X_{ij}$. A tighter interval confidence $\tilde{\mathcal{A}}(t)$ can then be obtained using $\tilde{X}, \tilde{X}_t$ in place of X, X_t. The confidence $1 - \delta$ itself is modified accordingly, and gets sharper. We omit the rather obvious details.
2. Suppose one must compute $\mathcal{A}(t)$ for $t \in \{t_1, ..., t_k\}$, rather than for a single point. By a slight modification of $\mathcal{A}$, it is possible to return confidence intervals for each of the $F(t_i|S^*)$ in a single run. The required modifications of the core

algorithm $\mathcal{A}$ are: first, the covering $\mathcal{C}$ is ensured to refine all the $\mathbb{R}^n_{\leq t_i}$'s; second, in step 3, all the variables $X_{t_1}, ..., X_{t_k}$ are defined, and the corresponding k confidence intervals are computed accordingly. We denote by $\mathcal{A}(t_1, ..., t_k)$ a call to this modifed algorithm. By a union bound, the probability that $F(t_i|S^*) \in \mathcal{A}(t_i)$ for *all* $i = 1, ..., k$ is at least $1 - k\delta$. This algorithm can be used to compute the confidence bands described in Sect. 2.

5   Confidence Intervals for Posterior Moments

We concentrate on the first marginal of the posterior distribution, denoted by $F(\lambda|S^*)$ according to the notation introduced in Sect. 2. The same arguments, of course, apply to the other marginals. In Sect. 2, we have seen how to compute stepwise approximations of $F(\lambda|S^*)$, starting from confidence intervals for $F(\lambda|S^*)$ for a set of node points, say $\mathcal{A}(\lambda_1), ..., \mathcal{A}(\lambda_k)$, see (2). The computed approximation forms a confidence band, (F^-, F^+). Starting from this band, we can compute confidence intervals for various statistics of $F(\lambda|S^*)$. We first examine in detail the case of the expected value, $E[\Theta_1|S^*]$.

In the result below, the hypothesis $\lambda_0 \geq 0$ can be removed by resorting to a slightly more complicated formula.

Corollary 1. *Let (F^-, F^+) be the confidence band defined in (2). Assume $\lambda_0 \geq 0$. Then $E[\Theta_1|S^*] \in [v^-, v^+]$ with probability at least $1 - (k-1)\delta$, where:*

$$v^- := \lambda_0 + \sum_{j=0}^{k-1}(1 - F^+(\lambda_j|S^*))(\lambda_{j+1} - \lambda_j) \quad v^+ := \lambda_0 + \sum_{j=0}^{k-1}(1 - F^-(\lambda_j|S^*))(\lambda_{j+1} - \lambda_j).$$

Proof. According to a well known formula [12], for a positively supported random variable Z with CDF F, one has $E[Z] = \int_0^{+\infty}(1 - F(z))dz$. Below, we apply this formula to $F(\lambda|S^*)$.

We have

$$E[\Theta_1|S^*] = \int_0^{+\infty}(1 - F(z|S^*))dz \qquad\qquad \leq \int_0^{+\infty}(1 - F^-(z|S^*))dz$$

$$= \int_0^{\lambda_0} dz + \int_{\lambda_0}^{\lambda_k}(1 - F^-(z|S^*))dz \quad = \lambda_0 + \sum_{j=0}^{k-1}\int_{\lambda_j}^{\lambda_{j+1}}(1 - F^-(z|S^*))dz$$

$$= \lambda_0 + \sum_{j=0}^{k-1}\int_{\lambda_j}^{\lambda_{j+1}}(1 - F^-(\lambda_j|S^*))dz \quad = \lambda_0 + \sum_{j=0}^{k-1}(1 - F^-(\lambda_j|S^*))(\lambda_{j+1} - \lambda_j)$$

$$= v^+$$

where, thanks to Proposition 1, the $\leq$ in the first row above holds true with probability at least $1 - (k-1)\delta$. The other inequality is proven similarly.

Example 3 (simple ball/3). Consider again the simple ball model introduced in Example 1. We build a confidence band (F^-, F^+) as specified in (2). We choose $k = 25$ evenly spaced points in R_0: $\lambda_0 = 7, \lambda_1, ..., \lambda_{25} = 12$. Then we compute $A(\lambda_1), ..., A(\lambda_{24})$ as specified in Remark 2(2), this time setting SIVIA's resolution to $\rho = 0.1 \times 5$, and choosing ϵ, N in such a way that $(k-1)\delta \leq 0.001$. The obtained F^- and F^+ are plotted in Fig. 1, left. Relying on (F^-, F^+), we apply Corollary 1, and obtain the confidence interval $[v^-, v^+] = [9.65, 9.85]$ for $E[\Theta|S^*]$.

A similar confidence interval can be obtained for the variance $\sigma^2 = \mathrm{Var}[\Theta_1|S^*]$, relying on the formula $E[Z^2] = 2\int_0^{+\infty} z(1 - F(z))dz$, which again holds for a nonnegative Z. One gets $\sigma^2 \in [\sigma^{2-}, \sigma^{2+}]$ with probability $\geq 1 - 2(k-1)\delta$, where $\sigma^{2-} = 2[\lambda_0 + \sum_{j=0}^{k-1}\lambda_j(1 - F^+(\lambda_j|S^*))(\lambda_{j+1} - \lambda_j)] - (v^+)^2$ and $\sigma^{2+} = 2[\lambda_0 + \sum_{j=0}^{k-1}\lambda_{j+1}(1 - F^-(\lambda_j|S^*))(\lambda_{j+1} - \lambda_j)] - (v^-)^2$.

6 Optimal Allocation of Computational Resources

Generally speaking, optimality in this section should be intended in the sense of algorithmic choices that minimize the expression of the Hoeffding bound. In the following analysis, we refer to the core algorithm of Sect. 4. Suppose that a covering $\mathcal{C}$ of $\mathcal{X}_0$ is given. The next result says how to optimally allocate a budget of N samples among the K rectangles of $\mathcal{C}$, that is a strategy that minimizes the quantity δ_0 defined in Theorem 1(1), which bounds (up to a factor of 2) the probability of error of $\mathcal{A}$.

Theorem 2. *Let* $\mathcal{C} = \{R_1, ..., R_K\}$ *be a covering of* $\mathcal{X}_0$, $A = \bigcup_{i=1}^{K} R_i$ *with* $\mu_0(A) > 0$, $N = N_1 + \cdots + N_K$ ($N_i \geq 1$) *and* δ_0 *as in Theorem 1(1). Then*

$$\delta_0 \geq 2\exp\left(-\frac{2N\epsilon^2}{\mu_0(A)^2}\right). \tag{7}$$

Equality in (7) holds if $N_i = N\frac{\mu_0(R_i)}{\mu_0(A)}$ *for* $i = 1, ..., K$.

Proof. Recall that $\delta_0 = 2\exp\left(-\frac{2\epsilon^2}{\sum_{i=1}^{K}\frac{\mu_0(R_i)^2}{N_i}}\right)$. Consider the denominator inside the exponential. We have

$$\sum_{i=1}^{K}\frac{\mu_0(R_i)^2}{N_i} = N\sum_{i=1}^{K}\frac{N_i}{N}\left(\frac{\mu_0(R_i)}{N_i}\right)^2 \geq N\left(\sum_{i=1}^{K}\frac{N_i}{N}\frac{\mu_0(R_i)}{N_i}\right)^2$$

$$= N\left(\sum_{i=1}^{K}\frac{1}{N}\mu_0(R_i)\right)^2 = \frac{\mu_0(A)^2}{N}$$

where in the second step we have applied Jensen's inequality to the convex function $\lambda \mapsto \lambda^2$. Now (7) is a direct consequence of the inequality just proved. Finally, by inspection, *equality* in (7) holds true under the stated condition.

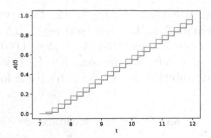
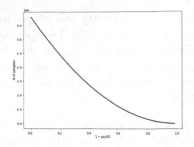

Fig. 1. Left: confidence band ($\delta \leq 0.001$) for the posterior CDF of the simple ball model in Example 3. **Right**: number N of samples necessary to achieve $\delta = 0.001$ as a function of $1 - \mu_0(A)$. Here $\epsilon = 0.001$.

With reference to the expression of δ_0 in Theorem 1, the preceding theorem elucidates two important facts. First, an optimal allocation of a budget of N samples can be obtained by drawing $N_i = N \frac{\mu_0(R_i)}{\mu_0(A)}$ samples from each rectangle R_i of the covering. As we have assumed we can sample efficiently from rectangles, we will adopt this strategy. Note that we will actually draw $N_i := \lceil N \frac{\mu_0(R_i)}{\mu_0(A)} \rceil$ samples per rectangle, leading to an actual number of samples slightly larger than the allocated budget of N, but still less than $N + K$. With the actual sampling strategy, we will still have

$$\delta_0 \leq 2 \exp\left(-\frac{2N\epsilon^2}{\mu_0(A)^2}\right). \tag{8}$$

Second, with the above optimal strategy, δ_0 will only depend on the volume of the set A that encloses $\mathcal{X}_0$: hence coverings that yield tighter enclosures A of $\mathcal{X}_0$ should be preferred. Letting $\delta = 2\delta_0$ (cf. Theorem 1(3)) and holding ϵ fixed, the number $N = \frac{\ln(1/2\delta)}{2\epsilon^2} \mu_0(A)^2$ of samples necessary to guarantee a confidence level δ decreases quadratically as $\mu_0(A)$ decreases: see the plot in Fig. 1. This part of the result also explains why a covering-based based algorithm is, typically, by far more convenient than a crude MC sampling from a rectangle R_0 containing $\mathcal{X}_0$: switching from R_0 to A, the number of samples drops from $N_0 = \frac{\ln(1/2\delta)}{2\epsilon^2}$ (recall that $\mu_0(R_0) = 1$) to $N_0 \cdot \mu_0(A)^2$.

The above discussion on budget allocation presupposes that a covering $\mathcal{C}$ is given. What if the cost of building $\mathcal{C}$ must explicitly be taken into account? The costs of refinement and of simulation are not easily comparable on the same scale, mainly because, depending on the function f, computing $\overline{f}(R)$ can be much more expensive than drawing a sample θ and computing $f(\theta)$. A practical strategy might be to allocate a time budget for the construction of $\mathcal{C}$ and stop the iterations of SIVIA as soon as this time expires, returning the current $\mathcal{C} = \mathcal{C}_{in} \cup \mathcal{C}_{out} \cup L$ as a covering. If the extraction policy in step 4 of SIVIA privileges rectangles with the largest width, this $\mathcal{C}$ is, practically speaking, the best covering that can be obtained with the allocated time budget.

Table 1. Comparison of $\mathcal{A}$, DSA and MCMC on the benchmarks from [5]. **Legend.** n_z: number of state variables of the ODE; n_θ: number of parameters of the ODE; θ^*: values of θ used to generate the observed data; $E_\mathcal{A}$, **confidence interval**: confidence interval for the posterior expectations $E[\Theta_i|S^*]$ ($i = 1, ..., n_\theta$), built via $\mathcal{A}$ (Corollary 1); $t_\mathcal{A}$: execution time (s) of $\mathcal{A}$, including SIVIA; γ: half-side of S^* (tolerance) for $\mathcal{A}$; $E_{DSA}, E_{MCMC}, E_{SMC}$: estimates of posterior expectations obtained via DSA, MCMC and SMC; t_{DSA}, t_{MCMC}: execution times (s) of DSA and MCMC.

Benchmark	n_z, n_θ	θ^*	$E_\mathcal{A}$ confidence interval	$t_\mathcal{A}$	γ	E_{DSA}	t_{DSA}	E_{MCMC}	t_{MCMC}	E_{SMC}
FITZHUGH-NAGUMO										
model-1	2, 2	0.3	0.308±7.0e-03	**11.46**	0.05	0.295	38	0.29	2011	0.29
		0.15	0.153±6.7e-03		0.05	0.16		0.16		0.16
model-2	2, 3	0.30	0.294±2.4e-02	**15.90**	0.05	0.26	1077	0.26	11410	0.26
		0.15	0.152±1.0e-03		0.05	0.10		0.09		0.09
		0.5	0.569±2.5e-02		0.01	0.41		0.39		0.40
LAUB-LOOMIS										
model-1	7, 3	1.8	1.810±2.8e-02	**16.81**	0.05	1.83	2612	1.52	1282	1.92
		0.8	0.797±2.3e-02		0.05	0.80		0.80		0.80
		0.3	0.297±2.3e-02		0.05	0.34		0.39		0.33
model-2	7, 3	0.9	0.912±5.8e-03	**24.55**	0.05	0.93	848	0.90	1251	0.89
		0.8	0.799±2.3e-02		0.05	0.77		0.82		0.82
		0.3	0.297±2.3e-02		0.05	0.27		0.28		0.28
model-3	7, 4	1.8	1.811±2.81e-02	**28.13**	0.07	1.85	557	1.64	4008	1.39
		0.8	0.797±2.3e-02		0.05	0.79		0.79		0.78
		0.3	0.298±2.3e-02		0.05	0.29		0.33		0.50
		2.5	2.556±2.9e-02		0.06	2.49		2.54		2.58
model-4	7, 4	0.9	0.912±5.8e-03	**41.93**	0.05	0.94	1794	0.94	3828	0.95
		0.8	0.797±2.3e-02		0.05	0.78		0.80		0.80
		0.3	0.297±2.3e-02		0.05	0.26		0.28		0.26
		1.4	1.457±5.6e-03		0.05	1.46		1.47		1.48
model-5	7, 5	0.9	0.90±5.7e-03	**118.25**	0.05	0.89	3974	0.86	4213	0.87
		0.8	0.839±4,0e-02		0.05	0.78		0.85		0.82
		0.3	0.339±4.0e-02		0.05	0.29		0.28		0.29
		2.5	2.594±4.6e-02		0.05	2.63		2.52		2.59
		1.3	1.297±2.3e-02		0.05	1.27		1.26		1.27
model-6	7, 5	1.8	1.847±5.1e-02	**185.66**	0.05	1.82	5239	1.62	3811	1.92
		0.8	0.839±4.0e-02		0.05	0.76		0.77		0.77
		0.3	0.296±2.3e-02		0.05	0.31		0.37		0.32
		2.5	2.529±8.7e-03		0.05	2.67		2.66		2.67
		1.3	1.297±2.3e-02		0.05	1.26		1.29		1.29
model-7	7, 6	0.9	0.909±5.79e-03	**386.70**	0.05	0.89	75166	0.85	4189	0.84
		0.8	0.797±2.3e-02		0.05	0.78		0.84		0.87
		0.3	0.297±2.3e-02		0.05	0.34		0.34		0.34
		2.5	2.509±6.4e-03		0.05	2.68		2.78		2.67
		1.3	1.308±2.3e-02		0.05	1.28		1.30		1.3
		1.8	1.799±2.8e-02		0.05	1.99		2.14		2.04
P53	6, 2	9.0e-04	9.0e-04±5.2e-07	**50.28**	0.04	8.9e-04	111	8.9e-04	5600	8.9e-04
		9.9e-06	9.6e-06±4.9e-08		0.04	9.9e-06		9.8e-06		9.8e-06
ROSSLER	3, 2	0.1	0.150±2.6e-02	**16.66**	0.05	0.12	34	0.11	1386	0.11
		0.1	0.170±6.7e-02		0.05	0.09		0.09		0.10
GENETIC	9, 2	50.0	50.013±1.1e-02	**10.26**	0.05	50	155	50	3120	50.0
		50.0	50.042±4.8e-01		0.05	49.1		49.1		49.0
DALLA-MAN	10, 2	0.0581	0.060±1.31e-03	**11.47**	0.05	0.05	9119	0.055	24000	0.055
		0.0871	0.088±9.75e-04		0.05	0.082		0.081		0.082

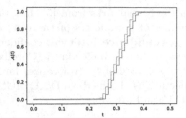 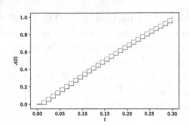

Fig. 2. Confidence bands ($\delta \leq 0.001$) for the marginal CDFs of the two parameter Fitzhugh-Nagumo ODE. **Left:** θ_1. **Right:** θ_2.

7   Experiments

We have put a proof-of-concept implementation[7] of $\mathcal{A}$ at work on a number of examples concerning ODEs models and neural network classifiers.

7.1   Discretized ODES

We have put $\mathcal{A}$ at work on the benchmarks[8] in [5]. We describe our experimental setting with reference to the notation introduced in Subsect. 3.2. In all cases, we apply $\mathcal{A}$ to an Euler-discretized version of the ODE. The timestep τ is chosen small enough to guarantee that, over the considered time horizon, the discretized solution is in very good agreement with the solution obtained via a traditional numerical ODE integrator—specifically, Python's odeint(). In all cases, the observed data y^* consists of a single measurement of the trajectory, taken at the end of the time horizon: multiple measurements do not bring any advantage in terms of accuracy, and introduce unnecessary computational burden. We consider an additive gaussian noise Ψ centered at the origin, with standard deviation as specified in [5,6], but truncated at ≥ 5 standard deviations (only in one case, 3) to the left and to the right of the origin. The set S^* is a hypercube centered at y^* of side 2γ chosen experimentally, with small volumes ranging from 10^{-1} to 10^{-10}. Initial conditions and true parameter values are fixed as specified in [5,6]. In all cases, the sampling budget N, the error threshold ϵ and the number $k+1$ of node points in the construction of the confidence bands via $\mathcal{A}$ (Proposition 1) are chosen so as to ensure a confidence $1 - \delta \geq 0.999$. We focus on computing confidence intervals for the posterior's marginal expectations, as per Corollary 1. The obtained results are reported in Table 1. By way of example, in Fig. 2 we also report the confidence bands for the marginal CDFs of an instance of the Fitzhugh-Nagumo model.

For comparison, we have also reported the results of the DSA algorithm of [5] and of classical MCMC and SMC estimation; all these figures are taken from

---

[7] Python code and examples available at https://github.com/Luisa-unifi/Posterior_estimates.

[8] With one exception, the ANALOG model, which has switching control features not easy to represent in our framework.

[5, Table 1]. In comparing theses results, one should be aware of the differences between the theoretical and experimental settings here and in [5]. First, while considering non discretized ODEs, [5] relies on sensitivity analysis for the estimation of rectangle measures, which can be regarded as a form of—in general, unsound—discretization of the parameter space; on the contrary, we start from the outset with a discretized ODE model, and do not introduce further levels of discretization or unsoundness. Second, [5] provides an estimate of the posterior density given an individual observation y^*; we consider the posterior probability distribution given an observation set S^* of small but positive measure. Third, the estimate intervals provided by [5] are certain—modulo the unsoundness discussed above; in our case, confidence intervals by definition introduce an extra level of aleatoric uncertainty, however small. The case of MCMC/SMC is still different, because no formal guarantee is provided. Despite these differences, we note that the estimates of the posteriors' expected value produced by the three approaches are overall remarkably similar. The execution time of $\mathcal{A}$ is up to *three orders of magnitude smaller* than DSA's and MCMC's[9]. A final caveat concerns the comparison with the true value θ^* used to generate the observation y^*. Although θ^* is often taken as a proxy of the exact posterior expectation $E[\Theta|S^*]$, one should remark that these two values need not coincide. Indeed, in a few experiments, it is observed that θ^* lies outside the returned confidence interval: this is not an indication that the computed interval is 'wrong'.

7.2   Feature Relevance in Neural Network Classifiers

We discuss an application of our algorithm to neural networks, in particular to the quantification of feature relevance in classifiers. We illustrate this methodology in the case of a classifier for images of the MNIST dataset [24]. Consider a classifier that maps items x consisting of p real valued features, say $x = (x_1, ..., x_p)$, into one of s categories. Possibly after normalization of domain and range, without loss of generality one can regard such a classifier as a function $C : [0,1]^p \rightarrow [0,1]^s$, where $C = (C_1, ..., C_s)$. Here it is understood that $x \in [0,1]^p$ is classified as $i \in \{1, ..., s\}$ if and only if $i = \text{argmax}_{j=1,...,s} C_j(x)$ *and* $C_i(x) > l_0$, for a chosen threshold $1 > l_0 \geq 1/2$—in particular, ties are ruled out. Assume a given $x^* = (x_1^*, ..., x_p^*) \in [0,1]^p$ is classified as i by C. One is often interested in assessing the relative importance of a certain feature, or set thereof, in obtaining such a classification: one speaks of *feature relevance*. Here we follow a recent proposal in [1]. Consider the k-th feature ($k \in \{1, ..., p\}$) and the function $C_{i,k} : [0,1] \rightarrow [0,1]$ defined by

$$C_{i,k}(\theta) := C_i(x_1^*, ..., x_{k-1}^*, \theta, x_{k+1}^*, ...x_p^*).$$

Let $\mu(\cdot)$ denote the uniform probability measure on $[0,1]$. The relevance of the k-th feature in classifying x^* to i is defined as:

$$\eta_k(x^*) := 1 - \mu\big(C_{i,k}^{-1}([l_0, 1])\big).$$

---

⁹ [5] does not report execution times for SMC, but we expect them to be in line with MCMC's.

Note that $\eta_k(x^*) \in [0, 1)$. According to [1], the value $\eta_k(x^*)$ reflects how sensitive the classification of x^* is to changes in the k-th feature, other features held constant: values closer to 1 indicate higher sensitivity.

We recast the problem of estimating $\eta_k(x^*)$ with guarantees in the framework of Sect. 2. In the notation of that section, we let $f = C_{i,k}$, hence let $n = m = 1$, and $\ell = 0$, that is no noise, which implies $f = g$ and $\mu = \mu_0$; moreover, we let $S^* = [l_0, 1]$. Then $\mu\big(C_{i,k}^{-1}([l_0, 1])\big) = \mu\big(f^{-1}(S^*)\big)$. Letting $R_0 = [0, 1]$, we define the variable X as in Algorithm 2 in Sect. 4. Then apply the first item of Theorem 1 to find that $\mu\big(C_{i,k}^{-1}([l_0, 1])\big) = E[X]$. The same theorem says that X approximates $E[X]$ with confidence $1 - \delta_0$. As a consequence, $1 - X$ approximates $\eta_k(x^*)$ with confidence $1 - \delta_0$.

We have applied the above outlined methodology to the estimation of the feature relevance for a classifier of the MNIST image dataset [24]. The considered classifier is implemented by a feedforward neural network with two hidden layers of respectively 30 and 10 neurons, using eLU (exponential LU) as an activation function for all layers. The trained network exhibits an accuracy of around 96% on a test subset of the MNIST dataset. For a few selected pictures from the dataset, we have computed the relevance of each of the $28 \times 28 = 784$ features (pixels) composing the image, applying the above methodology. Each pixel is represented by a grayscale value in the interval $[0, 1]$. This way, for each selected picture, we have obtained a 28×28 matrix of feature relevance values in $[0, 1)$. We have set $\delta_0 \leq 0.001$. The average computation time for each pixel is about 6 s. For a better visualization and interpretation of the results, a feature relevance matrix can be converted to a colormap, called *relevance map* in [1]: see Fig. 3. While, as expected, most pixels have relevance 0, there are a few clusters of highly relevant pixels, located approximately in the void zones of the original image. These empirical findings differ slightly from those reported in [1], where highly relevant features tend to reproduce the contours of the represented digit. The experimental settings here and in [1] are not exactly comparable though, as [1] considers a neural network with a different structure, with only one hidden layer of neurons.

Fig. 3. Example of relevance map. **Left**: original image from the MNIST dataset, correctly classified as '4'. **Center**: relevance map, with brighter colours representing more relevant features (black = 0, white = 1). **Right**: overlay of the first and second image. Clusters of highly relevant pixels tend to occupy void zones.

8   Conclusion

Assuming a functional relation between observations and parameters, we describe a method to estimate the Bayesian posterior parameters distribution, given that the observation belongs to a small set. Guarantees for the estimated a posteriori quantities, including CDFs and moments thereof, are given in the form of confidence bands or confidence intervals. We leverage IA to drastically reduce the computational cost of MC simulation. In terms of accuracy and execution time, the method compares very favourably to state-of-the-art techniques on benchmarks for noisy ODE parameter estimation. An application to relevance feature in neural networks has also been proposed.

As for future research, we would like to further investigate the scalability of the method, studying applications to more complex models and possibly to probabilistic programming. Another direction is the relation with Approximate Bayesian Computation (ABC) [25] and Importance Sampling [29], which, among well-established techniques, appear to be closest in spirit to our approach.

Aknowlegments. We wish to express our gratitude to Dr. Yi Chou and Prof. Sriram Sankaranarayanan for providing us with detailed data and assistance on their experimental setting [5,6]. Thanks to Prof. Fabio Corradi and Dr. Cecilia Viscardi for stimulating discussions on MC methods in Bayesian statistics.

References

1. Adam, S.P., Likas, A.C.: A Set Membership Approach to Discovering Feature Relevance and Explaining Neural Classifier Decisions (2022). arXiv:2204.02241
2. Althoff, M.: An introduction to CORA 2015. In: Proceedings of the Workshop on Applied Verification for Continuous and Hybrid Systems, pp. 120–151 (2015)
3. Boreale, M., Collodi, L.: Linearization, model reduction and reachability in nonlinear ODEs. In: Lin, A.W., Zetzsche, G., Potapov, I. (eds.) RP 2022. LNCS, vol. 13608, pp. 49–66. Springer, Cham (2022). https://doi.org/10.1007/978-3-031-19135-0_4
4. Chen, X., Ábrahám, E., Sankaranarayanan, S.: Flow*: an analyzer for non-linear hybrid systems. In: Sharygina, N., Veith, H. (eds.) CAV 2013. LNCS, vol. 8044, pp. 258–263. Springer, Heidelberg (2013). https://doi.org/10.1007/978-3-642-39799-8_18
5. Chou, Y., Sankaranarayanan, S.: Bayesian parameter estimation for nonlinear dynamics using sensitivity analysis. In: International Joint Conferences on Artificial Intelligence Organization, pp. 5708–5714 (2019)
6. Chou, Y.: Detailed experimental set up of [5]. Personal Communication (2022)
7. Coelho, F., Codeco, C., Gabriela, M., Gomes, M.: A Bayesian framework for parameter estimation in dynamical models. PLoS One **6** (2011)
8. Dempster, A.P.: Upper and lower probabilities induced by a multivalued mapping. Ann. Math. Stat. **38**(2), 325–339 (1967)
9. Doucet, A., De Freitas, N., Gordon, N.: Sequential Monte Carlo Methods in Practice. Springer, Heidelberg (2001). https://doi.org/10.1007/978-1-4757-3437-9

10. Dvoretzky, A., Kiefer, J., Wolfowitz, J.: Asymptotic minimax character of the sample distribution function and of the classical multinomial estimator. Ann. Math. Stat. **27**(3), 642–669 (1956). https://doi.org/10.1214/aoms/1177728174

11. Faes, M.G.R., Daub, M., Marelli, S., Patelli, E., Beer, M.: Engineering analysis with probability boxes: a review on computational methods. Struct. Saf. **93**, 102092 (2021)

12. Feller, W.: An Introduction to Probability Theory and Its Applications, vol. II (2nd edn. of 1966 Original edn.). Wiley, New York (1971). MR 0270403

13. Ferson, S., Kreinovich, V., Ginzburg, L., Myers, D.S., Sentz, K.: Constructing probability boxes and Dempster-Shafer structures. Technical report, SAND2002-4015, Sandia Laboratories (2003)

14. Fox, E.B.: Bayesian nonparametric learning of complex dynamical phenomena. Ph.D. thesis, MIT (2009)

15. Girolami, M.: Bayesian inference for differential equations. Theor. Comput. Sci. **408**(1), 4–16 (2008)

16. Goodfellow, I.J., Bengio, Y., Courville, A.: Deep Learning. MIT Press, Cambridge (2016)

17. Goodman, N.D., Stuhlmüller, A.: The Design and Implementation of Probabilistic Programming Languages (2014). http://dippl.org/

18. Hastings, W.: Monte Carlo sampling methods using Markov chains and their applications. Biometrika **57**, 97–109 (1970)

19. Hoeffding, W.: Probability inequalities for sums of bounded random variables. J. Am. Stat. Assoc. **58**(301) (1963)

20. Jaulin, L.: Computing minimal-volume credible sets using interval analysis; application to Bayesian estimation. IEEE Trans. Signal Process. **54**(9), 3632–3636 (2006). https://doi.org/10.1109/TSP.2006.877676

21. Jaulin, L., Walter, E.: Guaranteed nonlinear parameter estimation via interval computations. Interval Comput. **3**, 61–75 (1993)

22. Jaulin, L., Walter, E.: Set inversion via interval analysis for nonlinear bounded-error estimation. Automatica **29**(4), 10531064 (1993)

23. Kosorok, M.R.: Introduction to Empirical Processes and Semiparametric Inference. Springer, Heidelberg (2006). https://doi.org/10.1007/978-0-387-74978-5

24. LeCun, Y., Cortes, C., Burges, C.: MNIST handwritten digit database. https://yann.lecun.com/exdb/mnist/

25. Lintusaari, J., Gutmann, M.U., Dutta, R., Kaski, S., Corander, J.: Fundamentals and recent developments in approximate bayesian computation. Syst. Biol. **66**(1), 66–82 (2017)

26. Massart, P.: The tight constant in the Dvoretzky-Kiefer-Wolfowitz inequality. Ann. Probab. **18**(3), 1269–1283 (1990). https://doi.org/10.1214/aop/1176990746

27. Metropolis, N., Rosenbluth, A., Rosenbluth, M., Teller, A., Teller, E.: Equations of state calculations by fast computing machines. J. Chem. Phys. **21**, 1087–1092 (1953)

28. Moore, R.E.: Interval Analysis. Prentice-Hall, Englewood Cliffs (1966)

29. Robert, C.P., Casella, G.: Monte Carlo Statistical Methods. Springer, Heidelberg (1999)

30. Shafer, G.: A Mathematical Theory of Evidence. Princeton University Press, Princeton (1976)

31. Vanlier, J., Tiemann, C.A., Hilbers, P.A.J., van Riel, N.A.W.: Parameter uncertainty in biochemical models described by ordinary differential equations. Math. Biosci. **246**(2), 305–314 (2013)

A Pragmatic Approach to Stateful Partial Order Reduction

Berk Cirisci[1]([✉])[iD], Constantin Enea[2][iD], Azadeh Farzan[3][iD], and Suha Orhun Mutluergil[4][iD]

[1] IRIF, Université Paris Cité, Paris, France
cirisci@irif.fr
[2] LIX, Ecole Polytechnique, CNRS and Institut Polytechnique de Paris, Palaiseau, France
cenea@lix.polytechnique.fr
[3] University of Toronto, Toronto, Canada
azadeh@cs.toronto.edu
[4] Sabanci University, Tuzla, Turkey
suha.mutluergil@sabanciuniv.edu

Abstract. Partial order reduction (POR) is a classic technique for dealing with the state explosion problem in model checking of concurrent programs. Theoretical optimality, i.e., avoiding enumerating equivalent interleavings, does not necessarily guarantee optimal overall performance of the model checking algorithm. The computational overhead required to guarantee optimality may by far cancel out any benefits that an algorithm may have from exploring a smaller state space of interleavings. With a focus on overall performance, we propose new algorithms for stateful POR based on the recently proposed source sets, which are less precise but more efficient than the state of the art in practice. We evaluate efficiency using an implementation that extends Java Pathfinder in the context of verifying concurrent data structures.

1 Introduction

Concurrency results in insidious programming errors that are difficult to reproduce, locate, and fix. Therefore, verification techniques that can automatically detect and pinpoint errors in concurrent programs are invaluable. *Model checking* [7,37] explores the state space of a given program in a systematic manner and verifies that each reachable state satisfies a given property. It provides high coverage of program behavior, but it faces the infamous state explosion problem, i.e., the number of possible thread interleavings grows exponentially in the size of the source code. In this paper, we consider shared-memory programs running on a sequentially consistent memory model, for which interleavings of atomic steps in different threads are a precise model of concrete executions.

Partial order reduction (POR) [8,16,34,40] is an approach that limits the number of explored interleavings without sacrificing coverage. POR relies on an equivalence relation between interleavings, where two interleavings are equivalent

C. Dragoi et al. (Eds.): VMCAI 2023, LNCS 13881, pp. 129–154, 2023.
https://doi.org/10.1007/978-3-031-24950-1_7

if one can be obtained from the other by swapping consecutive independent (non-conflicting) execution steps. It guarantees that at least one interleaving from each equivalence class (called a Mazurkiewicz trace [30]) is explored. Optimal POR techniques explore exactly one interleaving from each equivalence class. Beyond this classic notion of optimality, POR techniques may aim for optimality by avoiding visiting states from which no optimal execution may pass. There is a large body of work on POR techniques that address its soundness when checking a certain class of specifications for a certain class of programs, or its theoretical optimality (see Sect. 6). The set of interleavings explored by some POR technique is defined by restricting the set of threads that are explored from each state (scheduling point). Depending on the class specifications, assumptions about programs, or optimality targets, there are various definitions for this set of processes, including stubborn sets [40], persistent sets [16], ample sets [8], and source sets [3].

The design of a model checking algorithm based on POR has to consider several computational tradeoffs. First, such an algorithm can be stateful or stateless [17], which corresponds to a tradeoff between memory consumption versus execution time. Stateful model checking records visited states, thereby consuming more memory, but stateless model checking performs redundant exploration from already visited states. Second, the computation of the set of threads that are explored from some state can be more or less complex. Focusing on theoretical optimality, e.g., exploring *exactly* one interleaving from each Mazurkiewicz trace, may make this computation more complex. This complexity in turn may diminish the overall performance when the potential for reducing the state space is not large, i.e., most Mazurkiewicz traces contain a small number of interleavings. In such a case, exploring more interleavings can take less time than computing more precise constraints on the explored schedules. Third, POR algorithms may compute the information they use for the purpose of reduction *statically*, by some kind of conservative static analysis of the source code, or *dynamically*, during the exploration of interleavings. Static computation is usually cheaper and less precise than dynamic computation.

In this work, we investigate the use of POR from a practical point of view. In the context of verifying concurrent data structures, we investigate the following research question: what tradeoffs in POR families of algorithms may lead to practical net gains in verification or bug-finding times? We focus on the application domain of verification of Java concurrent data structures using a tool like Violat [9]. Concurrent data structures provide implementations of common abstract data types (ADTs) like queues, key-value stores, and sets. Their correctness amounts to observational refinement [22,23,35] which captures the substitutability of an ADT with an implementation [29]: any combination of values admitted by a given implementation is also admitted by the given ADT specification. Violat can be used to generate tests of observational refinement, i.e., bounded-size clients of the concurrent data structure that include assertions to check that any combination of return values observed in an execution belongs to a statically precomputed set of ADT-admitted return-value outcomes. Violat is integrated with the Java Pathfinder (JPF) model checker [41], which enables complete systematic coverage of a given test program and outputting execution traces leading to consistency violations, thus facilitating diagnosis and repair. We investigate POR algorithms implementable in JPF.

We study several stateful model checking algorithms with POR in the context of Violat's test programs. This choice was inspired by experiments that demonstrated that it is much faster than the stateless variation (see Sect. 5). We introduce POR algorithms that combine *static and dynamic* computations of sets of threads to explore from a given state. In the context of stateful model checking, static techniques may seem like the better option. A dynamic computation usually requires re-traversing the state space starting in an already visited state which can be time consuming. Note however that re-traversing the state space that is already loaded in memory takes less time than generating that state space in the first place, which involves executing program statements.

Our starting point is a simple static POR algorithm, called S-POR, that makes use of *invisible* transitions. These transitions are independent of any transition of another concurrently-executing thread (they correspond to the safe actions introduced in [24]). Based on a syntactic analysis of the code, we identify shared and synchronization objects, and assume that every transition that does not access such an object is invisible. For clients of concurrent data structures, such objects correspond to class fields accessed in a method of the data structure. Invisible actions include starting and joining threads, and method calls and returns, for instance. The POR algorithm prioritizes the exploration of invisible transitions over visible ones, i.e., if an invisible transition is enabled in a given state then this is the only explored transition from that state, and otherwise, all enabled transitions are explored. We demonstrate that S-POR has a small overhead and the potential for substantial reductions, and therefore leads to significant speedups with respect to standard JPF which employs a very conservative heuristic for its POR (see Sect. 5).

S-POR is effective, but by the nature of being lightweight, does not always reduce the state space effectively. We introduce two new algorithms as extensions of S-POR, with the idea of performing a more aggressive reduction while keeping the overhead reasonably low. They *dynamically* compute *source sets*, which restrict the set of threads explored from a state with only *visible* enabled transitions. We focus on source sets since they are provably minimal, i.e., the set of explored threads from some state must be a source set in order to guarantee exploration of all Mazurkiewicz traces. Moreover, any superset of a source set is also a source set, which makes their computation less sensitive to the order in which transitions from a given state are enumerated. This property does not hold for other definitions such as stubborn sets, persistent sets, or ample sets.

The design principle behind our algorithms is to favor efficiency over theoretical optimality. Our algorithms are not theoretically optimal. However, we demonstrate that they are more efficient than the optimal algorithm [3] where the overhead of source set computation subsumes any gains from not exploring the redundant interleavings.

In general, a dynamic computation of source sets relies on tracking dependencies between actions in the already explored executions. Our two proposed algorithms differ in the way in which the tracking is performed: one is *eager* and called DE-S-POR, and the other is *lazy* and called DL-S-POR. Intuitively, DE-S-POR advances the computation of source sets for predecessors in the current execution in a style similar previous dynamic POR algorithms, e.g. [3,13], while DL-S-POR advances the computation of the source set in a given state only when the exploration backtracks to that state and one must decide if a new transition has to be explored.

The thesis of this paper is that when there is a big enough potential for reducing the state space of a concurrent program, i.e., many Mazurkiewicz traces are large enough, non-optimal but carefully customized algorithms, like DE-S-POR and DL-S-POR, can have the largest impact compared to the two extremes of the spectrum, that is, S-POR or theoretically optimal algorithms like [3]. If the potential for reduction is small, then a simple static algorithm like S-POR provides the best overhead-gain tradeoff.

To support this thesis, we implemented these algorithms in JPF and evaluated them on a number of clients of concurrent data structures from the Synchrobench repository [21]. Our evaluation shows that they outperform (1) their variations that are directly built on top of the standard setup of JPF, (2) their stateless variations, and (3) a best-effort implementation of a stateful variation of the optimal algorithm in [3]. The lazy algorithm DL-S-POR is more efficient than the eager DE-S-POR, and more efficient than S-POR on clients with a big enough potential for reducing the state space.

More details and experimental results can be found in a full version [6].

2   Preliminaries

We model a concurrent (multi-threaded) program with a bounded number of threads as a labeled transition system (LTS) $L = (\mathcal{S}, s_I, \Gamma)$. We assume that programs run under sequential consistency. A state in $\mathcal{S}$ represents a finite set of *shared* objects visible to all threads and a finite set of *local* objects visible to a single fixed thread, and a program counter for each thread. The state $s_I \in \mathcal{S}$ is the unique initial state. Γ is a set of labeled transitions (s, a, s') where $s, s' \in \mathcal{S}$ and a is an *action* (transition label) representing the execution of an *atomic* statement in the code. Action a records the executing thread, its program counter, and shared object accesses. There are two types of actions: (1) *invisible actions*: $a = (t, pc, \epsilon)$ where a thread t executes a statement at program counter pc that accesses no shared object, and (2) *visible actions*: $a = (t, pc, r/w, o)$ where t executes a statement at pc that reads (r) or writes (w) the shared object o. For an action a, $tid(a)$ is the thread id t, and $op(a)$ and $obj(a)$ refer to the third and fourth components when a is visible (otherwise they are undefined).

A transition labeled by a visible (resp. invisible) action is called visible (resp. invisible). In the context of a full-fledged programming language, invisible transitions are related to local computations, control-flow manipulations (e.g., starting/stopping threads and calling or returning from a method), or accesses to

"low-level" shared objects that are irrelevant for the intended (functional) specification. Visible transitions correspond to the execution of a single atomic statement that accesses a shared object followed by a maximal sequence of *local*
statements that only modify the local states of that thread.

We assume that LTSs are deterministic and acyclic. An action a is *enabled*
in state s if there exists s' such that $(s, a, s') \in \Gamma$. We use $next(s, t)$ to denote
the transition $(s, a, s') \in \Gamma$ for some a and s' with $tid(a) = t$, if it exists, and
$succ(s, t)$ to denote the successor s' in this transition. Otherwise, we say that t
is *blocked* in s. The set $enabled(s)$ is the set of threads that are not blocked in
s. A state s is *final* if $enabled(s) = \emptyset$.

Two actions a and a' of different threads are *independent* if they are both
enabled in a state s and either one of them is an invisible action, or they are
both visible and access different shared objects $(obj(a) \neq obj(a'))$, or they both
perform a read access $(op(a) = op(a') = r)$. The actions a and a' are called
dependent, denoted by $a \sim a'$, if they are not independent. We assume that if
an action a enables or disables another action a', then $a \sim a'$. Two transitions
are (in)dependent iff they contain actions that are (in)dependent.

An *execution from a state* s is a sequence of alternating states and actions
$E = s_0, a_0, s_1, a_1, \ldots, s_n$ with $s_0 = s$ and $(s_i, a_i, s_{i+1}) \in \Gamma$ for each $0 \leq i \leq n-1$.
The set of execution starting from s in the LTS L is denoted by $E(L, s)$. An
initialized execution is an execution from s_I. Initialized executions that end
with a final state are called *full* executions. We assume absence of deadlocks,
i.e., a full execution E contains every action enabled in a state of E.

The *happens-before* relation in an execution E, denoted by $\rightarrow_E$, captures the
causal relation among actions in E (the program order between actions of the
same thread and the order between actions accessing the same shared object
where at least one of them is a write). Given two actions a and a' labeling
transitions in E, $a \rightarrow_E a'$ holds iff $a \sim a'$ and the transition labeled by a occurs
before the transition labeled by a' in E. Two executions E and E' are called
equivalent if $\rightarrow_E = \rightarrow_{E'}$. For a full execution E, we use $[E]$ to denote the set of
full executions E' that are equivalent to E.

Given an LTS $L = (\mathcal{S}, s_I, \Gamma)$ that models a concurrent program, an LTS
$L_r = (\mathcal{S}_r, s_I, \Gamma_r)$ with $\mathcal{S}_r \subseteq \mathcal{S}$ and $\Gamma_r \subseteq \Gamma$ is called *sound for* L if for each full
execution E of L, there exists a full execution E' of L_r that is equivalent to E.

2.1 Partial Order Reduction

The set of executions explored by POR techniques is defined by restricting the
set of threads that are explored from each state. The algorithms discussed in this
paper fall into two categories in this respect: persistent sets and source sets. Both
guarantee soundness, i.e., at least one execution from each equivalence class is
explored.

Intuitively, a set of threads T is *persistent* for a state s if in any execution
starting from s, the first transition that is dependent on some transition starting
from s of some thread $t \in T$ is taken by some thread $t' \in T$ (t and t' may
be equal). A set of threads T is a *source* set for s if for any execution starting

from s, there is some thread in T that can take the first step, modulo reorderings of independent transitions. We define persistent and source sets as sets of threads, which correspond to sets of transitions in the classical sense, under the assumption of determinacy of individual threads.

Definition 1 (Persistent Set [16]). *A set of threads T is called a persistent set for a state s if for every execution E from s that contains only transitions from thread $t' \notin T$, every transition in E is independent of every transition $next(s, t)$ with $t \in T$.*

For an execution E from a state s that ends in a final state, a thread t is called a *weak initial* of E if there exists an execution E' that is equivalent to E and starts with a transition of t.

Definition 2 (Source Set [3]). *A set of threads T is called a source set for a state s if every execution from s that ends with a final state has a weak initial thread in T.*

An exploration where each state is expanded w.r.t. the threads in a persistent or source set is sound (when finished it produces an LTS which is sound for the "full" LTS of the program). However, source sets guarantee a stronger notion of optimality [3]. There exist programs where any persistent set (for the initial state) is strictly larger than a source set [2], but every persistent set is also a source set. Note that source sets are monotonic in the sense that any superset of a source set is also a source set, but this is not true for other definitions such as stubborn sets, persistent sets, or ample sets.

3   Eager Source Set POR (DE-S-POR)

We present a first stateful POR algorithm that selects a sufficient set of threads to expand a state based on two criteria: (1) a static criterion based on (in)visible actions, and (2) a dynamic criterion based on *source sets* computed on-the-fly during the exploration. Source sets are maintained *eagerly* for each new transition that is explored, in a style similar to previous algorithms, e.g. [3,13]. For presentation reasons, we start with a simplified version that includes only the static criterion and continue with the full version afterwards.

3.1   Safe Set POR (S-POR)

Algorithm 1 presents a stateful DFS traversal of a concurrent program, represented by an LTS, which restricts the traversal to so called *safe sets*. Figure 1 illustrates the core idea of this algorithm. The safe sets prioritize the exploration of *invisible* transitions over visible ones.

For a state s, if there is an enabled thread $t \in enabled(s)$ whose enabled transition is invisible, then $safeSet(s) = \{t\}$. Otherwise, $safeSet(s)$ contains all the threads enabled in s, and s is called an *irreducible state*. In Fig. 1, only state s' is irreducible since any other state has at least one enabled invisible transition, and all other states are reducible.

Algorithm 1: SAFE SET POR (S-POR)

Initialize: $Stack \leftarrow \emptyset$; $Stack.push(s_I)$; $L_r \leftarrow \emptyset$;

1 **Explore()**
2      $s \leftarrow Stack.\mathbf{top}$;
3   **if** $notVisited(s)$ **then**
4   **forall the** $t \in safeSet(s)$ **do**
5              $(s, a, s') \leftarrow next(s, t)$;
6              $Stack.\mathbf{push}(s')$; // transition (s, a, s') is added to L_r
7   **Explore()**;
8              $Stack.\mathbf{pop}()$;

$notVisited(s)$ holds if s is final in L_r but $enabled(s) \neq \emptyset$

$$safeSet(s) = \begin{cases} \{t\}, & \exists t \in enabled(s) : next(s, t) = (s, a, s') \text{ and } a \text{ is invisible} \\ enabled(s), & \text{otherwise} \end{cases}$$

In Algorithm 1, *Stack* represents the stack of the DFS traversal and it is considered to be a global variable, and L_r records transitions explored during the traversal. Note that the DFS traversal stops the exploration whenever it visits a state s that has been visited in the past (see the condition at line 3). The choice of safe sets then provides additional savings on top of the standard DFS traversal strategy. When the traversal ends, L_r is sound (for the "full" LTS of the program).

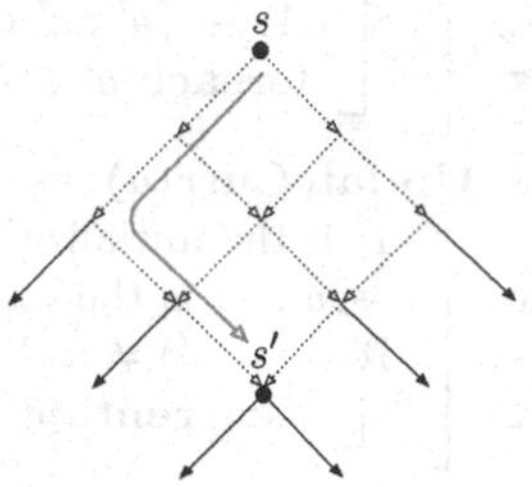

Observe that Algorithm 1 can reduce the number of visited states in a significant way. The diagram in Fig. 1 corresponds to a fully explored program LTS while the path marked by the blue arrow is the result

Fig. 1. Full traversal vs. partial S-POR (in blue). (Color figure online)

of Algorithm 1. It is easy to observe that one can obtain an exponential reduction (with the base of the number of consecutive invisible transitions and the exponent of the number of threads) with this algorithm.

3.2 Full Algorithm

Algorithm 2 builds on top of Algorithm 1 by computing on-the-fly source sets to limit exploration of transitions from the *irreducible* states. More precisely, reducible states are traversed according to the strategy of Algorithm 1 (i.e., only one enabled invisible transition is followed) and for irreducible states, source sets determine what transitions are followed. Since safe sets are also source sets, the overall algorithm remains sound if the new source sets are computed correctly.

Figure 2 provides a declarative description of the key components of Algorithm 2. For a state s in the current execution (stored on the stack), the s.**current** set may be updated every time a new visible transition is explored, and the s.**backtrack** set may be updated every time the exploration backtracks to s. The update of s.**backtrack** relies on the sets s.**current** computed while traversing successors of s.

Algorithm 2: EAGER SOURCE SET POR (DE-S-POR)

Initialize: $Stack \leftarrow \emptyset$; $Stack.\text{push}(s_I)$; $L_r \leftarrow \emptyset$;

```
1  Explore()
2      s ← Stack.top;
3      if notVisited(s) then
4          if ∃t ∈ safeSet(s)  then
5              s.backtrack ← {t}; s.current ← ∅; s.done ← ∅;
6              while ∃t' ∈ s.backtrack \ s.done  do
7                  (s, a, s') = next(s, t');
8                  Stack.push(s');
9                  s.done = s.done ∪ {t'};
10                 s.current[t'] ← {t'};
11                 if a is visible then UpdateCurr(a) ;
12                 Explore();
13                 s.backtrack ← UpdateBack(s, a);
14                 Stack.pop();
15         else
16             A_s ← {a' : a' occurs in an execution from E(L_r, s)};
17             foreach a' ∈ A_s do UpdateCurr(a');

18 UpdateCurr(a)
19     E is the initialized execution of L_r following states in Stack;
20     (s, a', s') is the last transition of E with a ≁ a' ∧ tid(a) ≠ tid(a')
21     if (s, a', s') ≠ null then
22         s.current[tid(a')] = s.current[tid(a')] ∪ {tid(a)};
```

$$UpdateBack(s,a) = \begin{cases} safeSet(s), & \exists t \in s.\textbf{current}[tid(a)] \setminus safeSet(s) \\ s.\textbf{done}, & \exists T \subset s.\textbf{done} : T = \bigcup_{t \in T} s.\textbf{current}[t] \\ \bigcup_{t \in s.\textbf{done}} s.\textbf{current}[t], & \textbf{otherwise} \end{cases}$$

$s.\textbf{current}[t]$: set of threads that execute a transition dependent on $next(s,t)$ which appears after it in an execution.

$s.\textbf{done}$: set of threads whose transitions have been fully explored from s.

$s.\textbf{backtrack}$: when equal to $s.\textbf{done}$, a source set for s.

Fig. 2. Description of important components in Algorithm 2.

When a new transition (s, a, s') from a state s is traversed, the active thread $tid(a)$ is added to the current set $s_l.\textbf{current}[t]$, where s_l is the last state from which the current execution performs a transition that is dependent on a such that $t \neq tid(a)$ is the thread of that transition. See line 11 and the **UpdateCurr** function. When a transition is followed to a visited state s, the same update is done for *every* transition that is reachable from s, as if these transitions are traversed again. See lines 16–17 and note that the declarative definition of

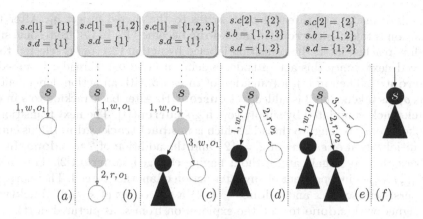

Fig. 3. An example for Algorithm 2. Solid grey circles represent states stored on the stack, hollow dotted circles represent states on the top of the stack, and solid black circles represent states from which the exploration has been completed, i.e. their **backtrack** sets are equal to their **done** sets. Transitions follow the same pattern: dotted transitions are the latest to have been explored, solid grey ones are between states on the stack, and solid black ones are the ones taken in the past. Solid black triangles represent completed explorations starting from some state. We omit program counters from actions. **backtrack**, **current**, and **done** are abbreviated by the first letter.

A_s at line 16 corresponds to a traversal of all the executions starting from s. This may be time-consuming, and yet, such updates are unavoidable in stateful POR algorithms because the current execution reaching s (stored on the stack) may belong to a different Mazurkiewicz trace compared to a previous execution reaching s (whose sequence of transitions leading to s was different).

When backtracking to a state s, the set $s.$**backtrack** is updated to take into account the transitions which are dependent on and occur after the last explored transition starting from s, called τ_s. If τ_s is a transition of thread t, the threads performing those dependent transitions are stored in $s.$**current**$[t]$. If there is a dependent transition τ performed by a thread t' that is not in the safe set of s, then $s.$**backtrack** is updated conservatively to contain the safe set of s. This situation occurs when τ becomes enabled after executing some other thread t'' enabled in s, and observing an execution where τ_s occurs after τ requires first executing the transition of t''. Otherwise, the algorithm checks to see if a subset T of threads enabled in s which have already been explored are sufficient to cover $s.$**current**$[t]$, and that T's transitions in s are independent from future transitions of threads not in T. In that case, $s.$**backtrack** is assigned with $s.$**done** and the exploration from s is halted. The subset of threads T defines a persistent set **and** a source set for s. Since source sets are monotonic, $s.$**done** is a source set for s. If none of the previous conditions hold, then $s.$**current**$[t]$ is simply added to $s.$**backtrack**. This computation is defined by the macro *UpdateBack* in Fig. 2.

We illustrate the algorithm using Fig. 3. In (a), s is reached for the first time and the transition labeled by $(1, w, o_1)$ is selected first to be executed. This is a visible transition of thread 1 that writes to the shared object o_1. After this transition is taken, $s.\mathbf{current}[1]$ and $s.\mathbf{done}$ become $\{1\}$. In (b), from some state which is reached later, the transition labeled by action $(2, r, o_1)$ is selected to be followed next. Since this action is dependent on $(1, w, o_1)$, thread 2 is added to $s.\mathbf{current}[1]$. Then in (c), a transition of thread 3 with an action dependent on $(1, w, o_1)$ is taken, and 3 is added to $s.\mathbf{current}[1]$. After backtracking to s in (d), $s.\mathbf{backtrack}$ is updated by simply copying $s.\mathbf{current}[1]$. The next transition to be taken from s belongs to thread 2 which is in $s.\mathbf{backtrack} \backslash s.\mathbf{done}$. This entails the initialization of $s.\mathbf{current}[2] = \{2\}$ and the addition of 2 to $s.\mathbf{done}$. In (e), we backtrack to s again and without having changed $s.\mathbf{current}[2]$. This means that $(2, r, o_2)$ is independent of any later action of another thread. Therefore, $\{2\}$ is a persistent set of s and $s.\mathbf{done} = \{1, 2\}$ a source set of s, and $s.\mathbf{backtrack}$ is assigned with $s.\mathbf{done}$ to stop the exploration from s, as pictured in (f).

This example shows that Algorithm 2 explores sets of transitions from a given state s that may correspond to a source set which is *not* a persistent set. The exploration in Fig. 3 stops when $s.\mathbf{backtrack} = s.\mathbf{done} = \{1, 2\}$, but the only persistent set that includes thread 1 is $\{1, 2, 3\}$.

Theorem 1. *Given a program represented by an LTS L, Algorithm 2 terminates with an LTS L_r that is sound for L.*

Proof. Based on the soundness of source sets (see Sect. 2.1), it is enough to show that for every state s in L_r,

$$s.\mathbf{backtrack} \text{ is a source set for } s \text{ in } L \text{ when it becomes equal to } s.\mathbf{done} \quad (1)$$

Due to the condition of **while** in line 6 of Algorithm 2, equality of $s.\mathbf{backtrack}$ and $s.\mathbf{done}$ is the only condition for stopping an exploration from a state s. Therefore, if some successor state s' of s is already explored and the search is backtracked to s, then $s'.\mathbf{backtrack} = s'.\mathbf{done}$, because otherwise, **while** loop in line 6 wouldn't be terminated for s'. Since $s.\mathbf{done}$ keep tracks of threads whose enabled transitions from s is already executed, the proof is reduced to showing that the following proposition holds:

$$\text{For any state } s, s.\mathbf{backtrack} \text{ is a source set when the exploration from} \quad (2)$$
$$s \text{ is finished}$$

If s is a reducible state, then only one transition is explored from s which is an invisible transition. The fact that the thread performing this invisible transition is a persistent set and hence a source set follows directly from definitions as every persistent set is also a source set. When $s.\mathbf{backtrack} = s.\mathbf{done}$ is different from $safeSet(s) = enabled(s)$ (which is trivially a source set), it must be the case that there exists $T \subset s.\mathbf{done}$ such that $T = \bigcup_{t \in T} s.\mathbf{current}[t]$ due to the definition of $UpdateBack$ method. Now we show that T is a persistent set for s in L. Assume by contradiction that this is not the case, then due to the definition of persistent set, L admits an execution E starting from s that contains only transitions of threads different from those in T and at least one of these transitions τ of a thread $t' \notin T$ is dependent on some transition $next(s,t)$ with $t \in T$. For every $t \in T$, the successor state s' of s reached by $next(s,t)$ must be in L_r. Due to deadlock freedom assumption, some transition that has the same transition label with τ must be enabled eventually in some successor state of s'. Let $E' \in L_r$ be that execution from s which starts with $next(s,t)$ and contains such transition that shares the same label with τ. Now we will move forward by showing that the following proposition is correct, which will be used in the rest of the proof:

$$\text{If } L_r \text{ admits an execution } E'' \text{ from } s \text{ whose last transition that} \qquad (3)$$
$$\text{depends on and occurs before } \tau' \text{ is } next(s,t) \text{ where } act(\tau') = act(\tau)$$
$$\text{and } tid(\tau') = tid(\tau) = t', \text{ then } t' \in T$$

When τ' is executed from some successor state of s, t' is added to $s.\mathbf{current}[t]$ (and eventually will be added to T due to *UpdateBack* method) by invoking the **UpdateCurr** function as $next(s,t)$ will be the transition in line 20 of Algorithm 2. This contradicts the assumption of $t' \notin T$ and therefore, it is enough to show that proposition below is correct for concluding the proof:

$$\text{If } L_r \text{ admits such an execution } E', \text{ then } L_r \text{ admits such an execution } E'' \quad (4)$$

To show that Proposition 4 holds, we proceed by induction on the order of $next(s,t)$ when we go backwards in E'. The base step is trivial since E'' can be E' when $next(s,t)$ is the first transition. Assuming by induction that $next(s,t)$ is the n-th transition that is dependent on and occurs before τ' in E' and L_r admits such execution E'', we show that this also holds when $next(s,t)$ is $(n+1)$-th transition with the same properties. Let s'' be the state that is reached from s by executing E'_p which is the prefix of E' until (not included) the last transition τ'' that is dependent on and occurs before τ' and hence, τ'' is enabled in s''. Using Proposition 3, as $tid(\tau')$ must be in T of s'', there must be another execution E''' from s'' such that τ''' occurs before τ'' where $act(\tau') - act(\tau''')$ and $tid(\tau') = tid(\tau''') = t'$. Since in the execution starting from s as E'_p and continues as E''', $next(s,t)$ is the n-th transition that is dependent on and occurs before τ''' (when we go backwards), Proposition 4 is correct by using induction assumption. As mentioned, this contradicts the assumption of $t' \notin T$ in proof of Proposition 2 and thus, T is a persistent set and a source set. By monotonicity of source sets, $s.\mathbf{backtrack}$ is also a source set. $\qquad \square$

4   Lazy Source Set POR (DL-S-POR)

Algorithm 2 tracks dependencies between transitions in an eager manner, i.e., every new transition leads to updates of **current** sets. In this section, we present a lazy variation that computes such dependencies only when the exploration backtracks to a state. The incentive is to compute such dependencies only when needed to decide if the exploration from a given state should continue or not. Also, this enables several optimizations when traversing the state space to compute such dependencies that are not possible in the eager version.

Algorithm 3: LAZY SOURCE SET POR (DL-S-POR)

Initialize: $Stack \leftarrow \emptyset$; $Stack.push(s_I)$; $L_r \leftarrow \emptyset$;

1   **Explore()**
2    $s \leftarrow Stack$.top;
3   s.**backtrack** $\leftarrow \emptyset$; s.**done** $\leftarrow \emptyset$; s.**current** $\leftarrow \emptyset$;
4   **while** $true$ **do**
5   **if** $\exists t_1 \in s$.**backtrack**$\backslash s$.**done** **then**
6        $t \leftarrow t_1$
7   **else**
8   choose $t \in safeSet(s)\backslash s$.**done**
9      $(s, a, s') = next(s, t)$;
10     $Stack.push(s')$;
11   **if** $notVisited(s')$ **then**
12      $\lfloor$ **Explore()**;
13   **if** **IsComplete**(s) **then**
14      $Stack.pop()$;
15   **return**
16     $Stack.pop()$;

17   **IsComplete(s)**
18   **forall the** $(s, a, s') \in \{s' \in L_r : t = tid(a) \notin s$.**done**$\}$ **do**
19   s.**done** $= s$.**done** $\cup \{t\}$;
20      $T \leftarrow safeSet(s)$;
21   **if** s.**done** $= T \vee (\forall t' \in T : isVisited(succ(s, t')) \vee t' \in s$.**done**$)$
      then
22   add transitions (s, a, s') to L_r;
23   s.**done** $\leftarrow T$;
24   s.**backtrack** $\leftarrow s$.**done**;
25   **return** true;
26   s.**current**$[t] \leftarrow \{t\}$;
27      $A_{s'} \leftarrow \{a' : a'$ occurs in an execution from $E(L_r, s')\}$;
28   s.**current**$[t] = s$.**current**$[t] \cup \{tid(a') : a' \in A_{s'}$ and $a \nsim a'\})$;
29   s.**backtrack** $\leftarrow UpdateBack(s, a)$;
30   **if** s.**backtrack** $= s$.**done** **then**
31   **return** true;
32   **return** false;

Algorithm 3 presents our POR algorithm based on a lazy computation of source sets. Rather than updating the **current** sets on-the-fly for states on the stack, this algorithm re-traverses part of the state space each time it backtracks to a state s in order to update just **current** sets of s. This is done in the function **IsComplete**. As a result, s.**done** is populated with a new thread t just before computing dependencies with t's transition in s and not after executing that transition in the style of Algorithm 2 (see line 19). For every transition τ

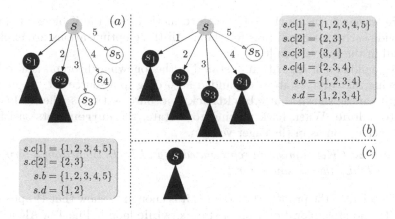

Fig. 4. An example exploration of Lazy Source Set POR. We use the same conventions as in Fig. 3.

of a thread t from s that has been followed since the last time the algorithm backtracks to s (i.e., t is not already in s.**done** – see line 18), the algorithm updates s.**current**$[t]$ to include all threads t' that later execute a transition that is dependent on τ (see lines 26–28). Subsequently, the s.**backtrack** set is updated exactly in the style of Algorithm 2 (see line 29). If s.**backtrack** becomes equal to s.**done** then **IsComplete** returns *true* and the exploration from s stops.

We explain how Algorithm 3 works by an example. Figure 4(a) illustrates a scenario in which the exploration backtracks to a state s for the second time. After the first backtrack to s, the state space starting from the successor s_1 (resulted from following a transition of thread 1) was re-traversed in order to compute s.**current**$[1]$. We assume that s.**current**$[1]$ is changed to $\{1,2,3,4,5\}$ due to the dependent transitions encountered during this traversal. The set s.**backtrack** is set to s.**current**$[1]$ as the latter contains all the enabled transitions. The exploration continues with a transition from s of thread 2 which is possible because thread 2 is in s.**backtrack**\s.**done**. After backtracking to s for the second time, the re-traversal of the state space starting in s_2 leads to s.**current**$[2] = \{2,3\}$. The set s.**backtrack** remains the same after this computation. Then, in Fig. 4(b), when backtracking to s for the fourth time, we assume that s.**current**$[3] = \{3,4\}$ and s.**current**$[4] = \{2,3,4\}$. Since transitions of threads 2, 3, and 4 starting in s are independent of transitions of other threads that occur later, we can conclude that $\{2,3,4\}$ is a persistent set and $\{1,2,3,4\}$ is a source set, and update s.**backtrack** to s.**done**. Therefore, the exploration from s stops, as pictured in Fig. 4(c). The set of transitions explored from s corresponds to a source set which is not a persistent set. The only persistent set that includes thread 1 is $\{1,2,3,4,5\}$.

In both DE-S-POR and DL-S-POR, we optimize re-traversals after some state s (computing A_s at line 16 and line 27, respectively) by not traversing all the executions after s but just traversing each transition after s only once. DL-S-POR is also amenable to other optimizations that are not possible or difficult to implement for DE-S-POR. These optimizations for DL-S-POR either prevent some re-traversals inside the **IsComplete** method (see the if block at line 21) or provide early exit conditions for them. All these optimizations are explained in detail in the full version [6].

The soundness of Algorithm 3, stated in the following theorem, is also based on proving that every state is expanded according to a source set. As in Theorem 1, it can be shown that s.**backtrack** is a source set for s when it becomes equal to s.**done**. When backtracking to a state, the **current** sets satisfy the same specification as in the eager version.

Theorem 2. *Given a program represented by an LTS L, Algorithm 3 terminates with an LTS L_r that is sound for L.*

Proof. Similar to the proof of Theorem 1, it is enough to show that Proposition 1 holds. To end an exploration from a state s, **while** loop in line 4 of Algorithm 3 must be terminated. For this, **return** statement in line 15 must be reached and therefore, **IsComplete** in line 13 should return true. First, we show that **IsComplete** method eventually returns true. Due to to method *UpdateBack*, we know that s.**backtrack** can not contain a thread $t \notin safeSet(s)$. Hence, all the transitions of s that can be executed are only from threads in $safeSet(s)$ because of lines 5–8. Each time a transition of s is executed and then the search backtracks to s, **IsComplete**(s) is initiated. By the **for** loop in line 18, every transition from s that is executed is considered and by line 19, we know that all these transitions will be added to s.**done**. Thus, s.**done** eventually becomes equal to $safeSet(s)$ which satisfies the condition in line 21 and as a result, **IsComplete** method returns true.

For **IsComplete** method to return true, either condition in line 21 or line 30 should be satisfied, where in both conditions s.**backtrack** must be equal to s.**done** before the return statement. That's why, equality of s.**done** and s.**backtrack** is the only condition for stopping an exploration from a state s. Similar to Algorithm 2, since s.**done** keep tracks of threads whose enabled transitions from s is already executed, the proof is deduced to showing that Proposition 2 holds for Algorithm 3 as well. The part between Proposition 2 and Proposition 3 in the proof of Theorem 1 applies totally the same and we show that T is a persistent set for s in L using the fact that L_r admits such an execution E' as it is concluded in the same proof. Assume by contradiction that T is not a persistent set for s. But as a result of backtracking to s after executing E' and invoking **IsComplete** method, t' will be added to s.**current**$[t]$ (and eventually to T due to *UpdateBack* method) in line 28 of Algorithm 3 since the transition label of τ' is an element of $A_{s'}$ ($t' \notin T$ and $act(\tau) \curlywedge act(\tau')$) in line 27. Since it contradicts the assumption, T (and also s.**backtrack** by monotonicity of source sets) is a persistent set and a source set. □

5   Experimental Evaluation

We evaluate an implementation of the three algorithms S-POR, DE-S-POR, and DL-S-POR, presented in Sect. 3 and Sect. 4, in the context of the Java Pathfinder (JPF) model checker. As benchmark, we use bounded-size clients of Java concurrent data structures.

Implementation. We implement our algorithms as an extension of the DFSHeuristic class in JPF. To identify (in)visible actions (for computing safe sets), the only manual input is a list of class names that constitute the implementation of the concurrent data structure. The (in)visible transitions are automatically inferred from these class names and Java synchronization-related native methods used to implement compare-and-swap (CAS) for instance, which are all known. Every action reading or writing a field of an object in one of these classes, or which corresponds to a native method call are marked as visible (JPF makes it possible to parse the Bytecode instructions executed in a transition and determine the read/written object fields). Calls to the lock and unlock methods of a lock object are both considered as writes to the lock object, and therefore, visible. Any other action is considered as invisible. The dependency relation between visible actions is defined as usual, i.e., two actions that access the same object field, one of them being a write, are considered dependent. The way we define (in)visible actions is sound because the clients we consider do not contain additional computation. They simply call methods of the data structure (from different threads), the verification goal being related to combinations of return values observed in their executions.

Benchmarks. Our benchmark consists of bounded-size clients of 7 concurrent data structures from JDK8 or Synchrobench [21]: two set implementations based on *coarse-grain* and *fine-grain* locking, respectively (RWLockCoarseGrainedList-IntSet and OptimisticListSortedSetWaitFreeContains), a set implementation based on a binary search tree and CAS, a wrapper on top of java.util.concurrent. ConcurrentLinkedQueue, java.util.concurrent.ConcurrentHashMap and a wrapper on top of it, and a hash map implementation based on coarse-grain locking. Since these implementations update shared memory using compare-and-swap or guarded by locks, they are data-race free and the restriction to sequential consistency is sound.

To evaluate our algorithms, we sampled 75 clients of these data structures where each client calls add and remove methods from 3 threads. Each thread contains up to 5 calls. We varied the contention on shared objects using less or more distinct inputs for add and remove methods.

We also use a number of buggy variations of the lock-based sets, RWLock-CoarseGrainedListIntSet and OptimisticListSortedSetWaitFreeContains. We used Violat to generate client programs of these variations that admit consistency violations. Violat generates these client programs in three steps. First, Violat enumerates arbitrary test programs of a given data structure based on other inputs such as number of threads, maximum number of programs and so on. Next, it computes expected (ADT-admitted return-value) outcomes for each test program by computing and then recording the outcomes of all possible sequential executions. Finally, it runs the threads of each test program in parallel (using a stress testing tool or JPF), checks if the results are as expected, and reports the test programs that violates linearizability which is witnessed by observing an unexpected outcome.

To introduce bugs in the selected data structures before inputting them to Violat, we modify the placement of locks dynamically under certain conditions in certain methods (e.g., when the set contains a specific element). These conditions make it possible to control the difficulty of a bug. We consider four different classes of clients based on the number of invocations to methods that lead to bugs: (1) all of the invocations, (2) half of the invocations, (3) just a single invocation and (4) none of the invocations. We sampled 310 clients of these buggy variations with 3 threads and up to 4 calls per thread using Violat.

Results. We use S-POR, DL-S-POR, DE-S-POR to denote the three algorithms presented in this paper. For the same algorithms, we use JPF, DL-JPF, DE-JPF to represent the standard setup of JPF, and variations of the DL-S-POR and DE-S-POR when the safe set of a state s contains all the enabled threads in s ($safeSet(s) = enabled(s)$). The latter are used to evaluate the performance of the eager and lazy approaches while disabling the benefit of the static S-POR method. We compare implementations of S-POR, DL-S-POR and DE-S-POR between them, with JPF, DL-JPF and DE-JPF, with their stateless variations, and with a stateful variation of the optimal source set algorithm in [3] (called O-DPOR). For a fair comparison, we implement O-DPOR on top of S-POR without wakeup trees as their operations are quite expensive. The experiments were run on a 2,3 GHz Dual-Core Intel Core i5 processor with 8GB of RAM. We consider a timeout of 30 min.

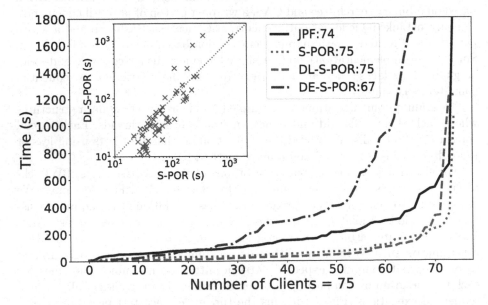

Fig. 5. Quantile plot of running times for S-POR, DL-S-POR, DE-S-POR and JPF (for each algorithm, clients are ordered w.r.t. time in ascending order). The top left part shows a scatter plot for comparing S-POR and DL-S-POR.

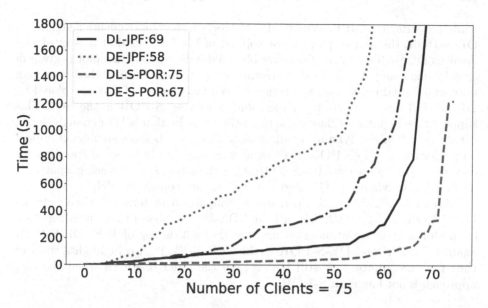

Fig. 6. Quantile plot of running times for DL-JPF, DE-JPF, DL-S-POR and DE-S-POR (for each algorithm, clients are ordered w.r.t. time in ascending order as in Fig. 5).

Execution Time Comparison. Figure 5 and Fig. 6 present a comparison in terms of execution time between different sets of algorithms. In Fig. 5, we compare JPF, S-POR, DL-S-POR and DE-S-POR to observe the advantages of using our algorithms against the standard setup of JPF. In Fig. 6, we compare DL-S-POR, DE-S-POR, DL-JPF and DE-JPF for investigating the gain by applying static filtering using S-POR as a baseline in dynamic algorithms. To ease the interpretation of the results, for each algorithm, we order clients according to execution time in ascending order. The numbers in the legend represent the number of clients on which a given algorithm terminates before the timeout. We omit O-DPOR because it times out for a large part of the benchmark, i.e., 39 out of the 46 clients on which it was run (our implementation of the algorithm in [3] does not support programs using locks which makes it inapplicable to the rest of the clients). This optimal algorithm manipulates happens-before constraints between steps in an execution, which results in a large overhead compared to our simpler tracking of pairwise dependencies. We also omit stateless variations of our algorithms since none of them finished before the timeout for any client. Note that stateless versions are obtained by disabling the state matching[1] in JPF, which also disables storing the full reachability graph.

Results based on Fig. 5 show that the lazy source set computation in DL-S-POR gives a significant speedup w.r.t. DE-S-POR (and intuitively O-DPOR)

---

[1] JPF uses hashing for state matching which is theoretically imperfect and can lead to incomplete results on rare occasions.

while outperforming JPF. While S-POR processes few more clients faster w.r.t. DL-S-POR, the scatter plot on the top-left of Fig. 5 shows that it is mostly in favor to DL-S-POR when clients are observed individually (this plot is given in logarithmic scale). DL-S-POR performs better than S-POR if there is a high potential for reduction, i.e., the ratio between the number of states explored by DL-S-POR over S-POR is smaller, and otherwise, S-POR is the best. This supports the hypothesis that if the potential for reduction is high enough then a carefully customized dynamic computation of source sets has a significant impact on performance. DL-S-POR gives an average speedup (average of speedups for each client) of 2.6 compared to S-POR. Overall picture suggests using a portfolio model checker where S-POR and DL-S-POR are run in parallel.

Similar to Fig. 5, Fig. 6 illustrates a comparison in terms of time between DL-S-POR, DE-S-POR, DL-JPF and DE-JPF. It shows that our algorithms outperforms their variations that are directly built on top of JPF (DL-S-POR against DL-JPF and DE-S-POR against DE-JPF). It also highlights the fact that the lazy approach is still better than the eager one even when the lazy approach is not based on S-POR.

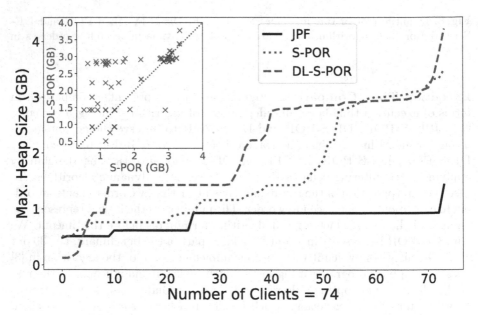

Fig. 7. Quantile plot of memory consumption for S-POR, DL-S-POR, and JPF (for each algorithm, clients are ordered w.r.t. memory in ascending order). The top left part shows a scatter plot for comparing S-POR and DL-S-POR.

Memory Consumption Comparison. Figure 7 presents a comparison in terms of memory consumption between S-POR and DL-S-POR, the most efficient algorithms according to Fig. 5 and Fig. 6, against the standard setup of JPF. We compared the maximum heap sizes using 74 clients that terminate before timeout for all algorithms. In all the experiments, the highest allocated

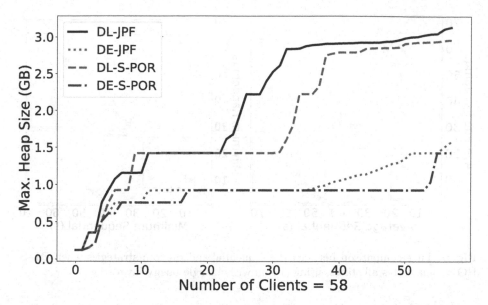

Fig. 8. Quantile plot of memory consumption for DL-JPF, DE-JPF, DL-S-POR and DE-S-POR (for each algorithm, clients are ordered w.r.t. memory in ascending order as in Fig. 7).

heap size is 4.2 GB. S-POR and DL-S-POR consume more memory than JPF because they have to store the transition labels which are used to reduce the explored state space. This overhead is unavoidable for any form of dynamic partial order reduction. However, this memory consumption overhead is counterbalanced by significant speedups in terms of time. There is some memory overhead also due to storing the sets of transition labels manipulated by the algorithms, e.g., s.**current**. But since these sets are maintained only for irreducible states and they are deleted for a state s when s.**done** equals s.**backtrack**, their effects are not significant as storing transition labels.

For 32% of the clients, S-POR and DL-S-POR consume at most twice the memory consumed by JPF. For these clients, the average memory overhead is 1.00 for S-POR and 1.34 for DL-S-POR while the average speedup against JPF is 2.54 and 6.67, respectively. For 50% of the clients, S-POR and DL-S-POR consume in between 2 and 4 times the memory used by JPF. The average memory overhead for these clients is 2.20 for S-POR and 2.63 for DL-S-POR while the average speedup is 2.86 and 7.81, respectively. For the rest of the clients, the memory overhead is at most 7.79 and in average 4.11 for S-POR and 5.39 for DL-S-POR while the average speedup 3.28 and 5.31, respectively.

The top-left part of Fig. 7 shows a pair-wise comparison of allocated maximum heap sizes in S-POR and DL-S-POR. These algorithms are incomparable in general. After investigating the clients individually, the results confirm that DL-S-POR consumes less memory than S-POR when there is a high potential for reduction. The memory consumed for computing source sets is compensated by the reduction in the state space.

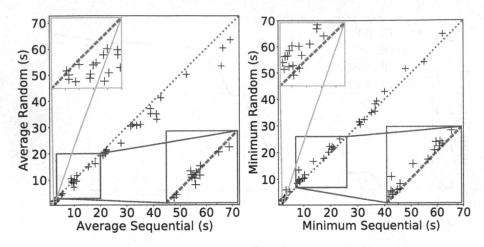

Fig. 9. Time comparison between the sequential and random strategies when DL-S-POR enumerates all states, using clients with a single buggy invocation.

When we take a look at to Fig. 8, it illustrates a comparison between DL-S-POR, DE-S-POR, DL-JPF and DE-JPF as in Fig. 5, but in terms of memory consumption. As in Fig. 7, we compared the maximum heap sizes of clients that finished before it timed out for all algorithms, which are 58 of them. It demonstrates that our algorithms are slightly better than their variations that are directly built on top of JPF. This memory overhead is mainly because of the additional transition labels that are not removed by the static filter. The overhead is also due to storing the sets of transition labels manipulated by the algorithms for all of the states rather than just for the irreducible ones but as mentioned previously, this overhead is negligible. This figure also shows that tracking dependencies with an eager approach does not increase the heap size as much as the lazy approach, although they explore the same state space. This difference in the memory overhead can not be explained by the memory that is used for storing the sets of transition labels or LTSs as they are all the same for both algorithms and they are kept in the same data structures. We suspect that this overhead can be due to low-level, internal details of JPF or due to the garbage collection process which might not keep up.

Transition Enumeration. The performance of dynamic POR algorithms is generally affected by the order in which transitions starting in a certain state are enumerated. This order influences the size of the computed persistent/source sets. This order is also important when enumerating states only until the first error is detected. We evaluate two strategies for defining this order, called *sequential* and *random*. For both of these strategies, the algorithm first selects a transition that leads to an already visited state, if one exists. We made this choice because it leads to better performance (this is adopted by the standard setup of JPF as well). In the sequential strategy, if there is no such transition, then the algorithm picks a transition by respecting a pre-defined order between the thread ids. In random, the next transition is selected uniformly at random.

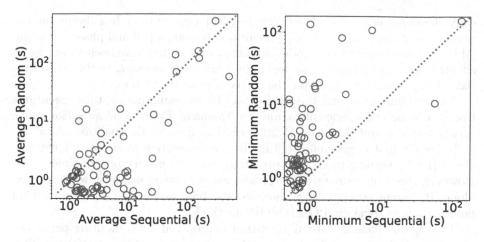

Fig. 10. Time comparison (log scale) between the sequential and random strategies when DL-S-POR enumerates states only until the first error, using clients in which all the invocations are buggy.

We ran S-POR and DL-S-POR, the best algorithms as shown above, with all 6 permutations of the 3 threads for sequential and 3 different seeds for random. For each strategy, we report the average and minimum time over different instances.

We report on the impact of these enumeration strategies for DL-S-POR when computing *all* reachable states (up to POR) in Fig. 9 and *only until the first error* in Fig. 10. For S-POR, the enumeration strategy is not important for the first case (since there is no dynamic computation of persistent/source sets) and it has a similar effect as for DL-S-POR in the second case. These figures consider clients of buggy libraries where a single or all method invocations are buggy. The rest of the cases are presented in the full version [6] and are similar.

The results show that the random strategy performs better in average, shown on the left of Fig. 9 and 10, but worse w.r.t. minima, shown on the right of Fig. 9 and 10.

The differences are more significant when enumerating states only until the first error. Figure 10 reporting on this case is given in logarithmic scale. Thus, the sequential strategy should be preferred when using a portfolio model checker, i.e., parallel runs for each permutation of thread ids, and otherwise, the random strategy is better. This follows also from the average standard deviation being 28 s for random and 60 s for sequential, where the means are 17 and 20 s, resp. Note that, there is no significant impact observed from changing the algorithms or the number of buggy method invocations in the clients.

6 Related Work

Over the years various different techniques have been introduced to deal with the state explosion problem in model checking. For concurrent programs specifically, depth bounding [17], delay bounding [10], context bounding (bounding the number of context switches) [36], preemption bounding [32] and phase bounding [5] bring tractability to the model checking problem and have been shown to be effective for bug finding. These techniques are all *incomplete*, in the sense that lack of bugs does not guarantee the correctness of the system.

POR techniques reduce the search space by not exploring multiple executions from the same equivalence class, and are *complete*. Early techniques like *ample sets* [7,24] and *stubborn sets* [15,20] were based on static analysis. *Sleep sets* [15] were the first to guarantee optimality (one execution from each equivalence class) [18] by keeping track of information from the history of the exploration. However, they only prune transitions and cannot eliminate any state when used alone. Persistent sets [19,25] generalized stubborn and ample sets and enabled development of dynamic POR (DPOR) methods.

In [13], an efficient stateful algorithm is proposed for computing persistent sets dynamically by considering currently explored parts of the state space. This algorithm needed large memory for keeping discovered states and the happens-before relation. The algorithm is improved in [42] with a more efficient state representation, and in [43] with a summary-based representation of the happens before.

In [28,39], stateless dynamic POR techniques were introduced. *Source sets* [3] were introduced in the context of dynamic POR techniques such that the state space can be reduced up to the limit that is theoretically possible. They are generalizations of persistent sets and their relation with persistent sets are investigated in [2]. Our DE-S-POR and DL-S-POR algorithms are relying on source sets but operate in the context of stateful model checking. The technique from [33] is similar to our S-POR algorithm for the GPU setting, but their choice of invisible actions is different than ours.

While we focus on shared-memory programs running on top of a sequential consistency memory model, POR techniques have been also investigated in the context of weak memory models such as TSO or C11, e.g., [1,4,26,27].

7   Conclusions

We proposed two algorithms for stateful model checking based on POR which build on the recently proposed source sets. Our algorithms focus on overall performance instead of theoretical optimality. Their evaluation in the context of JPF shows that they outperform a theoretically optimal algorithm [3], and a simple static POR algorithm when there is a big enough potential for reducing the state space. This suggests that an effective model checker would have to run S-POR and DL-S-POR in parallel, and depending on the amount of parallelism resources available, with different instances of the sequential or random strategies for enumerating transitions starting in a certain state.

Reductions based on Mazurkiewicz trace equivalence [30] has also been used in proof simplification for concurrent verification [12] hypersafety verification [11]. A relevant problem of interest is the automatic inference of inductive invariants [14,31,38] that prove the correctness of concurrent libraries. The sound

LTS's that are computed in this paper for the verification of individual instances, each provide a data point in what the inductive invariant for a most general client may look like. The key observation is that the inductive invariant for a most general client under some reduction scheme may be substantially simpler than an inductive invariant for all executions of the most general client. An interesting direction of future work is to investigate if the results of a sequence of individual client tests can be generalized to the discovery of a complete invariant (and hence a complete proof) under an appropriate reduction scheme. DL-S-POR provides an efficient way to produce data for a data-driven inference algorithm.

References

1. Abdulla, P.A., Aronis, S., Atig, M.F., Jonsson, B., Leonardsson, C., Sagonas, K.: Stateless model checking for TSO and PSO. Acta Inform. **54**(8), 789–818 (2017). https://doi.org/10.1007/s00236-016-0275-0

2. Abdulla, P., Aronis, S., Jonsson, B., Sagonas, K.: Comparing source sets and persistent sets for partial order reduction. In: Aceto, L., Bacci, G., Bacci, G., Ingólfsdóttir, A., Legay, A., Mardare, R. (eds.) Models, Algorithms, Logics and Tools. LNCS, vol. 10460, pp. 516–536. Springer, Cham (2017). https://doi.org/10.1007/978-3-319-63121-9_26

3. Abdulla, P.A., Aronis, S., Jonsson, B., Sagonas, K.: Source sets: a foundation for optimal dynamic partial order reduction. J. ACM **64**(4), 25:1–25:49 (2017). https://doi.org/10.1145/3073408

4. Abdulla, P.A., Atig, M.F., Jonsson, B., Ngo, T.P.: Optimal stateless model checking under the release-acquire semantics. Proc. ACM Program. Lang. **2**(OOPSLA), 135:1–135:29 (2018). https://doi.org/10.1145/3276505

5. Bouajjani, A., Emmi, M.: Bounded phase analysis of message-passing programs. In: Flanagan, C., König, B. (eds.) TACAS 2012. LNCS, vol. 7214, pp. 451–465. Springer, Heidelberg (2012). https://doi.org/10.1007/978-3-642-28756-5_31

6. Cirisci, B., Enea, C., Farzan, A., Mutluergil, S.O.: A pragmatic approach to stateful partial order reduction (2022). https://arxiv.org/abs/2211.11942

7. Clarke, E.M., Emerson, E.A., Sistla, A.P.: Automatic verification of finite state concurrent systems using temporal logic specifications: a practical approach. In: Wright, J.R., Landweber, L., Demers, A.J., Teitelbaum, T. (eds.) Conference Record of the Tenth Annual ACM Symposium on Principles of Programming Languages, Austin, Texas, USA, January 1983, pp. 117–126. ACM Press (1983). https://doi.org/10.1145/567067.567080

8. Clarke, E.M., Grumberg, O., Minea, M., Peled, D.A.: State space reduction using partial order techniques. Int. J. Softw. Tools Technol. Transf. **2**(3), 279–287 (1999)

9. Emmi, M., Enea, C.: Violat: generating tests of observational refinement for concurrent objects. In: Dillig, I., Tasiran, S. (eds.) CAV 2019, Part II. LNCS, vol. 11562, pp. 534–546. Springer, Cham (2019). https://doi.org/10.1007/978-3-030-25543-5_30

10. Emmi, M., Qadeer, S., Rakamaric, Z.: Delay-bounded scheduling. In: Ball, T., Sagiv, M. (eds.) Proceedings of the 38th ACM SIGPLAN-SIGACT Symposium on Principles of Programming Languages, POPL 2011, Austin, TX, USA, 26–28 January 2011, pp. 411–422. ACM (2011). https://doi.org/10.1145/1926385.1926432

11. Farzan, A., Vandikas, A.: Automated hypersafety verification. In: Dillig, I., Tasiran, S. (eds.) CAV 2019, Part I. LNCS, vol. 11561, pp. 200–218. Springer, Cham (2019). https://doi.org/10.1007/978-3-030-25540-4_11

12. Farzan, A., Vandikas, A.: Reductions for safety proofs. Proc. ACM Program. Lang. 4(POPL), 13:1–13:28 (2020). https://doi.org/10.1145/3371081

13. Flanagan, C., Godefroid, P.: Dynamic partial-order reduction for model checking software. In: Palsberg, J., Abadi, M. (eds.) Proceedings of the 32nd ACM SIGPLAN-SIGACT Symposium on Principles of Programming Languages, POPL 2005, Long Beach, California, USA, 12–14 January 2005, pp. 110–121. ACM (2005). https://doi.org/10.1145/1040305.1040315

14. Garg, P., Neider, D., Madhusudan, P., Roth, D.: Learning invariants using decision trees and implication counterexamples. In: Bodík, R., Majumdar, R. (eds.) Proceedings of the 43rd Annual ACM SIGPLAN-SIGACT Symposium on Principles of Programming Languages, POPL 2016, St. Petersburg, FL, USA, 20–22 January 2016, pp. 499–512. ACM (2016). https://doi.org/10.1145/2837614.2837664

15. Godefroid, P.: Using partial orders to improve automatic verification methods. In: Clarke, E.M., Kurshan, R.P. (eds.) Computer-Aided Verification, Proceedings of a DIMACS Workshop 1990, New Brunswick, New Jersey, USA, 18–21 June 1990. DIMACS Series in Discrete Mathematics and Theoretical Computer Science, vol. 3, pp. 321–340. DIMACS/AMS (1990). https://doi.org/10.1090/dimacs/003/21

16. Godefroid, P.: Partial-Order Methods for the Verification of Concurrent Systems - An Approach to the State-Explosion Problem. Lecture Notes in Computer Science, vol. 1032. Springer, Heidelberg (1996). https://doi.org/10.1007/3-540-60761-7_31

17. Godefroid, P.: Model checking for programming languages using verisoft. In: Lee, P., Henglein, F., Jones, N.D. (eds.) Conference Record of POPL 1997: The 24th ACM SIGPLAN-SIGACT Symposium on Principles of Programming Languages, Papers Presented at the Symposium, Paris, France, 15–17 January 1997, pp. 174–186. ACM Press (1997). https://doi.org/10.1145/263699.263717

18. Godefroid, P., Holzmann, G.J., Pirottin, D.: State-space caching revisited. Formal Methods Syst. Des. 7(3), 227–241 (1995). https://doi.org/10.1007/BF01384077

19. Godefroid, P., Pirottin, D.: Refining dependencies improves partial-order verification methods (extended abstract). In: Courcoubetis, C. (ed.) CAV 1993. LNCS, vol. 697, pp. 438–449. Springer, Heidelberg (1993). https://doi.org/10.1007/3-540-56922-7_36

20. Godefroid, P., Wolper, P.: Using partial orders for the efficient verification of deadlock freedom and safety properties. Formal Methods Syst. Des. 2(2), 149–164 (1993)

21. Gramoli, V.: More than you ever wanted to know about synchronization: synchrobench, measuring the impact of the synchronization on concurrent algorithms. In: Cohen, A., Grove, D. (eds.) Proceedings of the 20th ACM SIGPLAN Symposium on Principles and Practice of Parallel Programming, PPoPP 2015, San Francisco, CA, USA, 7–11 February 2015, pp. 1–10. ACM (2015). https://doi.org/10.1145/2688500.2688501

22. He, J., Hoare, C.A.R., Sanders, J.W.: Data refinement refined resume. In: Robinet, B., Wilhelm, R. (eds.) ESOP 1986. LNCS, vol. 213, pp. 187–196. Springer, Heidelberg (1986). https://doi.org/10.1007/3-540-16442-1_14

23. Hoare, C.A.R., He, J., Sanders, J.W.: Prespecification in data refinement. Inf. Process. Lett. 25(2), 71–76 (1987)

24. Holzmann, G.J., Peled, D.A.: An improvement in formal verification. In: Hogrefe, D., Leue, S. (eds.) Formal Description Techniques VII, Proceedings of the 7th

IFIP WG6.1 International Conference on Formal Description Techniques, Berne, Switzerland, 1994. IFIP Conference Proceedings, vol. 6, pp. 197–211. Chapman & Hall (1994)

25. Katz, S., Peled, D.A.: Verification of distributed programs using representative interleaving sequences. Distrib. Comput. **6**(2), 107–120 (1992)

26. Kokologiannakis, M., Vafeiadis, V.: HMC: model checking for hardware memory models. In: Larus, J.R., Ceze, L., Strauss, K. (eds.) ASPLOS 2020: Architectural Support for Programming Languages and Operating Systems, Lausanne, Switzerland, 16–20 March 2020, pp. 1157–1171. ACM (2020). https://doi.org/10.1145/3373376.3378480

27. Kokologiannakis, M., Vafeiadis, V.: GENMC: a model checker for weak memory models. In: Silva, A., Leino, K.R.M. (eds.) CAV 2021, Part I. LNCS, vol. 12759, pp. 427–440. Springer, Cham (2021). https://doi.org/10.1007/978-3-030-81685-8_20

28. Lauterburg, S., Karmani, R.K., Marinov, D., Agha, G.: Evaluating ordering heuristics for dynamic partial-order reduction techniques. In: Rosenblum, D.S., Taentzer, G. (eds.) FASE 2010. LNCS, vol. 6013, pp. 308–322. Springer, Heidelberg (2010). https://doi.org/10.1007/978-3-642-12029-9_22

29. Liskov, B., Wing, J.M.: A behavioral notion of subtyping. ACM Trans. Program. Lang. Syst. **16**(6), 1811–1841 (1994)

30. Mazurkiewicz, A.: Trace theory. In: Brauer, W., Reisig, W., Rozenberg, G. (eds.) ACPN 1986, Part II. LNCS, vol. 255, pp. 278–324. Springer, Heidelberg (1987). https://doi.org/10.1007/3-540-17906-2_30

31. Miltner, A., Padhi, S., Millstein, T.D., Walker, D.: Data-driven inference of representation invariants. In: Donaldson, A.F., Torlak, E. (eds.) Proceedings of the 41st ACM SIGPLAN International Conference on Programming Language Design and Implementation, PLDI 2020, London, UK, 15–20 June 2020, pp. 1–15. ACM (2020). https://doi.org/10.1145/3385412.3385967

32. Musuvathi, M., Qadeer, S.: Iterative context bounding for systematic testing of multithreaded programs. In: Ferrante, J., McKinley, K.S. (eds.) Proceedings of the ACM SIGPLAN 2007 Conference on Programming Language Design and Implementation, San Diego, California, USA, 10–13 June 2007, pp. 446–455. ACM (2007). https://doi.org/10.1145/1250734.1250785

33. Neele, T., Wijs, A., Bošnački, D., van de Pol, J.: Partial-order reduction for GPU model checking. In: Artho, C., Legay, A., Peled, D. (eds.) ATVA 2016. LNCS, vol. 9938, pp. 357–374. Springer, Cham (2016). https://doi.org/10.1007/978-3-319-46520-3_23

34. Peled, D.: All from one, one for all: on model checking using representatives. In: Courcoubetis, C. (ed.) CAV 1993. LNCS, vol. 697, pp. 409–423. Springer, Heidelberg (1993). https://doi.org/10.1007/3-540-56922-7_34

35. Plotkin, G.D.: LCF considered as a programming language. Theor. Comput. Sci. **5**(3), 223–255 (1977)

36. Qadeer, S., Rehof, J.: Context-bounded model checking of concurrent software. In: Halbwachs, N., Zuck, L.D. (eds.) TACAS 2005. LNCS, vol. 3440, pp. 93–107. Springer, Heidelberg (2005). https://doi.org/10.1007/978-3-540-31980-1_7

37. Queille, J.P., Sifakis, J.: Specification and verification of concurrent systems in CESAR. In: Dezani-Ciancaglini, M., Montanari, U. (eds.) Programming 1982. LNCS, vol. 137, pp. 337–351. Springer, Heidelberg (1982). https://doi.org/10.1007/3-540-11494-7_22

38. Sharma, R., Aiken, A.: From invariant checking to invariant inference using randomized search. Formal Methods Syst. Des. **48**(3), 235–256 (2016)

39. Tasharofi, S., Karmani, R.K., Lauterburg, S., Legay, A., Marinov, D., Agha, G.: TransDPOR: a novel dynamic partial-order reduction technique for testing actor programs. In: Giese, H., Rosu, G. (eds.) FMOODS/FORTE -2012. LNCS, vol. 7273, pp. 219–234. Springer, Heidelberg (2012). https://doi.org/10.1007/978-3-642-30793-5_14

40. Valmari, A.: Stubborn sets for reduced state space generation. In: Rozenberg, G. (ed.) ICATPN 1989. LNCS, vol. 483, pp. 491–515. Springer, Heidelberg (1991). https://doi.org/10.1007/3-540-53863-1_36

41. Visser, W., Pasareanu, C.S., Khurshid, S.: Test input generation with java pathfinder. In: Avrunin, G.S., Rothermel, G. (eds.) Proceedings of the ACM/SIGSOFT International Symposium on Software Testing and Analysis, ISSTA 2004, Boston, Massachusetts, USA, 11–14 July 2004, pp. 97–107. ACM (2004). https://doi.org/10.1145/1007512.1007526

42. Yang, Y., Chen, X., Gopalakrishnan, G., Kirby, R.M.: Efficient stateful dynamic partial order reduction. In: Havelund, K., Majumdar, R., Palsberg, J. (eds.) SPIN 2008. LNCS, vol. 5156, pp. 288–305. Springer, Heidelberg (2008). https://doi.org/10.1007/978-3-540-85114-1_20

43. Yi, X., Wang, J., Yang, X.: Stateful dynamic partial-order reduction. In: Liu, Z., He, J. (eds.) ICFEM 2006. LNCS, vol. 4260, pp. 149–167. Springer, Heidelberg (2006). https://doi.org/10.1007/11901433_9

Compositional Verification of Stigmergic Collective Systems

Luca Di Stefano[1,2]($\boxtimes$) (ID) and Frédéric Lang[1] (ID)

[1] Univ. Grenoble Alpes, Inria, CNRS, Grenoble INP, LIG, Grenoble, France
[2] University of Gothenburg, Gothenburg, Sweden
luca.di.stefano@gu.se

Abstract. Collective adaptive systems may be broadly defined as ensembles of autonomous agents, whose interaction may lead to the emergence of global features and patterns. Formal verification may provide strong guarantees about the emergence of these features, but may suffer from scalability issues caused by state space explosion. Compositional verification techniques, whereby the state space of a system is generated by combining (an abstraction of) those of its components, have shown to be a promising countermeasure to the state space explosion problem. Therefore, in this work we apply these techniques to the problem of verifying collective adaptive systems with stigmergic interaction. Specifically, we automatically encode these systems into networks of LNT processes, apply a static value analysis to prune the state space of individual agents, and then reuse compositional verification procedures provided by the CADP toolbox. We demonstrate the effectiveness of our approach by verifying a collection of representative systems.

1 Introduction

In a collective adaptive system, autonomous individuals or *agents* interact with each other according to simple local rules, which may lead to the emergence of global features and patterns despite the lack of centralized coordination [32]. Using these systems as a modelling framework to study complex phenomena, such as the spread of diseases through a social network [14], the role of spatial constraints in an economy [40], or the evolution of an ecosystem [30], appears to be a trending research methodology. Depending on the field of application, the resulting models are variously referred to as individual- or agent-based models, in silico cell models, or multi-agent systems, but they all share the essential traits of collective adaptive systems.

This increasing popularity owes to the fact that, under such a framework, one can easily specify heterogeneous agents with stateful, nonlinear or discontinuous behavioural rules [3]; additionally, one can easily refine these specifications if they turn up to be incomplete or incorrect, e.g., if undesired behaviour is

Work partially funded by ERC consolidator grant no. 772459 *D-SynMA* (Distributed Synthesis: from Single to Multiple Agents).

observed when simulating the model. However, simulations are unsuitable to achieve strong confidence in the *correctness* of a model of this kind. In fact, even small collective adaptive systems may evolve in a multitude of different ways, which increases exponentially in the number of agents and the complexity of their behaviour. This *state space explosion* problem means that simulations and testing may only cover a small portion of all feasible evolutions that such models can exhibit: therefore, attempts at uncovering unexpected or problematic behaviour by these means are likely to fail.

Formal verification techniques, in principle, may provide the correctness guarantees that are out of the reach of simulation-based analysis, but they also suffer from complexity issues related to the state space explosion problem. *Compositional* techniques, essentially based on a divide-and-conquer strategy to break down the analysis of large systems, appear to be a general, effective approach to mitigate the state space explosion problem [18]. To support this claim, in this work we introduce a fully-automated workflow to perform compositional verification of *stigmergic* collective systems specified in a high-level language called LAbS [7]. In these systems, agents do not interact directly with each other, but rather share information by manipulating a shared medium called a virtual stigmergy [41]. The concept of stigmergies originates from biology, where it has been used to explain the collective behaviour of social insects such as ants, termites, and bees [47], but appears well-suited to describe a much wider range of phenomena, including the creation and curation of content on the Wikipedia collaborative encyclopedia [6], or the development of open-source software [45]. The indirect and asynchronous nature of this interaction mechanism induces vast state spaces even in modestly-sized systems, making their verification challenging [10,12]. However, by combining compositional state space generation with a static value analysis that allows us to prune the state space of individual agents, we are able to verify a collection of example systems with significant gains over a non-compositional model checking procedure.

The rest of this paper is organized as follows. Section 2 outlines the specification language that we intend to verify, and provides the necessary background about verification tools, techniques, and abstract domains that are relevant to our work; it also includes an example of a stigmergic system that we will recurringly use to illustrate our approach. Section 3 discusses the encoding of LAbS agents into LNT processes, and how these processes are composed into a parallel program that emulates the agents' evolution and interactions. Section 4 introduces a static value analysis that helps us prevent state-space explosion as we generate the state space of individual agents. Section 5 describes the implementation of our approach and its experimental evaluation over a collection of LAbS examples. Section 6 discusses related work. Lastly, Sect. 7 contains our conclusions and potential directions for future work.

2   Background

In this section, we provide an overview of concepts that will be referred to in the rest of the paper. First, we will introduce the LAbS language for stigmergic collective systems, as well as a running example to demonstrate the peculiarities of these systems. Then, we will describe the *Intervals* and *Powerset of Intervals* abstract domains; the CADP analysis platform and the LNT process calculus; and some notions related to compositional state space generation.

Stigmergic Collective Systems and LAbS. The LAbS language [7] is a high-level formalism to specify stigmergic collective systems. Agents in a LAbS system cannot explicitly exchange messages with each other: rather, they assign values to specific local variables, which we call *stigmergic variables*. After an assignment to one of these variables, an agent will asynchronously diffuse the assigned value among its neighbours by sending out a *put-message*. All assignments are timestamped, and the receivers of a *put*-message with a newer value will update their own local copy of the variable to that value. Upon receiving a more recent value, agents also help propagate it by forwarding the *put*-message to their own neighbours. Similarly, after accessing the value of a stigmergic variable, an agent will asynchronously check whether someone among its neighbours has a newer value for that variable, by sending a *qry-message*. Neighbours react to the query by sending out a new *put*-message containing either their own value for the variable or the received one, depending on which is newer. These simple mechanisms allow local information to spread from one agent to the others, and new information to replace older data.

In LAbS, the definition of an agent's neighbourhood is not fixed: in fact, the language allows to equip stigmergic variables with *link predicates* to customize this concept. A link predicate is a Boolean function over the state of a sender and a (potential) receiver. Whenever an agent sends a message regarding a given variable x, this message will be received by all the agents that (together with the sender) satisfy x's link predicate. These agents are the *neighbours* of the sender with respect to x's predicate. This feature makes LAbS quite flexible, as it allows modelling different capabilities among the agents, such as their communication range, or having privileged access to some variables, and so on; it also induces neighbourhoods that may vary as the system evolves.

Running Example: Stigmergic Bully Election. A bully election is a simple protocol to elect a leader in a distributed system [24]. The protocol assumes that each node in the system has a fixed, unique numeric identifier (id) in the range $0..N - 1$, where N is the number of nodes. Intuitively, each node in the system initially considers itself the leader, and advertises this by broadcasting its id to the rest of the system. However, a node that receives a message with an id i lower than their own will instead regard node i as its new leader. When this happens, the node also stops advertising itself, but keeps changing the leader every time it receives a message with a lower id. This protocol eventually makes all nodes agree that the one with the lowest id is the leader.

Replicating such a protocol in a stigmergic system is not immediate, as agents have no primitive to explicitly exchange messages with one another. However, we can let them manipulate a stigmergic variable **leader** until they reach a consensus on its value. Essentially, each node only needs to check whether **leader** is currently higher than its own id. If it is, it means that the node still has a

Listing 1. A sketch of a stigmergic election system in LAbS.

```
1  system {                    8  agent Node {
2    spawn = Node: N            9    stigmergies = Election
3  }                          10    Behavior =
4  stigmergy Election {       11      leader > id ->
5    link = true              12        leader <~ id;
6    leader: N                13      Behavior
7  }                          14  }
```

chance of becoming the leader: so, the node assigns its own id to `leader`. As the
link predicate for `leader`, we use the one that is always satisfied: this induces
a broadcast communication model, i.e., every time a node assigns a value to
`leader`, this will be (asynchronously) diffused to every other node in the system.
Every time a value j gets diffused in this way, it immediately puts all nodes with
id $i > j$ out of the race. We can speculate that, eventually, every node gets out
of the election except the one with the lowest id, and all nodes have that id
assigned to `leader`.[1]

Listing 1 shows how such a protocol can be expressed in LAbS. The code
specifies that the system is composed of N agents of type `Node`; declares a
stigmergy (i.e., a collection of stigmergic variables with the same link predicate)
`Election`, equipped with the always-satisfied link predicate `true` and containing
a single variable `leader`, which is initialised to N; and finally specifies the `Node`
type. Namely, each `Node` participates in the `Election` stigmergy, meaning that
it will have a local copy of the `leader` variable. Its behaviour is a guarded
recursive process: a guard blocks the agent until the value of `leader` is greater
than its identifier `id`. When this is the case, the agent assigns `id` to `leader`
and then starts over. (The `<~` operator denotes an assignment to a stigmergic
variable).

Intervals and Their Powersets. For our purpose, an *interval* is either the empty
interval $\perp$ or a pair $[a, b]$, with $a \in \mathbb{R} \cup \{-\infty\}$, $b \in \mathbb{R} \cup \{\infty\}$, and $a \leq b$; we do
not need open-bounded intervals, which are excluded from our definition. Intu-
itively, an interval-based value analysis [5] starts from an initial *abstract state*
s_0 of the program under analysis, i.e., a mapping from program variables to
intervals. The precise way in which s_0 is computed depends on the semantics of
the language; generally, variables that are initialized to a constant κ are mapped
to the singleton interval $[\kappa, \kappa]$, while nondeterministic variables, e.g., those representing inputs to the program, are mapped to $[-\infty, \infty]$, meaning that they
may initially assume any value. The analysis then explores the abstract states
that are reachable from s_0 by performing an abstract interpretation of the program. As an example, Fig. 1a shows a function $[\![e]\!](s)$, defined by induction on
the structure of a very simple expression language, to evaluate an expression e
on an abstract state s. An integer constant κ evaluates to the single-element
interval $[\kappa, \kappa]$. A reference to a variable x evaluates to the interval $s(x)$. For a
binary operation $e_1 \circ e_2$, one evaluates e_1, e_2 to obtain two intervals, and then

---

[1] In Sect. 5, we will prove by model-checking that this speculation is correct.

$$[a, b] +^{\#} [c, d] = [a + c, b + d]$$

$$[\![x]\!](s) = s(x)$$

$$[a, b] -^{\#} [c, d] = [a - d, b - c]$$

$$[\![\kappa]\!](s) = [\kappa, \kappa], \quad \kappa \in \mathbb{Z}$$

$$[a, b] \times^{\#} [c, d] = [\min(ac, bc, ad, bd),$$

$$[\![e_1 \circ e_2]\!](s) = [\![e_1]\!](s) \circ^{\#} [\![e_2]\!](s)$$

$$\max(ac, bc, ad, bd)]$$

(a) Abstract evaluation of expressions.   (b) Examples of abstract operators.

$$[a, b] \sqcup [c, d] = [\min(a, c), \max(b, d)]$$
$$[a, b] \sqcup \perp = [a, b]$$
$$\perp \sqcup [a, b] = [a, b]$$

$$[\![x \leftarrow e]\!](s) = s [x \mapsto [\![e]\!](s)]$$

$$\perp \sqcup \perp = \perp$$

(c) Abstract evaluation of assignments.   (d) The join operator.

$$[a, b] \cap [c, d] = \begin{cases} \perp & \text{iff } a > d \text{ or } b < c \\ [\max(a, c), \min(b, d)] & \text{otherwise} \end{cases}$$

$$\perp \cap [a, b] = [a, b] \cap \perp = \perp$$
$$\perp \cap \perp = \perp$$

(e) The intersection operator.

Fig. 1. Definitions related to the interval abstraction.

composes the intervals according to an abstract version $\circ^{\#}$ of the operation. As a minimal example, in Fig. 1b we show the usual definition of abstract addition, subtraction, and multiplications over integers. Lastly, we can slightly abuse our notation and write $[\![x \leftarrow e]\!](s)$ to denote the abstract evaluation of an *assignment* statement on state s, where variable x will receive the result of expression e. This operation returns a new abstract state that is identical to s, except that x maps to the abstract evaluation of e on s (Fig. 1c).

Interval-based reasoning provides a rather coarse approximation of the concrete set of values that a variable may assume. For instance, interval $[0, 10]$ is a sound abstraction of the concrete set $\{0, 3, 10\}$, but includes several elements that do not belong to the set. To enjoy a tighter approximation while still relying on the (computationally cheap) domain of intervals, we consider the *powerset of intervals* [5,16] domain, commonly denoted by $P(I)$. Intuitively, an element in $P(I)$ is a set of disjoint intervals; we say that two intervals i, j are disjoint when their intersection $i \cap j$, as defined in Fig. 1e, is the empty interval. Given any set of intervals S, possibly including non-disjoint intervals, we can find its *normal form* $n(S) \in P(I)$, defined by Eq. 3 below, which replaces subsets of continuous (Eq. 1) intervals disjoint (Eq. 2) from the rest by their *join*, where the join of two intervals $i \sqcup j$ is the smallest interval that entirely contains both i and j, and is computed as shown in Fig. 1d. Lastly, we can use $P(I)$ as a domain for abstract interpretation of expressions by lifting the abstract operators already defined over intervals. Namely, if S_1, S_2 are elements of $P(I)$ and $\circ$ is a binary operator, one can soundly define $S_1 \circ^{\sharp} S_2$ as in Eq. 4 below, by evaluating the operation pairwise over the elements of S_1 and S_2, and then finding the normal form of the resulting set.

$$cont(S) = (\forall x \in \bigsqcup S)\,(\exists i \in S)\; x \in i \tag{1}$$

$$disj(S, S') = (\forall i \in S, j \in S')\; i \cap j = \bot \tag{2}$$

$$n(S) = \{\bigsqcup S' \mid S' \subseteq S \wedge cont(S') \wedge disj(S', S \setminus S')\} \tag{3}$$

$$S_1 \circ^{\sharp} S_2 = n(\{i_1 \circ^{\sharp} i_2 \mid i_1 \in S_1, i_2 \in S_2\}) \tag{4}$$

Compositional State Space Generation. The systems we are interested in analysing may be imagined as trees of parallel processes, branching out from a root parallel composition and whose leaves correspond to sequential processes. To generate the state space of these systems, we may apply a divide-and-conquer approach where we first generate the state spaces of each leaf, and then compose them together [49]. What makes this approach appealing is that, under appropriate assumptions and depending on our goals (e.g., on which properties we want to verify on the system), we can also perform hiding and minimization steps on the components, facilitating their composition.

Several compositional *strategies* have been put forward to outline the order in which these steps are carried out [18]. In this work we exploit one such strategy, namely *root leaf reduction*. Under this strategy, first, hiding operators are propagated as far down the tree as allowed; then, the state spaces of the leaves are generated and minimized modulo some equivalence relation R; lastly, the state spaces are composed together according to the structure of the tree and the resulting state space is further minimized modulo R.

The CADP Toolbox and LNT. CADP [19] is a software toolbox for the analysis of asynchronous concurrent systems. It provides a wide range of tools for simulation, test generation, verification, performance evaluation, etc., and accepts system descriptions in several languages whose semantics can be expressed in the form of an LTS (labelled transition system). CADP provides efficient model-checking procedures for a data-aware extension of the modal μ-calculus called MCL [37], and allows declaring complex verification tasks by means of an ad-hoc scripting language called SVL [17]. SVL natively supports several compositional strategies, including root leaf reduction.

In this work we will use LNT [21] to describe networks of *processes* that interact by means of *offers*. We will use the following subset of LNT *communication actions*: $G(x_1, \ldots, x_n)$ denotes an *output offer*, i.e., an action by which a process is willing to output the values $x_1, \ldots, x_n$ through a gate G; on the other hand, $G(?x_1, \ldots, ?x_n)$ denotes an *input offer*, i.e., an action where a process is willing to receive any n values from gate G and bind them to variables $x_1, \ldots, x_n$. Finally, $G(?x_1, \ldots, ?x_n)$ **where** $p(x_1, \ldots, x_n)$ is also an input offer, but the process is only willing to receive those values that satisfy a given Boolean predicate

p. The semantics of a process is an LTS in which each label corresponds to an offer.

To make LNT processes synchronize on offers, we have to compose them in parallel and specify a *synchronization set* for the composition. Specifically, the syntax **par** $G_1, \ldots, G_m$ **in** $P_1 \| \cdots \| P_n$ **end par** denotes a parallel composition of n processes where an offer on any of the gates $G_1, \ldots, G_m$ may only take place if *all* processes are willing to perform it simultaneously; all other offers may happen freely. *Partial* synchronization sets may also be defined by using the syntax **par** $\Gamma_1 {\rightarrow} P_1 \| \cdots \| \Gamma_n {\rightarrow} P_n$ **end par**, where $\Gamma_1, \ldots, \Gamma_n$ are sets of gates. In this case, synchronization over a gate G is only required among those processes that have G in their set of gates [23].

Notice that an input offer is semantically equivalent to an output offer of a nondeterministic value. For instance, if x is a Boolean variable, $G(?x)$ is the same as a choice between $G(\mathtt{true})$ and $G(\mathtt{false})$, followed by an assignment of the offered value to x. Thus, even though the LNT syntax might suggest asymmetrical interactions (*á la* CCS [39]), its synchronization semantics makes no difference between senders and receivers, and naturally supports multi-party rendezvous.

3 Parallel Emulation Programs

Given a specification $\mathbb{S}$ of a collective system, an emulation program $\mathbb{P}$ for it is a program, written in some target programming language, that may reproduce all feasible executions of $\mathbb{S}$ without introducing spurious ones. Thus, one may check whether a given temporal property holds in $\mathbb{S}$ by verifying an adequate encoding of the property against $\mathbb{P}$. A *sequential* emulation program replaces agent concurrency with nondeterminism, essentially applying sequentialization [43] to $\mathbb{S}$. Sequential emulation programs may be written in any imperative language and enable verification of collective systems by means of several analysis techniques for sequential programs [9]. In this section, instead, we show how we can exploit LNT's native constructs for parallelism to construct a *parallel* emulation program, where each agent is encoded into a separate process and communication is described via process synchronization. This encoding preserves the structure of the original system and enables compositional analysis.

The structure of an **Agent** LNT process is summarized in Fig. 2. We assume that each agent has a unique identifier, denoted by id. First, the agent performs an initialization (*init*), where its (potentially nondeterministic) internal state is set up according to the specifications. This state is entirely contained into two arrays L_{id} (the *local stigmergy* of id) and I_{id} (the *interface* of id), which respectively contain the values of stigmergic variables and of other internal variables.

After the initialization, the agent enters a loop during which it may repeatedly choose between six alternative behaviours. Namely, it may perform an individual action itself (*step*), signalled by an offer $\mathtt{tick}(id)$, or let another agent do the same (*yield*), signalled by an input offer $\mathtt{tick}(?j)$ for some $j \neq id$. It may also send one of its pending stigmergic messages, if any (*send-put, send-qry*); or it may react to a message sent by another agent (*react-put, react-qry*).

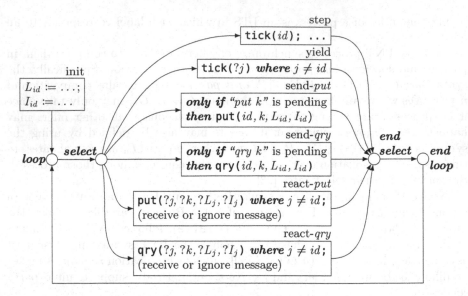

Fig. 2. Structure of an `Agent` process with identifier id.

By *reacting* to a message, we mean that the agent either accepts or ignores it, based on the semantic rules outlined in Sect. 2. To take this decision, we need the sender to evaluate the link predicate for the variable k within the message. To do so, it needs to know the state of the sender j, which is why L_j and I_j are part of the input offers (and L_{id}, I_{id} are part of the corresponding output offers).

Note that, while the definition of most blocks depends only on the variables of the input LAbS agent, the *step* block is the only one whose definition depends on its behaviour: generating the *step* block essentially amounts to transforming the LAbS behaviour into LNT fragments, by means of an automated procedure [9]. All other blocks simply implement the semantic rules of LAbS and are the same for every specification.

To model a LAbS system of n agents, we construct an LNT parallel emulation program following the structure shown in Listing 2. Specifically, we instantiate n `Agent` processes, each with a unique identifier from 0 to $n - 1$ (lines 3–7). These processes are composed in parallel, with a global synchronization set containing gates `tick`, `put`, and `qry`, so that all agents must synchronize in order to perform an input/output offer over these gates. This ensures that individual actions never overlap with each other nor with message-passing steps, and also that messages are always visible to all agents (which then decide whether to accept or ignore them). These restrictions are necessary to avoid spurious executions, i.e., computations of the programs that do not correspond to a trace in the original specifications. The program also features a `Timestamps` process (line 8) that tracks timestamping information about stigmergic variables. This information, in turn, determines how agents will react when they receive a stig-

Listing 2. A parallel emulation program for a system of n agents.

```
 1 process Main [...] is
 2 par refresh, request in
 3    par tick, put, qry in
 4       Agent [...] (0)
 5    || ...
 6    || Agent [...] (n − 1)
 7    end par
 8 || Timestamps [refresh, request]
 9 end par
10 end process
```

Listing 3. An emulation program with round-robin scheduling.

```
 1 process Sched [tick] is loop
 2    tick(0); tick(1); ... ;
 3    tick(n − 1)
 4 end loop end process
 5
 6 process Main [...] is
 7 par
 8    tick, refresh, request ->
 9       par tick, put, qry in
10          Agent [...] (0)
11       || ...
12       || Agent [...] (n − 1)
13       end par
14 || refresh, request ->
15       Timestamps [refresh, request]
16 || tick -> Sched [tick]
17 end par
```

Listing 4. Sketch of the LNT translation of the **Node** process from Listing 1.

```
 1 process Agent [...] (id: Nat) is
 2    -- init
 3    L[0] := N; -- leader
 4    pending := ∅;
 5    ...;
 6    loop
 7    select
 8       -- step
 9       tick(id);
10       if L[0] > id then
11          L[0] := id;
12          updateTimestamp(0);
13          addPendingMsg("put", 0)
14       else loop spurious end loop
15       end if
16    []
17       -- yield
18       tick(?j) where j <> id
19    []
20       -- other behaviours
21       -- (react-put, react-qry,
22       -- send-put, send-qry)
23    end select
24    end loop
25 end process
```

mergic message. Each agent may independently contact this process through two gates: namely, they can either **refresh** their timestamp for a variable (which they do after assigning a new value to it), or **request** a comparison between their timestamp for a variable and the one of another agent (which they do when they process an incoming message).

Round-Robin Scheduling. In the emulation program shown so far, agents can freely interleave their actions. In some cases, however, it makes sense to only consider *round-robin* executions, i.e., those where agents perform one *step* at a time, in a fixed sequence given by their identifiers. (Message passing actions can still happen at any time). We may enforce this restriction by adding to our program a *scheduler* process that constrains **tick** offers, so that the agent with identifier 0 is forced to act first, followed by the one with id 1, and so on (Listing 3).

Correctness of the Encoding. Our argument for the correctness of the LAbS-to-LNT encoding is essentially the same as the one we put forward for generic emulation programs [9], the main difference being that the sequentializing scheduler in that work is now replaced by multi-party synchronizations where the agents decide who should act next. First, we assume there exists a translation from each LAbS action α (e.g., as an assignment) to an LNT fragment that respects the operational semantics of α. By forcing synchronizations over the `tick` gate, we prevent agents from overlapping their actions with other actions or with message exchanges. Therefore, for every sequence of actions allowed by the original specification $\mathbb{S}$, the emulation program $\mathbb{P}$ features some execution in which the corresponding LNT code fragments are invoked in the correct order. Vice versa, if $\mathbb{P}$ allows a sequence of code fragments to be executed, then there is a feasible execution of $\mathbb{S}$ where agents perform the corresponding LAbS actions in the same order.

Running Example. Listing 4 contains a simplified sketch of how the `Node` agent from Listing 1 would get translated into LNT. First, the agent's stigmergic variable `leader` (stored in `L[0]`) and its set of pending messages are initialized to N and to the empty set, respectively. We omit the rest of the initialization. Then, the agent enters the loop we graphically depicted in Fig. 2. For sake of brevity, we only include a simplified version of the *step* and *yield* behaviours. We encode the guarded action of Listing 1 by means of an `if` statement. If the guard is not satisfied, we send the agent to a sink state where it repeatedly performs a special `spurious` action (line 14). We will use this action to detect and ignore invalid traces during our analysis. If the guard holds, the agent can proceed with the stigmergic assignments, which consists of three steps: updating the value, setting its timestamp to the current instant, and adding a *put*-message to the list of pending messages (lines 11–13).

Notice that, for an agent with a more elaborate behaviour, the `step` block would be a nondeterministic choice over the feasible actions that the agent can perform at the present time. This also requires keeping track of the agent's execution point: for instance, an agent whose behaviour is a sequence of two actions $a; b$ must necessarily perform a before it is able to perform b. To maintain track of this information, we constrain the agent's choice of actions by a dedicated variable that acts as a program counter, and is updated after every action [9].

4   Value Analysis of LAbS Specifications

As seen in Fig. 2, the exchange of stigmergic messages requires the sender to offer its local stigmergy and interface to all other agents, so that they can perform a corresponding input offer to receive this information and evaluate whether they should receive or ignore the message. These input offers make generating the state space of individual agents problematic. In fact, each agent may expect to receive *any* possible L_j and I_j over the `put` and `qry` gates, meaning that its transition system has to enumerate all potential offers. This easily makes the agent's state space explode, even when we assume that variables range over relatively modest intervals, such as the 8-bit representation range $(-128, \ldots, 127)$.

We work around this issue by observing that, in typical systems, agents will only ever see a rather small subset of those offers. We can over-approximate this subset by performing an automated value analysis on the input specifications,

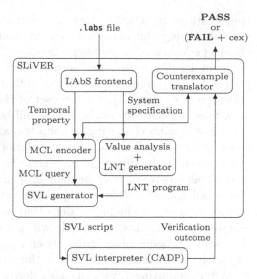

PASS
or
(**FAIL** + cex)

.labs file

```
1   ς₀ ← initial abstract state of S
2   A ← all assignments within S
3   σ ← {ς₀}
4   while true do
5   │   σ' ← {[[a]](ς) | ς ∈ σ, a ∈ A}
6   │   if σ' ⊆ σ then
7   │   │   break
8   │   else
9   │   │   σ ← σ ∪ σ'
10  │   end
11  end
12  return ⊔ σ
```

Fig. 3. Value analysis of a LAbS specification S.

Fig. 4. Our compositional verification workflow.

and then constrain input offers based on the result of this analysis. Notice that this over-approximation will not lead our procedure to produce spurious counterexamples: during composition, spurious input offers will find no matching output offers, and will therefore be pruned away.

Given a specification S, let us denote by $\mathcal{V}$ the set of its variables. From now on, we define an *abstract state* for S as a mapping from $\mathcal{V}$ to $P(I)$. Furthermore, we define the *merge* of two abstract states $ς_1 \sqcup ς_2$ as the state $ς$ such that, for every $x \in \mathcal{V}$, $ς(x) = n(ς_1(x) \cup ς_2(x))$. Our value analysis (Fig. 3) is straightforward. Initially, we compute the initial abstract state $ς_0$ for the given specification, and create a set $σ$ that only contains this state. Computing $ς_0$ is immediate, as every LAbS specification must specify one or more feasible initial values for each declared variable.[2] We also extract from S a set A of all assignment statements that appear in it (lines 1–3). Then, we run a loop in which we abstractly evaluate every assignment $a \in A$ on every state $ς \in σ$ and add the resulting states to $σ$ (lines 4–11). If, at some point, we fail to find any new states, then we break out of this loop, and return as the final value analysis the merge of all states in $σ$ (line 12).

The result of running this algorithm is an abstract state $\bar{ς}$, mapping every variable name x to a powerset of intervals $\bar{ς}(x)$. We can easily see that $\bar{ς}(x)$ over-approximates the set of all values that x may actually assume across all feasible executions of S. In fact, our analysis simply performs every possible assignment at every iteration, without considering the order in which they appear in the

---

[2] LAbS allows variables with an *undefined* initial value undef, but we currently do not support that feature in our analysis.

specifications, or whether they are guarded or not. Thus, we may say that we are considering the *chaos automaton* $chaos(A)$, i.e., the automaton that can always perform any of the assignments in $\mathbb{S}$. Every sequence of assignments from A, including those that are actual executions of $\mathbb{S}$, is a feasible execution in $chaos(A)$. Therefore, the set of values that a variable x can ever assume in the state space of $chaos(A)$ is a superset of those it may assume in the state space of $\mathbb{S}$, meaning that our analysis is sound but potentially over-approximating.

It is also nonterminating on infinite-state specifications, which are out of the scope of this work. This over-approximation also takes into account the exchange of values through stigmergic messages. To understand this, it suffices to notice that a value (say, κ) may be sent in a message only if it has been previously computed and stored in a stigmergic variable (say, x) by some agent. That is, messages cannot include values that are not the result of some sequence of assignments. But then, $chaos(A)$ will necessarily allow every agent to perform that same sequence of assignments and assign κ to x. Thus, there is no need to explicitly model message passing within our algorithm.

After computing $\bar{\varsigma}$, we can easily derive a Boolean function goodL that takes a local stigmergy L and returns true if and only if, for every stigmergic variable x, $L(x)$ is in $\bar{\varsigma}(x)$. Likewise, we can derive a function goodI that does the same for interfaces. Then, we force agents to only consider these objects as valid by constraining all their input offers, such as put($?j, ?k, ?L_j, ?I_j$) (as in the *recv-put* block of Fig. 2), by the predicate $j \neq id \wedge$ goodL$(L_j) \wedge$ goodI(I_j).

Running Example. Our bully election system (Listing 1) contains one variable leader, initialized to N for every agent. Plus, it refers to a special variable id that stores the agent's identifier. LAbS guarantees that identifiers are unique, contiguous and start at 0: so, the initial abstract state for our analysis will be $\varsigma_0 = \{$leader $\mapsto [N, N];$ id $\mapsto [0, N-1]\}$. Then, interpreting the assignment leader <~ id over ς_0 yields a new abstract state $\varsigma_1 = \{$leader $\mapsto [0, N-1];$ id $\mapsto [0, N-1]\}$. It is plain to see that our analysis cannot find any other states beyond this, since $[\![a]\!](\varsigma_1) = \varsigma_1$. Thus, the result of the analysis is just the merge $\varsigma_0 \sqcup \varsigma_1 = \{$leader $\mapsto \{[0, N-1], [N, N]\};$ id $\mapsto [0, N-1]\}$.

5   Compositional Verification Workflow

In this section, we describe how we combined the contributions described so far into an automated workflow for the compositional verification of LAbS systems. The workflow is implemented as a module within the SLiVER analysis tool,[3] and it is depicted in Fig. 4.

First, a frontend parses a LAbS file and extracts the temporal property to verify, as well as the system specification. The former is transformed into an equivalent MCL query [10]; the latter, instead, is fed to a code generator to construct a parallel emulation program as described in Sect. 3. The code generator also runs the value analysis described in Sect. 4 and uses the results to

---

[3] https://github.com/labs-lang/sliver.

constrain all input offers on gates put, qry. Then, we generate an SVL script that describes the compositional verification task, and submit it to CADP; when the task is completed, we interpret its verdict (e.g., if a counterexample is found, we translate it into the syntax of LABS) and show the result to the user.

Listing 5 shows the structure of a verification script generated by our workflow. Intuitively, we ask CADP to generate the state space of the parallel emulation program of Listing 2 by means of *root leaf* reduction, minimizing modulo divergence-preserving sharp bisimulation [35], and then to verify our MCL query against the resulting transition system using the *Evaluator4* model checker. Notice that the program is wrapped in a *hiding* and a *priority* operator.

Hiding (***hide** G **in** P **end hide***) replaces all offers over gate G that occur in P with internal actions, denoted by τ. When generating our script, we determine all labels that are relevant to our query, denoted as *gates*("query.mcl"), and then hide all other gates. This reduces the state space (as sharp bisimulation compresses sequences of τ-transitions) and thus accelerates model-checking.

The priority operator (***prio** \Omega **in** P **end prio***) allows to specify a partial order of labels so that, when the state space of P is generated, transitions with a low-priority label are cut from every state that also features at least one transition with a higher-priority label. We use priorities to prune some sections of our programs where agents are free to interleave their actions in any order (e.g., when they have to react to an incoming message). These sections are not part of the semantics of LAbS, where message exchanges are treated as atomic events, but are rather an artefact of the encoding into parallel LNT programs. Furthermore, in these sections, each agent only affects its internal state, so reordering their actions does not affect the satisfaction of properties we are interested in verifying. Thus, we can analyse all orderings by only considering a representative one. Specifically, we give decreasing priorities to offers over gates refresh, request, and l (which agents use to signal a new assignment to a stigmergic variable); additionally, when multiple agents are willing to perform an action over one of these gates, the agent with lowest *id* is prioritized. This prioritization is independent of the specification being analysed, as it concerns the LNT encoding of stigmergic messaging regardless of the actual data being exchanged.

Lastly, our choice of sharp bisimulation is motivated by our use of the priority operator. In fact, applying sharp minimization under an appropriate set of strong actions, as we do here, preserves priorities (like strong minimization does), but also results in smaller LTSs than the one obtained through strong minimization. Minimizing modulo divergence-preserving branching bisimulation [27] (also known as divbranching bisimulation, for short) or weaker equivalences could in principle lead to even smaller LTSs, but would not preserve the semantics of the system. In fact, divbranching bisimulation and weaker equivalences are not congruences for the priority operator [11]. To see this, it suffices to consider the process $\tau.a$ which is divbranching bisimilar to a (Eq. 5), and observe that we can easily find a context with priorities, e.g., $C[P] = $ ***prio** a>b **in** P \parallel b **end prio***, such that replacing P with either $\tau.a$ or a gives us non-bisimilar terms (Eq. 8).

Listing 5. Structure of an SVL script for our verification workflow

```
 1 "system.bcg" = root leaf divsharp reduction
 2 hold "refresh", "request", "l"
 3 in (
 4   hide all but gates("query.mcl") in
 5     prio
 6     "refresh" > "request" > "l"
 7     "refresh i .*" > "refresh j .*", i < j
 8     "request i .*" > "request j .*", i < j
 9     "l i .*" > "l j .*", i < j
10     in
11     ... (* Parallel emulation program (Listing 2 or 3) *)
12     end prio
13   end hide);
14
15 property CHECK is
16   verify "query.mcl" with evaluator4
17   in "system.bcg"
18   expected TRUE
19 end property;
```

$$\tau.a \sim_{db} a \tag{5}$$

$$C[\tau.a] = \textbf{\textit{prio}}\ a > b\ \textbf{\textit{in}}\ \tau.a \parallel b\ \textbf{\textit{end prio}} = (\tau.a.b + b.\tau.a) \tag{6}$$

$$C[a] = \textbf{\textit{prio}}\ a > b\ \textbf{\textit{in}}\ a \parallel b\ \textbf{\textit{end prio}} = a.b \tag{7}$$

$$(\tau.a.b + b.\tau.a) \not\sim_{db} a.b \tag{8}$$

Experimental Evaluation. To demonstrate our approach, we carry out a collection of verification tasks [10] in two different ways: first, we use a baseline workflow that generates a sequential LNT program, constructs its state space, minimizes it modulo divergence-preserving branching bisimulation,[4] and finally model-checks the reduced state space; then, we apply the compositional procedure proposed above. We then measure and compare the time and memory requirements of the two approaches.

We now provide a short overview of each system along with the properties to verify. The reader may refer to [10] for a detailed description. Systems whose name ends in -rr were verified assuming round-robin scheduling of agents. All properties are checked under fairness assumptions that exclude unfair loops from the verification [44].

---

[4] We use this relation because it preserves all the properties that we are interested in checking.

Table 1. Experimental results for compositional verification. Values in bold are better. $-^{a}$: theoretical, based on *Compositional* measurements.

System	Baseline [10]		Compositional		Parallel[a]	
	Time (s)	Memory (MiB)	Time (s)	Memory (MiB)	Time (s)	Memory (MiB)
flock-rr	**1875**	12000	4461	**11805**	4426	**11805**
flock	4787	30865	4071	**11113**	**4038**	**11113**
formation-rr	1670	**1657**	2511	1938	**1558**	5875
leader5	**10**	**41**	34	117	18	212
leader6	77	**147**	104	225	**65**	258
leader7	1901	2038	374	**404**	**326**	**404**
twophase2	**9**	**50**	67	93	34	210
twophase3	500	**209**	233	322	**131**	560

The `flock` and `flock-rr` systems describe a simplified flocking behaviour. The systems feature 3 agents in a 5×5 arena. Each agent is initially given a nondeterministic position and direction of movement, with the latter stored as a pair of stigmergic variables. The agents move by following this direction vector. When two agents are sufficiently close (5 spaces apart or fewer), one of them may imitate the other's direction by receiving a stigmergic message. We check that, eventually, all agents move in the same direction.

In the `formation-rr` system, 3 agents are placed on a line segment of length 10. They use stigmergic variables to signal their presence to nearby agents. If an agent detects that it is too close to another, it moves one step away from it, unless it is at either end of the segment. We check that, eventually, all agents are at least 2 spaces apart from each other.

The `leader<N>` systems are three instances of our running example (Listing 1), respectively with 5, 6, and 7 nodes. We verify that all nodes eventually choose the one with id 0 as the leader.

The `twophase<N>` systems describe a two-phase commit scenario [29] with *N* *workers* and one *coordinator*. The coordinator initiates a voting session where all workers must decide whether a transaction should be committed. If all workers agree, the coordinator commits the transaction and starts a new voting round. We implemented the workers so as they always agree to commit, and all communication happens through stigmergy variables. We check that the coordinator commits transactions infinitely often.

All the experiments were performed on the Grid'5000 testbed, specifically on a node of the *Dahu* cluster. The node is equipped with two Intel Xeon Gold 6130 CPUs and 192 GiB of physical memory, and runs Debian 11 with version 5.10.0 of the Linux kernel.[5] We used CADP version 2022-h, and set a timeout of 3 h and a memory limit of 32 GiB for all experiments. We collected the raw experimental data into a persistent replication package [13], which also includes the input LAbS specifications as well as binaries and scripts to facilitate reproducing the experiments.

---

[5] https://www.grid5000.fr/w/Grenoble:Hardware#dahu.

We summarize the experimental results in Table 1. Columns from left to right report the name of the system and the time and memory required to verify it by the baseline and compositional approaches, respectively. In the last column, called *Parallel*, we show the time and memory it would take to perform the compositional workflow if we generated the individual state spaces simultaneously, e.g., on separate machines, or on separate cores of a multi-core machine. These are hypothetical measurements, derived from the *Compositional* ones. Specifically, each compositional verification experiment is made of several *tasks*, namely: k tasks $T_1, \ldots, T_k$ that construct the individual state spaces of the n agents, plus those of the processes Timestamps and (for round-robin systems) Sched; a task $T_\mathbb{P}$ that assembles these state spaces into the one of the whole emulation program; and lastly, a model-checking task $T_\models$. Let us denote the time and memory required to execute a task T by $time(T)$ and $mem(T)$, respectively. We can gather these measurements by executing an experiment with the *Compositional* workflow. Under this workflow, tasks are carried out sequentially: thus, the time required by the experiment is the sum of $time(T)$ for each task T. For the same reason, the memory footprint is just the maximum of the memory requirements of every task. However, if the tasks T_i are carried out in parallel, then we would only have to wait for the task with the maximum $time(T_i)$ before we are able to begin $T_\mathbb{P}$. At the same time, we would need to satisfy the memory requirements of all individual tasks at the same time, so we have to take into account the *sum* of all $mem(T_i)$. We summarize these simple computations in Table 2. Notice that, on smaller systems, the memory requirements of SLiVER itself (around 400 MiB) would dominate that of the actual memory used for model checking. To better focus on comparing the performance of the two verification workflows, the table omits this overhead; we reserve the implementation of a more memory-efficient SLiVER for future work.

We can see that the baseline method is more time-efficient than the compositional one on some specific cases, e.g., when the overall system is rather small (leader5, leader6, twophase2) or round-robin scheduling has to be enforced (flock-rr, formation-rr). Full interleaving has an opposite effect: with the baseline procedure, verifying flock takes longer than flock-rr, whereas the compositional one can verify it faster. This may sound counterintuitive, since the former system only considers a subset of the latter's traces. Our explanation is that, for the compositional procedure, it is much easier to just freely compose agents rather than having to take the scheduling constraints into account. In other words, the scheduler acts as a sort of bottleneck to the compositional task, even though the resulting state space is smaller.

On small systems, namely leader5, leader6, and twophase2, the performance of the compositional procedure is likely affected by the overhead brought about by the component-wise state space generation. Furthermore, we are aware that CADP currently invokes the LNT compiler multiple times, i.e., for each component process, compounding this overhead. In conclusion, under specific conditions, the baseline approach may still produce a verdict faster than the compositional one. At the same time, the compositional approach appears to

Table 2. Time and memory requirements for the *Compositional* and *Parallel* workflows.

	Compositional	Parallel
Time	$\sum\limits_{Tasks} time(\mathcal{T})$	$\max\limits_{i}\{time(\mathcal{T}_i)\} + time(\mathcal{T}_\mathbb{P}) + time(\mathcal{T}_\models)$
Memory	$\max\limits_{Tasks} mem(\mathcal{T})$	$\max\left\{\sum\limits_{i} mem(\mathcal{T}_i), mem(\mathcal{T}_\mathbb{P}), mem(\mathcal{T}_\models)\right\}$

scale better than the baseline as the size of the systems grows. This is most evident in the `leader` systems, where every additional agent severely impacts the time and memory required by the baseline workflow; instead, the compositional approach shows a much less explosive, though still super-linear, progression.

The (theoretical) parallel procedure is, by definition, always faster than the compositional one. This speedup is most noticeable when the system involves many agents (`leader7`), or complex behavioural rules (`formation-rr`, `twophase3`). In some experiments, parallelization also incurs an increased memory usage, a rather obvious consequence of generating all individual state spaces at once. At the same time, it typically allows enjoying greater memory capacities (especially when done across multiple machines), so we do not expect this drawback to be significant. In others, however, both workflows have the same memory footprint, as the memory required to generate all state spaces simultaneously does not exceed the amount used by the other tasks ($\mathcal{T}_\mathbb{P}$ or $\mathcal{T}_\models$). Thus, in these cases the speedup from parallelization actually comes for free, i.e., it does not impact the overall memory usage.

6   Related Work

Compositional verification has been successfully applied in several domains, ranging from hardware systems to communication protocols and service choreographies [18,20]. From a recent, extensive experimental evaluation, it appears to be effective under diverse network topologies, and its benefits generally become more evident as the size of the system under verification grows [8].

In this work, we exploit compositionality of state space generation. A somewhat related approach to fight state space explosion is modular (or compositional) reasoning [26], whereby a program is analysed by splitting it into components, for instance according to rely-guarantee conditions [34]. This form of compositionality has proved effective in several use cases, such as multi-robot and multi-agent systems [4,33], railway networks [15], smart contracts [48], and authentication protocols [50]. All these applications, like our own work, exploit fully automated verification procedures; other frameworks, such as IVy [38], combine rely-guarantee reasoning with semi-automated procedures. LNT does provide constructs to express pre- and post- conditions on procedures (respectively denoted by `require` and `ensure`), but CADP does not use them in a compositional fashion yet.

Some classes of collective adaptive systems may be expressed in the form of population protocols [1], for which efficient parameterized verification procedures are known [2]. These may prove that a protocol satisfies a given property regardless of its size, but the properties of interest typically concern its eventual convergence to certain configurations, as the focus is to verify whether the protocol is able to carry out a desired computation. Our workflow checks systems of fixed size, but may support arbitrary branching-time temporal properties.

Preprocessing techniques to speed up program analysis by excluding invalid or infeasible values have also been proposed in the context of symbolic model checking. For instance, bounded model checking of programs featuring dynamic data structures may get more efficient by precomputing tight field bounds based on the structures' type invariants [42].

7   Conclusion and Future Work

In this work, we have argued that collective adaptive systems, being collections of autonomous and mutually interacting components, are naturally amenable to compositional techniques that can palliate state space explosion and thus aid in their verification. To support our claim, we have presented an encoding from high-level specifications into networks of LNT processes, introduced a simple value analysis to over-approximate the set of feasible offers between these processes, and demonstrated an automated workflow that exploits these ingredients to compositionally verify a collection of representative systems. Our experimental results do indicate that this procedure brings significant advantages over plain model checking. Besides evident gains in terms of absolute time and memory requirements, the proposed workflow appears to scale better in the number of agents, and can deal with freely-interleaved systems without particular effort compared to round-robin ones.

As future work, we intend to pursue several lines of research. For instance, the value analysis presented in this work is just a prototype and may be improved in several ways. Its approximation may be tightened by preserving some of the original behavioural structure and adding sensitivity to LAbS control constructs, such as guards. In general, powerset domains have well-known scalability issues that we could overcome by switching to more advanced abstract domains, such as BOXES [31] or donut domains [25], which may also track relations between variables. Our analysis only exploits data restriction; interfaces [28] could complement that with *behavioural* constraints, allowing to prune the state space of agents by cutting sequences of actions that are impossible under a given context. Thus, synthesizing such interfaces could enhance our compositional approach.

We also plan to actually implement the *Parallel* workflow theorized in Sect. 5, so that the generation of individual state spaces is distributed across multiple machines. This could be integrated with existing procedures for distributed state space generation [22], to further exploit parallelism; it would also allow us to measure how other factors, e.g., networked storage latency and transfer times, may affect the theoretical measurements presented in this work. An implementation of lighter-weight formal techniques, such as runtime verification [36] or statistical model-checking [46], could also provide some degree of assurance about the behaviour of very large collective systems.

Acknowledgements. Experiments presented in this paper were carried out using the Grid'5000 testbed, supported by a scientific interest group hosted by Inria and including CNRS, RENATER and several Universities as well as other organizations (see https://www.grid5000.fr). Preliminary experiments were enabled by resources provided by the Swedish National Infrastructure for Computing (SNIC) at Umeå University partially funded by the Swedish Research Council through grant agreement no. 2018-05973. The authors wish to thank Wendelin Serwe and Hubert Garavel for their assistance in installing and running CADP on these machines.

References

1. Angluin, D., Aspnes, J., Eisenstat, D., Ruppert, E.: The computational power of population protocols. Distrib. Comput. **20**(4), 279–304 (2007). https://doi.org/10.1007/s00446-007-0040-2
2. Blondin, M., Esparza, J., Jaax, S.: PEREGRINE: a tool for the analysis of population protocols. In: Chockler, H., Weissenbacher, G. (eds.) CAV 2018. LNCS, vol. 10981, pp. 604–611. Springer, Cham (2018). https://doi.org/10.1007/978-3-319-96145-3_34
3. Bonabeau, E.: Agent-based modeling: methods and techniques for simulating human systems. Proc. Natl. Acad. Sci. **99**(Suppl. 3), 7280–7287 (2002). https://doi.org/10.1073/pnas.082080899
4. Cardoso, R.C., Dennis, L.A., Farrell, M., Fisher, M., Luckcuck, M.: Towards compositional verification for modular robotic systems. In: 2nd Workshop on Formal Methods for Autonomous Systems (FMAS). EPTCS, vol. 329, pp. 15–22 (2020). https://doi.org/10.4204/EPTCS.329.2
5. Cousot, P., Cousot, R.: Static determination of dynamic properties of programs. In: 2nd International Symposium on Programming, pp. 106–130. Dunod (1976)
6. Crowston, K., Rezgui, A.: Effects of stigmergic and explicit coordination on Wikipedia article quality. In: 53rd Hawaii International Conference on System Sciences (HICSS), pp. 1–10. ScholarSpace (2020)
7. De Nicola, R., Di Stefano, L., Inverso, O.: Multi-agent systems with virtual stigmergy. Sci. Comput. Program. **187** (2020). https://doi.org/10.1016/j.scico.2019.102345
8. de Putter, S., Wijs, A.: To compose, or not to compose, that is the question: an analysis of compositional state space generation. In: Havelund, K., Peleska, J., Roscoe, B., de Vink, E. (eds.) FM 2018. LNCS, vol. 10951, pp. 485–504. Springer, Cham (2018). https://doi.org/10.1007/978-3-319-95582-7_29
9. Di Stefano, L., De Nicola, R., Inverso, O.: Verification of distributed systems via sequential emulation. ACM Trans. Softw. Eng. Methodol. **31**(3) (2022). https://doi.org/10.1145/3490387
10. Di Stefano, L., Lang, F.: Verifying temporal properties of stigmergic collective systems using CADP. In: Margaria, T., Steffen, B. (eds.) ISoLA 2021. LNCS, vol. 13036, pp. 473–489. Springer, Cham (2021). https://doi.org/10.1007/978-3-030-89159-6_29
11. Di Stefano, L., Lang, F.: Compositional verification of priority systems using sharp bisimulation. Research report, INRIA (2022). https://hal.inria.fr/hal-03640683

12. Di Stefano, L., Lang, F., Serwe, W.: Combining SLiVER with CADP to analyze multi-agent systems. In: Bliudze, S., Bocchi, L. (eds.) COORDINATION 2020. LNCS, vol. 12134, pp. 370–385. Springer, Cham (2020). https://doi.org/10.1007/978-3-030-50029-0_23

13. Di Stefano, L., Lang, F.: Replication Package for the paper: Compositional Verification of Stigmergic Collective Systems (2022). https://doi.org/10.5281/zenodo.7043353

14. El-Sayed, A.M., Scarborough, P., Seemann, L., Galea, S.: Social network analysis and agent-based modeling in social epidemiology. Epidemiol. Perspect. Innov. **9** (2012). https://doi.org/10.1186/1742-5573-9-1

15. Fantechi, A., Haxthausen, A.E., Macedo, H.D.: Compositional verification of interlocking systems for large stations. In: Cimatti, A., Sirjani, M. (eds.) SEFM 2017. LNCS, vol. 10469, pp. 236–252. Springer, Cham (2017). https://doi.org/10.1007/978-3-319-66197-1_15

16. Filé, G., Ranzato, F.: The powerset operator on abstract interpretations. Theor. Comput. Sci. **222**(1–2), 77–111 (1999). https://doi.org/10.1016/S0304-3975(98)00007-3

17. Garavel, H., Lang, F.: SVL: a scripting language for compositional verification. In: Kim, M., Chin, B., Kang, S., Lee, D. (eds.) FORTE 2001. IIFIP, vol. 69, pp. 377–392. Springer, Boston (2002). https://doi.org/10.1007/0-306-47003-9_24

18. Garavel, H., Lang, F., Mateescu, R.: Compositional verification of asynchronous concurrent systems using CADP. Acta Informatica **52** (2015). https://doi.org/10.1007/s00236-015-0226-1

19. Garavel, H., Lang, F., Mateescu, R., Serwe, W.: CADP 2011: a toolbox for the construction and analysis of distributed processes. Softw. Tools Technol. Transfer **15** (2013). https://doi.org/10.1007/s10009-012-0244-z

20. Garavel, H., Lang, F., Mounier, L.: Compositional verification in action. In: Howar, F., Barnat, J. (eds.) FMICS 2018. LNCS, vol. 11119, pp. 189–210. Springer, Cham (2018). https://doi.org/10.1007/978-3-030-00244-2_13

21. Garavel, H., Lang, F., Serwe, W.: From LOTOS to LNT. In: Katoen, J.-P., Langerak, R., Rensink, A. (eds.) ModelEd, TestEd, TrustEd. LNCS, vol. 10500, pp. 3–26. Springer, Cham (2017). https://doi.org/10.1007/978-3-319-68270-9_1

22. Garavel, H., et al.: DISTRIBUTOR and BCG_MERGE: tools for distributed explicit state space generation. In: Hermanns, H., Palsberg, J. (eds.) TACAS 2006. LNCS, vol. 3920, pp. 445–449. Springer, Heidelberg (2006). https://doi.org/10.1007/11691372_30

23. Garavel, H., Sighireanu, M.: A graphical parallel composition operator for process algebras. In: Joint International Conference on Formal Description Techniques for Distributed Systems and Communication Protocols (FORTE) and Protocol Specification, Testing and Verification (PSTV). IFIPAICT, vol. 156, pp. 185–202. Kluwer (1999)

24. Garcia-Molina, H.: Elections in a distributed computing system. IEEE Trans. Comput. **31**(1), 48–59 (1982). https://doi.org/10.1109/TC.1982.1675885

25. Ghorbal, K., Ivančić, F., Balakrishnan, G., Maeda, N., Gupta, A.: Donut domains: efficient non-convex domains for abstract interpretation. In: Kuncak, V., Rybalchenko, A. (eds.) VMCAI 2012. LNCS, vol. 7148, pp. 235–250. Springer, Heidelberg (2012). https://doi.org/10.1007/978-3-642-27940-9_16

26. Giannakopoulou, D., Namjoshi, K.S., Păsăreanu, C.S.: Compositional reasoning. In: Clarke, E., Henzinger, T., Veith, H., Bloem, R. (eds.) Handbook of Model Checking, pp. 345–383. Springer, Cham (2018). https://doi.org/10.1007/978-3-319-10575-8_12

27. van Glabbeek, R.J., Weijland, W.P.: Branching time and abstraction in bisimulation semantics. J. ACM **43** (1996)
28. Graf, S., Steffen, B.: Compositional minimization of finite state systems. In: Clarke, E.M., Kurshan, R.P. (eds.) CAV 1990. LNCS, vol. 531, pp. 186–196. Springer, Heidelberg (1991). https://doi.org/10.1007/BFb0023732
29. Gray, J.N.: Notes on data base operating systems. In: Bayer, R., Graham, R.M., Seegmüller, G. (eds.) Operating Systems. LNCS, vol. 60, pp. 393–481. Springer, Heidelberg (1978). https://doi.org/10.1007/3-540-08755-9_9
30. Grimm, V., Railsback, S.F.: Agent-based models in ecology: patterns and alternative theories of adaptive behaviour. In: Billari, F.C., Fent, T., Prskawetz, A., Scheffran, J. (eds.) Agent-Based Computational Modelling: Applications in Demography, Social, Economic and Environmental Sciences, pp. 139–152. Physica-Verlag, Heidelberg (2006). https://doi.org/10.1007/3-7908-1721-X_7
31. Gurfinkel, A., Chaki, S.: BOXES: a symbolic abstract domain of boxes. In: Cousot, R., Martel, M. (eds.) SAS 2010. LNCS, vol. 6337, pp. 287–303. Springer, Heidelberg (2010). https://doi.org/10.1007/978-3-642-15769-1_18
32. Hillston, J.: Challenges for quantitative analysis of collective adaptive systems. In: Abadi, M., Lluch Lafuente, A. (eds.) TGC 2013. LNCS, vol. 8358, pp. 14–21. Springer, Cham (2014). https://doi.org/10.1007/978-3-319-05119-2_2
33. Jones, A.V.: Model checking and compositional reasoning for multi-agent systems. Ph.D. thesis, Imperial College London, UK (2014). https://doi.org/10.25560/32695
34. Jones, C.B.: Tentative steps toward a development method for interfering programs. ACM Trans. Program. Lang. Syst. **5** (1983). https://doi.org/10.1145/69575.69577
35. Lang, F., Mateescu, R., Mazzanti, F.: Sharp congruences adequate with temporal logics combining weak and strong modalities. In: Biere, A., Parker, D. (eds.) TACAS 2020. LNCS, vol. 12079, pp. 57–76. Springer, Cham (2020). https://doi.org/10.1007/978-3-030-45237-7_4
36. Leucker, M., Schallhart, C.: A brief account of runtime verification. J. Logic Algebraic Program. **78** (2009). https://doi.org/10.1016/j.jlap.2008.08.004
37. Mateescu, R., Thivolle, D.: A model checking language for concurrent value-passing systems. In: Cuellar, J., Maibaum, T., Sere, K. (eds.) FM 2008. LNCS, vol. 5014, pp. 148–164. Springer, Heidelberg (2008). https://doi.org/10.1007/978-3-540-68237-0_12
38. McMillan, K.L., Padon, O.: Ivy: a multi-modal verification tool for distributed algorithms. In: Lahiri, S.K., Wang, C. (eds.) CAV 2020. LNCS, vol. 12225, pp. 190–202. Springer, Cham (2020). https://doi.org/10.1007/978-3-030-53291-8_12
39. Milner, R. (ed.): A Calculus of Communicating Systems. LNCS, vol. 92. Springer, Heidelberg (1980). https://doi.org/10.1007/3-540-10235-3
40. Olner, D., Evans, A.J., Heppenstall, A.J.: An agent model of urban economics: digging into emergence. Comput. Environ. Urban Syst. **54** (2015)
41. Pinciroli, C., Beltrame, G.: Buzz: an extensible programming language for heterogeneous swarm robotics. In: IEEE/RSJ International Conference on Intelligent Robots and Systems (IROS), pp. 3794–3800. IEEE (2016)
42. Ponzio, P., Godio, A., Rosner, N., Arroyo, M., Aguirre, N., Frias, M.F.: Efficient bounded model checking of heap-manipulating programs using tight field bounds. In: Guerra, E., Stoelinga, M. (eds.) FASE 2021. LNCS, vol. 12649, pp. 218–239. Springer, Cham (2021). https://doi.org/10.1007/978-3-030-71500-7_11
43. Qadeer, S., Wu, D.: KISS: keep it simple and sequential. In: Conference on Programming Language Design and Implementation (PLDI), pp. 14–24. ACM (2004). https://doi.org/10.1145/996841.996845

44. Queille, J.P., Sifakis, J.: Fairness and related properties in transition systems - a temporal logic to deal with fairness. Acta Informatica **19** (1983). https://doi.org/10.1007/BF00265555
45. Robles, G., Merelo, J.J., Gonzales-Barahona, J.M.: Self-organized development in libre software: a model based on the stigmergy concept. In: 6th International Workshop on Software Process Simulation and Modeling (ProSim). Fraunhofer (2005)
46. Sen, K., Viswanathan, M., Agha, G.: Statistical model checking of black-box probabilistic systems. In: Alur, R., Peled, D.A. (eds.) CAV 2004. LNCS, vol. 3114, pp. 202–215. Springer, Heidelberg (2004). https://doi.org/10.1007/978-3-540-27813-9_16
47. Theraulaz, G., Bonabeau, E.: A brief history of stigmergy. Artif. Life **5** (1999). https://doi.org/10.1162/106454699568700
48. Wesley, S., Christakis, M., Navas, J.A., Trefler, R., Wüstholz, V., Gurfinkel, A.: Compositional verification of smart contracts through communication abstraction. In: Drăgoi, C., Mukherjee, S., Namjoshi, K. (eds.) SAS 2021. LNCS, vol. 12913, pp. 429–452. Springer, Cham (2021). https://doi.org/10.1007/978-3-030-88806-0_21
49. Yeh, W.J., Young, M.: Compositional reachability analysis using process algebra. In: Symposium on Testing, Analysis, and Verification (TAV), pp. 49–59. ACM (1991). https://doi.org/10.1145/120807.120812
50. Zhang, Z., de Amorim, A.A., Jia, L., Pasareanu, C.S.: Automating compositional analysis of authentication protocols. In: 20th Conference on Formal Methods in Computer Aided Design (FMCAD), pp. 113–118. IEEE (2020). https://doi.org/10.34727/2020/isbn.978-3-85448-042-6_18

Efficient Interprocedural Data-Flow Analysis Using Treedepth and Treewidth

Amir Kafshdar Goharshady[✉] and Ahmed Khaled Zaher[✉]

Hong Kong University of Science and Technology, Clear Water Bay, Hong Kong
{goharshady,akazaher}@cse.ust.hk

Abstract. We consider interprocedural data-flow analysis as formalized by the standard IFDS framework, which can express many widely-used static analyses such as reaching definitions, live variables, and null-pointer. We focus on the well-studied on-demand setting in which queries arrive one-by-one in a stream and each query should be answered as fast as possible. While the classical IFDS algorithm provides a polynomial-time solution for this problem, it is not scalable in practice. More specifically, it will either require a quadratic-time preprocessing phase or takes linear time per query, both of which are untenable for modern huge code-bases with hundreds of thousands of lines. Previous works have already shown that parameterizing the problem by the treewidth of the program's control-flow graph is promising and can lead to significant gains in efficiency. Unfortunately, these results were only applicable to the limited special case of same-context queries.

In this work, we obtain significant speedups for the general case of on-demand IFDS with queries that are not necessarily same-context. This is achieved by exploiting a new graph sparsity parameter, namely the treedepth of the program's call graph. Our approach is the first to exploit the sparsity of control-flow graphs and call graphs at the same time and parameterize by both the treewidth and the treedepth. We obtain an algorithm with a linear preprocessing phase that can answer each query in constant time wrt the size of the input. Finally, our experimental results demonstrate that our approach significantly outperforms the classical IFDS and its on-demand variant.

Keywords: Static analysis · Data-flow analysis · IFDS · Parameterized algorithms

C. Dragoi et al. (Eds.): VMCAI 2023, LNCS 13881, pp. 177–202, 2023.
https://doi.org/10.1007/978-3-031-24950-1_9

1 Introduction

Data-Flow. Data-flow analysis is a catch-all term for a wide and expressive variety of static program analyses that include common tasks such as reaching definitions [30], points-to and alias analysis [74, 76, 77, 80–82], null-pointer dereferencing [32, 55, 58], uninitialized variables [61] and dead code elimination [43], as well as several other standard frameworks, e.g. gen-kill and bit-vector problems [47, 49, 50]. The common thread among data-flow analyses is that they consider certain "data facts" at each line of the code and then try to ascertain which data facts may/must hold at any given point [69]. This is often achieved by a worklist algorithm that keeps discovering new data facts until it reaches a fixed point and converges to the final solution [48, 69]. Variants of data-flow analysis are already included in most IDEs and compilers. For example, Eclipse has support for various data-flow analyses, such as unused variables and dead code elimination, both natively [35] and through plugins [31, 63]. Data-flow analyses have also been applied in the context of compiler optimization, e.g. for register allocation [52] and constant propagation analysis [17, 42, 73]. Additionally, they have found important use-cases in security [18], including in taint analysis [4] and detection of SQL injection attacks [41]. Due to their apparent importance, data-flow analyses have been widely studied by the verification, compilers, security and programming languages communities over the past five decades and are also included in program analysis frameworks such as Soot [9] and WALA [1].

Intraprocedural vs Interprocedural Analysis. Traditionally, data-flow analyses are divided into two general groups [46]:

- *Intraprocedural* approaches analyze each function/procedure of the code in isolation [31, 47]. This enables modularity and helps with efficiency, but the tradeoff is that the call-context and interactions between the different procedures are not accounted for, hence leading to relatively lower precision.
- In contrast, *interprocedural* analyses consider the entirety of the program, i.e. all the procedures, at the same time. They are often sensitive to call context and only focus on execution paths that respect function invocation and return rules, i.e. when a function ends, control has to return to the correct site of the last call to that function [29, 69]. Unsurprisingly, interprocedural analyses are much more accurate but also have higher complexity than their intraprocedural counterparts [66, 69, 72, 75].

IFDS. One of the most classical and widely-used frameworks for interprocedural data-flow analysis is that of *Interprocedural Finite Distributive Subset problems* (IFDS) [68, 69]. IFDS is an expressive framework that can perform all the analyses enumerated above by assigning a set D of data facts to each line of the program and then applying a reduction to a variant of graph reachability with side conditions ensuring that function call and return rules are enforced. For example, in a null-pointer analysis, each data fact d_i in D is of the form "the pointer p_i might be null". See Sect. 2 for details. Given a program with n lines, the original IFDS algorithm in [69] solves the data-flow problem *for a fixed starting point* in time $O(n \cdot |D|^3)$. Due to its elegance and generality, this framework has been thoroughly studied by the community. It has been extended to various platforms and settings [4,10,57], notably the on-demand setting [45] and in presence of correlated method calls [65], and has been implemented in standard static analysis tools [1,9].

On-Demand Data-Flow Analysis. Due to the expensiveness of exhaustive data-flow analysis, i.e. an analysis that considers every possible starting point, many works in the literature have turned their focus to on-demand analysis [6,22, 34,45,67,77,81,82]. In this setting, the algorithm can first run a preprocessing phase in which it collects some information about the program and produces summaries that can be used to speedup the query phase. Then, in the query phase, the algorithm is provided with a series of queries and should answer each one as efficiently as possible. Each query is of the form $(\ell_1, d_1, \ell_2, d_2)$ and asks whether it is possible to reach line ℓ_2 of the program, with the data fact d_2 holding at that line, assuming that we are currently at line ℓ_1 and data fact d_1 holds[1]. It is also noteworthy that on-demand algorithms commonly use information found in previous queries to handle the current query more efficiently. On-demand analyses are especially important in just-in-time compilers and their speculative optimizations [7,22,28,37,53], in which having dynamic information about the current state of the program can dramatically decrease the overhead for the compiler. In addition, on-demand analyses have the following merits (quoted from [45,68]):

- narrowing down the focus to specific points of interest,
- narrowing down the focus to specific data-flow facts of interest,
- reducing the work in preliminary phases,
- side-stepping incremental updating problems, and
- offering on-demand analysis as a user-level operation that helps programmers with debugging.

On-Demand IFDS. An on-demand variant of the IFDS algorithm was first provided in [45]. This method has no preprocessing but memoizes the information obtained in each query to help answer future queries more efficiently. It outperforms the classical IFDS algorithm of [69] in practice, but the only theoretical guarantee is that of same worst-case complexity, i.e. the on-demand version will never be any worse than running a new instance of the IFDS algorithm for each query. Hence, the worst-case runtime on m queries is $O(n \cdot m \cdot |D|^3)$. Recall that n is the number of lines in the program and $|D|$ is the number of data facts at each line. Alternatively, one can push all the complexity to the preprocessing phase, running the IFDS algorithm exhaustively for each possible starting point, and then answering queries by a simple table lookup. In this case, the preprocessing will take $O(n^2 \cdot |D|^3)$. Unfortunately, none of these two variants are scalable

---

[1] Instead of single data facts d_1 and d_2, we can also use a set of data facts at each of ℓ_1 and ℓ_2, but as we will see in Sect. 2, this does not affect the generality.

enough to handle codebases with hundreds of thousands of lines, e.g. standard utilities in the DaCapo benchmark suite [8] such as Eclipse or Jython.

Same-Context On-Demand IFDS. The work [22] provides a parameterized algorithm for a special case of the on-demand IFDS problem. The main idea in [22] is to observe that control-flow graphs of real-world programs are sparse and tree-like and that this sparsity can be exploited to find faster algorithms for *same-context* IFDS analysis. More specifically, the sparsity is formalized by a graph parameter called treewidth [70,71]. Intuitively speaking, treewidth is a measure of how much a given graph resembles a tree, i.e. more tree-like graphs have smaller treewidth. See Sect. 3 for a formal definition. It is proven that structured programs in several languages, such as C, have bounded treewidth [78] and there are experimental works that establish small bounds on the treewidth of control-flow graphs of real-world programs written in other languages, such as Java [44], Ada [16] and Solidity [19]. Using these facts, [22] provides an on-demand algorithm with $O(n \cdot |D|^3)$ preprocessing time and $O\left(\lceil \frac{|D|}{\lg n} \rceil\right)$ time per query[2]. In practice, $|D|$ is often tiny in comparison with n and hence this algorithm is considered to have linear preprocessing and constant query time. Unfortunately, the algorithm in [22] is not applicable to the general case of IFDS and can only handle *same-context* queries. Specifically, the queries in [22] provide a tuple $(\ell_1, d_1, \ell_2, d_2)$ just as in standard IFDS queries but they ask whether it is possible to reach (ℓ_2, d_2) from (ℓ_1, d_1) by an execution path that *preserves the state of the stack*, i.e. ℓ_1 and ℓ_2 are limited to being in the same function and the algorithm only considers execution paths in which every function call returns before reaching ℓ_2.

Our Contribution. In this work, we present a novel algorithm for the general case of on-demand IFDS analysis. Our contributions are as follows:

- We identify a new sparsity parameter, namely the treedepth of the program's call graph, and use it to find more efficient parameterized algorithms for IFDS data-flow analysis. Hence, our approach exploits the sparsity of both call graphs and control-flow graphs and bounds both the treedepth and the treewidth. Treedepth [14,60] is a well-studied graph sparsity parameter. It intuitively measures how much the graph resembles a star, i.e. a shallow tree [59, Chapter 6].
- We provide a scalable algorithm that is not limited to same-context queries as in [22] and is much more efficient than the classical on-demand IFDS algorithm of [45]. Specifically, after a lightweight preprocessing that takes $O(n \cdot |D|^3 \cdot \text{treedepth}^2)$ time, our algorithm is able to answer each query in $O(|D|^3 \cdot \text{treedepth})$. Thus, this is the first algorithm that can solve the general case of on-demand IFDS scalably and handle codebases and programs with hundreds of thousands or even millions of lines of code.
- We provide experimental results on the standard DaCapo benchmarks [8] illustrating that:

---

[2] This algorithm uses the Word-RAM model of computation. The division by $\lg n$ is obtained by encoding $\lg n$ bits in one word.

- our assumption of the sparsity of call graphs and low treedepth holds in practice in real-world programs; and
- our approach comfortably beats the runtimes of exhaustive and on-demand IFDS algorithms [45,69] by two orders of magnitude.

Novelty. Our approach is novel in several directions:

- Unlike previous optimizations for IFDS that only focused on control-flow graphs, we exploit the sparsity of both control-flow graphs and call graphs.
- To the best of our knowledge, this is the first time that the treedepth parameter is exploited in a static analysis or program verification setting. While this parameter is well-known in the graph theory community and we argue that it is a natural candidate for formalizing the sparsity of call graphs (See Sect. 3), this is the first work that considers it in this context.
- We provide the first theoretical improvements in the runtime of general on-demand data-flow analysis since [45], which was published in 1995. Previous improvements were either heuristics without a theoretical guarantee of improvement or only applicable to the special case of same-context queries.
- Our algorithm is much faster than [45] in practice and is the first to enable on-demand interprocedural data-flow analysis for programs with hundreds of thousands or even millions of lines of code. Previously, for such large programs, the only choices were to either apply the data-flow analysis intraprocedurally, which would significantly decrease the precision, or to limit ourselves to the very special case of same-context queries [22].

Limitation. The primary limitation of our algorithm is that it relies on the assumption of bounded treewidth for control-flow graphs and bounded treedepth for call graphs. In both cases, it is theoretically possible to generate pathological programs that have arbitrarily large width/depth: [44] shows that it is possible to write Java programs whose control-flow graphs have any arbitrary treewidth. However, such programs are highly unrealistic, e.g. they require a huge number of labeled nested while loops with a large nesting depth and break/continue statements that reference a while loop that is many levels above in the nesting order. Similarly, in Sect. 3, we construct a pathological example program whose call graph has a large treedepth. Nevertheless, this is also unrealistic and real-world programs, such as those in the DaCapo benchmark suite, have both small treewidth and small treedepth, as shown in Sect. 5 and [16,19,44,78].

Organization. In Sect. 2, we present the standard IFDS framework and formally define our problem. This is followed by a presentation of the graph sparsity parameters we will use, i.e. treewidth and treedepth, in Sect. 3. Our algorithm is then presented in Sect. 4, followed by experimental results in Sect. 5.

2 The IFDS Framework

In this section, we provide an overview of the IFDS framework following the notation and presentation of [22,69] and formally define the interprocedural data-flow problem considered in this work.

Model of Computation. Throughout this paper, we consider the standard word RAM model of computation in which every word is of length $w = \Theta(\lg n)$, where n is the length of the input. We assume that common operations, such as addition, shift and bitwise logic between a pair of words, take $O(1)$ time. Note that this has no effect on the implementation of our algorithms since most modern computers have a word size of at least 64 and we are not aware of any possible real-world input to our problems whose size can potentially exceed 2^{64}. We need this assumption since we use the algorithm of [22] as a black box. Our own contribution does not rely on the word RAM model.

Control-Flow Graphs. In IFDS, a program with k functions $f_1, f_2, \ldots, f_k$ is modeled by k control-flow graphs $G_1, G_2, \ldots, G_k$, one for each function, as well as certain interprocedural edges that model function calls and returns. The graphs G_i are standard control-flow graphs, having a dedicated *start vertex* s_i modeling the beginning point of f_i, another dedicated *end vertex* e_i modeling its end point, one vertex for every line of code in f_i, and a directed edge from u to v, if line v can potentially be reached right after line u in some execution of the program. The only exception is that function call statements are modeled by two vertices: a *call* vertex c_l and a *return site* vertex r_l. The vertex c_l has only incoming edges, whereas r_l has only outgoing edges. There is also an edge from c_l to r_l, which is called a *call-to-return-site* edge. This edge is used to pass local information, e.g. information about the variables in f_i that are unaffected by the function call, from c_l to r_l.

Supergraphs. The entire program is modeled by a *supergraph* G, consisting of all the control-flow graphs G_i, as well as interprocedural edges between them. If a function call statement in f_i, corresponding to vertices c_l and r_l in G_i, calls the function f_j, then the supergraph contains the following interprocedural edges:

- a *call-to-start* edge from the call vertex c_l to the start vertex s_j of the called function f_j, and
- an *exit-to-return-site* edge from the endpoint e_j of the called function f_j back to the return site r_l.

Call Graphs. Given a supergraph G as above, a call graph is a directed graph C whose vertices are the functions $f_1, \ldots, f_k$ of the program and there is an edge from f_i to f_j iff there is a function call statement in f_i that calls f_j. In other words, the call graph models the interprocedural edges in the supergraph and the supergraph can be seen as a combination of the control-flow and call graphs.

Example. Figure 1 shows a program consisting of two functions (left) and its supergraph (right).

Valid Paths. The supergraph G potentially contains invalid paths, i.e. paths that are not realizable by an actual run of the underlying program. The IFDS framework only considers *interprocedurally valid* paths in G. These are the paths that respect the rules for function invocation and return. More concretely, when a function's execution ends, control should return to the correct return-site vertex

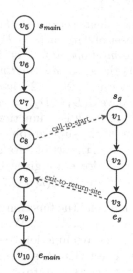

```
1  void g(int *&a, int *&b) {
2      b = a;
3  }
4
5  int main() {
6      int *a, *b;
7      a = new int(42);
8      g(a, b);
9      *b = 0;
10 }
```

Fig. 1. A C++ program (left) and its supergraph (right).

in its parent function. Formally, consider a path Π in G and let Π^* be the subsequence of Π that is obtained by removing any vertex that was not a call vertex c_l or a return-site vertex r_l. Then, Π is called a *same-context interprocedurally valid* path if Π^* can be generated from the non-terminal S in the following grammar:

$$S \rightarrow \ \epsilon \ | \ c_l \ S \ r_l \ S.$$

In other words, any function call in Π that was invoked in line c_l should end by returning to its corresponding return-site r_l. A same-context valid path preserves the state of the function call stack. In contrast, the path Π is *interprocedurally valid* or simply *valid* if Π^* is generated by the non-terminal S' in the following grammar:

$$S' \rightarrow \ S \ | \ S' \ c_l \ S.$$

A valid path has to respect the rules for returning to the right return-site vertex after the end of each function, but it does not necessarily keep the function call stack intact and it is allowed ot have function calls that do not necessarily end by the end of the path. Let u and v be vertices in the supergraph G. We denote the set of all same-context valid paths from u to v by SCVP(u, v) and the set of all interprocedurally valid paths from u to v by IVP(u, v). In IFDS, we only focus on valid paths and hence the problem is to compute a *meet-over-all-valid-paths* solution to data-flow facts, instead of the *meet-over-all-paths* approach that is usually taken in intraprocedural data-flow analysis [69].

IFDS Arena [69]. An *arena* of the IFDS data-flow analysis is a tuple $(G, D, \Phi, M, \sqcap)$ wherein:

- $G = (V, E)$ is a supergraph consisting of control-flow graphs and interprocedural edges, as illustrated above.

- D is a finite set of *data facts*. Intuitively, we would like to keep track of which subset of data facts in D hold at any vertex of G (line of the program).
- The *meet operator* $\sqcap$ is either union or intersection, i.e. $\sqcap \in \{\cup, \cap\}$.
- Φ is the set of *distributive flow functions* over $\sqcap$. Every function $\varphi \in \Phi$ is of the form $\varphi : 2^D \to 2^D$ and for every pair of subsets of data facts $D_1, D_2 \subseteq D$, we have $\varphi(D_1 \sqcap D_2) = \varphi(D_1) \sqcap \varphi(D_2)$.
- $M : E \to \Phi$ is a function that assigns a distributive flow function to every edge of the supergraph. Informally, $M(e)$ models the effect of executing the edge e on the set of data facts. If the data facts that held before the execution of the edge e are given by a subset $D' \subseteq D$, then the data facts that hold after e are $M(e)(D') \subseteq D$.

We can extend the function M to any path Π in G. Let Π be a path consisting of the edges $e_1, e_2, \ldots, e_\pi$. We define $M(\Pi) := M(e_\pi) \circ M(e_{\pi-1}) \circ \cdots \circ M(e_1)$. Here, $\circ$ denotes function composition. According to this definition, $M(\Pi)$ models the effect that Π's execution has on the set of data facts that hold in the program's current state.

Problem Formalization. Consider an initial state $(u, D_1) \in V \times 2^D$ of the program, i.e. we are at line u of the program and we know that the data facts in D_1 hold. Let $v \in V$ be another line, we define

$$\text{MIVP}(u, D_1, v) := \bigsqcap_{\Pi \in \text{IVP}(u,v)} M(\Pi)(D_1).$$

We simplify the notation to $\text{MIVP}(v)$, when the initial state is clear from the context. Our goal is to compute the MIVP values. Intuitively, MIVP corresponds to *meet-over-all-valid-paths*. If the meet operator is intersection, then $\text{MIVP}(v)$ models the data facts that *must* hold whenever we reach v. Conversely, if we use union as our meet operator, then $\text{MIVP}(v)$ is the set of data facts that *may* hold when reaching v. The work [69] provides an algorithm to compute $\text{MIVP}(v)$ for every end vertex v in $O(n \cdot |D|^3)$, in which $n = |V|$.

Same-Context IFDS. We can also define a same-context variant of MIVP as follows:

$$\text{MSCVP}(v) := \bigsqcap_{\Pi \in \text{SCVP}(u,v)} M(\Pi)(D_1).$$

The intuition is similar to MIVP, except that in MSCVP we only consider same-context valid paths that preserve the function call stack's status and ignore other valid paths. The work [22] uses parameterization by treewidth of the control-flow graphs to obtain faster algorithms for computing MSCVP. However, its algorithms are limited to the same-context setting. In contrast, in this work, we follow the original IFDS formulation of [69] and focus on MIVP, not MSCVP. Our main contribution is that we present the first theoretical improvement for computing MIVP since [45,69].

Dualization. In this work, we only consider the cases in which the meet operator is union. In other words, we focus on *may* analyses. IFDS instances in which the meet operator is intersection, also known as *must* analyses, can be reduced to union instances by a simple dualization. See [48,69] for details.

Data Fact Domain. In our presentation, we are assuming that there is a fixed global data fact domain D. In practice, the domain D can differ in every function of the program. For example, in a null-pointer analysis, the data facts in each function keep track of nullness of the pointers that are either global or local to that particular function. However, having different D sets would reduce the elegance of the presentation and has no real effect on any of the algorithms. So, we follow [22,69] and consider a single domain D in the sequel. Our implementation in Sect. 5 supports different domains for each function.

Graph Representation of Functions [69]. Every union-distributive function $\varphi : 2^D \to 2^D$ can be succinctly represented by the following relation $R_\varphi \subseteq (D \cup \{0\}) \times (D \cup \{0\})$:

$$R_\varphi := \{(0,0)\} \ \cup \ \{(0,d) \mid d \in \varphi(\emptyset)\} \ \cup \ \{(d_1,d_2) \mid d_2 \in \varphi(\{d_1\}) \setminus \varphi(\emptyset)\}.$$

The intuition is that, in order to specify the union-distributive function φ, it suffices to fix $\varphi(\emptyset)$ and $\varphi(\{d\})$ for every $d \in D$. Then, we always have

$$\varphi(\{d_1, d_2, \ldots, d_r\}) = \varphi(\{d_1\}) \cup \varphi(\{d_2\}) \cup \cdots \cup \varphi(\{d_r\}).$$

We use a new item 0 to model $\varphi(\emptyset)$, i.e. $0 \ R_\varphi \ d \Leftrightarrow d \in \varphi(\emptyset)$. To specify $\varphi(\{d\})$, we first note that $\varphi(\emptyset) \subseteq \varphi(\{d\})$, so we only need to specify the elements that are in $\varphi(\{d\})$ but not $\varphi(\emptyset)$. These are precisely the elements that are in relation with d. In other words, $\varphi(\{d\}) = \varphi(\emptyset) \cup \{d' \mid d \ R_\varphi \ d'\}$. We can further represent the relation R_φ as a bipartite graph H_φ in which each part consists of the vertices $D \cup \{0\}$ and R_φ defines the edges.

Example. Figure 2 shows the graph representation of several union-distributive functions.

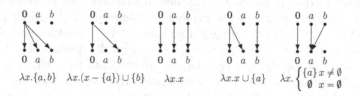

Fig. 2. Graph representation of union-distributive functions with $D = \{a, b\}$ [22].

Composition of Graph Representations [69]. What makes this graph representation particularly elegant is that we can compose two functions by a simple reachability computation. Specifically, if φ_1 and φ_2 are distributive, then so is $\varphi_2 \circ \varphi_1$. By definition chasing, we can see that $R_{\varphi_2 \circ \varphi_1} = R_{\varphi_1}; R_{\varphi_2} = \{(d_1,d_2) \mid \exists d_3 \ (d_1,d_3) \in R_{\varphi_1} \wedge (d_3,d_2) \in R_{\varphi_2}\}$. Thus, to compute the graph representation $H_{\varphi_2 \circ \varphi_1}$, we simply merge the bottom part of H_{φ_1} with the top part of

H_{φ_2} and then compute reachability from the top-most layer to the bottom-most layer.

Example. Figure 3 illustrates how the composition of two distributive functions can be obtained using their graph representations. Note that this process sometimes leads to superfluous edges. For example, since we have the edge $(\mathbf{0}, a)$ in the result, the edge (b, a) is not necessary. However, having it has no negative side-effects, either.

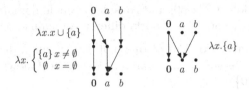

Fig. 3. Composing two distributive functions using reachability [22].

Exploded Supergraph [69]. Consider an IFDS arena $(G = (V, E), D, \Phi, M, \cup)$ as above and let $D^* := D \cup \{\mathbf{0}\}$. The *exploded supergraph* of this arena is a directed graph $\overline{G} = (\overline{V}, \overline{E})$ in which:

- $\overline{V} = V \times D^*$, i.e. we take $|D^*|$ copies of each vertex in the supergraph G, one corresponding to each data fact in D^*.
- $\overline{E} = \{(u_1, d_1, u_2, d_2) \in \overline{V} \times \overline{V} \mid (u_1, u_2) \in E \ \wedge \ (d_1, d_2) \in R_{M(u_1, u_2)}\}$. In other words, every edge between vertices u_1 and u_2 in the supergraph G is now replaced by the graphic representation of its corresponding distributive flow function $M(u_1, u_2)$.

Naturally, we say a path $\overline{\Pi}$ in $\overline{G}$ is interprocedurally (same-context) valid, if the path Π in G, obtained by ignoring the second component of every vertex in $\overline{\Pi}$, is interprocedurally (same-context) valid.

Reduction to Reachability. We can now reformulate our problem based on reachability by valid paths in the exploded supergraph $\overline{G}$. Consider an initial state $(u, D_1) \in V \times 2^D$ of the program and let $v \in V$ be another line. Since the exploded supergraph contains representations of all distributive flow functions, it already encodes the changes that happen to the data facts when we execute one step of the program. Thus, it is straightforward to see that for any data fact d_2, we have $d_2 \in \texttt{MIVP}(u, D_1, v)$ if and only if there exist a data fact $d_1 \in D_1$ such that the vertex (v, d_2) in $\overline{G}$ is reachable from the vertex (u, d_1) using an interprocedurally valid path [69]. Hence, our data-flow analysis is now reduced to reachability by valid paths. Moreover, instead of computing $\texttt{MIVP}$ values, we can simplify our query structure so that each query provides two vertices (u, d_1) and (v, d_2) in the exploded supergraph $\overline{G}$ and asks whether there is a valid path from (u, d_1) to (v, d_2).

Example. Figure 4 shows the same program as in Fig. 1, together with its exploded supergraph for null-pointer analysis. Here, we have two data facts: d_1 models the fact "the pointer a may be null" and d_2 does the same for b. Starting

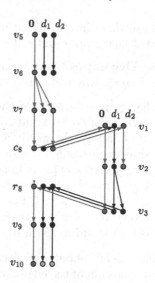

```
1  void g(int *&a, int *&b) {
2      b = a;
3  }
4
5  int main() {
6      int *a, *b;
7      a = new int(42);
8      g(a, b);
9      *b = 0;
10 }
```

Fig. 4. A program (left) and its exploded supergraph (right).

from line 5, i.e. the beginning of the main function, and knowing no data facts, i.e. $D_1 = \{\mathbf{0}\}$, we would like to see if either a or b might be null at the end of the main function. Using a reachability analysis on the exploded supergraph, we can identify all vertices that can be reached by a valid path (green) and conclude that neither a nor b may be null by the end of the program.

On-Demand Analysis. As mentioned in Sect. 1, we focus on on-demand analysis and distinguish between a preprocessing phase in which the algorithm can perform a lightweight pass over the input and a query phase in which the algorithm has to respond to a large number of queries. The queries appear in a stream and the algorithm has to handle each query as fast as possible. Based on the discussion above, each query is of the form $(u_1, d_1, u_2, d_2) \in V \times D^* \times V \times D^*$ and the algorithm should report whether there exist an interprocedurally valid path from (u_1, d_1) to (u_2, d_2) in the exploded supergraph $\overline{G}$.

Bounded Bandwidth Assumption. Following previous works such as [22, 45, 69], we assume that the "bandwidth" in function calls and returns is bounded. More concretely, we assume there exists a small constant β such that for every interprocedural call-to-start or exit-to-return-site edge e in our supergraph G, the degree of each vertex in the graph representation $H_{M(e)}$ is at most β. This is a classical assumption made in IFDS and all of its extensions. Intuitively, it models the idea that every parameter in a called function depends on only a few variables in the call site line c of the callee, and conversely, that the return value of a function is only dependent on a few variables at its last line.

3   Treewidth and Treedepth

In this section, we provide a short overview of the concepts of treewidth and treedepth. Treewidth and treedepth are both graph sparsity parameters and we will use them in our algorithm in the next section to formalize the sparsity of control-flow graphs and call graphs, respectively.

Tree Decompositions [13,70,71]. Given an undirected graph $G = (V, E)$, a *tree decomposition* of G is a rooted tree $T = (\mathfrak{B}, E_T)$ such that:

i. Every node $b \in \mathfrak{B}$ of the tree T has a corresponding subset $V_b \subseteq V$ of vertices of G. To avoid confusion, we reserve the word *"vertex"* for vertices of G and use the word *"bag"* to refer to nodes of the tree T. This is natural, since each bag b has a subset V_b of vertices.

ii. Every vertex appears in some bag, i.e. $\bigcup_{b \in \mathfrak{B}} V_b = V$.

iii. For every edge $\{u, v\} \in E$, there is a bag that contains both of its endpoints, i.e. $\exists b \in \mathfrak{B} \ \{u, v\} \subseteq V_b$.

iv. Every vertex $v \in V$ appears in a connected subtree of T. Equivalently, if b is on the unique path from b' to b'' in T, then $V_b \supseteq V_{b'} \cap V_{b''}$.

When talking about tree decompositions of directed graphs, we simply ignore the orientation of the edges and consider decompositions of the underlying undirected graph. Intuitively, a tree decomposition covers the graph G by a number of bags[3] that are connected to each other in a tree-like manner. If the bags are small, we are then able to perform dynamic programming on G in a very similar manner to trees [11,21,39,40,54]. This is the motivation behind the following definition.

Treewidth [71]. The *width* of a tree decomposition is defined as the size of its largest bag minus 1, i.e. $w(T) := \max_{b \in \mathfrak{B}} |V_b| - 1$. The *treewidth* of a graph G is the smallest width amongst all of its tree decompositions. Informally speaking, treewidth is a measure of tree-likeness. Only trees and forests have a treewidth of 1, and, if a graph G has treewidth k, then it can be decomposed into bags of size at most $k + 1$ that are connected to each other in a tree-like manner.

Example. Figure 5 shows a graph G on the left and a tree decomposition of width 2 for G on the right. In the tree decomposition, we have highlighted the connected subtree of each vertex by dotted lines. This tree decomposition is optimal and hence the treewidth of G is 2.

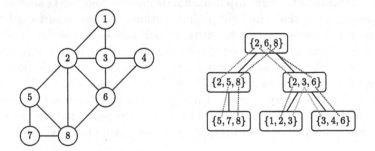

Fig. 5. A graph G (left) and one of its tree decompositions (right).

---

[3] The bags do not have to be disjoint.

Computing Treewidth. In general, it is NP-hard to compute the treewidth of a given graph. However, for any constant k, there is a linear-time algorithm that decides whether the graph has treewidth at most k and, if so, also computes an optimal tree decomposition [12]. As such, most treewidth-based algorithms assume that an optimal tree decomposition is given as part of the input.

Treewidth of Control-Flow Graphs. In [78], it was shown that the control-flow graphs of goto-free programs in a number of languages such as C and Pascal have a treewidth of at most 7. Moreover, [78] also provides a linear-time algorithm that, while not necessarily optimal, always outputs a tree decomposition of width at most 7 for the control-flow graph of programs in these languages by a single pass over the parse tree of the program. This algorithm is implemented in [24], and is the algorithm we use for obtaining our tree decompositions in Sect. 5. Alternatively, one can use the algorithm of [12] to ensure that an optimal decomposition is used at all times. The theoretical bound of [78] does not apply to Java, but the work [44] showed that the treewidth of control-flow graphs in real-world Java programs is also bounded. This bounded-treewidth property has been used in a variety of static analysis and compiler optimization tasks to speed up the underlying algorithms [2,3,5,20,23,25–27,36,38,62]. Nevertheless, one can theoretically construct pathological examples with high treewidth.

Balancing Tree Decompositions. The runtime of our algorithm in Sect. 4 depends on the height of the tree decomposition. However, [15] provides a linear-time algorithm that, given a graph G and a tree decomposition of constant width t, produces a binary tree decomposition of height $O(\lg n)$ and width $O(t)$. Combining this with the algorithms of [78] and [12] for computing low-width tree decompositions allows us to assume that we are always given a balanced and binary tree decomposition of bounded width for each one of our control-flow graphs as part of our IFDS input.

We now switch our focus to the second parameter that appears in our algorithms, namely treedepth.

Partial Order Trees [60]. Let $G = (V, E)$ be an undirected connected graph. A *partial order tree* (POT)[4] over G is a rooted tree $T = (V, E_T)$ on the same set of vertices as G that additionally satisfies the following property:

– For every edge $\{u, v\} \in E$ of G, either u is an ancestor of v in T or v is ancestor of u in T.

The intuition is quite straightforward: T defines a partial order $\prec_T$ over the vertices V in which every element u is assumed to be smaller than its parent p_u, i.e. $u \prec_T p_u$. For T to be a valid POT, every pair of vertices that are connected by an edge in G should be comparable in $\prec_T$. If G is not connected, then we will have a partial order forest, consisting of a partial order tree for each connected component of G. With a slight abuse of notation, we call this a POT, too.

---

[4] The name *partial order tree* is not standard in this context, but we use it throughout this work since it provides a good intuition about the nature of T. Usually, the term "treedepth decomposition" is used instead.

Example. Figure 6 shows a graph G (left) together with a POT of depth 4 for G (right). In the POT, the edges of the original graph G are shown by dotted red lines. Every edge of G goes from a node in T to one of its ancestors.

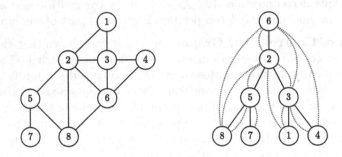

Fig. 6. A graph G (left) and a POT of depth 4 for G (right).

Treedepth [60]. The treedepth of an undirected graph G is the smallest depth among all POTs of G.

Path Property of POTs [60]. Let $T = (V, E_T)$ be a POT for a graph $G = (V, E)$ and u and v two vertices in V. Define A_u as the set of ancestors of u in T and define A_v similarly. Let $A := A_u \cap A_v$ be the set of common ancestors of u and v. Then, any path that goes from u to v in the graph G has to intersect A, i.e. it has to go through a common ancestor.

Sparsity Assumption. In the sequel, our algorithm is going to assume that call graphs of real-world programs have small treedepth. We establish this experimentally in Sect. 5. However, there is also a natural reason why this assumption is likely to hold in practice. Consider the functions in a program. It is natural to assume that they were developed in a chronological order, starting with base (phase 1) functions, and then each phase of the project used the functions developed in the previous phases as libraries. Thus, the call graph can be partitioned to a small number of layers based on the development phase of each function. Moreover, each function typically calls only a small number of previous functions. So, an ordering based on development phase is likely to give us a POT with small depth. The depth would typically depend on the number of phases and the degree of each function in the call graph, but these are both small parameters in practice.

Pathological Example. It is possible in theory to write a program whose call graph has an arbitrarily large treedepth. However, such a program is not realistic. Suppose that we want a program with treedepth n. We can create n functions $f_1, f_2, \ldots, f_n$ and then ensure that each function f_i calls every other function f_j $(j \neq i)$. In this strange program, our call graph will simply be a complete graph on n vertices. Since every two vertices in this graph have to be comparable, its POT will be a path with depth n. So, its treedepth is $\Theta(n)$.

Computing Treedepth. As in the case of treewidth, it is NP-hard to compute the treedepth of a given graph [64]. However, for any fixed constant k, there is a linear-time algorithm that decides whether a given graph has treedepth at most k and, if so, produces an optimal POT [56]. Thus, in the sequel, we assume that all inputs include a POT of the call graph with bounded depth.

4 Our Parameterized Algorithm

In this section, we present our parameterized algorithm for solving the general case of IFDS data-flow analysis, assuming that the control-flow graphs have bounded treewidth and the call graph has bounded treedepth. Throughout this section, we fix an IFDS arena $(G, D, \Phi, M, \cup)$ given by an exploded supergraph $\overline{G}$ and assume that every control-flow graph comes with a balanced binary tree decomposition of width at most k_1. We also assume that a POT of depth k_2 over the call graph is given as part of the input. All these assumptions are without loss of generality since the tree decompositions and POT can be computed in linear time using the algorithms mentioned in Sect. 3. Before presenting our algorithm, we should first define a few useful notions.

Algorithm for Same-Context IFDS. The work [22] provides an on-demand parameterized algorithm for same-context IFDS. This algorithm requires a balanced and binary tree decomposition of constant width for every control-flow graph and provides a preprocessing runtime of $O(n \cdot |D|^3)$, after which it can answer *same-context* queries in time $O\left(\lceil \frac{|D|}{\lg n} \rceil\right)$. A same-context query is a query of the form $(u_1, d_1, u_2, d_2) \in V \times D^* \times V \times D^*$ which asks whether there exists a *same-context* valid path from (u_1, d_1) to (u_2, d_2) in the exploded supergraph $\overline{G}$. Below, we use [22]'s algorithm for same-context queries as a black box.

Stack States. Let F be the set of functions in our program. A *stack state* is simply a finite sequence of functions $\xi = \langle \xi_i \rangle_{i=1}^s \in F^s$. We use a stack state to keep track of the set of functions that have been called but have not finished their execution and returned yet.

Persistence. Consider an interprocedurally valid path $\Pi = \langle \pi_i \rangle_{i=1}^p$ in the supergraph G and let $\Pi^* = \langle \pi_i^* \rangle_{i=1}^s$ be the sub-sequence of Π that only includes call vertices c_l and return vertices r_l. For each π_i^* that is a call vertex, let f_i be the function called by π_i^*. We say the function call to f_i is *temporary* if π_i^* is matched by a corresponding return-site vertex π_j^* in Π^* with $j > i$. Otherwise, f_i is a *persistent* function call. In other words, temporary function calls are the ones who return before the end of the path Π and persistent ones are those that are added to the stack but never popped. So, if the stack is at state ξ before executing Π, it will be in state $\xi \cdot \langle f_{i_1} \cdot f_{i_2} \cdots f_{i_r} \rangle$ after Π's execution, in which the f_{i_j}'s are our persistent function calls. Moreover, we can break down the path Π as follows:

$$\Pi = \Sigma_0 \cdot \Sigma_1 \cdot \pi_{i_1} \cdot \Sigma_2 \cdot \pi_{i_2} \cdots \Sigma_r \cdot \pi_{i_r} \cdot \Sigma_{r+1} \tag{1}$$

in which Σ_0 is an *intraprocedural* path, i.e. the part of Π that does not leave the initial function. Note that we either have $\Pi = \Sigma_0$ or Σ_0 should end with a function call. For every $i \neq 0$, Σ_i is a same-context valid path from the starting point of a function and π_{i_j} is a call vertex that calls the next persistent function f_{i_j}. We call (1) the *canonical partition* of the path Π.

Exploded Call Graph. Let $C = (F, E_C)$ be the call graph of our IFDS instance, in which F is the set of functions in the program. We define the *exploded call graph* $\overline{C} = (\overline{F}, \overline{E_C})$ as follows:

- Our vertex set $\overline{F}$ is simply $F \times D^*$. Recall that $D^* := D \cup \{\mathbf{0}\}$.
- There is an edge from the vertex (f_1, d_1) to the vertex (f_2, d_2) in $\overline{E_C}$ iff:
  - There is a call statement $c \in V$ in the function f_1 that calls f_2;
  - There exist a data fact $d_3 \in D^*$ such that (i) there is a *same-context* valid path from (s_{f_1}, d_1) to (c, d_3) in the exploded supergraph $\overline{G}$, and (ii) there is an edge from (c, d_3) to (s_{f_2}, d_2) in the exploded supergraph $\overline{G}$.

The edges of the exploded call graph model the effect of a valid path that starts at s_{f_1}, i.e. the first line of f_1, when the function call stack is empty and reaches s_{f_2}, with stack state $\langle f_2 \rangle$. Informally, this corresponds to executing the program starting form f_1, potentially calling any number of temporary functions, then waiting for all of these temporary functions and their children to return so that we again have an empty stack, and then finally calling f_2 from the call-site c, hence reaching stack state $\langle f_2 \rangle$. Intuitively, this whole process models the substring $\Sigma \cdot c$ in the canonical partition of a valid path, in which Σ is a same-context valid path, and f_2 is the next persistent function, which was called at c. Hence, going forward, we do not plan to pop f_2 from the stack.

Treedepth of $\overline{C}$. Recall that we have a POT T of depth k_2 for the call graph C. In $\overline{C}$, every $f \in C$ is replaced by $|D^*|$ vertices $(f, \mathbf{0}), (f, d_1), \ldots, (f, d_{|D|})$. We can obtain a valid POT $\overline{T}$ for $\overline{C}$ by processing the tree in a top-down order and replacing every vertex that corresponds to a function f with a path of length $|D^*|$, as shown in Fig. 7. It is straightforward to verify that $\overline{T}$ is a valid POT of depth $k_2 \cdot |D^*|$ for $\overline{C}$.

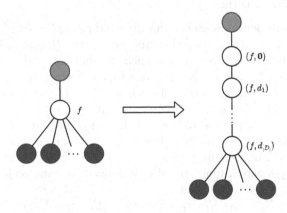

Fig. 7. Obtaining $\overline{T}$ from T by expanding each vertex to a path.

Preprocessing. The preprocessing phase of our algorithm consists of the following four steps:

1. *Same-context Preprocessing:* Our algorithm runs the preprocessing phase of [22]'s algorithm for same-context IFDS. This is done as a black box. See [22] for the details of this step.

2. *Intraprocedural Preprocessing:* For every vertex $(u, d) \in \overline{G}$, for which u is a line of the program in the function f, our algorithm performs an *intraprocedural* reachability analysis and finds a list of all the vertices of the form (c, d') such that:

   - c is a call-site vertex in the same function f.
   - There is an *intraprocedural* path from (u, d) to (c, d') that always remains within f and does not cause any function calls.

   Our algorithm computes this by a simple reverse DFS from every (c, d'). Intuitively, this is so that we can later handle the first part, i.e. Σ_0, in the canonical partition in 1. Note that this step is entirely intraprocedural and our reverse DFS is equivalent to the classical algorithms of [48]. Moreover, we can consider Σ_0 to be a same-context path instead of merely an intraprocedural path. In this case, we can rely on queries to [22] to do this step of our preprocessing.

3. *Computing Exploded Call Graph:* Our algorithm generates the exploded call graph $\overline{C}$ using its definition above. It iterates over every function f_1 and call site c in f_1. Let f_2 be the function called at c. For every pair $(d_1, d_3) \in D^* \times D^*$, our algorithm queries the same-context IFDS algorithm of [22] to see if there is a same-context valid path from (s_{f_1}, d_1) to (c, d_3). Note that we can make such queries since we have already performed the required same-context preprocessing in Step 1 above. If the query's result is positive, the algorithm iterates over every $d_2 \in D^*$ such that (c, d_3, s_{f_2}, d_2) is an edge in the exploded supergraph $\overline{G}$, and adds an edge from (f_1, d_1) to (f_2, d_2) in $\overline{C}$. The algorithm also computes the POT $\overline{T}$ as mentioned above. Intuitively, this step allows us to summarize the effects of each function call in the call graph so that we can later handle the control-flow graphs and the call graph separately.

4. *Computing Ancestral Reachability in $\overline{T}$:* For every vertex u in $\overline{T}$, let $\overline{T}_u^{\downarrow}$ be the subtree of $\overline{T}$ rooted at u and $\overline{F}_u^{\downarrow}$ be the set of descendants of u. For every u and every $v \in \overline{F}_u^{\downarrow}$, our algorithm precomputes $reach(u, v)$, i.e. whether u is reachable from v in $\overline{C}$ and also $reach(v, u)$. Informally, the idea is that every path from a vertex a in our exploded call graph to a vertex b has to go through one of the ancestors of a and b (See Sect. 3). Thus, it is natural to precompute the reachability relations between every vertex and all of its ancestors.

   To compute this, for every vertex u and every descendant v of u, we define:

$$up[u, v] := \begin{cases} 1 & \text{there is a path from } v \text{ to } u \text{ in } \overline{C}[\overline{F}_u^{\downarrow}] \\ 0 & \text{otherwise} \end{cases},$$

$$down[u, v] := \begin{cases} 1 & \text{there is a path from } u \text{ to } v \text{ in } \overline{C}[\overline{F}_u^{\downarrow}] \\ 0 & \text{otherwise} \end{cases}.$$

Note that in the definition above, we are only considering paths whose every internal vertex is in the subtree of u. We can find the values of $down[u, v]$ by simply running a DFS from u but ignoring all the edges that leave the subtree $\overline{T}_u^{\downarrow}$. Similarly, we can find the values of $up[u, v]$ by a similar DFS in which the orientation of all edges are reversed.

By the path property of POTs, every path ρ from v to u in $\overline{C}$ either has all of its vertices in the subtree $\overline{T}_u^{\downarrow}$ or visits some ancestors of u as internal vertices. Let w be the highest ancestor of u that is visited by ρ. Then, we must have $up[w, v] = down[w, u] = 1$. Similarly, if there is a path from u to v, we must have $up[w, u] = down[w, v] = 1$. Our algorithm simply sets:

$$reach(u, v) = \bigvee_w (up[w, u] \wedge down[w, v]),$$

and

$$reach(v, u) = \bigvee_w (up[w, v] \wedge down[w, u]).$$

Query. After the end of the preprocessing phase, our algorithm is ready to accept queries. Suppose that a query q asks whether there exists a valid interprocedural path from (u_1, d_1) to (u_2, d_2) in $\overline{G}$. Suppose that $\overline{\Pi}$ is such a valid path and Π is its trace on the supergraph G, i.e. the path obtained from $\overline{\Pi}$ by ignoring the second component of every vertex. We consider the canonical partition of Π as

$$\Pi = \Sigma_0 \cdot (\Sigma_1 \cdot \pi_{i_1}) \cdot (\Sigma_2 \cdot \pi_{i_2}) \cdots (\Sigma_r \cdot \pi_{i_r}) \cdot \Sigma_{r+1}$$

and its counterpart in $\overline{\Pi}$ as

$$\overline{\Pi} = \overline{\Sigma_0} \cdot \left(\overline{\Sigma_1} \cdot \overline{\pi_{i_1}}\right) \cdot \left(\overline{\Sigma_2} \cdot \overline{\pi_{i_2}}\right) \cdots \left(\overline{\Sigma_r} \cdot \overline{\pi_{i_r}}\right) \cdot \overline{\Sigma_{r+1}}.$$

Let $\overline{\Sigma_j}[1]$ be the first vertex in $\overline{\Sigma_j}$. For every $j \geq 1$, consider the subpath

$$\overline{\Sigma_j} \cdot \overline{\pi_{i_j}} \cdot \overline{\Sigma_{j+1}}[1].$$

This subpath starts at the starting point s_f of some function f and ends at the starting point $s_{f'}$ of the function f' called in $\overline{\pi_{i_j}}$. Thus, it goes from a vertex of the form (s_f, d_1) to a vertex of the form $(s_{f'}, d_2)$. However, by the definition of our exploded call graph $\overline{C}$, we must have an edge $\overline{e_j}$ in $\overline{C}$ going from (f, d_1) to (f', d_2). With a minor abuse of notation, we do not differentiate between f and s_f and replace this subpath with $\overline{e_j}$. Hence, every interprocedurally valid $\overline{\Pi}$ can be partitioned in the following format:

$$\overline{\Pi} = \overline{\Sigma_0} \cdot \overline{e_1} \cdot \overline{e_2} \cdots \overline{e_r} \cdot \overline{\Sigma_{r+1}}.$$

In other words, to obtain an interprocedurally valid path, we should first take an intraprocedural path $\overline{\Sigma_0}$ in our initial function, followed by a path $\overline{e_1} \cdot \overline{e_2} \cdots \overline{e_r}$ in the exploded call graph $\overline{C}$, and then a same-context valid path $\overline{\Sigma_{r+1}}$ in our target function. Note that $\overline{\Sigma_{r+1}}$ begins at the starting point of our target function.

Our algorithm uses the observation above to answer the queries. Recall that the query q is asking whether there exists a path from (u_1, d_1) to (u_2, d_2) in $\overline{G}$. Let f_1 be the function of u_1 and f_2 the function containing u_2. Our algorithm performs the following steps to answer the query:

1. Take all vertices of the form (c, d_3) such that c is a call vertex in f_1 and (c, d_3) is intraprocedurally reachable from (u_1, d_1). Note that this was precomputed in Step 2 of our preprocessing.
2. Find all successors of the vertices in Step 1 in $\overline{G}$. These successors are all of the form $(s_{f'}, d_4)$ for some function f', and their corresponding nodes in the exploded call graph are of the form (f', d_4).
3. Compute the set of all (f_2, d_5) vertices in $\overline{C}$ that are reachable from one of the (f', d_4) vertices obtained in the previous step. In this case, the algorithm uses the path property of POTs and tries all possible common ancestors of (f_2, d_5) and (f', d_4) as potential internal vertices in the path.
4. For each (f_2, d_5) found in the previous step, ask the same-context query from (s_{f_2}, d_5) to (f_2, d_2). For these same-context queries, our algorithm uses the method of [22] as a black box.
5. If any of the same-context queries in the previous step return true, then our algorithm also answers true to the query q. Otherwise, it answers false.

Intuition. Figure 8 provides an overview of how our query phase breaks an interprocedurally valid path down between $\overline{G}$ (red) and $\overline{C}$ (blue). Note that we do not distinguish between the vertex (f_2, d_5) of $\overline{C}$ and vertex (s_{f_2}, d_5) of $\overline{G}$. Explicitly, any path from (u_1, d_1) to (u_2, d_2) should first begin with an intraprocedural segment in the original function f_1. This part is precomputed and shown in red. Then, it switches from the exploded control-flow graph to the exploded call graph and follows a series of function calls. This is shown in blue. We have already precomputed the effect of each edge in the call graph and encoded this effect in the exploded call graph. Hence, the blue part of the path is simply a reachability query, which we can answer efficiently using our POT. We would like to see whether there is a path from $a = (f', d_4)$ to $b = (f_2, d_5)$. However, any such path should certainly go through one of the common ancestors of a and b in the POT. Since the treedepth is bounded, a and b have only a few ancestors. Moreover, we have already computed the reachability between any vertex and all of its ancestors. So, a few table lookups can tell us whether there is a path from a to b. Finally, when we reach the beginning of our target function f_2, we have to take a same-context valid path to our target state (u_2, d_2). To check if such a path exists, we simply rely on [22] as a black box.

Runtime Analysis of the Preprocessing Phase. Our algorithm is much faster than the classical IFDS algorithm of [69]. More specifically, for the pre-processing, we have:

– Step 1 is a black box from [22] and takes $O(n \cdot |D|^3)$.

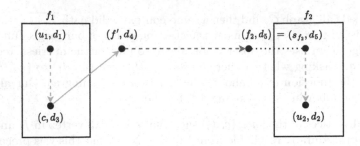

Fig. 8. An overview of the query phase. (Color figure online)

- Step 2 is a simple intraprocedural analysis that runs a reverse DFS from every node (c, d) in any function f. Assuming that the function f has α lines of code and a total of δ function call statements, this will take $O(\alpha \cdot \delta \cdot |D|^3)$. Assuming that δ is a small constant, this leads to an overall runtime of $O(n \cdot |D|^3)$. Note that this assumption is in line with reality since we rarely, if ever, encounter functions that call more than a constant number of other functions.
- In Step 3, we have at most $O(n \cdot |D|)$ call nodes of the form (c, d_3). Based on the bounded bandwidth assumption, each such node leads to constantly many possibilities for d_2. So, we perform at most $O(n \cdot |D|^2)$ calls to the same-context query procedure. Each same-context query takes $O(\lceil |D| / \lg n \rceil)$, so the overall runtime of this step is $O(n \cdot |D|^3 / \lg n)$.
- In Step 4, the total time for computing all the *up* and *down* values is $O(n \cdot |D|^3 \cdot k_2)$. This is because $\overline{C}$ has at most $O(n \cdot |D|)$ vertices and $O(n \cdot |D|^2)$ edges and each edge can be traversed at most $O(|D| \cdot k_2)$ times in the DFS, where k_2 is the depth of our POT for C. Note that the treedepth of $\overline{C}$ is a factor $|D|$ larger than that of C. Finally, computing the *reach* values takes $O(n \cdot |D|^3 \cdot k_2^2)$ time.

Therefore, the total runtime of our preprocessing phase is $O(n \cdot |D|^3 \cdot k_2^2)$, which has only linear dependence on the number of lines, n.

Runtime Analysis of the Query Phase. To analyze the runtime of a query, note that there are $O(\delta \cdot |D|)$ different possibilities for (c, d_3). Due to the bounded bandwidth assumption, each of these correspond to a constant number of (f', d_4)'s. For each (f', d_4) and (f_2, d_5), we should perform a reachability query using the POT $\overline{T}$. So, we might have to try up to $O(k_2 \cdot |D|)$ common ancestors. So, the total runtime for finding all the (f_2, d_5)'s is $O(|D|^3 \cdot k_2 \cdot \delta)$. Finally, we have to perform a same-context query from every (s_{f_2}, d_5) to (u_2, d_2). So, we do a total of at most $O(|D|)$ queries, each of which take $O(|D|)$. So, the total runtime is $O(|D|^3 \cdot k_2 \cdot \delta)$, which is $O(|D|^3)$ in virtually all real-world scenarios where k_2 and δ are small constants.

5   Experimental Results

Implementation and Machine. We implemented our algorithm, as well as the algorithms of [69] and [45], in a combination of C++ and Java, and used the Soot [79] framework to obtain the control-flow and call graphs. Specifically, we use the SPARK call graph created by Soot for the intermediate Jimple representation. To compute treewidth and treedepth, we used the winning open-source tools submitted to past PACE challenges [33,51]. All experiments were run on an Intel i7-11800H machine (2.30 GHz, 8 cores, 16 threads) with 12 GB of RAM.

Benchmarks and Experimental Setup. We compare the performance of our method against the standard IFDS algorithm [69] and its on-demand variant [45] and use the standard DaCapo benchmarks [8] as input programs. These are real-world programs with hundreds of thousands of lines of code. For each benchmark, we consider three different classical data-flow analyses: (i) reachability analysis for dead-code elimination, (ii) null-pointer analysis, and (iii) possibly-uninitialized variables analysis. For each analysis, we gave each of the algorithms 10 min time over each benchmark and recorded the number of queries that the algorithm successfully handled in this time. The queries themselves were randomly generated[5] and the number of queries was also limited to n, i.e. the number of lines in the code. We then report the average cost of each query, i.e. each algorithm's total runtime divided by the number of queries it could handle. The reason for this particular setup is that [69] and [45] do not distinguish between preprocessing and query. So, to avoid giving our own method any undue advantage, we have to include both our preprocessing and our query time in the mix.

Treewidth and Treedepth. In our experiments, the maximum encountered treewidth was 10, whereas the average was 9.1. Moreover, the maximum treedepth was 135 and the average was 43.8. Hence, our central hypothesis that real-world programs have small treewidth and treedepth holds in practice and the widths and depths are much smaller than the number of lines in the program.

Results. Figure 9 provides the average query time for each analysis. Each dot corresponds to one benchmark. We use PARAM, IFDS and DEM to refer to our algorithm, the IFDS algorithm in [69], and the on-demand IFDS algorithm in [45], respectively. The reported instance sizes are the number of edges in $\overline{G}$.

Discussion. As shown in Sect. 4, our algorithm's preprocessing has only linear dependence on the number n of lines and our query time is completely independent of n. Thus, our algorithm has successfully pushed most of the time complexity on the small parameters such as the treewidth k_1, treedepth k_2, bandwidth b and maximum number of function calls in each function, i.e. δ. All these parameters are small constants in practice. Specifically, the two most important ones

---

[5] For generating each query, we randomly and uniformly picked two points in the exploded supergraph. Note that none of our queries are same-context. Even when the two points of the query are in the same function, we are asking for reachability using interprocedurally valid paths that are not necessarily same-context.

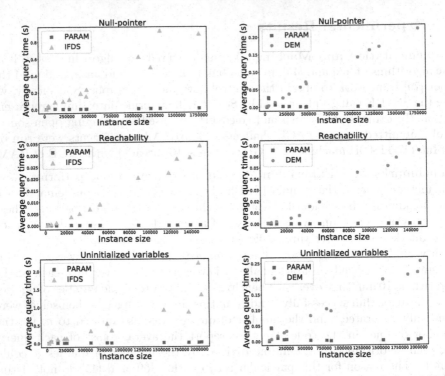

Fig. 9. Comparison of the average cost per query for our algorithm vs [69] and [45].

are always small: The treewidth in DaCapo benchmarks never exceeds 10 and the treedepth is at most 135. This is in contrast to n which is the hundreds of thousands and the instance size, which can be up to around $2 \cdot 10^6$. In contrast, both [69] and [45] have a quadratic dependence on n. Unsurprisingly, this leads to a huge gap in the practical runtimes and our algorithm is on average faster than the best among [69] and [45] by a factor of 158, i.e. more than two orders of magnitude. Moreover, the difference is much starker on larger benchmarks, in which the ratio of our parameters to n is close to 0. On the other hand, in a few small instances, simply computing the treewidth and treedepth is more time-consuming than the previous approaches and they outperform us.

6   Conclusion

In this work, we provided a parameterized algorithm for the general case of the on-demand data-flow analysis as formalized by the IFDS framework. We exploited a novel parameter, i.e. the treedepth of call graphs, to reduce the runtime dependence on the number of lines of code from quadratic to linear. This led to significant practical improvements of more than two orders of magnitude in the runtime of the IFDS data-flow analysis as demonstrated by our experimental results. Moreover, this is the first theoretical improvement in the runtime of the *general case* of IFDS since the original algorithm of [69], which was published in 1995.

Acknowledgments. The research was partially supported by the Hong Kong Research Grants Council ECS Project Number 26208122, the HKUST-Kaisa Joint Research Institute Project Grant HKJRI3A-055 and the HKUST Startup Grant R9272. Author names are ordered alphabetically.

References

1. T.J. Watson libraries for analysis, with frontends for Java, Android, and JavaScript, and many common static program analyses. https://github.com/wala/WALA
2. Ahmadi, A., Daliri, M., Goharshady, A.K., Pavlogiannis, A.: Efficient approximations for cache-conscious data placement. In: PLDI, pp. 857–871 (2022)
3. Aiswarya, C.: How treewidth helps in verification. ACM SIGLOG News **9**(1), 6–21 (2022)
4. Arzt, S., et al.: FlowDroid: precise context, flow, field, object-sensitive and lifecycle-aware taint analysis for Android apps. In: PLDI, pp. 259–269 (2014)
5. Asadi, A., Chatterjee, K., Goharshady, A.K., Mohammadi, K., Pavlogiannis, A.: Faster algorithms for quantitative analysis of MCs and MDPs with small treewidth. In: Hung, D.V., Sokolsky, O. (eds.) ATVA 2020. LNCS, vol. 12302, pp. 253–270. Springer, Cham (2020). https://doi.org/10.1007/978-3-030-59152-6_14
6. Babich, W.A., Jazayeri, M.: The method of attributes for data flow analysis: Part II. Demand analysis. Acta Informatica **10**, 265–272 (1978)
7. Bebenita, M., et al.: SPUR: a trace-based JIT compiler for CIL. In: OOPSLA, pp. 708–725 (2010)
8. Blackburn, S.M., et al.: The DaCapo benchmarks: Java benchmarking development and analysis. In: OOPSLA, pp. 169–190 (2006)
9. Bodden, E.: Inter-procedural data-flow analysis with IFDS/IDE and soot. In: SOAP, pp. 3–8 (2012)
10. Bodden, E., Tolêdo, T., Ribeiro, M., Brabrand, C., Borba, P., Mezini, M.: SPLLIFT: statically analyzing software product lines in minutes instead of years. In: PLDI, pp. 355–364 (2013)
11. Bodlaender, H.L.: Dynamic programming on graphs with bounded treewidth. In: ICALP, pp. 105–118 (1988)
12. Bodlaender, H.L.: A linear time algorithm for finding tree-decompositions of small treewidth. In: STOC, pp. 226–234 (1993)
13. Bodlaender, H.L.: A tourist guide through treewidth. Acta Cybern. **11**(1–2), 1–21 (1993)
14. Bodlaender, H.L., et al.: Rankings of graphs. SIAM J. Discret. Math. **11**(1), 168–181 (1998)
15. Bodlaender, H.L., Hagerup, T.: Parallel algorithms with optimal speedup for bounded treewidth. SIAM J. Comput. **27**(6), 1725–1746 (1998)
16. Burgstaller, B., Blieberger, J., Scholz, B.: On the tree width of Ada programs. In: Llamosí, A., Strohmeier, A. (eds.) Ada-Europe 2004. LNCS, vol. 3063, pp. 78–90. Springer, Heidelberg (2004). https://doi.org/10.1007/978-3-540-24841-5_6
17. Callahan, D., Cooper, K.D., Kennedy, K., Torczon, L.: Interprocedural constant propagation. In: CC, pp. 152–161 (1986)
18. Chang, W., Streiff, B., Lin, C.: Efficient and extensible security enforcement using dynamic data flow analysis. In: CCS, pp. 39–50 (2008)
19. Chatterjee, K., Goharshady, A.K., Goharshady, E.K.: The treewidth of smart contracts. In: SAC, pp. 400–408 (2019)

20. Chatterjee, K., Goharshady, A.K., Goyal, P., Ibsen-Jensen, R., Pavlogiannis, A.: Faster algorithms for dynamic algebraic queries in basic RSMs with constant treewidth. TOPLAS **41**(4), 23:1–23:46 (2019)
21. Chatterjee, K., Goharshady, A.K., Ibsen-Jensen, R., Pavlogiannis, A.: Algorithms for algebraic path properties in concurrent systems of constant treewidth components. In: POPL, pp. 733–747 (2016)
22. Chatterjee, K., Goharshady, A.K., Ibsen-Jensen, R., Pavlogiannis, A.: Optimal and perfectly parallel algorithms for on-demand data-flow analysis. In: ESOP, pp. 112–140 (2020)
23. Chatterjee, K., Goharshady, A.K., Okati, N., Pavlogiannis, A.: Efficient parameterized algorithms for data packing. In: POPL, pp. 53:1–53:28 (2019)
24. Chatterjee, K., Goharshady, A.K., Pavlogiannis, A.: JTDec: a tool for tree decompositions in soot. In: D'Souza, D., Narayan Kumar, K. (eds.) ATVA 2017. LNCS, vol. 10482, pp. 59–66. Springer, Cham (2017). https://doi.org/10.1007/978-3-319-68167-2_4
25. Chatterjee, K., Ibsen-Jensen, R., Goharshady, A.K., Pavlogiannis, A.: Algorithms for algebraic path properties in concurrent systems of constant treewidth components. TOPLAS **40**(3), 9:1–9:43 (2018)
26. Chatterjee, K., Ibsen-Jensen, R., Pavlogiannis, A.: Faster algorithms for quantitative verification in constant treewidth graphs. In: Kroening, D., Păsăreanu, C.S. (eds.) CAV 2015. LNCS, vol. 9206, pp. 140–157. Springer, Cham (2015). https://doi.org/10.1007/978-3-319-21690-4_9
27. Chatterjee, K., Ibsen-Jensen, R., Pavlogiannis, A.: Quantitative verification on product graphs of small treewidth. In: FSTTCS, pp. 42:1–42:23 (2021)
28. Chen, T., Lin, J., Dai, X., Hsu, W.-C., Yew, P.-C.: Data dependence profiling for speculative optimizations. In: Duesterwald, E. (ed.) CC 2004. LNCS, vol. 2985, pp. 57–72. Springer, Heidelberg (2004). https://doi.org/10.1007/978-3-540-24723-4_5
29. Chow, A.L., Rudmik, A.: The design of a data flow analyzer. In: CC, pp. 106–113 (1982)
30. Collard, J.F., Knoop, J.: A comparative study of reaching-definitions analyses (1998)
31. Dangel, A., Fournier, C., et al.: PMD Eclipse plugin. https://github.com/pmd/pmd-eclipse-plugin
32. Das, A., Lal, A.: Precise null pointer analysis through global value numbering. In: D'Souza, D., Narayan Kumar, K. (eds.) ATVA 2017. LNCS, vol. 10482, pp. 25–41. Springer, Cham (2017). https://doi.org/10.1007/978-3-319-68167-2_2
33. Dell, H., Komusiewicz, C., Talmon, N., Weller, M.: The PACE 2017 parameterized algorithms and computational experiments challenge: the second iteration. In: IPEC, pp. 30:1–30:12 (2018)
34. Duesterwald, E., Gupta, R., Soffa, M.L.: Demand-driven computation of interprocedural data flow. In: POPL, pp. 37–48 (1995)
35. Eclipse Foundation: Eclipse documentation, Java development user guide. http://help.eclipse.org/2022-06/index.jsp?topic=/org.eclipse.jdt.doc.user/reference/preferences/java/compiler/ref-preferences-errors-warnings.htm
36. Ferrara, A., Pan, G., Vardi, M.Y.: Treewidth in verification: local vs. global. In: Sutcliffe, G., Voronkov, A. (eds.) LPAR 2005. LNCS (LNAI), vol. 3835, pp. 489–503. Springer, Heidelberg (2005). https://doi.org/10.1007/11591191_34
37. Flückiger, O., Scherer, G., Yee, M., Goel, A., Ahmed, A., Vitek, J.: Correctness of speculative optimizations with dynamic deoptimization. In: POPL, pp. 49:1–49:28 (2018)

38. Goharshady, A.K.: Parameterized and algebro-geometric advances in static program analysis. Ph.D. thesis, Institute of Science and Technology Austria, Klosterneuburg, Austria (2020)
39. Goharshady, A.K., Hooshmandasl, M.R., Meybodi, M.A.: [1, 2]-sets and [1, 2]-total sets in trees with algorithms. Discret. Appl. Math. **198**, 136–146 (2016)
40. Goharshady, A.K., Mohammadi, F.: An efficient algorithm for computing network reliability in small treewidth. Reliab. Eng. Syst. Saf. **193**, 106665 (2020)
41. Gould, C., Su, Z., Devanbu, P.T.: JDBC checker: a static analysis tool for SQL/JDBC applications. In: ICSE, pp. 697–698 (2004)
42. Grove, D., Torczon, L.: Interprocedural constant propagation: a study of jump function implementations. In: PLDI, pp. 90–99 (1993)
43. Gupta, R., Benson, D., Fang, J.Z.: Path profile guided partial dead code elimination using predication. In: PACT, pp. 102–113 (1997)
44. Gustedt, J., Mæhle, O.A., Telle, J.A.: The treewidth of Java programs. In: Mount, D.M., Stein, C. (eds.) ALENEX 2002. LNCS, vol. 2409, pp. 86–97. Springer, Heidelberg (2002). https://doi.org/10.1007/3-540-45643-0_7
45. Horwitz, S., Reps, T.W., Sagiv, S.: Demand interprocedural dataflow analysis. In: FSE, pp. 104–115 (1995)
46. Khedker, U., Sanyal, A., Sathe, B.: Data Flow Analysis: Theory and Practice. CRC Press, Boca Raton (2017)
47. Kildall, G.A.: A unified approach to global program optimization. In: POPL, pp. 194–206 (1973)
48. Kildall, G.A.: Global Expression Optimization During Compilation. University of Washington (1972)
49. Knoop, J., Steffen, B.: Efficient and optimal bit-vector data flow analyses: a uniform interprocedural framework. Institut für Informatik und Praktische Mathematik Kiel, Bericht (1993)
50. Knoop, J., Steffen, B., Vollmer, J.: Parallelism for free: efficient and optimal bitvector analyses for parallel programs. TOPLAS **18**(3), 268–299 (1996)
51. Kowalik, Ł., Mucha, M., Nadara, W., Pilipczuk, M., Sorge, M., Wygocki, P.: The PACE 2020 parameterized algorithms and computational experiments challenge: treedepth. In: IPEC, pp. 37:1–37:18 (2020)
52. Kurdahi, F.J., Parker, A.C.: REAL: a program for register allocation. In: DAC, pp. 210–215 (1987)
53. Lin, J., et al.: A compiler framework for speculative optimizations. TACO (3), 247–271 (2004)
54. Meybodi, M.A., Goharshady, A.K., Hooshmandasl, M.R., Shakiba, A.: Optimal mining: maximizing Bitcoin miners' revenues from transaction fees. In: Blockchain, pp. 266–273. IEEE (2022)
55. Meyer, B.: Ending null pointer crashes. Commun. ACM **60**(5), 8–9 (2017)
56. Nadara, W., Pilipczuk, M., Smulewicz, M.: Computing treedepth in polynomial space and linear FPT time. CoRR abs/2205.02656 (2022)
57. Naeem, N.A., Lhoták, O., Rodriguez, J.: Practical extensions to the IFDS algorithm. In: Gupta, R. (ed.) CC 2010. LNCS, vol. 6011, pp. 124–144. Springer, Heidelberg (2010). https://doi.org/10.1007/978-3-642-11970-5_8
58. Nanda, M.G., Sinha, S.: Accurate interprocedural null-dereference analysis for Java. In: ICSE, pp. 133–143. IEEE (2009)
59. Nešetřil, J., De Mendez, P.O.: Sparsity: Graphs, Structures, and Algorithms. Springer, Cham (2012)
60. Nesetril, J., de Mendez, P.O.: Tree-depth, subgraph coloring and homomorphism bounds. Eur. J. Comb. **27**(6), 1022–1041 (2006)

61. Nguyen, T.V.N., Irigoin, F., Ancourt, C., Coelho, F.: Automatic detection of uninitialized variables. In: Hedin, G. (ed.) CC 2003. LNCS, vol. 2622, pp. 217–231. Springer, Heidelberg (2003). https://doi.org/10.1007/3-540-36579-6_16

62. Obdržálek, J.: Fast mu-calculus model checking when tree-width is bounded. In: Hunt, W.A., Somenzi, F. (eds.) CAV 2003. LNCS, vol. 2725, pp. 80–92. Springer, Heidelberg (2003). https://doi.org/10.1007/978-3-540-45069-6_7

63. Pessoa, T., Monteiro, M.P., Bryton, S., et al.: An eclipse plugin to support code smells detection. arXiv preprint arXiv:1204.6492 (2012)

64. Pothen, A.: The complexity of optimal elimination trees. Technical report (1988)

65. Rapoport, M., Lhoták, O., Tip, F.: Precise data flow analysis in the presence of correlated method calls. In: Blazy, S., Jensen, T. (eds.) SAS 2015. LNCS, vol. 9291, pp. 54–71. Springer, Heidelberg (2015). https://doi.org/10.1007/978-3-662-48288-9_4

66. Reps, T.: Undecidability of context-sensitive data-dependence analysis. TOPLAS 22(1), 162–186 (2000)

67. Reps, T.W.: Demand interprocedural program analysis using logic databases. In: Ramakrishnan, R. (ed.) Applications of Logic Databases. SECS, pp. 163–196. Springer, Boston (1993). https://doi.org/10.1007/978-1-4615-2207-2_8

68. Reps, T.W.: Program analysis via graph reachability. Inf. Softw. Technol. 40(11–12), 701–726 (1998)

69. Reps, T.W., Horwitz, S., Sagiv, S.: Precise interprocedural dataflow analysis via graph reachability. In: POPL, pp. 49–61 (1995)

70. Robertson, N., Seymour, P.D.: Graph minors. III. Planar tree-width. J. Comb. Theory Ser. B 36(1), 49–64 (1984)

71. Robertson, N., Seymour, P.D.: Graph minors. II. Algorithmic aspects of tree-width. J. Algorithms 7(3), 309–322 (1986)

72. Rountev, A., Kagan, S., Marlowe, T.: Interprocedural dataflow analysis in the presence of large libraries. In: Mycroft, A., Zeller, A. (eds.) CC 2006. LNCS, vol. 3923, pp. 2–16. Springer, Heidelberg (2006). https://doi.org/10.1007/11688839_2

73. Sagiv, S., Reps, T.W., Horwitz, S.: Precise interprocedural dataflow analysis with applications to constant propagation. Theor. Comput. Sci. 167, 131–170 (1996)

74. Shang, L., Xie, X., Xue, J.: On-demand dynamic summary-based points-to analysis. In: CGO, pp. 264–274 (2012)

75. Späth, J., Ali, K., Bodden, E.: Context-, flow-, and field-sensitive data-flow analysis using synchronized pushdown systems. In: POPL, pp. 48:1–48:29 (2019)

76. Sridharan, M., Bodík, R.: Refinement-based context-sensitive points-to analysis for Java. In: PLDI, pp. 387–400 (2006)

77. Sridharan, M., Gopan, D., Shan, L., Bodík, R.: Demand-driven points-to analysis for Java. In: OOPSLA, pp. 59–76 (2005)

78. Thorup, M.: All structured programs have small tree-width and good register allocation. Inf. Comput. 142(2), 159–181 (1998)

79. Vallée-Rai, R., Co, P., Gagnon, E., Hendren, L.J., Lam, P., Sundaresan, V.: Soot - a Java bytecode optimization framework. In: CASCON, p. 13. IBM (1999)

80. Xu, G., Rountev, A., Sridharan, M.: Scaling CFL-reachability-based points-to analysis using context-sensitive must-not-alias analysis. In: Drossopoulou, S. (ed.) ECOOP 2009. LNCS, vol. 5653, pp. 98–122. Springer, Heidelberg (2009). https://doi.org/10.1007/978-3-642-03013-0_6

81. Yan, D., Xu, G., Rountev, A.: Demand-driven context-sensitive alias analysis for Java. In: ISSTA, pp. 155–165 (2011)

82. Zheng, X., Rugina, R.: Demand-driven alias analysis for C. In: POPL, pp. 197–208 (2008)

Maximal Robust Neural Network Specifications via Oracle-Guided Numerical Optimization

Anan Kabaha[✉][iD] and Dana Drachsler-Cohen[iD]

Technion, Haifa, Israel
anan.kabaha@campus.technion.ac.il, ddana@ee.technion.ac.il

Abstract. Analyzing the robustness of neural networks is crucial for trusting them. The vast majority of existing works focus on networks' robustness in ϵ-ball neighborhoods, but these cannot capture complex robustness specifications. We propose MaRVeL, a system for computing maximal non-uniform robust specifications that maximize a target norm. The main idea is to employ *oracle-guided numerical optimization*, thereby leveraging the efficiency of a numerical optimizer as well as the accuracy of a non-differentiable robustness verifier, acting as the oracle. The optimizer iteratively submits to the verifier candidate specifications, which in turn returns the closest inputs to the decision boundaries. The optimizer then computes their gradients to guide its search in the directions the specification can expand while remaining robust. We evaluate MaRVeL on several datasets and classifiers and show that its specifications are larger by 5.1x than prior works. On a two-dimensional dataset, we show that the average diameter of its specifications is 93% of the optimal average diameter, whereas the diameter of prior works' specifications is only 26%.

1 Introduction

Neural networks are susceptible to adversarial examples [14,15,21,37,46,48]. To understand the robustness level of neural networks, many works verify local robustness [3,18,28,31,34,38,41,43]. These works focus on analyzing the network's robustness at an ϵ-ball centered at a given input, where every input entry can be perturbed by up to $\pm\epsilon$. However, focusing only on this kind of neighborhood hinders the overall robustness level of the network. To illustrate, consider Fig. 1 showing the decision boundaries of a small network (the black curves), taking two-dimensional inputs, and an input (the red dot). Its maximal ϵ-ball is bounded by the closest decision boundary and thus it is quite small (the blue square). This is because an ϵ-ball uniformly bounds all perturbations by the same ϵ.

This gave rise to works that compute maximal non-uniform robust neighborhoods [25,26]. These neighborhoods are defined by interval specifications, generalizing ϵ-balls, where each input entry is bounded by an interval. A robust interval specification is maximal if expanding any interval results in including

C. Dragoi et al. (Eds.): VMCAI 2023, LNCS 13881, pp. 203–227, 2023.
https://doi.org/10.1007/978-3-031-24950-1_10

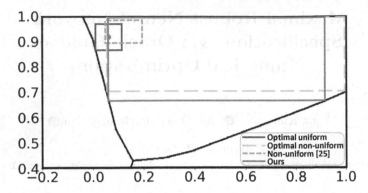

Fig. 1. A comparison of MaRVeL's maximal non-uniform specification to the maximal (uniform) ϵ-ball, the optimal non-uniform specification (computed by a naive approach), and the maximal non-uniform specification computed by [25].

an adversarial example. In other words, every interval is approaching a decision boundary. To pick among the multiple maximal robust specifications, it is common to maximize a given size metric (e.g., the L_1 or L_2 norm). Computing maximal non-uniform specifications is challenging because (1) the search space is exponentially large and (2) determining whether an interval specification belongs to this space, i.e., whether it is robust, requires to call a robustness verifier, which takes non-negligible time. A naive optimal approach begins by computing all decision boundaries around the given input using a grid search. Accordingly, it computes all maximal robust interval specifications and returns the specification maximizing the size metric. However, this approach is highly time-consuming and impractical if the input dimension is high. Figure 1 shows an optimal non-uniform specification (the dashed yellow rectangle).

Existing works propose efficient approaches to compute maximal non-uniform specifications [25,26]. These approaches rely on numerical optimization, to search in the large space, and on an incomplete robustness analysis, to determine robustness of candidate specifications. This analysis overapproximates the network's computation with differentiable linear functions and thus it scales well to large networks and amenable to first-order optimization. However, incomplete analysis suffers from precision loss. Hence, their specifications are not always maximal and are quite small. Figure 1 shows the maximal non-uniform specification computed by [25] (the dashed green rectangle). It is significantly smaller than the optimal specification and it does not reach any decision boundary. This raises the question: *Can we efficiently compute optimal maximal robust specifications?*

We present MaRVeL (**Ma**ximal **R**obustness **Ve**rification of Interva**L** specifications). Like prior works, MaRVeL relies on a numerical optimizer to look for a robust specification maximizing a given size metric. Unlike prior works, it relies on a MILP robustness verifier [38], which provides a more accurate analysis but is not differentiable. To employ first-order optimization, we propose a novel way to compute the analysis' gradient from the set of *weakest points*. These are inputs contained in the specification that are the closest to the decision boundaries, and they are computed during the robustness analysis. Based on this idea, MaRVeL employs *oracle-guided numerical optimization*. At each iteration, it submits a specification to the robustness MILP verifier to obtain the weakest points. Accordingly, it computes the gradient and constructs the next specification. MaRVeL also employs counterexample-guided synthesis (CEGIS) to prune the search space based on non-robust specifications. If the verifier determines a specification is non-robust (i.e., it contains an adversarial example), then MaRVeL prunes the search space by restricting the relevant interval bounds. Figure 1 shows the specification computed by MaRVeL (the red rectangle), which is maximal and its average diameter is only 7% smaller than that of the optimal specification.

We evaluate MaRVeL on several benchmarks and compare to prior works [25, 26]. First, we consider the two-dimensional synthetic dataset of [26] and show that MaRVeL's specifications are maximal and their average diameters is 93% of the average diameters of the optimal specifications, whereas the average diameters of prior works' specifications are at most 26%. Second, we consider popular datasets (MNIST, Fashion-MNIST, CIFAR-10, and Contagio/Virustotal [7,40]) and several networks, including convolutional networks. Results show that the average diameter of MaRVeL's specifications is 5.1x larger than that of prior works' specifications. We further show that the CEGIS component leads to 1.8x larger average diameters. The execution time of MaRVeL is 19.9x longer than that of prior works and 6.5x longer if MaRVeL terminates upon the first non-robust specification. The longer execution time is mostly because MaRVeL relies on a more accurate verifier. Lastly, we show that MaRVeL's specifications identify robustness attributes of the networks that ϵ-balls cannot identify and even prior works' specifications do not identify.

2   Preliminaries

In this section, we provide background on network classifiers and local robustness.

Neural Network Classifiers. Given an input domain $\mathbb{R}^d$ and a set of classes $C = \{1, \ldots, c\}$, a classifier maps inputs to a score vector over the possible classes $D : \mathbb{R}^d \to \mathbb{R}^c$. We focus on classifiers in which every input entry has a minimum and maximum domain value. A fully-connected network consists of L layers. The first layer z_0 takes as input a vector from $\mathbb{R}^d$, denoted x, and it passes the input as is to the next layer (i.e., $z_{0,k} = x_k$). The last layer outputs a vector, denoted $D(x)$, consisting of a score for each class in C. The classification of the network for input x is the class with the highest score, $c' = \mathrm{argmax}(D(x))$. The layers are functions, denoted $h_1, h_2, \ldots, h_L$, each taking as input the output of the preceding layer. The network's function is the composition of the layers: $D(x) = h_L(h_{L-1}(\cdots(h_1(x))))$. The function of layer m is defined by a set of processing units called neurons, denoted $z_{m,1}, \ldots, z_{m,k_m}$. Each neuron takes as input the outputs of all neurons in the preceding layer and outputs a real number. The output of layer m is the vector $(z_{m,1}, \ldots, z_{m,k_m})^T$ consisting of all its neurons' outputs. A neuron $z_{m,k}$ has a weight for each input $w_{m,k,k'}$ and a single bias $b_{m,k}$. Its function is computed by first computing the sum of the bias and the multiplication of every input by its respective weight: $\hat{z}_{m,k} = b_{m,k} + \sum_{k'=1}^{k_{m-1}} w_{m,k,k'} \cdot z_{m-1,k'}$. This output is then passed to an activation function σ to produce the output $z_{m,k} = \sigma(\hat{z}_{m,k})$. Activation functions are typically non-linear functions. In this work, we focus on the ReLU activation function, $\text{ReLU}(\hat{z}) = \max(0, \hat{z})$. We note that, for simplicity's sake, we explain our approach for fully-connected networks, but it extends to other architectures, e.g., convolutional networks.

Local Robustness. A safety property for neural networks that has drawn a lot of interest is *local robustness*. A network is locally robust at a given input if it does not change the classification under a given type of perturbation. Formally, given a classifier D, an input x and a neighborhood containing x, $I(x) \subseteq \mathbb{R}^d$, we say D is robust at $I(x)$ if: $\forall x' \in I(x)$. $\text{argmax}(D(x')) = \text{argmax}(D(x))$. There are many robustness verifiers for neural networks. Most of them can analyze hyperrectangular neighborhoods, where each input entry (i.e., pixel, if the input is an image) is bounded in an interval $[l, u]$, where $l, u \in \mathbb{R}$. These neighborhoods capture popular robustness neighborhoods, e.g., ϵ-balls. Among the robustness verifiers, some are complete, i.e., for every neighborhood, they return *robust* or *non-robust*, while others are incomplete, i.e., they may also return *unknown*. Many complete verifiers rely on constraint solvers, e.g., SAT-solvers [10], SMT-solvers [18], or mixed-integer linear programming (MILP) solvers [38]. Incomplete verifiers often employ linear or convex relaxations to the network's non-linear computations to scale the analysis [1,3,13,28,29,31–34,41,43]. While complete verifiers tend to be slower than incomplete verifiers, today's MILP solvers are very efficient and can reason about relatively large networks. They also provide a natural way to trade-off accuracy with scalability, as we explain in Sect. 5.1.

3   Problem Definition

In this section, we define the problem of maximal robust specifications for neural networks. We then discuss the challenges, prior work and the current gap.

Robustness Specifications. We focus on interval specifications defining hyperrectangular neighborhoods. An interval specification is a sequence of intervals, each corresponding to an input entry and constraining its possible values. Formally, interval specifications are parameterized by an input x and take the form of: $I_{l_1,u_1,\ldots,l_d,u_d}(x) = [l_1, u_1], \ldots, [l_d, u_d]$, where $l_i \leq x_i \leq u_i$, for every $i \in [d]$. The specification's neighborhood contains all inputs bounded by the intervals: $N_{I_{l_1,u_1,\ldots,l_d,u_d}}(x) = \{x' \mid \forall i \in [d]. \ x'_i \in [l_i, u_i]\}$. When it is clear from the context, we write I. If $x' \in N_I(x)$, we write $x' \in I$ and say x' is contained in I. We say I is a *robustness specification* for a classifier D if D is robust at $N_I(x)$. Our goal is to compute maximal robust specifications maximizing a given norm. Formally:

Definition 1 (Problem Definition). *Given a classifier D, a correctly classified input x and its class c_x, and a differentiable almost everywhere p-norm $||\cdot||_p$ (e.g., $p = 1, 2, \ldots$), the goal is to compute a specification $I_{l_1,u_1,\ldots,l_d,u_d}(x)$ satisfying:*

1. D is robust at $N_{I_{l_1,u_1,\ldots,l_d,u_d}}(x)$.
2. For every interval specification I' expanding I, D is not robust at $N_{I'}(x)$.
3. $I_{l_1,u_1,\ldots,l_d,u_d}(x)$ maximizes $||\cdot||_p$, among specifications meeting 1 & 2.

This problem is challenging for several reasons. First, it involves searching in high-dimensional space: a specification is a vector in $\mathbb{R}^{2d}$. Second, determining whether a specification belongs to the search space (namely, whether it is robust) involves querying a robustness verifier, which takes non-negligible time. Third, it involves identifying the decision boundaries of the classifier to determine that the specification is maximal. We note that for (uniform) ϵ-ball neighborhoods, computing the maximal neighborhood is significantly simpler. An ϵ-ball specification allows perturbations of each input entry by up to a given ϵ: $B_\epsilon(x) = [x_1 - \epsilon, x_1 + \epsilon], \ldots, [x_d - \epsilon, x_d + \epsilon]$. Namely, an ϵ-ball is defined by a real number ϵ. Thus, computing the maximal ϵ-ball of a given x is a search in a one-dimensional space. It can be done using a binary search, where each candidate ϵ' is submitted to a robustness verifier. Determining whether an ϵ-ball specification is maximal is also simpler and does not require estimating the decision boundaries: the maximal robust ϵ-ball is the one maximizing ϵ. However, as we demonstrate later, considering the more expressive interval specifications leads to revealing a more accurate perspective on the classifier's robustness level.

Prior Work and Current Gap. Two works address the problem of maximal robust interval specifications [25,26]. These works assume an incomplete robustness verifier that relies on linear relaxations to bound each neuron by linear bounds. They leverage the linear bounds to overapproximate the classifier's function $D(x)$ as a linear function of the inputs $\tilde{D}(x)$. This allows them to search for a maximal specification using numerical optimization guided by the gradient of $\tilde{D}$. While these approaches compute larger specifications than their counterpart maximal ϵ-ball specifications (as we show in Sect. 6), they suffer from precision loss. The precision loss stems both from the accumulated overapproximation error of the incomplete verifier's analysis and the inaccuracy of computing the gradient based on $\tilde{D}(x)$ and not the actual classifier's function $D(x)$. As a result, the computed specifications are not maximal. As demonstrated in Fig. 1, existing approaches compute non-maximal specifications, which are also significantly smaller than the optimal specification. We note that although Fig. 1 demonstrates one of the existing approaches, similar results are obtained for the other one. In this work, we propose a new approach for computing maximal robust specifications.

4 Key Idea: An Oracle-Guided Numerical Optimization

In this section, we present our key idea for computing maximal robust specifications, on which we later build to design MaRVeL. Our goal is to compute a maximal robust specification $I(x)$ maximizing a given norm $||\cdot||_p$. To this end, we rely on a MILP verifier, which loses less precision than verifiers relying on linear relaxations. However, the computation of this verifier is not differentiable and thus not amenable to numerical optimization, as proposed by prior works [25, 26]. On the other hand, numerical optimization is very efficient for (differentiable) maximization problems, and thus we wish to leverage it for searching for candidate specifications. We draw inspiration from program synthesis and propose to rely on *oracle-guided numerical optimization*.

In oracle-guided numerical optimization, we have two entities: the numerical optimizer and the verifier, which interact iteratively. At every iteration, the numerical optimizer computes a new candidate specification and then submits it to the verifier. The verifier checks whether the specification defines a robust neighborhood and returns information to the optimizer that guides it in which directions the current specification can expand (if it is robust) or should shrink (if it is not robust). The process terminates when the optimizer does not have more directions to expand. It then returns the last candidate specification that is robust, according to the verifier. We next formalize the optimization problem that the optimizer solves to compute a maximal robust specification. We then explain at a high-level how the optimizer solves the optimization problem and describe the information provided by the verifier to guide the optimization.

4.1 The Optimization Problem

Ideally, we would like the optimizer to solve a constrained optimization problem over specifications, where the maximization function is the p-norm of the specification and the constraints are that the specification is valid (i.e., contains x) and robust. We note that, in this section, we ignore the domain constraints, bounding the input entries by minimum and maximum values, because they are enforced differently (explained in Sect. 5.3). Expressing that the specification is valid is straightforward: we require $x_i \geq l_i$ and $x_i \leq u_i$, for every $i \in [d]$. Expressing that the specification is robust is more subtle because it requires to enforce that the network classifies every input contained in the specification as c_x (i.e., x's class):

$$\forall x' \in I(x).\ \text{class}(D(x')) = c_x$$

However, this constraint is not differentiable, because we rely on a MILP-based encoding of the network's computation, to avoid precision loss. Thus, we rewrite this constraint into a term, which is easier for differentiation, preserving the constraint's semantics. We begin with an equivalent constraint requiring that the difference between c_x's score and the maximal score of any other class is positive:

$$\forall x' \in I(x).\ D(x')_{c_x} - \max\{D(x')_{c'} \mid c' \neq c_x\} > 0$$

Next, to eliminate the for-all operator, which is generally not supported by numerical optimizers, we rewrite this constraint by requiring that the minimum value of the above difference is positive:

$$\min\{D(x')_{c_x} - \max\{D(x')_{c'} \mid c' \neq c_x\} \mid x' \in I(x)\} > 0$$

This constraint has the same semantics: if the minimal value of this difference is positive, then the specification is robust, and otherwise, it is not robust. We call the minimal difference the *robustness level*. We next define it formally.

Definition 2 (Robustness Level). *Given a classifier D, a correctly classified input x and its class c_x, and a specification $I(x)$, the robustness level of $I(x)$ is*
$$RL(I(x)) = \min\{D(x')_{c_x} - \max\{D(x')_{c'} \mid c' \neq c_x\} \mid x' \in I(x)\}.$$

Lastly, since such constraint is challenging for a numerical optimizer, we relax it by adding the robustness level as an additional term to the maximization function (such relaxation is common, for example, to compute adversarial examples [4,6,37,39]). By aiming to maximize the robustness level, the optimizer guides its search towards robust specifications. Overall, the optimizer computes a maximal robust specification by solving the following optimization problem:

$$\max_{I_{l_1,u_1,\ldots,l_d,u_d}} \|I_{l_1,u_1,\ldots,l_d,u_d}\|_p + \lambda \cdot RL(I_{l_1,u_1,\ldots,l_d,u_d})$$

subject to

$$x_i \geq l_i \quad \forall i \in [d]$$
$$x_i \leq u_i \quad \forall i \in [d]$$

(1)

Here, λ is the balancing term, which we define in Sect. 5.2. This constrained problem aims to maximize both the specification's size and its robustness level.

We note that although an optimal solution to this problem may be a non-robust specification, this does not affect the overall soundness of our approach. This is because every candidate is submitted to a sound verifier, and eventually we return the maximal specification that is robust, according to the verifier.

4.2 Solving the Optimization Problem

To solve the optimization problem, the optimizer runs stochastic gradient descent (SGD). At every SGD iteration, the optimizer computes the gradient of the maximization problem and accordingly updates the current specification by a small step: $I \mapsto I + \eta \cdot \nabla(\|I\|_p + \lambda \cdot RL(I))$. Afterwards, it clips the specification to respect the validity constraints $x_i \geq l_i$ and $x_i \leq u_i$. The main question is how to compute the gradient of the maximization function. Since the norm is differentiable almost everywhere, the challenge is only in computing the gradient of $RL(I)$. Our idea is to rely on the robustness level that the MILP verifier computes as part of its analysis, and in particular on the *inputs* defining the robustness level. That is, the inputs minimizing the difference between the score of c_x and the maximal score of any other class. We call these inputs the *weakest points*. We next define them formally.

Definition 3 (Weakest Points). *Given a classifier D, a correctly classified input x and its class c_x, and a specification $I(x)$, the weakest points of $I(x)$ is the following set of inputs $\hat{W} \subseteq \mathbb{R}^d$: $\hat{W} = \{x' \in I(x) \mid RL(x') = RL(I(x))\}$, where for every $x' \in \mathbb{R}^d$, we define $RL(x') = D(x')_{c_x} - \max\{D(x')_{c'} \mid c' \neq c_x\}$.*

We next explain how the weakest points enable the optimizer to compute the gradient of $RL(I)$. By the definition of robustness level, we have $RL(I) = RL(\hat{W})$. In particular, their gradients are equal: $\nabla RL(I) = \nabla RL(\hat{W})$. Thus, to compute the gradient of $RL(I)$, the optimizer computes the gradients of the weakest points, which is typically a very small set. Computing the gradient at a single point $x' \in \hat{W}$ is a simple standard computation involving a forward pass and a backward pass over the classifier D. The gradient of $\hat{W}$ is the average of the weakest points' gradients. In practice, the gradient of the weakest points may direct to other decision boundaries, which are close to inputs with a low robustness level, but not the lowest. To avoid it, we identify the set of classes with a low robustness level $C' \subseteq C \setminus \{c_x\}$ and obtain from the verifier's analysis the weakest points of every class $c' \in C'$, namely the inputs minimizing the robustness level for every class in C'. Formally, given the set of classes with a low robustness level C', its set of weakest points is $W = \{W_{c'} \mid c' \in C'\}$, where for $c' \in C'$, $W_{c'} = \{x' \in I(x) \mid D(x')_{c_x} - D(x')_{c'} = RL_{c'}(I(x))\}$ and $RL_{c'}(I(x)) = \min\{D(x')_{c_x} - D(x')_{c'} \mid x' \in I(x)\}$. The optimizer constructs a weighted gradient from all the points in $\bigcup W$ (defined in Sect. 5.2).

5   MaRVeL: Computing Maximal Robust Specifications

In this section, we present MaRVeL, our algorithm to compute maximal robust interval specifications. MaRVeL builds on oracle-guided numerical optimization (Sect. 4). Figure 2 shows its operation. MaRVeL takes as arguments a classifier D, an input x and its class c_x. Throughout execution, it maintains three specifications: the current specification I, the last verified robust specification I^r, and the termination specification I^f, keeping the maximal bounds. Initially, I is the specification containing only x, and I^r and I^f are undefined. MaRVeL operates iteratively to maximize the optimization problem of Eq. (1). It begins at the Verify step. This step begins with a call to a fast incomplete verifier to identify the set of classes with low robustness levels C'. For every $c' \in C'$, it encodes the verification task as a MILP and submits it to a MILP solver. The solver returns, for every class $c' \in C'$, the weakest points and their robustness level $W_{c'}$. MaRVeL then continues to the Progress step to decide how to advance the computation. If I is robust, I^r is updated. Otherwise, MaRVeL resets I to the previous I^r and updates I^f using CEGIS, to prevent expanding in the maximal directions. It further updates the balancing factor λ_0 (described later), if I is not sufficiently larger than the previous I^r or if I is not robust. Then, MaRVeL checks the termination conditions. MaRVeL terminates in one of the following cases: (1) if x is misclassified (in which case, I is set to an undefined I^r and thus has $\bot$), (2) all bounds are maximal (I^f has no $\bot$), or (3) the balancing factor λ_0 is below a predetermined threshold ($\lambda_0 < \lambda_{min}$). If MaRVeL does not terminate, it continues to the Optimize step. This step first computes the specification size's gradient, the robustness level's gradient (from the weakest points), and the value of λ. Accordingly, it updates I. Lastly, it employs clipping to I based on the validity constraints (i.e.,

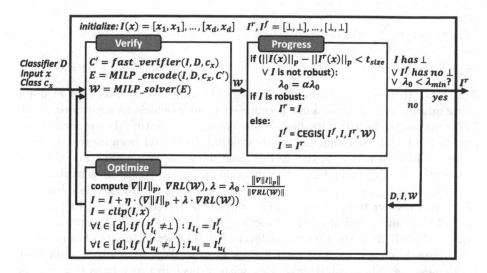

Fig. 2. MaRVeL: system description.

$l_i \le x_i \le u_i$) and enforces the bounds in I^f. When one of the termination conditions is true, the last robust specification I^r is returned. We next explain these steps, show an example, and discuss correctness.

5.1 The Verify Step

For the verifier, MaRVeL relies on the MILP encoding of an existing MILP-based robustness verifier [38]. We begin with a short background on its encoding, which is necessary to understand MaRVeL's encoding and optimizations, and then describe MaRVeL's call to the fast incomplete verifier and the optimizations that MaRVeL employs.

Background: A MILP Robustness Verifier. The MILP verifier [38] encodes the robustness analysis as MILPs, which are then submitted to a MILP solver. We begin by describing the encoding of the network's computation. Given a network, the encoding associates to each neuron $z_{m,k}$ the following (we abuse notation, for simplicity's sake): (1) a real-valued variable $\hat{z}_{m,k}$ for the affine computation, (2) a real-valued variable $z_{m,k}$ for the ReLU computation, (3) concrete lower and upper bounds $l_{m,k}, u_{m,k} \in \mathbb{R}$, and (4) a boolean variable $a_{m,k} \in \{0,1\}$. For every neuron, it adds the following constraints, capturing the neuron's computation:

$$\hat{z}_{m,k} = b_{m,k} + \sum_{k'=1}^{k_{m-1}} w_{m,k,k'} \cdot z_{m-1,k'} \qquad z_{m,k} \geq 0 \qquad z_{m,k} \geq \hat{z}_{m,k}$$

$$z_{m,k} \leq u_{m,k} \cdot a_{m,k} \qquad\qquad\qquad z_{m,k} \leq \hat{z}_{m,k} - l_{m,k}(1 - a_{m,k})$$

The concrete bounds are computed before encoding the network, for example using interval arithmetic or a fast incomplete verifier (which, as mentioned, MaR-VeL runs before the MILP verifier). These constraints capture the network's computation precisely, without any approximation. Note that because the encoding relies on boolean variables, the overall function is a step function, and thus not differentiable. Given an interval specification $I_{l_1,u_1,\ldots,l_d,u_d}(x)$, the encoding adds the constraints: $z_{0,k} \geq l_k$ and $z_{0,k} \leq u_k$, for every $k \in [d]$. To encode robustness of c_x with respect to class $c' \neq c_x$, i.e., show that the score of c_x is higher than that of c', it adds a minimization function: $\min(z_{L,c_x} - z_{L,c'})$. An optimal solution to this MILP is the robustness level of c'. The set of inputs obtaining this robustness level is the set of weakest points of c', $\mathcal{W}_{c'}$. Overall, to show local robustness, i.e., show robustness of c_x with respect to every class $c' \neq c_x$, the verifier submits to a MILP solver $|C| - 1$ MILPs. If all optimal solutions (i.e., robustness levels) are positive, the classifier is locally robust at the given interval specification. Otherwise, it is not robust. That is, this encoding provides a sound and complete local robustness analysis [38]. In addition to the robustness levels, the MILP solver can also return the sets of weakest points.

The Fast Verifier. The MILP verifier is generally an efficient approach for exact analysis. However, since MaRVeL invokes it at every iteration, it becomes highly time consuming, especially for large networks. Thus, at every iteration, MaRVeL attempts to reduce the number of MILPs by pruning classes whose robustness level is not low. Thus, before the MILP encoding, MaRVeL runs DeepPoly [33]. Deep-Poly is an incomplete robustness verifier, relying on linear relaxations to scale the analysis. As part of its analysis, DeepPoly computes, for every output neuron, real-valued lower and upper bounds bounding the possible values. MaRVeL relies on these bounds to compute, for every class $c' \neq c_x$, a lower bound on the robustness level. Then, it constructs the set C' of all classes whose robustness level's lower bound is not positive. Note that the MILP verifier need not check the other classes to determine local robustness. If $C' = \emptyset$ (it may happen for very small specifications), it adds to C' the class with the minimal robustness level.

The MILP Verifier. MaRVeL encodes a MILP for each class in C', as described, and submits them to a MILP solver. To reduce execution times, it employs two optimizations. First, it employs a *partial* MILP encoding. That is, it encodes only part of the neurons using boolean variables, and the rest are overapproximated with DeepPoly's linear constraints (which lose precision). This allows MaRVeL to trade-off precision with scalability. Specifically, MaRVeL limits the number of neurons that are encoded precisely at every layer to n_m (a hyper-parameter). There are several heuristics to determine which neurons require precise encoding [34,41–43]. MaRVeL employs a common heuristic. It picks for every layer the

n_m neurons with the largest overapproximation error, i.e., the largest difference between their upper bound and lower bound, as determined by DeepPoly. Second, many MILP solvers support anytime computations. Thus, MaRVeL runs the MILP solver with a predetermined timeout T_{MILP}. If the solver reaches this timeout, it returns the current optimal solution. We note that these optimizations do not affect the soundness of MaRVeL, but may reduce its accuracy.

5.2   The Optimize Step

The Optimize step begins by computing the gradient of the maximization function: $||I_{l_1,u_1,\dots,l_d,u_d}||_p + \lambda \cdot RL(I_{l_1,u_1,\dots,l_d,u_d})$. This computation follows the description at Sect. 4.2, and we next explain it in detail. The gradient is a vector in $\mathbb{R}^{2d}$, defining for every l_i and u_i its derivative. Computing the gradient of $\nabla||I||_p$ is straightforward, for $p \geq 1$. For example, if $p = 2$, namely $||I||_2 = \sum_{i=1}^{d}(u_i - l_i)^2$, then $\nabla||I||_2 = (2l_1 - 2u_1, 2u_1 - 2l_1, \dots, 2l_d - 2u_d, 2u_d - 2l_d)^T$. To compute the gradient of the robustness level $\nabla RL(I)$, MaRVeL computes $\nabla RL(W)$. To this end, it computes for each input $x' \in W_{c'}$, where $W_{c'} \in W$, the gradient of $RL(x')$. This gradient is $\nabla_{x'}(z_{L,c_x} - z_{L,c'})$, and it can be computed as standard, using a forward pass and a backward pass over the classifier D. Given the gradient of $RL(x')$, which is a vector $(\dot{x}'_1, \dots, \dot{x}'_d)^T \in \mathbb{R}^d$, MaRVeL defines for every $i \in [d]$ the derivative of l_i and u_i to be $\dot{x}'_i$. Theoretically, the gradient of the points in $\bigcup W$ is the component-wise average. However, the weakest points of different classes have different robustness levels. Taking the average gradient assigns the same importance to all points, even if some are closer to a decision boundary than others. Instead, MaRVeL computes a weighted average: $\nabla RL(W) = \frac{1}{|\bigcup W|} \sum_{x' \in \bigcup W} \frac{exp(-RL(x'))}{\sum_{\bar{x} \in \bigcup W} exp(-RL(\bar{x}))} \cdot \nabla_{x'} RL(x')$. This weighted average assigns higher weights to the gradients of inputs with lower robustness levels. Having defined the gradients of each component, the overall gradient is the sum $\nabla||I||_p + \lambda \cdot \nabla RL(I)$. This is then normalized, as standard, by dividing it by its norm: $||\nabla||$. Our balancing term λ is a function of the gradients' ratio: $\lambda = \lambda_0 \cdot \frac{||(\nabla||I||_p)||}{||(\nabla RL(I))||}$, where λ_0 is initialized to a predetermined factor and decreases during the optimization. To allow the specifications expand at a reasonable rate, if I is not robust or $||I||_p - ||I^r||_p < t_{\text{size}}$, for a threshold t_{size}, MaRVeL multiplies λ_0 by a constant $\alpha \in (0,1)$. This update directs the optimizer to assign more weight to the specification's size term in the next iterations.

Specification Update. After computing the gradient, the specification update is a standard SGD step: $I \mapsto I + \eta \cdot (\nabla||I||_p + \lambda \cdot \nabla RL(I))$, where η is a small constant. The intuitive meaning of a single step is that MaRVeL updates the specification with the goal of increasing its size while expanding the bounds away from the current weakest points. After that, the specification is clipped to satisfy the validity constraints. Namely, every u_i that is smaller than x_i is set to x_i, and every l_i that is greater than x_i is set to x_i. Then, the specification is aligned with the maximal and minimal bounds in I^f. Namely, every u_i or l_i

that has a value in I^f is set to its value in I^f. The domain constraints bounding the input entries by minimum and maximum values are enforced through I^f, as we next explain.

5.3   CEGIS at the Progress Step

Lastly, we explain the CEGIS operation at the Progress step. Its goal is to leverage non-robust specifications to identify when MaRVeL reaches maximal bounds and thereby prevent the optimizer from proposing non-robust specifications. This operation draws inspiration from counterexample-guided inductive synthesis (CEGIS), where a program synthesizer prunes its search space after obtaining a counterexample from the oracle or user [17,36]. As described, if a specification is not robust, MaRVeL discards it and continues from the last robust specification. However, before discarding it, MaRVeL computes a set of maximal bounds and updates I^f accordingly. The maximal bounds are computed from the weakest points that are adversarial examples (if a specification is not robust, some of the weakest points are adversarial examples). By restricting these bounds, future specifications will not include these adversarial examples. Moreover, a clever restriction will also eliminate very close adversarial examples that otherwise will be discovered in the following iterations, thereby slowing down the computation.

We begin with several observations and afterward explain how MaRVeL leverages them to construct I^f. Consider two consecutive specifications $I^r_{l^r_1,u^r_1,...,l^r_d,u^r_d}$ and $I_{l_1,u_1,...,l_d,u_d}$, where I^r is robust and I is not. The MILP verifier computes for I the set of weakest points. Because I is not robust, at least one of them is an adversarial example, denoted $x' = (x'_1,...,x'_d)^T$. For every $i \in [d]$, one of the following holds: (1) $x'_i \in (u^r_i, u_i]$, (2) $x'_i \in [l_i, l^r_i)$, or (3) $x'_i \in [l^r_i, u^r_i]$. Because x' is an adversarial example and I^r is robust, there exists $i \in [d]$, for which cases (1) or (2) hold. We define $B_{x'}$ to be the set of bounds satisfying cases (1) or (2):

$$B_{x'} = \{U_i \mid x'_i \in (u^r_i, u_i]\} \cup \{L_i \mid x'_i \in [l_i, l^r_i)\}$$

Our first observation is that if, for future specifications, we prohibit *any* bound in $B_{x'}$ from reaching its respective value in I, then x' is not part of future specifications' neighborhoods. To eliminate all weakest points, it is sufficient to eliminate a single bound for each of them. Thus, the most permissive restriction on future specifications is a minimal hitting set over all $B_{x'}$-s. Namely, $B = \text{argmin}_{B \in \mathcal{B}} |B|$, where $\mathcal{B} = \{B \subseteq \{L_1, U_1,...,L_d, U_d\} \mid \forall x' \in \bigcup \mathcal{W}. B_{x'} = \emptyset \vee B_{x'} \cap B \neq \emptyset\}$. We can prove that if MaRVeL removes the most permissive restriction B at every iteration in which I is not robust, then MaRVeL returns a maximal robust specification (Sect. 5.5, Theorem 1).

Our second observation is that, in practice, the most permissive restriction results in high execution times, especially for high-dimensional specifications. This is because adversarial examples are not sporadic and often multiple adversarial examples appear in the same region [8]. Thus, while eliminating a single bound removes a particular adversarial example, it does not eliminate the adversarial region. Although eventually all adversarial examples in this region are removed, it requires many iterations in which I^r is not updated, causing a time waste.

Computing a minimal set of bounds defining adversarial regions is not trivial. Instead, we overapproximate it with the union of the bounds: $\tilde{B} = \bigcup_{x' \in \cup W} B_{x'}$. While this is the most restrictive approach, we empirically observe that among all approaches we experimented with, it leads to a minimal number of iterations until the optimizer again computes a candidate specification which is discovered as robust. We believe the reason is that the SGD's step size is very small, and thus if a bound is included in any $B_{x'}$, it should be restricted. When experimenting with less restrictive approaches (e.g., computing a minimal hitting set or restricting bounds based on their frequencies), MaRVeL required many more iterations to restrict all necessary bounds. During these iterations, I^r remains the same, because the specifications are not robust. Consequently, when limiting MaRVeL with a one hour timeout, the average diameter of the less restrictive approaches is at best 80% of the average diameter of the specifications computed with $\tilde{B}$.

MaRVeL builds on these observations to compute the maximal bounds. As described, it maintains a termination specification I^f, keeping for each bound a maximal or a minimal value. Initially, all bounds in I^f are undefined. At every iteration in which the verifier determines that I is not robust, I^f is updated based on the weakest points that are adversarial examples. To this end, MaRVeL first computes $\tilde{B}$. Then, for every $U_i \in \tilde{B}$, it sets in I^f at index u_i the value $I^r_{u_i}$, and for every $L_i \in \tilde{B}$, it sets in I^f at index l_i the value $I^r_{l_i}$. While we could set the bounds in I^f to the respective values in I minus a small constant, in practice this does not eliminate the adversarial region. Additionally, because MaRVeL advances I by small steps, the difference between $I^r_{u_i}$ and I_{u_i} is very small.

We note that I^f is also updated when bounds reach their maximal or minimal domain value. For example, assume the input domain is $[0, 1]^d$. If the verifier determines a specification I is robust, then for every $u_i = 1$ and $l_i = 0$ in I, their respective value in I^f is updated to 1 or 0 (respectively). This is required to guarantee termination (Sect. 5.5, Lemma 1).

5.4   An End-to-End Example

We next exemplify MaRVeL for the specification presented at Fig. 1. In this example, the classifier D is a fully-connected network, taking two-dimensional inputs in the range $[-1, 1]$ and consisting of three layers, each with ten neurons. The input is $x = (0.075, 0.93)^T$ and its class is $c_x = 7$. Given these arguments, MaRVeL computes a maximal robust specification with respect to the L_1 norm ($p = 1$). Figure 3 visualizes the key steps in MaRVeL's verification process. Every figure shows the following. The black curves show the decision boundaries. The blue square shows the input x. The red rectangle shows the current specification I. The green dots show the weakest points, and the light blue stars show the weakest points that are adversarial examples. The gradient is shown as two arrows (to simplify its visualization): the green arrow shows the gradient of the upper bounds and the dashed red arrow shows the gradient of the lower bounds.

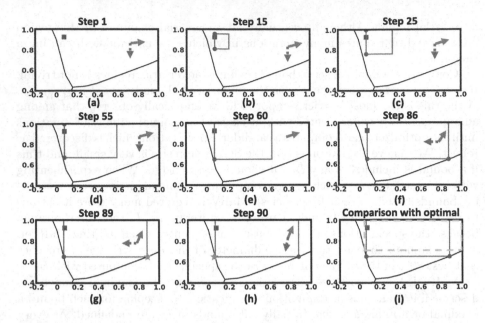

Fig. 3. A running example for computing MaRVeL's specification at Fig. 1.

At step 1, MaRVeL initializes the specification I to contain only the input x: $I_1 = [0.075, 0.075], [0.93, 0.93]$. It also initializes $I^r = I^f = [\bot, \bot], [\bot, \bot]$. The Verify step runs DeepPoly and determines that only $c' = 8$ has to be checked by the (MILP) verifier. That is, instead of nine MILPs, only one is encoded and submitted to the verifier. The verifier returns $\mathcal{W}_8 = \{x\}$ and its robustness level, which is positive. Namely, I is robust. No termination condition is true, and so MaRVeL computes the gradient of $\mathcal{W}_8 = \{x\}$. Accordingly, it expands the specification. Then, it clips to ensure that the specification contains x. Step 15 shows a very similar scenario only that there are two weakest points for $c' = 8$.

At step 25, the verifier returns that $I = [0.075, 0.32], [0.75, 1]$ is robust. Because one of the bounds reaches its maximal domain value, its respective value in I^f is updated: $I^f = [\bot, \bot], [\bot, 1]$. The specification is expanded as before.

At step 55, $I = [0.075, 0.61], [0.65, 1]$ and it approaches the decision boundary of class 8. The weakest point is $(0.075, 0.66)^T$. The gradient of this point directs to expand I only in the right-up direction (demonstrated also in step 60).

At step 86, I approaches the decision boundary of class 6. The verifier returns $\mathcal{W} = \{\{(0.065, 0.65)^T\}_8, \{(0.87, 0.64)^T\}_6\}$. The first point corresponds to $c' = 8$ and the second point to $c' = 6$. At step 89, I is not robust and one of the weakest points is an adversarial example: $(0.88, 0.649)^T$. MaRVeL constructs $\tilde{B} = \{U_1, L_2\}$ and updates their respective values in I^f based on their values in the last I^r: $I^f = [\bot, 0.878], [0.649, 1]$. At step 90, I is not robust and one of the weakest points is an adversarial example: $(0.055, 0.649)^T$. MaRVeL constructs $\tilde{B} = \{L_1\}$ and updates its respective bound in I^f based on the last I^r:

$I^f = [0.065, 0.878], [0.649, 1]$. At this point, there is no direction that MaRVeL can expand. Thus, MaRVeL terminates and returns the last robust specification $I^r = [0.065, 0.878], [0.649, 1]$. The figure shows that the specification is maximal: expanding any bound results in including an adversarial example. Figure 3(i) compares MaRVeL's specification with the optimal one (the dashed yellow rectangle), whose average diameter is larger by only 7%.

5.5 Correctness and Running Time

In this section, we discuss correctness and running time analysis.

Correctness. By MaRVeL's operation, it is sound because if it returns a defined specification (without $\perp$), it must have been verified by the MILP verifier, which provides a sound robustness analysis [38]. Under the following conditions the returned specification is maximal: (1) MaRVeL relies on the minimal hitting set B, (2) the step sizes are small enough (as standard in numerical optimization), (3) the MILP verifier is precise (i.e., the optimizations of the partial MILP and the anytime computations do not reduce its accuracy), and (4) MaRVeL terminates because I^f has no $\perp$. We next formalize this in a theorem.

Theorem 1. *Let D, x and c_x be arguments to MaRVeL. If MaRVeL relies on the minimal hitting set, the step sizes are small, the MILP verifier is precise, and MaRVeL completes because I^f has no $\perp$, then its specification is maximal: expanding any bound that has not reached its maximal or minimal possible value results in including an adversarial example.*

Proof (Sketch). Assume MaRVeL's maximization function was $\max \|I\|_p$. Then, at every iteration, the gradient is positive for any upper bound and negative for any lower bound, because an L_p norm is a monotonically increasing function[1]. Thus, at every iteration, the SGD step updates the current specification by increasing every u_i and decreasing every l_i that are not limited by I^f or the validity constraints. Thus, if a bound u_i stops increasing (or a bound l_i stops decreasing) and if it is not because of the validity constraints or because u_i has reached its maximal domain value, then it is because u_i prevents an adversarial example. This is guaranteed since the MILP verifier is precise. Because the step sizes are small, the bound of u_i is maximal (or the bound of l_i is minimal). Because MaRVeL relies on the minimal hitting set B, every adversarial example is prevented by limiting a single bound and no bound can be omitted from B without including an adversarial example. Thus, if the optimization is completed, it must be that every bound is preventing an adversarial example or has reached the maximal or minimal possible value. A similar reasoning applies to our maximization function with the robustness level, thanks to the adaptive

---

[1] There is an edge case where $l_i = u_i$, in which case the gradient is zero. There are standard corrections to guarantee that the gradient is monotonically increasing. For example, for the L_1 norm, which is the one currently supported in our implementation, the correction replaces the zero gradient by 1 (for u_i) or -1 (for l_i).

definition of λ_0. Recall that if the specification size increases too slowly or the specification is not robust, then λ_0 decreases. Thereby, MaRVeL assigns more weight to the specification's size term. Thus, the optimization process cannot terminate without attempting to increase every bound, due to the gradient of $\|I\|_p$. Hence, if the optimization is completed, it must be that every bound is preventing an adversarial example or has reached the maximal or minimal possible value.

If MaRVeL relies on $\tilde{B}$, we provide a lower bound on the number of dimensions in which the specification is maximal. Every time the specification I^f is updated, at least one of I^r's bounds is maximal, because $\tilde{B}$ is a hitting set. Given all (disjoint) sets $\tilde{B}_1, \ldots, \tilde{B}_k$ throughout the execution, the number of maximal bounds is at least k. This is a very loose lower bound, since in practice, several bounds tend to be maximal together, thereby inducing an adversarial region.

Running Time. Next, we analyze the running time of MaRVeL. We start with a lemma guaranteeing termination. Then, we analyze the running time of a single iteration of MaRVeL.

Lemma 1. *For every D, x and c_x, MaRVeL terminates.*

Proof (Sketch). At every iteration, one of the following holds:

- The current specification is robust: In this case, in most iterations, the size of the current specification is larger than previous specifications. We note that it may be that for a small number of iterations it is not the case, but then λ_0 decreases until the specification's size becomes large enough.
- The current specification is not robust: In this case, at least one of the bounds is set to a value (if it was $\bot$) or is tightened by the respective value in I^r. We note that a bound can be tightened in case MaRVeL relies on a more permissive set of bounds than $\tilde{B}$.

Because at every update of I^f at least one bound is set or tightened and because the step size is a discrete number, the number of iterations in which I^f is updated is finite. If at some iteration, I^f has no $\bot$, then MaRVeL terminates. Otherwise, it must be that at least one bound can continue increasing or decreasing. In this case, MaRVeL continues expanding the specification (by the definition of λ_0). If MaRVeL does not terminate because of I^f, even though every input entry is bounded by a minimum and maximum values, it must be that the specification's size increases too slowly, even when λ_0 continues decreasing. In this case, at some iteration, λ_0 decreases below λ_{min} and MaRVeL terminates.

The maximal running time of a single iteration of MaRVeL is the sum of $T_{\text{DeepPoly}} + |C - 1| \cdot T_{\text{MILP}} + |\bigcup \mathcal{W}| \cdot T_D$, where T_{DeepPoly} is the execution time of the incomplete verifier DeepPoly, T_{MILP} is the execution time of the MILP verifier (recall that MaRVeL sets a timeout to the solver), and $|\bigcup \mathcal{W}| \cdot T_D$ is the time to compute the gradient of the weakest points, involving a forward pass and a backward pass to each over the classifier D. The other computations take a negligible time. Note that because $\mathcal{W}$ is computed by a MILP solver, $\bigcup \mathcal{W}$ is finite. The dominant factor of the running time is T_{MILP} (under reasonable choices). To mitigate it, MaRVeL solves the MILPs parallelly. Naturally, advances in complete robustness verification or MILP solvers can reduce MaRVeL's execution time.

6 Evaluation

In this section, we evaluate MaRVeL. We begin by describing our experiment setup and baselines and then present our experiments.

Experiment Setup. We implemented MaRVeL[2] in Python, as a module in ERAN[3], to easily integrate with DeepPoly and the MILP-based verification. MaRVeL leverages ERAN's RefinePoly domain that runs DeepPoly and then the MILP verifier as described in Sect. 5.1. The MILP solver is Gurobi. Experiments ran on an Ubuntu 20.04.1 OS on a dual AMD EPYC 7713 server with 2TB RAM. We evaluated MaRVeL over several datasets. First, image datasets: MNIST [24] and Fashion-MNIST [44], consisting of 28×28 gray-scale images, and CIFAR-10 [20], consisting of $32 \times 32 \times 3$ colored images. Second, Contagio/Virustotal [7,40], a malware dataset consisting of malicious and benign PDF files, each with 135 features. Table 1 shows the different networks we used. Their activation function was ReLU. The *Conv2* architecture comprised of two convolutional layers followed by two fully-connected layers, while *Conv3* comprised of three convolutional layers followed by three fully-connected layers. We also used a toy synthetic dataset consisting of two-dimensional inputs [26], described later, to visualize the size of the specifications with respect to the decision boundaries. In our experiments, the norm is L_1 ($p = 1$), the balancing factor is $\lambda_0 = 0.99$, the number of precise neurons is $n_m = 200$, and the MILP timeout is $T_{\mathrm{MILP}} = 100$ seconds.

Baselines. We compare MaRVeL to existing works on computing maximal robust specifications [25,26]. Both approaches rely on CROWN [49], an incomplete robustness verifier, which overapproximates ReLU with linear constraints. They differ in the kind of specifications they compute. Liu et al. [26] compute non-uniform, symmetric robust specifications, defined by a non-negative vector ϵ: $I_\epsilon(x) = [x_1 - \epsilon_1, x_1 + \epsilon_1], \ldots, [x_d - \epsilon_d, x_d + \epsilon_d]$. Li et al. [25] build on [26] to compute non-uniform, asymmetric robust specifications, like our specifications. We used Liu et al.'s code[4], which supports only fully-connected networks, and extended their code to support Li et al.'s approach. We compare MaRVeL and these works by measuring the specifications' average diameter $\epsilon_{avg} = \frac{\sum_{i=1}^{d} u_i - l_i}{d}$. Since MaRVeL and these works rely on different robustness verifiers, for a fair comparison, we also report the diameter ϵ_u of the maximal robust (uniform) ϵ-ball of CROWN and the MILP verifier. This is computed by a binary search running 15 steps and starting from $\epsilon_0 = 0.1$.

---

[2] https://github.com/ananmkabaha/MaRVeL.git.
[3] https://github.com/eth-sri/eran.
[4] https://github.com/liuchen11/CertifyNonuniformBounds.

Table 1. The networks used in our experiments.

Dataset	Name	Architecture	#Neurons
MNIST	3×50	Fully-connected	100
	3×100	Fully-connected	200
	$Conv2$	Convolutional	2948
Fashion-MNIST	3×50	Fully-connected	100
	3×250	Fully-connected	500
	$Conv2$	Convolutional	2948
	$Conv3$	Convolutional	3664
CIFAR-10	3×400	Fully-connected	800
	$Conv2$	Convolutional	1188
	$Conv3$	Convolutional	4368
Contagio/virustotal	3×50	Fully-connected	100
	3×100	Fully-connected	200

Synthetic Dataset. We start by considering the toy synthetic dataset, presented by [26]. This dataset consists of two-dimensional inputs $x_1, x_2 \in [-1, 1]$ and ten classes $C = [1, \ldots, 10]$. We create training and test sets by randomly generating 9000 inputs and 1000 inputs, respectively. We consider a fully-connected network with two hidden layers, each with 10 ReLU neurons. After training, it reaches over 99% accuracy on the test set. Figure 4(a) shows the network's decision boundaries (the black curves). We run MaRVeL and both baselines on ten inputs. We further compare to a (highly impractical) optimal approach that computes all decision boundaries around the given input using a grid search, accordingly computes all maximal robust specifications, and returns the specification maximizing the L_1 norm. Figure 4(a) shows the maximal robust specifications of each approach. It shows that MaRVeL's specifications reach the decision boundaries and cannot be expanded in any dimension. In contrast, both prior works compute significantly smaller specifications. Part of the difference is attributed to the underlying verifier (MILP-based vs. CROWN). To illustrate this, Fig. 4(b) shows the average diameter, over 100 inputs, of the specifications computed by the non-uniform approaches and of the maximal uniform ϵ-ball computed by the verifiers. The results show that the average diameter of MaRVeL is 93% of the optimal approach's average diameter, while the diameter of the prior works is only 26%. The average diameter of the maximal uniform ϵ-balls computed by CROWN is only 50% of that computed by the MILP verifier. On average, the execution time of MaRVeL is 12.1 s and the execution time of prior works is 7.8 s.

Real Datasets. We next evaluate MaRVeL over the image datasets and the malware dataset. We compare MaRVeL to our two baselines over the fully-connected networks, because their code does not support other architectures. For every network, we run each approach over 50 inputs with a one hour timeout. We measure

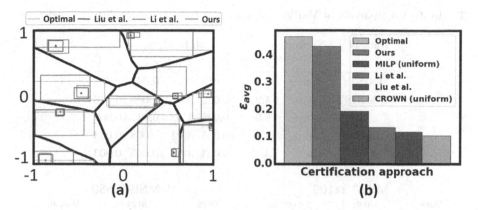

Fig. 4. A comparison of the maximal specifications computed by MaRVeL, by prior works, and by an impractical optimal approach, over a 2D synthetic dataset.

Table 2. A comparison of MaRVeL to the baselines over fully-connected networks.

Dataset	Network	MaRVeL					Li et al.		Liu et al.		ϵ_u
		$T^f[m]$	ϵ^f_{avg}	$T[m]$	ϵ_{avg}	ϵ_u	$T[m]$	ϵ_{avg}	$T[m]$	ϵ_{avg}	
Contagio	3×50	23.1	0.74	41.8	0.91	0.35	0.14	0.17	0.11	0.16	0.16
	3×100	11.9	0.31	28.2	0.40	0.26	0.24	0.15	0.22	0.15	0.14
MNIST	3×50	14.0	0.172	42.2	0.196	0.073	0.52	0.027	0.48	0.026	0.024
	3×100	12.4	0.093	32.6	0.10	0.068	4.2	0.025	3.12	0.022	0.020
F-MNIST	3×50	10.2	0.147	47.0	0.191	0.066	3.9	0.029	2.40	0.0295	0.024
	3×250	4.9	0.031	21.7	0.037	0.028	1.3	0.015	0.9	0.014	0.010
CIFAR-10	3×400	0.7	0.015	21.3	0.047	0.015	3.2	0.007	2.8	0.007	0.002

the execution time in minutes T and the average diameter ϵ_{avg}. To understand the advantage of the CEGIS step, we compare to a variant of MaRVeL that terminates at the first iteration that I is not robust (its results are denoted by ^{f}). We note that the variant that simply removes the CEGIS step and runs MaRVeL as described (in particular, it continues to run even if it encounters non-robust specifications) obtains very close results to the variant we consider, given a one hour timeout, but it always reaches the timeout. We also report, for each approach's verifier, the average diameter of the maximal ϵ-ball ϵ_u. Table 2 shows the results. The results indicate that MaRVeL's average diameter is larger by 5.2x compared to Liu et al. and by 5x compared to Li et al. Without the CEGIS step, MaRVeL's average diameter is larger by 3.8x compared to Liu et al. and by 3.7x compared to Li et al. The average diameter of the maximal ϵ-ball is 3.3x larger for the MILP verifier. MaRVeL's average execution time is 34 min, and if it terminates upon encountering the first non-robust specification, it is 8 min. Table 3 shows the results for the convolutional networks. The results show that MaRVeL's average diameter is larger by 3.1x than the average diameter of the maximal ϵ-balls, and without CEGIS, it is larger by 1.4x.

Table 3. A comparison of MaRVeL to uniform ϵ-balls over convolutional networks.

Dataset	Network	MaRVeL				
		$T^f[m]$	ϵ_{avg}^f	$T[m]$	ϵ_{avg}	ϵ_u
MNIST	$Conv_2$	2.7	0.010	51.9	0.041	0.007
F-MNIST	$Conv_2$	4.2	0.032	30.1	0.063	0.016
	$Conv_3$	6.0	0.008	25.0	0.011	0.009
CIFAR-10	$Conv_2$	4.6	0.005	39.2	0.008	0.003
	$Conv_3$	3.45	0.003	32.2	0.005	0.003

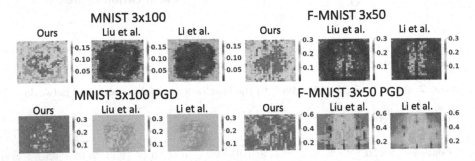

Fig. 5. The heatmaps of the specifications computed by MaRVeL and the baselines.

Robustness Interpretability. We next show how our maximal specifications can expose interpretable robustness attributes of the network. We focus on networks for MNIST and Fashion-MNIST, consisting of gray-scale images with a single centered object (a digit or a fashion item). For each network, we run MaRVeL and both baselines on 50 images of different classes. For each approach, given the 50 specifications $\Im$, we generate a heatmap image $y^{\Im} \in \mathbb{R}^d$. A pixel in $y^{\Im}$ is the average diameter of its interval: $y_i^{\Im} = \frac{\sum_{I \in \Im} u_i - l_i}{|\Im|}$. A heatmap shows which pixels are more robust: the brighter the pixel the larger its average interval. Figure 5 shows the heatmaps of four networks, the first two are trained without defense and the other two are trained with the PGD defense [27]. The heatmaps demonstrate the following. First, MaRVeL computes larger diameters than the baselines, for all pixels. Second, the baselines' heatmaps suggest that the maximal robust diameters are obtained for the background pixels. In contrast, MaRVeL's heatmaps suggest that some object pixels have the largest diameters. This shows that MaRVeL's maximal specifications provide a better perspective on the network's robustness. Third, all approaches show that, as expected, networks trained with PGD are more robust than their undefended counterparts. However, MaRVeL provides a clearer distinction between the robustness of object pixels and background pixels.

Hyper-parameters. Lastly, we discuss the effect of the hyper-parameters: (1) the balancing factor λ_0 and (2) n_m, the number of precise neurons at every layer, affecting the verifier's accuracy. In these experiments, we focus on

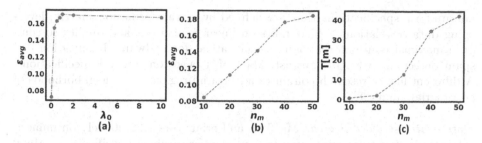

Fig. 6. The effect of the hyper-parameters of MaRVeL.

the 3×50 MNIST network. We begin with the effect of λ_0. Recall that a very small value leads to ignoring the robustness level, while a very large value leads to ignoring the specification size. We consider the values: $\lambda_0 \in \{0.01, 0.25, 0.5, 0.75, 0.99, 2, 10\}$. For each value, we run MaRVeL on 50 images. Figure 6(a) shows the average diameter as a function of λ_0. The figure shows that the largest diameters are obtained for $\lambda_0 \approx 1$. Next, we study the effect of the number of precise neurons n_m. The larger its value, the more accurate the verifier, but the execution time is longer. We consider the values: $n_m \in \{10, 20, 30, 40, 50\}$ (note that for $n_m = 50$ all neurons are precisely encoded). For each value, we run MaRVeL on 50 images. Figure 6(b) and (c) show the average diameter and execution time as a function of n_m. The results show that the higher the value of n_m, the larger the average diameter. Naturally, the higher the value of n_m, the longer the execution time.

7 Related Work

In this section, we discuss the closest related work to ours.

Robustness Verifiers and Specifications. Most existing robustness verifiers focus on ϵ-ball neighborhoods but can be easily extended to analyze interval specifications. These verifiers rely on various techniques, including overapproximation analysis [1,29,41] and in particular overapproximation by linear relaxations [3,13,28,31–33,43], simplex [11,18,19], mixed-integer linear programming (MILP) [23,34,38], and duality [9,30]. The closest works to ours present algorithms for computing non-uniform robust neighborhoods [25,26]. These works focus on image classification tasks, where one work computes non-uniform, symmetric robust specifications [26], and the other work computes non-uniform, asymmetric robust specifications [25]. Both works rely on the CROWN verifier [49], employing linear relaxations to the network's computation. Based on CROWN's linear constraints, they define a gradient to search for maximal robust specifications. The gradient computation is integrated as part of CROWN's analysis. In contrast, MaRVeL relies on a MILP verifier, providing a more accurate analysis, and the optimizer and verifier take turns. Another work focuses on non-uniform specifications that are defined by a transformation matrix, leveraging data correlations [12]. It relies on linear relaxations and duality to compute maximal non-uniform robust specifications for malware classification and spam detection tasks. In contrast, MaRVeL focuses on interval specifications. A different line of research computes adversarial regions, i.e., neighborhoods of adversarial inputs [8].

Optimization-Guided Search. MaRVeL and prior works [25,26] rely on numerical optimization to compute maximal non-uniform robust specifications. Many works rely on optimization to solve other robustness-related tasks. For example, for computing adversarial examples with uniform perturbation budgets [4–6,22,27] or non-uniform perturbation budgets [12,47], defending a network by adversarial training [27,35,45], or improving a verifier's precision by looking for spurious adversarial examples added during analysis [1,2].

Program Synthesis. MaRVeL builds on two common program synthesis techniques. First, it relies on oracle-guided synthesis [16,17], introduced for synthesizing programs by interaction with an oracle (e.g., a solver). Second, it leverages counterexample-guided inductive synthesis (CEGIS) [17,36], where a program synthesizer checks candidates with a solver, which either confirms or provides a counterexample. In the latter case, the synthesizer prunes the search space.

8   Conclusion

We present MaRVeL, an approach and a system for computing maximal non-uniform robust interval specifications, maximizing a target norm. The key idea is to rely on oracle-guided numerical optimization. This allows MaRVeL to rely on an accurate MILP verifier and thereby compute larger specifications than prior works. At each iteration, the optimizer submits a candidate specification to the MILP verifier. The verifier returns the *weakest points*, inputs minimizing the robustness level. Accordingly, the optimizer defines a gradient and computes a new candidate specification. To avoid wasting time on non-robust candidates, MaRVeL relies on CEGIS and restricts bounds that are part of non-robust specifications. We evaluate MaRVeL on several datasets and networks, including fully-connected and convolutional networks. We show it computes specifications with 5.1x larger average diameters compared to prior works, for fully-connected networks, and 2.6x larger average diameters compared to uniform ϵ-balls, computed with an accurate MILP verifier. We further show that CEGIS allows MaRVeL to compute specifications with 1.8x larger average diameters. We demonstrate that MaRVeL's specifications can identify input regions for which networks tend to be more robust or vulnerable as well as compare the robustness of different networks.

Acknowledgements. We thank the reviewers for their feedback. This research was supported by the Israel Science Foundation (grant No. 2605/20).

References

1. Anderson, G., Pailoor, S., Dillig, I., Chaudhuri., S.: Optimization and abstraction: a synergistic approach for analyzing neural network robustness. In: PLDI, pp. 731–744 (2019)
2. Balunovic, M., Vechev, M.T.: Adversarial training and provable defenses: bridging the gap. In: ICLR, pp. 1–18 (2020)
3. Boopathy, A., Weng, T., Chen, P., Liu, S., Dani., L.: Cnn-cert: an efficient framework for certifying robustness of convolutional neural networks. In: AAAI, pp. 3240–3247 (2019)
4. Carlini, N., Wagner., D.A.: Towards evaluating the robustness of neural networks. In: SP, pp. 39–57 (2017)
5. Chen, P., Sharma, Y., Zhang, H., Yi, J., Hsieh, C.: EAD: elastic-net attacks to deep neural networks via adversarial examples. In: AAAI (2018)
6. Chen, P., Zhang, H., Sharma, Y., Yi, J., Hsieh, C.: ZOO: zeroth order optimization based black-box attacks to deep neural networks without training substitute models. In: AISec Workshop, pp. 15–26 (2017)
7. Contagio: Contagio, pdf malware dump (2010). http://contagiodump.blogspot.de/2010/08/malicious-documents-archivefor.html
8. Dimitrov, D.I., Singh, G., Gehr, T., Vechev, M.T.: Provably robust adversarial examples. In: ICLR (2022)
9. Dvijotham, K., Stanforth, R., Gowal, S., Mann, T.A., Kohli, P.: A dual approach to scalable verification of deep networks. In: UAI, p. 3 (2018)
10. Ehlers, R.: Formal verification of piece-wise linear feed-forward neural networks. In: D'Souza, D., Narayan Kumar, K. (eds.) ATVA 2017. LNCS, vol. 10482, pp. 269–286. Springer, Cham (2017). https://doi.org/10.1007/978-3-319-68167-2_19
11. Elboher, Y.Y., Gottschlich, J., Katz, G.: An abstraction-based framework for neural network verification. In: Lahiri, S.K., Wang, C. (eds.) CAV 2020. LNCS, vol. 12224, pp. 43–65. Springer, Cham (2020). https://doi.org/10.1007/978-3-030-53288-8_3
12. Erdemir, E., Bickford, J., Melis, L., Aydöre, S.: Adversarial robustness with non-uniform perturbations. In: NeurIPS (2021)
13. Gehr, T., Mirman, M., Drachsler-Cohen, D., Tsankov, P., Chaudhuri, S., Vechev, M.T.: AI2: safety and robustness certification of neural networks with abstract interpretation. In: SP, pp. 3–18 (2018)
14. Goodfellow, I.J., Shlens, J., Szegedy, C.: Explaining and harnessing adversarial examples. In: ICLR (2015)
15. Ilyas, A., Santurkar, S., Tsipras, D., Engstrom, L., Tran, B., Madry, A.: Adversarial examples are not bugs, they are features. In: NeurIPS (2019)
16. Jha, S., Gulwani, S., Seshia, S.A., Tiwari, A.: Oracle-guided component-based program synthesis. In: ICSE, pp. 215–224 (2010)
17. Jha, S., Seshia, S.A.: A theory of formal synthesis via inductive learning. Acta Informatica 54(7), 693–726 (2017). https://doi.org/10.1007/s00236-017-0294-5
18. Katz, G., Barrett, C., Dill, D.L., Julian, K., Kochenderfer, M.J.: Reluplex: an efficient SMT solver for verifying deep neural networks. In: Majumdar, R., Kunčak, V. (eds.) CAV 2017. LNCS, vol. 10426, pp. 97–117. Springer, Cham (2017). https://doi.org/10.1007/978-3-319-63387-9_5
19. Katz, G., et al.: The marabou framework for verification and analysis of deep neural networks. In: Dillig, I., Tasiran, S. (eds.) CAV 2019. LNCS, vol. 11561, pp. 443–452. Springer, Cham (2019). https://doi.org/10.1007/978-3-030-25540-4_26

20. Krizhevsky, A.: Learning multiple layers of features from tiny images. In: CoRR, abs/1708.07747 (2009)
21. Kurakin, A., Goodfellow, I.J., Bengio, S.: Adversarial examples in the physical world. In: ICLR Workshop, pp. 99–112 (2017)
22. Kurakin, A., Goodfellow, I.J., Bengio, S.: Adversarial machine learning at scale. In: ICLR, pp. 99–112 (2017)
23. Lazarus, C., Kochenderfer, M.J.: A mixed integer programming approach for verifying properties of binarized neural networks. In: IJCAI Workshop (2021)
24. Lecun, Y., Bottou, L., Bengio, Y., Haffner, P.: Gradient-based learning applied to document recognition. Proc. IEEE **86**(11), 2278–2324 (1998)
25. Li, C., et al.: Towards certifying the asymmetric robustness for neural networks: quantification and applications. In: TDSC (2021)
26. Liu, C., Tomioka, R., Cevher, V.: On certifying non-uniform bounds against adversarial attacks. In: ICML, pp. 4072–4081 (2019)
27. Madry, A., Makelov, A., Schmidt, L., Tsipras, D., Vladu, A.: Towards deep learning models resistant to adversarial attacks. In: ICLR (2018)
28. Müller, C., Serre, F., Singh, G., Püschel, M., Vechev, M.: Scaling polyhedral neural network verification on GPUs. In: MLSys (2021)
29. Qin, C., et al.: Verification of non-linear specifications for neural networks. In: ICLR (2019)
30. Raghunathan, A., Steinhardt, J., Liang, P.: Certified defenses against adversarial examples. In: ICLR (2018)
31. Salman, H., Yang, G., Zhang, H., Hsieh, C., Zhang, P.: A convex relaxation barrier to tight robustness verification of neural networks. In: NeurIPS (2019)
32. Singh, G., Ganvir, R., Püschel, M., Vechev, M.T.: Beyond the single neuron convex barrier for neural network certification. In: NeurIPS (2019)
33. Singh, G., Gehr, T., Püschel, M., Vechev, M.T.: An abstract domain for certifying neural networks. In: POPL, pp. 1–30 (2019)
34. Singh, G., Gehr, T., Püschel, M., Vechev, M.T.: Boosting robustness certification of neural networks. In: ICLR (2019)
35. Sinha, A., Namkoong, H., Duchi, J.C.: Certifying some distributional robustness with principled adversarial training. In: ICLR (2019)
36. Solar-Lezama, A., Tancau, L., Bodík, R., Seshia, S.A., Saraswat, V.A.: Combinatorial sketching for finite programs. In: ASPLOS, pp. 404–415 (2006)
37. Szegedy, C., et al.: Intriguing properties of neural networks. In: ICLR (2014)
38. Tjeng, V., Xiao, K.Y., Tedrake, R.: Evaluating robustness of neural networks with mixed integer programming. In: ICLR (2019)
39. Tu, C., et al.: Autozoom: autoencoder-based zeroth order optimization method for attacking black-box neural networks. In: AAAI, pp. 742–749 (2019)
40. VirusTotal: Virustotal, a free service that analyzes suspicious files and urls and facilitates the quick detection of viruses, worms, trojans, and all kinds of malware (2004). https://www.virustotal.com/
41. Wang, S., Pei, K., Whitehouse, J., Yang, J., Jana, S.: Efficient formal safety analysis of neural networks. In: NeurIPS (2018)
42. Wang, S., Pei, K., Whitehouse, J., Yang, J., Jana, S.: Formal security analysis of neural networks using symbolic intervals. In: USENIX, pp. 1599–1614 (2018)
43. Wang, S., et al.: Beta-crown: efficient bound propagation with per-neuron split constraints for neural network robustness verification. In: NeurIPS (2021)
44. Xiao, H., Rasul, K., Vollgraf, R.: Fashion-mnist: a novel image dataset for benchmarking machine learning algorithms. arXiv preprint arXiv:1708.07747 (2017)

45. Xie, C., Wu, Y., van der Maaten, L., Yuille, A.L., He, K.: Feature denoising for improving adversarial robustness. In: CVPR, pp. 501–509 (2019)
46. Yuan, X., He, P., Zhu, Q., Li, X.: Adversarial examples: attacks and defenses for deep learning. IEEE Trans. Neural Netw. Learn. Syst. **30**(9), 2805–2824 (2019)
47. Zeng, H., Zhu, C., Goldstein, T., Huang, F.: Are adversarial examples created equal? A learnable weighted minimax risk for robustness under non-uniform attacks. In: AAAI, pp. 10815–10823 (2021)
48. Zhang, C., Benz, P., Imtiaz, T., Kweon, I.S.: Understanding adversarial examples from the mutual influence of images and perturbations. In: CVPR, pp. 14521–14530 (2020)
49. Zhang, H., Weng, T., Chen, P., Hsieh, C., Daniel, L.: Efficient neural network robustness certification with general activation functions. In: NeurIPS (2018)

A Generic Framework to Coarse-Grain Stochastic Reaction Networks by Abstract Interpretation

Jérôme Feret[1,2] and Albin Salazar[1,2(✉)]

[1] DI ENS, École normale supérieure, Université PSL, CNRS, INRIA,
75005 Paris, France
`albin.salazar@ens.fr`
[2] INRIA, Paris, France

Abstract. In the last decades, logical or discrete models have emerged as a successful paradigm for capturing and predicting the behaviors of systems of molecular interactions. Intuitively, they consist in sampling the abundance of each kind of biochemical entity within finite sets of intervals and deriving transitions accordingly. On one hand, formally-proven sound derivation from more precise descriptions (such as from reaction networks) may include many fictitious behaviors. On the other hand, direct modeling usually favors dominant interactions with no guarantee on the behaviors that are neglected.

In this paper, we formalize a sound coarse-graining approach for stochastic reaction networks. Its originality relies on two main ingredients. Firstly, we abstract values by intervals that overlap in order to introduce a minimal effort for the system to go back to the previous interval, hence limiting fictitious oscillations in the coarse-grained models. Secondly, we compute for pairs of transitions (in the coarse-grained model) bounds on the probabilities on which one will occur first.

We illustrate our ideas on two case studies and demonstrate how techniques from Abstract Interpretation can be used to design more precise discretization methods, while providing a framework to further investigate the underlying structure of logical and discrete models.

Keywords: Abstract interpretation · Stochastic reaction networks · Logical modeling · Coarse-graining

1 Introduction

The field of Systems Biology is driven by the development of tools to investigate emergent behaviors in populations of biological molecules from single entity interactions. Among these tools, mathematical models of systems of interactions have been critical in identifying hidden mechanisms by perturbation studies, drivers of disease phenotypes and in generating new hypotheses. Consequently, a major interest underlies the ability for modeling tools to recapitulate biological phenomenon and its repurposing for predictive studies.

© The Author(s), under exclusive license to Springer Nature Switzerland AG 2023
C. Dragoi et al. (Eds.): VMCAI 2023, LNCS 13881, pp. 228–251, 2023.
https://doi.org/10.1007/978-3-031-24950-1_11

In retrospect, developing modeling tools is a continuous field of investigation as it provides means to gain understanding of biological systems through *in silico* studies. A common battle to derive an ideal representation for a system process is the descriptive trade-off between the simplicity and accuracy. On one hand, too simple models are prone to reproduce only *a priori* knowledge. On the other hand, too descriptive models result in behaviors too difficult to parse, or even yet compute. In both cases, gaining new insights is hampered. Thus, it is arguably important in generating tools to measure the impact model selection has in capturing biological phenomena, especially those which can be verified. For example, an overview of various formal modeling frameworks for Systems Biology have been reviewed in [4]. In this paper, we assess the impact of discretization in the study of biological systems.

Logical models are a popular class of discretized models. Recent development have made it an ideal modeling tool to perform perturbation studies for a myriad of biochemical interactions. For example, some logical models have been developed to study iron metabolism in breast cancer cells [7] and a cell division process in mammalian cells [11]. A key feature of logical models is in obtaining knowledge about a process while only partial information is available about some particular interactions and their kinetics. Albeit their success in recapitulating experimental observations and predicting local system properties, their underlying modeling assumptions are often kept implicit. There is indeed a gap between hand-written logical models and the models that can be formally derived from a more concrete level of representation (such as a stochastic reaction network).

One popular approach to design logical models has been proposed by René Thomas [16,19]. With this method, each dimension is associated with a unique attractor state (or focal point), which may depend on the current state of the system. Then the transitions of the system are obtained by assuming that on each dimension, the system may get closer to its focal point. Reverse transitions may also occur (but at low probabilities), but in practice they are neglected. Such simplifications of the model is usually justified by some time-scale separation principles [11] (that are mainly asymptotic reasonings providing convergence results when scales are infinitely separated). Two critical observations emerge about this modeling process. Firstly, it is unclear to which extent these simplifications actually impact the behaviors of the systems they try to represent. Secondly, considering all reverse transitions, without any information about their potential likelihood, would lead to inaccurate models with many fictitious non-deterministic behaviors.

In order to increase confidence in the modeling process, we would like to derive formally discrete models from more precise representations (such as reaction networks). For this purpose we use the abstract interpretation framework to coarse-grain reaction networks into discrete models of abstract regions of states while preserving formal relationships between the respective behaviors of both models. Yet we have to face several issues. Firstly, several behaviors that are usually neglected in the logical models may occur with a low probability in the initial reaction network. Thus a non-deterministic abstraction would be unhelpful, because non-deterministic models provide no means to distinguish rare events from more the common ones. Instead we propose to propagate the probabilities of transitions from the reference reaction network to the coarse-grained model, so that bounds to the probability of unlikely behaviors can be indeed computed. For this purpose, we equip each transition in the abstract model with an interval for their probabilities which is formally derived from the underlying reaction network. Secondly, even with probabilities, naive abstractions lead to very imprecise models. This means that we have to adapt our abstraction in order to highlight the main behaviors of interest. Consequently, we obtain a Discrete-time Markov chain (DTMC) whereby the exclusion of probabilities would structurally mimic logical models, yet providing an accessible tool to quantify differences between logical models and the discrete models obtained from our formalizations. One may be tempted to use Continuous-time Markov chains (CTMC) rather than DTMCs. Yet, the additional information provided by the continuous setting is not relevant in our context. On one hand, we want to obtain models comparable to logical models where the notion of time has already been abstracted away. On the other hand, the exact moment when each event occurs does not affect the computation of the probabilities of transitions between abstract regions in our coarse-grained models. What matters is only the relative order between these events, not the exact moment when they have happened. Yet, abstracting away the notion of time of a CTMC while only keeping the relative order between events induces a DTMC, which justifies our choice thoroughly. Lastly, using DTMCs instead of CTMCs deeply simplifies the underlying mathematics. For instance, in the continuous setting, probability density functions are required to define when events are likely to occur and a topology is necessary to define the probability of which set of model executions can be computed [13]. In contrast, in the discrete setting, only discrete probabilities are necessary and the probability of each execution with a finite amount of steps can be computed.

Ideally, the upper bounds computed for the probabilities of transitions that correspond to less likely behaviors should be very low. To achieve this goal, it is important to refine the abstraction process and to distinguish the abstract regions of states according to which transitions have been taken to enter them. Furthermore, it is also important that every transition between abstract regions corresponds to sequences composed of at least a few concrete transitions. This motivates the use of overlapping intervals to coarse-grain models. In the formal discretization process, an interval is composed of a pair of boundary values that enclose a concrete value. Thus, for each concrete value, an abstract interval computation is sensitive to whether a concrete value has exited a visited interval. In the abstraction, we consider that a value changes to another interval only when it actually leaves its current interval (hence leaving the overlapping region between its previous and current intervals). This way, when entering a new region of states, going in the reverse direction requires crossing through the overlap between two consecutive intervals, which is likely to have low probability when it is against the main trend of the dynamics of the system. The so-obtained abstraction ignores small fluctuations while strengthening the sequences of transitions that follow the main trend of the system dynamics.

Related Work. Value sampling is widely used to simplify dynamical systems. In piece-wise linear systems [9], the dynamics are approximated by one system of linear equations per equivalence class of states. In [15], an abstract interpretation based on support functions makes the computation of their trajectories scale while ensuring a sound over-approximation of any potential behavior. In the context of discrete modeling, the Boolean semantics of BIOCHAM [10] is also an abstract interpretation of the stochastic semantics, but it is too conservative. In [1,2], value sampling is refined by exploiting some formal properties of the initial model. But this abstraction relies on some informal time-scale separation arguments. We propose to compute conservative bounds on the probabilities on unlikely transitions in the coarse-grained model rather than ignoring them.

Our abstraction of states is history sensitive. In [3], the abstraction of each state of a reaction network also depends on the previous state in its trajectory.

As noticed in [6], refining the sampling intervals further in a discretization process does not necessarily reduce the amount of potential behaviors. This is why a new update policy has been introduced to recover this lack of monotonicity—at the cost of considering more fictitious behaviors. We do not think that non-monotonicity is an issue: when discretizing a system, refining sampling intervals introduces new check-points. Thus it provides the obtained model more opportunities to change its trajectories. We think that it is more important to relate the behaviors of the discretized models to the ones of the initial system, that is to say to ensure that every behavior of the initial system is reflected in the abstraction, would they be some additional fictitious behaviors. This is why, we prove formally that the potential behaviors of each coarse-grain model over-approximate the ones of a reference system (would it be known or not).

Lastly, our goal is also to compare different modeling paradigms in order to understand their underlying assumptions. Building a landscape of different semantics is one of the initial motivations behind the abstract interpretation framework [8].

Outline. The rest of the paper is organized as follows. In Sect. 2, we introduce a unidimensional case study to motivate and illustrate our general framework. Section 3 generalizes this approach to coarse-grain arbitrary stochastic reaction networks. In Sect. 4, we apply this framework on a tridimensional case study. We conclude in Sect. 5.

2 First Case Study: Birth and Death Model

In order to motivate our framework, we introduce a well-studied unidimensional system: a birth and death (BD) model.

Directed graph	Logical function	Transition system
A (self-loop)	$f_{x_v} = \begin{cases} \{0,1,2\} & \to & \{0,1,2\} \\ x_v & \mapsto & \begin{cases} x_v - 1 & \text{if } x_v > 1 \\ x_v + 1 & \text{if } x_v < 1 \\ 1 & \text{otherwise.} \end{cases} \end{cases}$	$\begin{array}{ccc} x_A(t) & \to & x_A(t+1) \\ 0 & \to & 1 \\ 1 & \to & 1 \\ 2 & \to & 1 \end{array}$

Fig. 1. A logical BD model. A directed graph (left) displays A as a self-regulator, while the logical function (center) is derived to reflect the expected BD system behaviors. The transition system (right) is the result of applying the logical function to a state $x_A \in \{0, 1, 2\}$, each representing an interval of values of molecule A: low (0), medium (1) and high (2). Note that the notion of time is discrete.

2.1   Reaction Network

The BD system is characterized by two reactions having opposite behaviors. Both reactions are given as follows:

$$r_1 : \emptyset \xrightarrow{k_A} A \qquad\qquad r_2 : A \xrightarrow{k_{A'}} \emptyset.$$

The reactions represent production and consumption events of molecules of A. The first reaction, r_1, is a birth event with kinetic constant k_A, while the second, r_2, is a death event with kinetic constant $k_{A'}$. We denote as q the state of the system. The state of the system maps the components of the system to their copy numbers. In this model, A is the unique component.

By assuming stochastic mass-action kinetics law, we can obtain the propensity k_A for the production event and the propensity $k_{A'} \cdot q(A)$ for the death event. Then, the probabilities $\lambda_{r_1}(q)$ and $\lambda_{r_2}(q)$ that the next event is an instance of the reaction r_1 or an instance of the reaction r_2 are defined as follows:

$$\lambda_{r_1}(q) = \frac{k_A}{k_A + k_{A'} \cdot q(A)} \text{ and } \lambda_{r_2}(q) = \frac{k_{A'} \cdot q(A)}{k_A + k_{A'} \cdot q(A)}.$$

A probability for an event type is the ratio between its propensity and the sum of all possible propensities for this system. We can observe that these probabilities are non-constant, as the quantity of A, $q(A)$, may vary.

2.2   Logical Model

A logical model can be provided regardless of the exact structure of the reactions and the effective values of the kinetic parameters. Following René Thomas's principles, what matters is the general trend for the evolution of the copy numbers of each kind of components. In our case study, we can indeed distinguish three kinds of states: when the amount of A is such that the state of the system is likely to be stable; when (below this amount) the quantity of A is likely to increase; and when (above this amount) the quantity of A is likely to decrease.

These observations lead to the logical model that is described in Fig. 1. Firstly, a directed graph summarizes the potential regulations (or dependencies) between the components. Here a self arrow on the component A stipulates that the component A auto-regulates itself. We assume that the potential quantities of A are partitioned into three intervals, denoted as 0, 1, and 2. They stand respectively for below the steady state (low/(0)), for around the steady state (medium/(1)), and for above the steady state (high/(2)). Consequently, we derive a logical function that reflects how the system is expected to evolve: below the steady state, the amount of A increases; around the steady state, it remains constant; above the steady state, it decreases. Then, the logical function induces a transition system that captures the integrated process. Note, however, that at this level of abstraction stochastic fluctuations cannot be observed since reversible interval transitions are not permitted. Thus, the logical model is capable of capturing only the most expected behaviors.

2.3   Formal Derivation of a Coarse-Grained Model

Now that a logical model for the BD system has been proposed, we would like to compare its behaviors to those obtained by a formal discretization. We use the same BD reaction network and formally discretize its state space. Then we show that it is possible to restore information on probabilities in this new model.

A common discretization method uses a non-overlapping interval schema. This means that the state space of chemical values are partitioned into intervals that do not share any common values. For example, in the logical BD model interval partitioning are qualitative states with implicit meaning. We can obtain a similar representation in the formally derived model by explicitly choosing a sequence of contiguous intervals (with no intersecting values). It is worth noting that we expect this new model to cope with many more behaviors than the logical model. A question then rises as to whether these behaviors are consequence of the imprecision of the formal abstraction, or whether they reveal important behaviors that are missing in the logical models.

To answer this question, we use information about the stochastic behaviors of the reaction network to recover the probabilities to navigate between intervals (see Fig. 2). More precisely, when entering an interval, we compute the probability that the process will cross this interval or go back to the previous one. It is worth noting that whether an interval is entered from below or from above matters. So we duplicate each interval accordingly. We expect to observe the convergence of the process towards the interval containing the steady state, which is a stable behavior observed by a system. This is the main qualitative behavior of the reaction network and it should be reflected in the discretize model independently of the choices of the discretization intervals.

We call the transitions between intervals macro-transitions. Macro-transition probabilities are computed as follows. Given an interval with lower bound $l \in \mathbb{N}$ and upper bound $u \in \mathbb{N}$, we consider for every state q such that $l \leq q(A) \leq u$, the probability $P(q)$ such that the quantity of A will reach the upper bound u before reaching the lower bound l, knowing that the system starts in the state q. By reasoning on the potential BD events stemming from the state q and their probabilities, we obtain the following relation:

$$P(q) = \begin{cases} 0 \text{ whenever } q(A) = l, \\ 1 \text{ whenever } q(A) = u, \\ \lambda_{r_1}(q) \cdot P([A \mapsto q(A) + 1]) + (1 - \lambda_{r_1}(q)) \cdot P([A \mapsto q(A) - 1]) \text{ otherwise.} \end{cases}$$

As boundary conditions, the probabilities $P([A \mapsto l])$ and $P([A \mapsto u])$ are set respectively to 0 and 1. The last case combines the contribution of two processes: the potential increase in the amount of A with probability $\lambda_{r_1}(q)$ and the potential decrease with probability $1 - \lambda_{r_1}(q)$.

Finite unfolding of the previous recurrence relation converges to a lower bound on the probability $P(q)$. In Sect. 3.3, we discuss how one can exploit recurrence relations to bound an exact probability by an interval of probabilities. Yet, in the BD model, a closed form equation can be derived. The probability $P(q)$ is indeed defined by the following equality:

$$P(q) = \frac{Aux(q(A))}{Aux(u)} \tag{1}$$

where $Aux(j) = \sum_{l \le s' < j} \left(\prod_{l < s \le s'} \left(\frac{k_{A'} \cdot s}{k_A} \right) \right)$, for each $j \in \{u, q(A)\}$. We can use Eq. 1 to compute the probability to reach first an upper (and by complement, first a lower) bound.

We show in Fig. 2 the macro-transition system that we derive this way. Underlying each rectangular region are the dynamics stemming from the BD process with kinetic constants $k_A = 20$ and $k_{A'} = 1$. Thus, the intervals displayed are chosen according to the steady state of the system, which is when the quantity of A is equal to 20, or $q(A) = 20$, and in this example it is contained in the interval $[20, 24]$. Each macro-transition is composed of a source interval, an edge labeled with a probability and a target interval. To trigger a macro-transition type, a BD event (birth or death) must push the value $q(A)$ through an interval upper or lower bound value. Thus, it is possible to enter a target interval from below (via an upper bound) or above (via a lower bound). Intervals are duplicated accordingly. The one on the left side of the transition system denotes those entered from below and the one on the right side, those entered from above. Also the exact position of the source (resp. target) of each macro-transition indicates from which border of the interval the macro-transition starts (resp. ends).

We then want to observe whether the trend of the system will proceed upwards or downwards. Since abstraction loses all information about the probabilities of individual reactions, we recover them using Eq. 1. As a result, the probability to exit from the upper bound 19 when starting from the value $q(A) = 15$ is equal to 0.27, and its complement 0.73 is the probability to exit from the lower bound 15. Putting these pieces of information together results in the macro-transition $[15, 19] \xrightarrow{0.73} [10, 14]$ in Fig. 2. Namely, after going up to the interval $[15, 19]$, the system has a higher tendency to return to a lower interval. Note that this behavior is opposite in direction to the stable interval $[20, 24]$. Actually, the general trend of a majority of macro-transitions opposes the direction of the interval containing the steady state point. Mainly this is because it

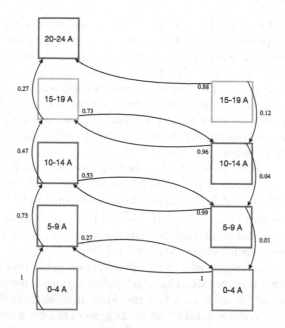

Fig. 2. A non-overlapping macro-transition system for the BD system. Each rectangle is a range of values for the molecule A and a labeled edge is a transition from a source to a target interval. Intervals are duplicated to distinguish whether they are entered from below or above. (Color figure online)

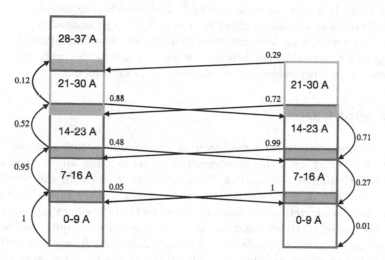

Fig. 3. An overlapping macro-transition for the BD system. The interpretation is similar to the non-overlapping case in 2 with the exception that intervals overlap (denoted by a gray region) (Color figure online)

requires only one transition in the initial reaction network to go back to the previous interval, whereas several ones are required to cross an interval entirely. Hence border effects give too much importance to backwards macro-transitions.

To cope with this artifact of the abstraction, we introduce a minimal effort for the system to perform fluctuations between consecutive intervals by using overlapping intervals instead. In Fig. 3, we compute a macro-transition system for the same BD system as in Fig. 2 but with overlapping intervals (overlaps are indicated in gray). The meaning of macro-transitions has to be defined carefully: we consider a macro-transition between a first interval and a second one, only when the system leaves the first interval (hence crossing the overlapping region). We adjust the position of the source and target of the macro-transitions accordingly in the drawing.

Now, the stable interval is $[14, 23]$. Starting from the value $q(A) = 10$, the probability to exit from the upper bound 16 of the interval $[7, 16]$ is equal to 0.95 while the probability to exit from its lower bound 7 is equal to 0.05. Consequently, we obtained the macro-transition $[7, 16] \xrightarrow{0.95} [14, 23]$ which shows the tendency to move towards the stable interval. Similar observations can be made about the other macro-transitions leading to the interval $[14, 23]$.

We notice two major features from our overlapping interval design. Firstly, quantities of molecule A contained in overlapping regions must surpass a buffer region to be able to go back into the previous interval, thus limiting border effects. And, secondly, the general trend of the system reflects a greater likelihood towards the stable interval which was not the case for non-overlapping intervals.

Altogether, we show how our framework is capable of coarse-graining the behaviors emerging from the BD reaction network using a more formal approach. Whereas discretization with non-overlapping intervals was not enough to keep only the likely behaviors as done in the logical models, the use of overlapping intervals introduces a minimal effort for the system to go back after entering an interval. The result is an abstract transition system when unnatural behaviors are assigned low probabilities, hence allowing to quantify the probability of the behaviors that are neglected in the hand-written logical model.

3   General Case

In the previous section, we introduced as an example a logical model of the BD system and compared this hand-written model to a formally derived coarse-grained model from the same underlying reaction network. In this section, we generalize this approach to coarse-grain arbitrary reaction networks. More specifically, we build a concrete semantics to capture all the behaviors that may emerge from a system of reactions, and an abstract semantics to approximate these behaviors with intervals (overlapping or not). After which, we bridge the two semantics to restore information on probabilities to the abstract semantics.

3.1   Concrete Semantics

Firstly we define the syntax for reactions and reaction networks.

Definition 1 (Chemical Reaction). *Given a finite set $\mathbb{V}$ of chemical species, a reaction over the set of species $\mathbb{V}$ is defined as a triple $r = (M, V, k)$ such that:*

1. $M : \mathbb{V} \to \mathbb{N}$,
2. $V : \mathbb{V} \to \mathbb{Z}$,
3. $k : \mathbb{V}^{\mathbb{N}} \to \mathbb{R}_{\geq 0}$.

In Definition 1, the function M stands for the (multi-)set of reactants, V denotes the reaction vector (which cumulates the production (positively) and the consumption (negatively) of each chemical species), and k is a function mapping a vector of chemical quantities to a real number which denotes a kinetic term.

Definition 2 (Chemical Reaction Network). *A reaction network R is defined as a pair $(\mathbb{V}, (r_j)_{1 \leq j \leq n})$ such that:*

1. $\mathbb{V}$ *is a set of chemical species;*
2. $(r_j)_{1 \leq j \leq n}$ *is a set of n reactions over the set $\mathbb{V}$ indexed with an integer j between 1 and n.*

For each integer j between 1 and n, the reaction r_j is also denoted as $(M_{r_j}, V_{r_j}, k_{r_j})$.

A chemical state encodes the values of each chemical species.

Definition 3 (Chemical State). *A chemical state is defined as a function $q : \mathbb{V} \to \mathbb{N}$. The set of all the chemical states is denoted as $\mathcal{Q}$.*

Additionally, a chemical state is an input to a kinetic function to obtain a kinetic term for each reaction that involves this state.

For example, we apply our definitions to the reaction network made of both following reactions:

$$r_1 : A \xrightarrow{k_B} B \qquad\qquad r_2 : 2A \xrightarrow{k_C} C.$$

Here, the set of chemical species is $\{A, B, C\}$. This reaction network is made of two reactions r_1 and r_2, with respective multiplicity vectors $M_{r_1} = [A \mapsto 1, B \mapsto 0, C \mapsto 0]$ and $M_{r_2} = [A \mapsto 2, B \mapsto 0, C \mapsto 0]$ and with respective reaction vectors $V_{r_1} = [A \mapsto -1, B \mapsto 1, C \mapsto 0]$ and $V_{r_2} = [A \mapsto -2, B \mapsto 0, C \mapsto 1]$. Furthermore, assuming the stochastic mass-action kinetics law, the kinetic functions are defined as $k_{r_1} = [q \mapsto k_B \cdot q(A)]$ for the reaction r_1 and $k_{r_2} = \left[q \mapsto \frac{k_C \cdot q(A) \cdot (q(A)-1)}{2}\right]$ for the reaction r_2.

Until the rest of Sect. 3, we assume that we are given $(\mathbb{V}, (r_j)_{1 \leq j \leq n})$ a generic reaction network that we also denote as R. The set $\{r_1, \ldots, r_n\}$ is also written as $\mathcal{R}$. This is the set of the reactions of the network R.

Furthermore, a system is updated via a reaction application, which is called a chemical transition. This is formalized in the following definition.

Definition 4 (Chemical Transition). *A chemical transition is a triple $(q, r, q') \in \mathcal{Q} \times \mathcal{R} \times \mathcal{Q}$ relating two chemical states $q, q' \in \mathcal{Q}$ by a reaction r such that for all chemical species $v \in \mathbb{V}$:*

1. $M_r(v) \leq q(v)$
2. $q'(v) = q(v) + V_r(v)$.

The set of all the chemical transitions is denoted as T.

A chemical transition captures an application of a reaction rule. Criterion 1 ensures that there are enough reactants available for a reaction to occur, while criterion 2 applies a reaction rule to update a predecessor value to obtain a successor value. Additionally, each chemical transition is given a probability. The probability that the transition will be the next one, given the current state, is defined as follows:

Definition 5 (Transition probability). *Let $(q, r, q') \in T$ be a chemical transition. The probability $\lambda_r(q)$ for a chemical state $q \in Q$ involved in a chemical transition is defined as:*

$$\frac{k_r(q)}{\sum_{r' \in R} k_{r'}(q)}.$$

In Definition 5 the probability that a given chemical transition is applied next, is equal to the ratio of its kinetic term to all kinetic terms of the reaction network.

For example, an instantiation of this definition used on the previous reaction network results in the following transition probabilities:

$$\lambda_{r_1}(q) = \frac{2 \cdot k_B \cdot q(A)}{q(A) \cdot (2 \cdot k_B + k_C \cdot (q(A) - 1))} \text{ and } \lambda_{r_2}(q) = \frac{k_C \cdot q(A) \cdot (q(A) - 1)}{q(A) \cdot (2 \cdot k_B + k_C \cdot (q(A) - 1))}$$

whenever $q(A) > 0$. Note that when $q(A) = 0$, no reaction is enabled and the system is deadlocked.

Starting from an initial chemical state, it is possible to chain chemical transitions. This leads to the notion of a chemical trace.

Definition 6 (Chemical Trace). *A trace of length $k \in \mathbb{N}$ is a pair $(q'_0, ((q_i, r_i, q'_i), \mu_i)_{1 \leq i \leq k}) \in Q \times (T \times [0, 1])^k$ that satisfies both conditions:*

1. *for every integer i between 0 and $k - 1$, we have $q'_i = q_{i+1}$;*
2. *for every integer i between 1 and k, we have $\mu_i = \lambda_{r_i}(q_i)$.*

Such a trace is usually written as $q_1 \xrightarrow[\mu_1]{r_1} \dots \xrightarrow[\mu_k]{r_k} q'_k$.

The set of all the chemical traces of a reaction network defines all the potential long-term behaviors of its underlying system. Given an initial chemical state $q'_0 = [A \mapsto 6, B \mapsto 0, C \mapsto 0]$ and the kinetic constants $k_B = 20$ and $k_C = 1$, an example of a chemical trace for the previous reaction network is given as follows:

$$(6, 0, 0) \xrightarrow[0.8]{r_1} (5, 1, 0) \xrightarrow[0.83]{r_1} (4, 2, 0) \xrightarrow[0.13]{r_2} (2, 2, 1) \xrightarrow[0.95]{r_1} (1, 3, 1) \xrightarrow[1]{r_1} (0, 4, 1),$$

where a state q is denoted as the triple $(q(A), q(B), q(C))$. At the end of this trace, no transition is available.

Finally, the following definition associates a probability to each chemical trace.

Definition 7. *Let* $(q_0', ((q_i, r_i, q_i'), \mu_i)_{1 \le i \le k})$ *be a chemical trace that we denote as* τ. *The probability* $P(\tau \mid q_0')$ *of the chemical trace* τ, *knowing that the system starts in the state* q_0' *is defined as* $\prod_{1 \le i \le k} \mu_i$.

For example, the probability for the previous trace is: $P(\tau \mid (6, 0, 0)) = 0.08$ (since $0.80 \cdot 0.83 \cdot 0.13 \cdot 0.95 \cdot 1 = 0.08$).

3.2   Abstract Semantics

The goal of the abstract semantics is to over-approximate the behaviors emerging from chemical reaction networks. Namely, it is obtained by sampling the value domains by the means of a set of intervals.

Definition 8 (Intervals). *We consider a family* $(\underline{q}_p^\sharp, \overline{q}_p^\sharp)_{1 \le p \le n}$ *of* n *pairs of values in* $\mathbb{N} \cup \{+\infty\}$ *(where* n *is a natural number in* $\mathbb{N}$*) such that both of the following properties are satisfied:*

1. *for every natural number* p *between* 2 *and* n, $\underline{q}_{p-1}^\sharp < \underline{q}_p^\sharp \le \overline{q}_{p-1}^\sharp < \overline{q}_p^\sharp$;
2. $\overline{q}_n^\sharp = +\infty$.

We denote by $D^\sharp$ *the set of intervals* $\{(\underline{q}_p^\sharp, \overline{q}_p^\sharp) \mid 1 \le p \le n\}$.

An interval $(\underline{q}_p^\sharp, \overline{q}_p^\sharp)$ denotes the set of values $\{k \in \mathbb{N} \mid \underline{q}_p^\sharp \le k < \overline{q}_p^\sharp\}$. There are finitely many of them. Each of them is well-formed. Their lower bounds form an increasing sequence, as well as their upper bounds. Also every natural number occurs in at least one of them.

Conversely, an abstraction of a value is an interval in the domain $D^\sharp$ that contains this value. There may be several such intervals. To decide which one, the abstraction function is parameterized by a context made of a reference interval. The following definition specifies that the so-contextualized abstraction function selects the interval nearest to the reference one among the potential ones.

Definition 9 (Value Abstraction Function). *Let* $(\underline{q}_{p_\star}^\sharp, \overline{q}_{p_\star}^\sharp)$ *be an interval in* $D^\sharp$. *The value abstraction function* $\beta_{(\underline{q}_{p_\star}^\sharp, \overline{q}_{p_\star}^\sharp)}^{\mathcal{D}} : \mathbb{N} \to D^\sharp$ *maps each value* $k \in \mathbb{N}$ *to the unique interval* $(\underline{q}_p^\sharp, \overline{q}_p^\sharp) \in D^\sharp$, *such that both following properties are satisfied:*

1. $\underline{q}_p^\sharp \le k < \overline{q}_p^\sharp$;
2. *for any* $(\underline{q}_{p'}^\sharp, \overline{q}_{p'}^\sharp) \in D^\sharp$ *such that* $\underline{q}_{p'}^\sharp \le k < \overline{q}_{p'}^\sharp$, *we have:* $|p_\star - p| \le |p_\star - p'|$.

Definition 9 is well-formed thanks to the hypotheses in Definition 8. More precisely, the existence of an interval in $D^\sharp$ containing a given value follows from the fact that the elements of $D^\sharp$ forms a covering of $\mathbb{N}$, then thanks to the monotonicity of the lower and upper bounds of the intervals, the set of intervals that contain a given value are contiguous elements in the domain. It follows the uniqueness of the interval that is the closest to the reference interval.

Now we lift the notions of values and value abstraction to all the chemical species of a reaction network.

Definition 10 (Abstract State). *An abstract state is a function* $q^\sharp : \mathbb{V} \to D^\sharp$. *The set of all abstract states is denoted* $\mathcal{Q}^\sharp$.

An abstract state contains all the interval values approximating the quantities of each chemical species.

Definition 11 (State Abstraction Function). *Let* $q_*^\sharp$ *be an abstract state in* $\mathcal{Q}^\sharp$. *The abstract state function* $\beta_{q_*^\sharp}^{\mathcal{S}} : \mathcal{Q} \to \mathcal{Q}^\sharp$ *maps each chemical state* q *to the abstract state* $\left[v \in \mathbb{V} \mapsto \beta_{q_*^\sharp(v)}^{D}(q(v)) \in D^\sharp \right]$.

Similarly to the value abstraction function, the state abstraction function is parameterized by a reference abstract state. An equivalent definition can be obtained by interpreting each abstract state as the box delimited on each chemical species in $\mathbb{V}$ by its corresponding intervals, and then by abstracting each concrete state by the unique box that contains this concrete state and that is at minimal Gaussian distance from the reference abstract state.

It is worth noting that we have used the same family of intervals to abstract the quantity of every chemical species. Using different families of intervals is also possible and it would have raised no further technical difficulties. Indeed, it would have been even useful in practice since there is no reason why the quantity of each component of the system should be abstracted the same way. Yet making this simplification deeply lighten the presentation of the framework and this is why we have proceeded this way.

We can now define the abstraction of a chemical trace. Each abstract trace is obtained by lifting point-wise the state abstraction function to each chemical state along a chemical trace, taking respectively for reference the previous abstract state. For the moment, we discard probabilities. Restoring information about the probabilities of abstract transitions is the purpose of Sect. 3.3.

Definition 12 (Abstract Trace). *An abstract trace is an element of the set* $\mathcal{Q}^\sharp \times (\mathcal{Q}^\sharp \times \mathcal{R} \times \mathcal{Q}^\sharp)^\star$.

Definition 13 (Trace Abstraction Function). *The trace abstraction function* $\beta^{\mathcal{T}}$ *maps each chemical trace* $(q_0', ((q_i, r_i, q_i'), \mu_i)_{1 \leq i \leq k})$ *to the abstract trace that is defined inductively as follows:*

1. $\beta^{\mathcal{T}}(q_0', ()) = (\beta_{q_0^\sharp}^{\mathcal{S}}(q_0'), ())$;

2. *By induction, if* $\beta^{\mathcal{T}}(q_0', ((q_i, r_i, q_i'), \mu_i)_{1 \leq i < k}) = (q_0^{\sharp\prime}, (q_i^\sharp, r_i^\sharp, q_i')_{1 \leq i < k})$, *the abstract trace* $\beta^{\mathcal{T}}(q_0', ((q_i, r_i, q_i'), \mu_i)_{1 \leq i \leq k})$ *is defined as* $(q_0^{\sharp\prime}, (q_i^\sharp, r_i^\sharp, q_i')_{1 \leq i \leq k})$ *where* $(q_k^\sharp, r_k^\sharp, q_k^{\sharp\prime}) = (q_{k-1}^{\sharp\prime}, r_k, \beta_{q_{k-1}^{\sharp\prime}}^{\mathcal{S}}(q_k'))$.

where $q_0^\sharp$ *is the abstract state mapping every component to the interval* $(\underline{q_0^\sharp}, \overline{q_0^\sharp})$.

The trace abstraction function starts by abstracting the initial state of a chemical trace by using the abstract state $q_0^\sharp$ as a reference; then it abstracts each chemical transition by abstracting the successor chemical state of the corresponding chemical transition while referencing the last encountered abstract state. For example, we apply our definition on the following chemical trace:

$$3 \xrightarrow[0.87]{r_1} 4 \xrightarrow[0.83]{r_1} 5 \xrightarrow[0.20]{r_2} 4 \xrightarrow[0.83]{r_1} 5 \xrightarrow[0.20]{r_2} 4 \xrightarrow[0.83]{r_1} 5 \xrightarrow[0.80]{r_1} 6$$

for the BD model introduced in Sect. 2.3 with several interval samplings.

With the following choice of non-overlapping intervals: $((0,4),(5,9))$, we obtain the following abstract trace:

$$(0,4) \xrightarrow{r_1}^{\sharp} (0,4) \xrightarrow{r_1}^{\sharp} (5,9) \xrightarrow{r_2}^{\sharp} (0,4) \xrightarrow{r_1}^{\sharp} (5,9) \xrightarrow{r_2}^{\sharp} (0,4) \xrightarrow{r_1}^{\sharp} (5,9) \xrightarrow{r_1}^{\sharp} (5,9),$$

whereas with the following choice of overlapping intervals: $((0,5),(2,7))$, we obtain the following one:

$$(0,5) \xrightarrow{r_1}^{\sharp} (0,5) \xrightarrow{r_1}^{\sharp} (0,5) \xrightarrow{r_2}^{\sharp} (0,5) \xrightarrow{r_1}^{\sharp} (0,5) \xrightarrow{r_2}^{\sharp} (0,5) \xrightarrow{r_1}^{\sharp} (0,5) \xrightarrow{r_1}^{\sharp} (2,7).$$

We notice that with the second choice, the system remains in the first interval until finally exiting via its final chemical state of the chemical trace. By emphasizing the effort to go back and forth between consecutive intervals, the abstraction has abstracted away the fluctuations. This is not the case with the first choice of intervals. Please observe that no concrete behavior has been lost in the process. Then, we will see in Sect. 3.3, how to recover some information about the probabilities of the behaviors of the initial chemical networks. In that context, not only, as in a non-deterministic setting, any potential behavior in the concrete is reflected in the abstraction, but also the probability attached to each potential concrete behavior is over-approximated in the abstraction: transitions between abstract regions come with an upper bound on their probability to occur. In particular, in case of a rare event that may occur in the concrete only at a very small probability, by construction, this event is taken into account in the abstraction, but its likelihood may be over-estimated.

Furthermore, each choice of sampling intervals comes with a different interpretation for the abstract traces, which may highlight or hide different information accordingly. When the initial system is complex, the process of interval parameterization may proceed heuristically. Our framework provides a tool to tune granularity: too coarse interval abstractions can mask the underlying dynamics; however, upon refining the intervals one will be able to identify the behavioral trend of a system. In practice, it is enough to pick intervals that are fine-grained enough to separate the main regimes of the system.

3.3   Recovering Information About Transition Probabilities

In Sect. 3.3, we refine the abstract semantics with some quantitative information to compare the likelihood of transitions between abstract states, the macro-transitions. This basically means that knowing that the system has just entered a new abstract state, and a given pair of potential macro-transitions, we would like to know with which probability the first macro-transition (in the pair) will occur before the second one.

By construction of our abstract domain, macro-transitions are triggered when a given chemical species reaches a particular copy number. We introduce target regions accordingly in the following definition.

Definition 14 (Target region). *A target region is a set of concrete states of the form* $\{q \in \mathcal{Q} \mid q(v) \ \square \ b\}$*, where* v *is a chemical species in* $\mathbb{V}$*,* $\square$ *a binary relation in the set* $\{\leq, \geq\}$*, and* b *a natural number in* $\mathbb{N}$*.*

This target region is denoted as $g_{v,\square,b}$*.*

Until the end of Sect. 3.3, we consider $q_\bullet \in \mathcal{Q}$ a state, $\mathcal{G}$ a set of target regions, and g a specific target region in the set $\mathcal{G}$. We want to define, the probability that the system when starting from the state $q_\bullet$, will enter the specific region g before entering any other target regions of the set $\mathcal{G}$. In order not to overcount the probabilities, we cut chemical traces as soon as they enter the region g and we ignore the traces that enter another region in the set $\mathcal{G}$ before.

Definition 15 (Minimum successful traces). *We denote as* $\chi_{(q_\bullet, \mathcal{G}, g)}$ *the set of the chemical traces* $(q_0', ((q_i, r_i, q_i'), \mu_i)_{1 \leq i \leq k})$ *such that the following conditions are satisfied:*

1. $q_0' = q_\bullet$*;*
2. $q_k' \in g$*;*
3. $\forall i \in \mathbb{N}$*, such that* $0 \leq i < k$*,* $q_i' \notin \bigcup \mathcal{G}$*.*

In Definition 15, the set $\chi_{(q_\bullet, \mathcal{G}, g)}$ contains all the chemical traces that start in the state $q_\bullet$ (Cond. 1), reach the target region g in their final state (Cond. 2), and have reached no other target regions before (Cond. 3).

We are now ready to integrate the probability of minimal successful traces.

Definition 16 (Probability to reach a specific goal first). *The probability* $P_g^{\mathcal{G}}(q_\bullet)$ *that the system reaches the target region* g *before any other target regions in* $\mathcal{G}$ *when starting in the state* $q_\bullet$ *is defined as:* $\sum_{\tau \in \chi_{(q_\bullet, \mathcal{G}, g)}} P(\tau \mid q_\bullet)$*.*

Namely, the probability to reach a specific goal first is computed by summing the probability of all corresponding minimum successful traces.

The following proposition provides an easier way to compute this probability.

Proposition 1 (Inductive definition). *The probabilities* $P_g^{\mathcal{G}}(q)$ *for every state* $q \in \mathcal{Q}$ *are related by the following three conditions:*

1. $P_g^{\mathcal{G}}(q) = 1$ *whenever* $q \in g$*;*
2. $P_g^{\mathcal{G}}(q) = 0$ *whenever* $q \in \bigcup \mathcal{G} \setminus g$*;*
3. $P_g^{\mathcal{G}}(q) = \sum_{q \xrightarrow{r_i} q'} \lambda_{r_i}(q) \cdot P_g^{\mathcal{G}}(q')$ *whenever* $q \notin \bigcup \mathcal{G}$*.*

Proposition 1 provides an iterative scheme that computes for every state q a sequence of values that converges from below to the value of $P_g^{\mathcal{G}}(q)$. By complementing, we can also obtain an upper bound to this probability.

We can go further by the means of matrix computations. We consider one dimension for each potential state. Each function that maps states to real numbers is interpreted as a vector, whereas each function that maps pairs of states to real numbers is interpreted as a matrix. We define the vector B, and both matrices I and A as follows:

$$B(q) = \begin{cases} 1 & \text{whenever } q \in g, \\ 0 & \text{otherwise;} \end{cases} \qquad I(q, q') = \begin{cases} 1 & \text{whenever } q = q', \\ 0 & \text{otherwise;} \end{cases}$$

$$A(q, q') = \begin{cases} 0 & \text{when } q \in \bigcup \mathcal{G}, \\ \sum_{q \xrightarrow{r_i} q'} \lambda_{r_i}(q) \cdot P_g^{\mathcal{G}}(q') & \text{otherwise.} \end{cases}$$

Then, the sequence $(X_k)_{k \in \mathbb{N}}$ of vectors that is defined by:

1. $X_0 = B$;
2. $X_{k+1} = A \cdot X_k + B$ for every $k \in \mathbb{N}$;

converges component-wise to the probability $P_g^{\mathcal{G}}(q)$. It follows that $P_g^{\mathcal{G}} = (\sum_{j \in \mathbb{N}} A^j) \cdot B$. Or even, $P_g^{\mathcal{G}} = (I - A)^{-1} \cdot B$, whenever the matrix $(I - A)$ is invertible.

The computation of the probabilities $P_g^{\mathcal{G}}(q)$ can be proceeded by using any available linear algebra library. This is indeed what the model checker PRISM [14,17] is doing. Yet, having unfolded the computation offers several advantages. For instance, in our setting, the probabilities can be approximated from below by finitely approximating the formal expansion of the sums of the powers of the sparse matrix A. Secondly, when dealing with high dimensional models, expressions with scalar coefficients can be symbolically simplified into expressions over interval coefficients [18] in order to eliminate some dimensions and to tune the trade-off between accuracy and efficiency. This would not be possible with a black box approach.

In Sect. 3, we have refined our non-deterministic abstract semantics with probabilities, hence providing information about the general trend for the dynamics of models. In Sect. 4, we apply this approach on a case study taken from [2], which consists of two reactions competing for a common resource at different time-scales according to the availability of this resource. In [2] a precise coarse-grained system has been derived, but at the cost of neglecting some slow reactions. We would like to assess this assumption from a formal perspective.

4 Second Case Study: Competition for Resources

In Sect. 2, we compared a logical model and a formally discretized one of the BD process. In this section, we will follow a similar strategy for a model of a system of reactions which compete for a common resource taken from [2].

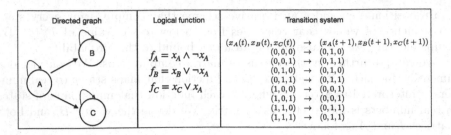

Fig. 4. A logical model for a system of resource competition. A directed graph (left) displays how molecule B and C have a common regulator, molecule A. Moreover, each kind of molecules auto-regulates it-self. Each logical function (center) is a Boolean rule which reflects the update scheme for a given chemical species. They can take values in the domain $\{0, 1\}$. The value 0 stands for low, 1 for high. The transition system (right) reflects the system dynamics from the logical functions.

4.1   Reaction Network

The second case study is composed of two reactions:

$$r_1 \; : \; A \xrightarrow{k_B} B \qquad\qquad r_2 \; : \; 2A \xrightarrow{k_C} C$$

where two chemical species, B and C are produced and each consume a common resource, A. In reaction r_1, a quantity of B is produced with a kinetic constant k_B. In reaction r_2, a quantity of C is produced with kinetic constant k_C. The production of molecule B consumes a quantity of A, while C requires two.

As in the first case study, we assume stochastic mass-action kinetics law to obtain the propensities $k_B \cdot q(A)$ and $\frac{k_C \cdot q(A) \cdot (q(A)-1)}{2}$ for, respectively, the reactions r_1 and r_2. Consequently, we derive a probability function for each reaction:

$$\lambda_{r_1}(q) = \frac{2 \cdot k_B \cdot q(A)}{q(A) \cdot (2 \cdot k_B + k_C \cdot (q(A) - 1))} \;\; \text{and} \; \lambda_{r_2}(q) = \frac{k_C \cdot q(A) \cdot (q(A) - 1)}{q(A) \cdot (2 \cdot k_B + k_C \cdot (q(A) - 1))}$$

Contrary to the first case study, this reaction system does not have reversible reactions and the quantities of B and C are strictly increasing up to the point at which all resources become depleted.

4.2   Logical Model

As in the first case study, we can write by hand a logical model for this second example. The property of interest is the competition between the production of the molecules B and C. Namely, depending on the quantity of the resource A, either B or C is produced more abundantly. When the quantity of A is under a certain value, the quantity of B increases more; and when it is above this value, the quantity of C increases more.

Using this knowledge we obtain the logical model that is described in Fig. 4. The directed graph shows different kinds of regulations. Firstly, A regulates B and C because it may be consumed to produce them. Secondly, each component auto-regulates itself since whatever it increases or decreases, its quantity at the next time step depends on the one at the current state. Lastly, A auto-regulates itself negatively (since it is consumed to produce B and C). We can abstract quantities by Boolean values. Namely, the state is a triple of Boolean variables $(x_A, x_B, x_C) \in \{0, 1\}^3$, here 0 stands for low quantity and 1 for high quantity. The logical (Boolean) functions are derived to capture the main feature of the reaction network. A gets consumed, while either B or C is produced according to the qualitative abundance of A.

Several update policies exist to define the operational semantics. Our transition system is derived by assuming the synchronous one, where the value of each chemical species is updated at each time step. It induces the eight transitions that are described in Fig. 4. It is worth mentioning that similar results can be achieved in the asynchronous mode by taking into account mass preservation of invariants and using priorities [2]. The model indicates that starting with a low amount of the molecule A, the system produces some B, but no C; whereas with a high amount, the system produces firstly some C, then some B. One may wonder whether more behaviors may occur in the underlying reaction system that have been discarded by the simplification into a logical model.

4.3   Formal Discretization of the Reaction Network

This motivates the formal discretization of the reaction network with overlapping intervals to compare the behaviors of the so-obtained model to the ones of the logical model. Specifically, we wonder whether our framework is capable of highlighting both main behaviors (production of B, or production of C followed by production of B), and provides low upper bounds for the probability of behaving differently.

Firstly, in order to simplify the computation, we would like to eliminate the variable $q(A)$ which stands for the quantity of A in the system. We denote as $q \in \mathcal{Q}$ the chemical state, which contains copy numbers of molecules A, B, and C. The system is constrained by the following mass invariant $q(A) = q_0(A) - (q(B) - q_0(B)) - 2 \cdot (q(C) - q_0(C))$, where q_0 stands for the initial state and is fixed once for all. We can safely replace each occurrence of the variable $q(A)$ with the right hand side of this equality in any expression.

As in the first case study, the domain of values can be sampled into overlapping intervals. This way, chemical states are gathered into rectangular regions (we are left with only two dimensions since we have eliminated the quantity of A), that are called abstract states. Our goal is then to derive some quantitative information about the macro-transitions in the so-obtained discretized model. More specifically, when a chemical state enters a new abstract state, the goal is to compute the probability that the system will cross the corresponding rectangular region and exit along the same axis, or via the alternate axis. Note that when entering a new abstract state, we do not know precisely the chemical state of the system. Thus, any potential position on the entering side must be considered (e.g., the system may be arbitrary close to the corner of the rectangular region so that exiting the rectangle via the alternate axis may require only one step in the concrete). Consequently, in order to retain a minimal effort strategy, we do not consider the next consecutive interval in the alternate axis, but the subsequent one. For instance, when an event drives molecule B into a new abstract interval, we consider as target goals the next consecutive interval for molecule B and the next two consecutive abstract values for molecule C, since we do not know precisely the concrete amount of C). The initial abstract state receives particular treatment: we can safely compute by which rectangular face a chemical state will exit, since the initial state is known perfectly.

The general framework described in Sect. 3 provides for any pair of thresholds for the quantities of molecules B and C and for any chemical state $q \in \mathcal{Q}$, the probability that the quantity of the molecule B reaches its threshold before the molecule C, and conversely. We denote by (m_B, M_B) (resp. (m_C, M_C)) an interval for the quantity of the molecule B (resp. C). We introduce $P_{g_1}^{\mathcal{G}}(q)$ (resp. $P_{g_2}^{\mathcal{G}}(q)$) as the probability for a chemical state where the quantity of B (resp. C) reaches the threshold M_B (resp. M_C) before that the quantity of C (resp. B) reaches the threshold M_C (resp. M_B) when starting from a state with $q(B)$ and $q(C)$ instances of the molecule B and C. Therefore, the probability $P_{g_1}^{\mathcal{G}}(q)$ satisfies the following relation:

$$
P_{g_1}^{\mathcal{G}}(q) = \begin{cases} 1 \text{ whenever } q(B) = M_B, \\ 0 \text{ whenever } q(B) < M_B \text{ and } q(C) = M_C, \\ \lambda_{r_1}(q) \cdot P_{g_1}^{\mathcal{G}}(q') + \lambda_{r_2}(q) \cdot P_{g_1}^{\mathcal{G}}(q'') \text{ otherwise.} \end{cases} \tag{2}
$$

for every $q(B), q(C) \in \mathbb{N}$, such that $m_B \leq q(B) \leq M_B$ and $m_C \leq q(C) \leq M_C$. A similar expression can be obtained for $P_{g_2}^{\mathcal{G}}(q)$ by switching the base cases for the alternate axis. First two cases stand for the boundary conditions (where thresholds are reached) whereas the third cases captures an increase in molecule B with probability $\lambda_{r_1}(q)$ or C with probability $\lambda_{r_2}(q)$.

As seen in Sect. 3.3, in general, Eq. 2 can be computed exactly by means of inverting a matrix (or equivalently solving a linear system of equations) or approximated, from below, by using a finite expansion of the sequence of the powers of a sparse matrix. Here, since the quantities of the molecules B and C never decrease, the recurrence relation can be solved exactly (up to rounding errors) in a finite amount of iterations.

In the logical version of the reaction network, the unlikely behaviors when C is produced at low abundance of A and when B is produced at high abundance of A have been discarded. We thus test our framework in capturing a low upper bound on the probability of the corresponding macro-transitions in the formal discretization of the underlying reaction network. As a result, in Fig. 5, we computed a macro-transition system for two scenarios: when the copy number of A

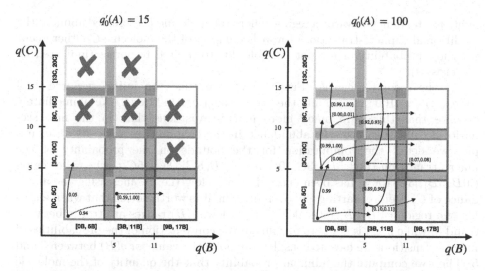

Fig. 5. The derived coarse-grained transition systems for different initial quantities of the molecule A. In the first case (left), the initial amount of A is equal to 15 whereas it is equal to 100 in the second case (right). The states of both systems are related by edges with one source and two targets, with the following meaning: upon entering a new state (the source), the range of probabilities reflects the probability to reach a target before the other one. (Color figure online)

is low (Fig. 5, left) or high (Fig. 5, right). As kinetic constant, we took $k_B = 20$ and $k_C = 1$ and parameterize intervals to reflect the system steady state, which is again when $q(A) = 20$. We tune the initial values $q_0(A)$ respectively to 15 and 100. In Fig. 5, a rectangular region represents an abstract state and is composed each of a respective range of values for molecules B and C. A labeled edge connects a source abstract state to a target abstract state. Each time, a dashed edge, that describes an increase in the quantity of the molecule B, competes with a solid edge, that describes an increase in the quantity of the molecule C.

Let us take an example by considering, in the first scenario, when $q_0(A) = 15$, the sequence of chemical reactions that starting from the initial state (where $q(B) = 0$ and $q(C) = 0$) drives the system out of the region $([0B, 5B], [0C, 5C]$ into the region $([3B, 11B], [0C, 5C]$. To recover the probability for this macro-transition, by Eq. 2 we can compute the probabilities to reach each next target abstract state. As such, the probability to exceed, from the initial state, the upper bound $q(B) = 5$ (resp. upper bound $q(C) = 5$) first is equal to 0.94 (resp. 0.05). The remaining 0.01 probability corresponds to the case where the system reaches the state where $q(B) = 5$ and $q(C) = 5$ (that is to say that the molecule A is completely depleted before having left the initial abstract state). This describes precisely the directional tendency of the behavior of the underlying reaction network. Combining these elements results in the macro-transition $([0B, 5B], [0C, 5C]) \xrightarrow{0.94} ([3B, 11B], [0C, 5C])$ (indicated by a dashed edge on the left in Fig. 5). Given a low resource environment, this macro-transition highlights the tendency towards a regime where the molecule B is created abundantly. Additionally, macro-transitions towards creation of the molecule C either occur with low probability or are not possible due to limited resources (indicated by red crosses).

In the second scenario (Fig. 5, right) we tune the initial resource pool to $q_0(A) = 100$ and retain the same kinetic conditions. We immediately observe that there are many more macro-transitions than in the first scenario, a consequence to the abundance in the common resource. As an example, we detail the computation for the bounds on the probability of the macro-transition $([0B, 5B], [3C, 10C]) \to ([0B, 5B], [8C, 15C])$, when the region $([0B, 5B], [3C, 10C])$ has been entered from below (i.e. by increasing the abundance of C). Before starting any computation, it is worth noting that when entering the region $([0B, 5B], [3C, 10C])$ from below, $q(B)$ ranges arbitrarily between 0 and 5, while $q(C)$ is equal to 6. Thus in the computation of the probabilities of macro-transitions we have to consider any potential value for $q(B)$ between 0 and 5. Then we compute the minimal probability that the quantity of the molecule C exceeds 10 before the quantity of the molecule B exceeds the quantity 11. By applying Eq. 2, we obtain 0.99 (it is indeed obtained when entering the region for $([0B, 5B], [3C, 10C])$ with the state $[B \mapsto 5, C \mapsto 6]$, the maximal probability that we obtain is 1.00 (when entering this region with the state $[B \mapsto 0, C \mapsto 6]$). Indeed the value is not exactly 1 but it is conservatively rounded to 1 because of floating point arithmetics. This highlights that the molecule C is abundantly created with very high probability at the begin of the system execution, and then eventually some B is synthesized.

Reflecting on the two scenarios, it becomes clear that one can bound accurately the probabilities on the likelihood for macro-transitions. Our coarse-graining approach has the following benefits. Firstly, our framework provides a means to compute lower and upper bounds on the probabilities of the transitions between abstract states by formally relating the semantics of reaction networks to its abstract counterpart. Hence providing formal confidence in the formally derived discretized models. Secondly, by enforcing a minimal effort for the system to perform any transition between abstract states, we were able to observe the expected dynamics, the same as in the logical model but without relying on arguments on concentration- and time-scale separation. This has been obtained by twisting the abstraction function instead. This way, the interpretation of the behaviors of the abstract model is less intuitive, but it is still rigorously formally specified. Finally, it is possible to assess the validity of logical models by providing upper bounds to the probabilities of the transitions that has been discarded during the modeling process without any formal justification.

5   Conclusion

We have proposed a generic framework to coarse-grain stochastic reaction networks by sampling the quantity of each kind of molecules within a set of intervals. Instead of neglecting unlikely transitions between abstract regions of states, we compute conservative bounds on their probability. Our goal is indeed to check whether we can derive as accurate models as hand-written ones while ensuring a formal relationship between the potential behaviors in the initial and the derived models. We expect to gain new insights to understand the underlying assumptions behind logical modeling.

Getting formal—but accurate—coarse-grained models requires a specific treatment of boundary effects. It is indeed important not to amplify the importance of some unlikely behaviors. In particular we ensure that every chemical transition between abstract regions of states corresponds to a minimal number of steps. For transitions induced by reverse reactions, we use overlapping intervals. When a chemical state can be arbitrary close to the boundary of an interval, we examine the capacity to cross the next interval instead of just entering it. This induces a non-standard interpretation of the dynamics of the coarse-grained systems. Fewer trajectories are considered in the abstract, while soundness is still ensured (by construction). As in a non-deterministic setting, every concrete behavior is reflected in the abstract, but additionally an upper bound on their probability is computed. Hopefully, rare events are assigned a small upper bound on their probability to occur.

An additional advantage of our framework is the ability to perform and combine numerical abstractions, such as finite expansions of infinite increasing series and include their overall impact as a unique bound on the numerical errors made on the computation of probability values. However, scaling the framework to more complex systems would require one to formally parse intricate relations between numerous variables. For example, to deal with higher dimensional models, symbolic simplification of expressions [18] is possible. Another avenue of thought is to use exact model reduction methods based on the structure of the components of an initial reaction network [12]. Yet, in practice, exact model reduction techniques are not very efficient, especially in a stochastic setting [13]. Still, in our context, we can be more optimistic since, on the one hand not all the properties of the underlying stochastic system have to be preserved and because on the other hand, we can admit numerical approximations on probability values as any other sources of numerical imprecision and include them in the computation of sound over-approximations.

In our paper, we have dealt only with very small case studies. Our motivation was to be able to explain them thoroughly and to focus on minimal difficulties that occur pervasively in models. Ideally we would like to target bigger—but still reasonable—models such as the one for the early events of the EGFR cascade presented in [5]. These models already cope with around three hundred kinds of molecular species. To scale up to this kind of model, we will restrict our study to the competition between pairs of macro-transitions. Yet special care will have to be taken to deal with the denominator of probability functions. These denominators involve the sum over the propensities of each potential event which make them particularly tricky to abstract. Instead of using numerical approaches, we plan to use marginalization to isolate independent subnetworks and reduce the number of terms in denominators accordingly. Yet, here again perfectly independent reaction sub-networks are very unlikely to occur, thus we plan to propose a relaxed version, at the cost of including an additional component in the computation of bounds of probability values.

References

1. Abou-Jaoudé, W., Feret, J., Thieffry, D.: Derivation of qualitative dynamical models from biochemical networks. In: Roux, O., Bourdon, J. (eds.) CMSB 2015. LNCS, vol. 9308, pp. 195–207. Springer, Cham (2015). https://doi.org/10.1007/978-3-319-23401-4_17
2. Abou-Jaoudé, W., Thieffry, D., Feret, J.: Formal derivation of qualitative dynamical models from biochemical networks. Biosystems **149**, 70–112 (2016). https://doi.org/10.1016/j.biosystems.2016.09.001
3. Adélaïde, M., Sutre, G.: Parametric analysis and abstraction of genetic regulatory networks. In: Proc. 2nd Workshop on Concurrent Models in Molecular Biology (BioCONCUR 2004), London. Electronic Notes in Theor. Comp. Sci., Elsevier (2004). http://www.labri.fr/~sutre/Publications/Documents/Adelaide:2004:BioCONCUR.ps.gz
4. Bartocci, E., Lió, P.: Computational modeling, formal analysis, and tools for systems biology. PLoS Comput. Biol. **12**(1), 1–22 (2016). https://doi.org/10.1371/journal.pcbi.1004591
5. Blinov, M.L., Faeder, J.R., Goldstein, B., Hlavacek, W.S.: A network model of early events in epidermal growth factor receptor signaling that accounts for combinatorial complexity. Biosystems **83**(2), 136–151 (2006). https://doi.org/10.1016/j.biosystems.2005.06.014
6. Chatain, T., Haar, S., Paulevé, L.: Boolean networks: beyond generalized asynchronicity. In: Baetens, J.M., Kutrib, M. (eds.) AUTOMATA 2018. LNCS, vol. 10875, pp. 29–42. Springer, Cham (2018). https://doi.org/10.1007/978-3-319-92675-9_3
7. Chifman, J., et al.: Activated oncogenic pathway modifies iron network in breast epithelial cells: a dynamic modeling perspective. PLoS Comput. Biol. **13**(2), e1005352 (2017). https://doi.org/10.1371/journal.pcbi.1005352
8. Cousot, P.: Constructive design of a hierarchy of semantics of a transition system by abstract interpretation. Theor. Comput. Sci. **277**(1–2), 47–103 (2002). https://doi.org/10.1016/S0304-3975(00)00313-3
9. de Jong, H., Gouzé, J.L., Hernandez, C., Page, M., Sari, T., Geiselmann, J.: Qualitative simulation of genetic regulatory networks using piecewise-linear models. Bull. Math. Biol. **66**(2), 301–340 (2004). https://doi.org/10.1016/j.bulm.2003.08.010
10. Fages, F., Soliman, S.: Formal cell biology in biocham. In: Bernardo, M., Degano, P., Zavattaro, G. (eds.) SFM 2008. LNCS, vol. 5016, pp. 54–80. Springer, Heidelberg (2008). https://doi.org/10.1007/978-3-540-68894-5_3
11. Faure, A., Naldi, A., Chaouiya, C., Thieffry, D.: Dynamical analysis of a generic Boolean model for the control of the mammalian cell cycle. Bioinformatics **22**(14), e124–e131 (2006). https://doi.org/10.1093/bioinformatics/btl210
12. Feret, J., Danos, V., Krivine, J., Harmer, R., Fontana, W.: Internal coarse-graining of molecular systems. Proc. Natl. Acad. Sci. **106**(16), 6453–6458 (2009). https://doi.org/10.1073/pnas.0809908106

13. Feret, J., Koeppl, H., Petrov, T.: Stochastic fragments: a framework for the exact reduction of the stochastic semantics of rule-based models. Int. J. Softw. Inform. **7**(4), 527–604 (2013). http://www.ijsi.org/ch/reader/view_abstract.aspx? file_no=i173

14. Forejt, V., Kwiatkowska, M., Norman, G., Parker, D.: Automated verification techniques for probabilistic systems. In: Bernardo, M., Issarny, V. (eds.) SFM 2011. LNCS, vol. 6659, pp. 53–113. Springer, Heidelberg (2011). https://doi.org/10.1007/978-3-642-21455-4_3

15. Grosu, R., et al.: From cardiac cells to genetic regulatory networks. In: Gopalakrishnan, G., Qadeer, S. (eds.) CAV 2011. LNCS, vol. 6806, pp. 396–411. Springer, Heidelberg (2011). https://doi.org/10.1007/978-3-642-22110-1_31

16. Kauffman, S.: Metabolic stability and epigenesis in randomly constructed genetic nets. J. Theor. Biol. **22**(3), 437–467 (1969). https://doi.org/10.1016/0022-5193(69)90015-0

17. Kwiatkowska, M., Norman, G., Parker, D.: PRISM 4.0: verification of probabilistic real-time systems. In: Gopalakrishnan, G., Qadeer, S. (eds.) CAV 2011. LNCS, vol. 6806, pp. 585–591. Springer, Heidelberg (2011). https://doi.org/10.1007/978-3-642-22110-1_47

18. Miné, A.: Symbolic methods to enhance the precision of numerical abstract domains. In: Emerson, E.A., Namjoshi, K.S. (eds.) VMCAI 2006. LNCS, vol. 3855, pp. 348–363. Springer, Heidelberg (2005). https://doi.org/10.1007/11609773_23

19. Thomas, R.: Boolean formalization of genetic control circuits. J. Theor. Biol. **42**(3), 563–585 (1973). https://doi.org/10.1016/0022-5193(73)90247-6

CosySEL: Improving SAT Solving Using Local Symmetries

Sabrine Saouli[1]([✉]), Souheib Baarir[2], Claude Dutheillet[1], and Jo Devriendt[3]

[1] Sorbonne Université, CNRS, LIP6, 75005 Paris, France
{Sabrine.Saouli, Claude.Dutheillet}@lip6.fr
[2] Université Paris Nanterre (now at EPITA, LRE), 92000 Nanterre, France
Souheib.Baarir@lip6.fr
[3] KU Leuven, Department of Computer Science, Celestijnenlaan 200A,
3001 Heverlee, Belgium
jo.devriendt@kuleuven.be

Abstract. Many satisfiability problems exhibit symmetry properties. Thus, the development of symmetry exploitation techniques seems a natural way to try to improve the efficiency of solvers by preventing them from exploring isomorphic parts of the search space. These techniques can be classified into two categories: dynamic and static symmetry breaking. Static approaches have often appeared to be more effective than dynamic ones. But although these approaches can be considered as complementary, very few works have tried to combine them.

In this paper, we present a new tool, CoSySEL, that implements a composition of the static Effective Symmetry Breaking Predicates (ESBP) technique with the dynamic Symmetric Explanation Learning (SEL). ESBP exploits symmetries to prune the search tree and SEL uses symmetries to speed up the tree traversal. These two accelerations are complementary and their combination was made possible by the introduction of *Local symmetries*.

We conduct our experiments on instances issued from the last ten SAT competitions and the results show that our tool outperforms the existing tools on highly symmetrical problems.

Keywords: Boolean satisfiability · Symmetry · Dynamic symmetry breaking · Static symmetry breaking · Local symmetries

1 Introduction

The Boolean satisfiability (SAT) problem is the problem of determining whether or not a solution that satisfies a Boolean formula exists, i.e., by assigning *true*

C. Dragoi et al. (Eds.): VMCAI 2023, LNCS 13881, pp. 252–266, 2023.
https://doi.org/10.1007/978-3-031-24950-1_12

or *false* values to the variables of a given Boolean formula, the latter can be evaluated as *true*. If such a solution exists, it is called a *model*.

Boolean satisfiability is a research area with application in fields such as cryptology [22], modal logic [15], decision planning [19], and hardware and software verification. Actually, SAT-based verification techniques have been widely explored [7,26–28,31].

Since SAT problems often exhibit symmetries, developing techniques to handle them prevents solving algorithms from needlessly exploring isomorphic parts of the search space. One common method to exploit symmetries is the *static symmetry breaking method* [1,10]. It consists in precomputing Symmetry Breaking Predicates (SBPs) and adding them to the original problem before starting the search process. These SBPs invalidate symmetrical solutions, so that the solver avoids exploring branches of the search tree symmetrical to the already explored ones. This method has been implemented in tools such as SHATTER [2] and BREAKID [12]. Even though these approaches are the most efficient on many symmetrical problems, highly symmetrical problems generate a large number of SBPs and this can affect the performance of the used solver.

Dynamic symmetry breaking techniques operate during the search process. Most of them are based on learning symmetric images of already learned clauses. The main such approaches are *Symmetric Learning Scheme (SLS)* [6], *Symmetry Propagation (SP)* [13] and *Symmetric Explanation Learning (SEL)* [11]. Even if these techniques are less effective than the static ones in general, they perform very well on some problems that static approaches fail to solve.

Hence the question of combining both approaches arises naturally, and has already been tackled in some studies: *Effective Symmetry Breaking Predicates method (ESBP)* [23], that uses the same principle as static methods, but operates dynamically, has been combined with SP in [24]. In [29], the authors generate SBPs in the preprocessing phase and apply the SEL method afterwards.

The tool we present in this paper combines ESBP with SEL. Our experiments show that it improves the capacity of the conflict-driven clause learning (CDCL) like algorithm to handle some classes of symmetrical SAT problems.

The paper is structured as follows: Sect. 2 gives the basic definitions relevant to this work. Section 3 recalls the notion of local symmetries and presents the combo algorithm. In Sect. 4, we discuss the implementation of the tool and the experimental results.

2 State of the Art and Some Definitions

We recall here some basic definitions and the main ideas of ESBP and SEL.

2.1 Basics on Boolean Satisfiability

Boolean satisfiability aims at checking whether a Boolean formula φ is satisfiable or not, i.e., whether there exists an assignment α of the Boolean variables for which the formula is true. If so, φ is said to be *satisfiable* (SAT), otherwise φ is *unsatisfiable* (UNSAT).

A formula φ in Conjunctive Normal Form (CNF) is a finite conjunction of *clauses*, each clause being a disjunction of (possibly negated) variables. The set of variables of a formula φ is denoted by $\mathcal{V}_\varphi$.

An assignment α is a function $\alpha\colon \mathcal{V}_\varphi \to \{\top, \bot\}$ and can be represented by the subset of its true literals. We call a true literal x if $\alpha(x) = \top$ or $\neg x$ if $\alpha(x) = \bot$. An extension of α is any α' such that $\alpha \subset \alpha'$. Assignment α is said to be *complete* if it contains one literal over each variable in $\mathcal{V}_\varphi$; it is partial otherwise. The set of all (possibly partial) assignments to $\mathcal{V}_\varphi$ is denoted $Ass(\mathcal{V}_\varphi)$.

An assignment α *satisfies* a clause ω, denoted $\alpha \models \omega$, if α contains at least one true literal from ω. An assignment α satisfies a formula φ, denoted $\alpha \models \varphi$, if α satisfies all the clauses in φ. Such an assignment α is said to be a model of φ. The formula φ is *unsatisfiable* (UNSAT) otherwise. For more details, the interested reader can refer to the very complete handbook [8].

Example. Let $\varphi = \{\{x_1, x_2, x_3\}, \{x_1, x_2\}, \{\neg x_1, x_3\}\}$ be a formula. The partial assignments $\{\neg x_1, x_2\}$ and $\{x_1, x_3\}$ satisfy φ, so φ is satisfiable. Extending φ with the unit clauses $\{\neg x_2\}$ and $\{\neg x_3\}$ would make it unsatisfiable.

2.2 Symmetry Group of a Formula

Let φ be a formula and let $\mathfrak{S}(\mathcal{V}_\varphi)$ be the group of permutations of $\mathcal{V}_\varphi$ under composition. We say that $g \in \mathfrak{S}(\mathcal{V}_\varphi)$ is a symmetry of φ if and only if for every complete assignment α such that $\alpha \models \varphi$, $g.\alpha \models \varphi$, with $g.\alpha = \{g(x) \mid x \in \alpha\} \cup \{\neg g(x) \mid \neg x \in \alpha\}$. We denote $S(\varphi) \subseteq \mathfrak{S}(\mathcal{V}_\varphi)$ the symmetry group of φ and we call *generator* the elements of a generating set of $S(\varphi)$. A variable x is said to appear in a generator g if $g(x) \neq x$.

Let G be a subgroup of $\mathfrak{S}(\mathcal{V}_\varphi)$. The *orbit of α under G* is the set $[\alpha]_G = \{g.\alpha \mid g \in G\}$. The set of orbits $\{[\alpha]_G \mid \alpha \in Ass(\mathcal{V}_\varphi)\}$ partitions $Ass(\mathcal{V}_\varphi)$ into equivalence classes, called *symmetry classes* of φ when $G = S(\varphi)$. We introduce an ordering relation between assignments in order to identify a unique representative for each symmetry class.

Definition 1. [23] *We assume a total order $\prec$ on $\mathcal{V}_\varphi$. Given two assignments $\alpha, \beta \in Ass(\mathcal{V}_\varphi)$, $\alpha \prec \beta$, if there exists a variable $v \in \mathcal{V}_\varphi$ such that:*

- *for all $v' \prec v$, either $v' \in \alpha \cap \beta$ or $\neg v' \in \alpha \cap \beta$,*
- *$\neg v \in \alpha$ and $v \in \beta$.*

Moreover, $\prec$ is a total order on complete assignments. For a complete assignment α we define the lexicographic leader (*lex-leader*) of an orbit $[\alpha]_G$ as the minimum of $[\alpha]_G$ w.r.t. $\prec$.

2.3 (Effective) Symmetry Breaking

From the above presentation, it is clear that either all the assignments within the same symmetry class satisfy the formula, or none do. Adding a *Symmetry Breaking Predicate* (SBP) to a symmetric SAT formula aims at limiting the search tree exploration to only one assignment per symmetry class, e.g., the *lex-leader*. However, finding the *lex-leader* of a class is computationally hard [20] and best-effort approaches are commonly used [2,12].

SBPs were first introduced as pre-generated predicates (i.e., in a *static* approach) but they required auxiliary variables, making the size of the formulas often intractable in practice. *Effective* SBPs (ESBPs) were then proposed to tackle the problem with a *dynamic* approach [23], where the solver detects on-the-fly when the current assignment cannot be extended to a *lex-leader*. Actually, assignment ordering is monotonic, i.e., whenever $\alpha < \beta$, any extension α' of α (resp. β' of β) are such that $\alpha' < \beta'$. Hence, if $g.\alpha < \alpha$, any possible extension α' of α is such that $g.\alpha' < \alpha'$, because $g.\alpha'$ is an extension of $g.\alpha$. In this case, we can define a predicate contradicting α that still preserves the satisfiability of the formula. Such a predicate will be used to discard α and all its extensions from further exploration, thus pruning the search tree.

Definition 2. [23] *Let $\alpha \in Ass(\mathcal{V}_\varphi)$, and $g \in \mathfrak{S}(\mathcal{V}_\varphi)$. We say that the formula ψ is an* Effective Symmetry Breaking Predicate *(ESBP) for α under g if:*

$$\alpha \not\models \psi \text{ and for all } \beta \in Ass(\mathcal{V}_\varphi), \beta \not\models \psi \Rightarrow g.\beta < \beta$$

The equi-satisfiability of φ and $\varphi \cup \psi$ is guaranteed by the fact that ψ will not prune the branch of the lex-leader. This approach avoids the pre-generation of a large SBP that could have a negative effect on the overall performance of the classical static symmetry breaking approaches. The extensive experiments conducted in [23] show that it outperforms other state-of-the-art symmetry breaking techniques, both dynamic and static, when considering the total number of solved instances. However, this technique fails to solve some problems that have been trivially solved by other dynamic symmetry breaking techniques such as SEL developed in [13]. We give an overview of SEL in the following section.

2.4 Symmetric (Explanation) Learning

An orthogonal approach to symmetry breaking is *symmetric learning*. The idea here is not to remove the symmetric assignments by posting extra constraints, but to add implied symmetric clauses to a SAT solver's internal *learned clause database*. Symmetric learning hinges on the following theorem:

Theorem 1. [6] *Let φ be a formula, $g \in S(\varphi)$ a symmetry for the formula, and ω a clause. Then, $\varphi \models \omega$ implies $\varphi \models g.\omega$.*

As a result, for any implied clause derived by a SAT solver, any of its symmetric images can safely be derived as well, which, through unit propagation, discourages the solver from visiting symmetric search branches [13]. The crucial question when implementing symmetric learning is how to avoid overloading the solver with exponentially many symmetric clauses, while retaining effective pruning of the search tree. One answer is *symmetric explanation learning* (SEL), which was shown to be competitive to (but not better than) state-of-the-art static symmetry breaking [11]. The idea behind SEL, given a small set[1] of symmetries G, is to keep track of all clauses $\{g.\omega \mid g \in G\}$ symmetric to the clauses ω that triggered a currently propagated literal. Only the symmetric clauses that propagate in turn will be added to the learned clause database. When they propagate, all their symmetric images will be tracked, in effect composing the symmetries in G. If the propagation explained by a clause is cancelled (when backtracking the search), SEL will quickly forget the symmetric images that did not propagate.

3   The Proposed Technique

Our first attempt to combine static and dynamic approaches was proposed in [24], where we combined ESBP with SP. However, as it appears that SEL is theoretically more effective than SP, we decided to investigate the integration of ESBP with SEL. This work can also be considered as a generalization of [29], where the combination was purely static and did not take advantage of the upcoming notion of local symmetries [24].

The idea of the SEL approach is to derive and efficiently use symmetrical clauses using a subset of $S(\varphi)$, the symmetry group of φ (the correctness is thus guaranteed by Theorem 1). If φ is extended by a set of clauses ψ, preserving equisatisfiability, then SEL can be applied, as long as $S(\varphi \cup \psi)$ is known. Therefore, the effectiveness of the composition of SEL and ESBP strongly depends on how hard it is to compute the elements of $S(\varphi \cup \psi)$.

The notion of *local symmetries* was introduced as the theoretical framework materializing the computation of the aforementioned elements in [24]. In this section, we recall the definition and properties of *local symmetries* and invite the interested reader to consult the original work [24] for more details.

3.1   Theoretical Foundations and Practical Considerations

Local symmetries of a clause of a formula are defined as follows.

Definition 3. *Let φ be a formula. We define $L_{\omega,\varphi}$, the set of local symmetries for a clause ω with respect to a formula φ, as follows:*

$$L_{\omega,\varphi} = \{g \in \mathfrak{S}(\mathcal{V}) \mid \varphi \models g.\omega\}$$

Through this definition, it is straightforward to derive the next proposition.

Proposition 1. *Let φ be a formula. Then, $\bigcap_{\omega \in \varphi} L_{\omega,\varphi} \subseteq S(\varphi)$.*

---

[1] Such a small set typically does not form a group, i.e., is not closed under composition, but closing it under composition *generates* a detected symmetry group for the formula.

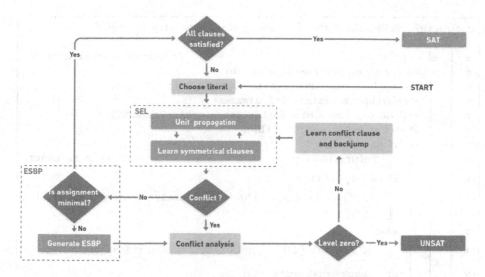

Fig. 1. Workflow of the ESBP_SEL algorithm.

A direct consequence is that the intersection of the sets of *local symmetries* of all the clauses of a formula φ are symmetries of $S(\varphi)$. Hence, when adding a symmetry breaking predicate ω to φ, a set of valid symmetries for $\varphi \cup \{\omega\}$ can be computed on-the-fly as the intersection of $L_{\omega,\varphi}$ and $\bigcap_{\omega' \in \varphi} L_{\omega',\varphi}$.

However, full $L_{\omega,\varphi}$ sets are hard to compute in general, hence our tool only computes subsets based on the following considerations. While solving a formula based on a symmetry breaking approach, three sets of clauses are manipulated: the original formula φ, the set of SBP clauses φ_e and the set φ_d of clauses derived from $\varphi \cup \varphi_e$. Our computation of the local symmetries of a clause ω takes into account the fact that symmetries $S(\varphi)$ of φ are already known, and depends on which of the three sets φ, φ_e, or φ_d, ω belongs to.

Let $\varphi' = \varphi \cup \varphi_e \cup \varphi_d$. There are three cases:

1. if $\omega \in \varphi$, then by definition $S(\varphi) \subseteq L_{\omega,\varphi'}$, so we take $S(\varphi)$ as a representative for $L_{\omega,\varphi'}$.
2. if $\omega \in \varphi_e$, this is an ESBP clause, and we choose the set of stabilizing symmetries: $Stab(\omega) = \{g \in \mathfrak{S}(\mathcal{V}) \mid \omega = g.\omega\} \subseteq L_{\omega,\varphi'}$.
3. if $\omega \in \varphi_d$, this is a derived clause, and we choose the set $(\bigcap_{\omega' \in \varphi_1} L_{\omega',\varphi'})$, where φ_1 is the set of clauses that derived ω.

3.2 Algorithm

In this section, we describe how we combined SEL and ESBP.

Figure 1 gives an overview of the integration of ESBP and SEL in the CDCL algorithm.

```
 1  function  ESBP_SEL(φ: CNF formula, symCtrl: symmetry controller)
 2
 3  │  dl ← 0 ;                                     // Current decision level
 4  │  while not all variables are assigned do
 5  │  │   isConflict ← unitPropagation() ∧ selPropagation();
 6  │  │   symCtrl.updateAssign(crtAssignment());
 7  │  │   isReduced ← symCtrl.isNotLexLeader(crtAssignment());
 8  │  │   if isConflict ∨ isReduced then
 9  │  │   │   if dl = 0 then
10  │  │   │   └   return UNSAT;                      // φ is UNSAT
11  │  │   │   if isConflict then
12  │  │   │   │   ⟨ω, L = ⋂      L_{ω',φ₁} ⋃ Stab(ω)⟩ ← analyzeConflictEsbpSel();
       │  │   │   │          ω'∈φ₁
13  │  │   │   else
14  │  │   │   │   ⟨ω, L = Stab(ω)⟩ ← symCtrl.genEsbpSel(crtAssignment());
15  │  │   │   dl ← backjumpOrRestart();
16  │  │   │   φ ← φ ∪ {ω} ;
17  │  │   └   symCtrl.updateCancel(crtAssignment());
18  │  │   else
19  │  │   │   assignDecisionLiteral();
20  │  │   └   dl ← dl + 1;
21  │  return SAT;                                   // φ is SAT
```

Algorithm 1: The ESBP_SEL algorithm. Instructions derived from ESBP and SEL algorithms are reported in blue and red (respectively). Instructions derived from the combination are reported with a grey background.

The integration of SEL in the CDCL algorithm operates the same way as a basic CDCL, except for the unit propagation function (SEL in Fig. 1). With SEL, the algorithm keeps symmetrical versions of propagated literals' explanation clauses in a different database and in addition to regular unit propagation over the regular clauses, when a symmetrical clause is asserting, SEL adds it to the learnt clauses and the asserting symmetrical literal is propagated.

The integration of ESBP then consists in controlling the behaviour of the previous algorithm by introducing a *symmetry controller component* that operates all symmetry-based actions. It inspects all partial assignments and detects non-minimal ones as soon as possible. In this case, it generates an ESBP clause and injects it into the original problem (see ESBP in Fig. 1).

The details of the aforementioned approach are given in Algorithm 1. The algorithm first executes the *unitPropagation()* and *selPropagation* functions (line 5). In propagation phase, regular and symmetrical unit clauses are propagated until a conflict is detected or fixed point is reached. Next, the symmetry controller updates the current assignment and checks if it can still be extended to a lex-leader (lines 6 − 7). When a conflict is detected, function *analyzeCon-*

flictEsbpSel() (line 12) analyses the conflict and generates a *learnt clause* ω. With respect to a classical analysis function of a basic CDCL algorithm, *analyzeConflictEsbpSel()* will generate the set of local symmetries associated with ω. This is done by computing the intersection of the sets of symmetries of all the clauses used to derive ω (as explained in Sect. 3.1), augmented with the stabilizers set. If the current assignment is conflict free but can not be extended to a lex-leader, function *genEsbpSel()* (line14) is called. It generates the ESBP clause to inject, along with its set of stabilizers. Function *updateCancel()* (line17) is the counterpart of function *backjumpOrRestart* but for *symCtrl*, the symmetry controller.

4   Tooling and Evaluation

In this section, we first present the tooling support of our combined approach ESBP_SEL combining ESBP and SEL. Then, we compare it to the vanilla SAT solver GLUCOSE[2] [5] and to the implementations of ESBP, SEL on top of GLUCOSE and discuss the results.

4.1   Tool Usage

COSY[3] is a C++ library offering all the functionalities necessary for the implementation of the ESBP method. This library can easily be integrated to any CDCL-like solver.

The implementation of ESBP approach on top of GLUCOSE is available at https://github.com/lip6/cosy/tree/master/solvers/glucose-3.0. In the remaining of this paper, this implementation is simply called COSY. The implementation of SEL approach on top of GLUCOSE is available at https://bitbucket.org/krr/glucose-sel. Our implementation, COSYSEL, of the combined approach ESBP_SEL is available at https://github.com/sabrinesaouli/CosySEL. It integrates COSY in the already mentioned implementation of SEL according to Algorithm 1.

COSYSEL can be used with two symmetry generator tools: BLISS [17] or SAUCY [18]. These are two of the best graph automorphism tools, that compute a set of symmetries for a given graph.

In our tool-chain, a given CNF formula is first encoded as a colored graph into a file that is given to BLISS or SAUCY to obtain the set of symmetry generators as a file in the corresponding format (`.bliss` or `.sym` respectively). The obtained file is then given along with the `.cnf` file to the COSYSEL solver (Fig. 2).

This workflow is encapsulated in a script `cosysel.sh` that we can execute with **bliss** or **saucy** options.

```
$./cosysel.sh bliss <\cnf>
$./cosysel.sh saucy <\cnf>
```

---

[2] https://www.labri.fr/perso/lsimon/downloads/softwares/glucose-syrup.tgz.
[3] Cosy library is released under GPL v3 license at https://github.com/lip6/cosy.

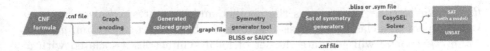

Fig. 2. Workflow of the COSYSEL tool.

4.2 Evaluation

Among all the instances from the last ten SAT competitions (from 2012 to 2021) [16], we selected the 1362 for which BLISS detects at least one symmetry generator in at most 1000 s of CPU time.

All experiments were conducted using the following settings: each solver ran once on each problem, with a memory limit of 15 GB and a time-out of 7200 s seconds (this time limit includes symmetry detection time for all the solvers except for GLUCOSE which does not compute symmetries). Experiments were executed on a computer with an Intel(R) Xeon(R) Gold 6148 CPU @ 2.40 GHz and 1500 GB of memory, running Linux 5.0.16.

All the approaches dealing with symmetries are built on top of a CDCL-like solver. Therefore, for a fair comparison, the solver must not introduce side effects. As our tool combines SEL and ESBP, we must at least prove that it outperforms both. SEL is built on top of GLUCOSE, and there is also a version of ESBP on GLUCOSE, so we chose GLUCOSE [5] to avoid solver-induced side effects.

To be complete in our study, it seems natural to compare our new approach with the combination of ESBP and SP presented in [24] and implemented on top of MINISAT (in a tool referred here as COSYSP). However, although it was shown in [11] that SEL approach is theoretically more efficient than SP, no available implementation supports this claim. Nevertheless, as SP is built on top of MINISAT, we decided to implement SEL on top of MINISAT and compare the approaches on the whole benchmark. The results are given in Table 1 and confirm that SEL is by far better than SP. Therefore, considering that implementing a combination of ESBP and SP on top of GLUCOSE is not simple and would require a great effort, we relied on these results to consider it is not relevant.

We computed the symmetries of each instance with BLISS and SAUCY. BLISS is known to compute a larger number of generators for the symmetry group compared to SAUCY. As shown in [23], in an SBP-like approach, this influences the results since it allows to cut branches of the search tree early.

Table 2 compares the use of SAUCY and BLISS for computing symmetry generators. The values represent the number of SAT, UNSAT, and the total number

Table 1. Comparison of the number of instances solved by MINISAT-SEL and MINISAT-SP using BLISS.

	MINISAT-SEL	MINISAT-SP
SAT	304	271
UNSAT	402	378
TOTAL(1362)	706	649

Table 2. Comparison of different approaches when using SAUCY and BLISS.

	None	SAUCY			BLISS		
	GLUCOSE	COSY	SEL	CosySEL	COSY	SEL	CosySEL
SAT	238	227	255	229	235	253	240
UNSAT	473	452	505	474	497	503	553
TOTAL(1362)	711	679	760	703	732	756	793

Table 3. Comparison of instances solved by each approach according to the percentage of the variables in the symmetries, using BLISS to detect symmetry (the table is restricted to instances solved by at least one solver).

(a) SAT instances					(b) UNSAT instances				
% sym vars	GLUCOSE	COSY	SEL	CosySEL	% sym vars	GLUCOSE	COSY	SEL	CosySEL
0% - 25% (195)	152	149	174	165	0% - 25% (250)	242	211	233	218
25% - 50% (28)	19	23	19	16	25% - 50% (21)	20	19	21	20
50% - 75% (14)	14	14	13	12	50% - 75% (11)	8	7	9	9
75% - 100% (62)	53	49	47	47	75% - 100% (330)	203	260	240	306
Total (299)	238	235	253	240	Total (612)	473	497	503	553

of instances solved. It shows that our tool performs poorly with SAUCY. Actually it computes too few symmetries to allow CosySEL to detect early non-lex-leader assignments, hence the overhead of keeping track of local symmetries is not counterbalanced. The effectiveness of COSY, and thus the combined tool, is largely relying on the number of observed (tracked) literals while solving the problem and because of the reduced number of generators given by SAUCY, this can be an issue for these approaches.

The results in Table 2 confirm that CosySEL is more effective using BLISS than SAUCY. Globally, we notice that CosySEL, when used with BLISS, is the most effective, especially when considering UNSAT instances. It solves 50 more UNSAT and 13 fewer SAT instances than the second-best method (SEL).

After establishing that our tool works better with BLISS, we compared its effectiveness against the three others in each class of problems. In Table 3, results are split according to the percentage of variables appearing in at least one generator computed by BLISS, with the first column giving the intervals of percentages. Table 3a (respectively 3b) shows the number of SAT (respectively UNSAT) instances solved by each approach in each interval.

UNSAT problems are exceptionally difficult to solve as they require traversing the entire search space, but our tool is particularly effective for highly symmetrical UNSAT problems. As far as SAT problems are concerned, the loss of performance can be explained by the fact that COSY can stop the exploration of a satisfying branch of the search space because it is not a lex-leader, even though this branch could still contain a non-lex-leader solution.

It is essential to mention that (as shown in Table 4a), CosySEL increased the VBS(Virtual Best Solver)—which represents the best performances combined, i.e. the number of instances that at least one solver can solve—with **35** problems

Table 4. Virtual Best Solver (VBS) results with and without COSYSEL.

(a) Comparing the VBS when using GLUCOSE, COSY and SEL only and when adding COSYSEL to the set of solvers.

	without COSYSEL	with COSYSEL
SAT	297	299 (**+2**)
UNSAT	579	612 (**+33**)
TOTAL(1362)	876	911 (**+35**)

(b) Comparing the VBS when using KISSAT-MAB only and when adding COSYSEL.

	KISSAT-MAB	KISSAT-MAB+COSYSEL
SAT	436	443 (**+7**)
UNSAT	570	680 (**+110**)
TOTAL(1362)	1006	1123 (**+117**)

that the other methods failed to solve. We also compared it to the best solver of the SAT contest 2021 KISSAT-MAB [9] (see Table 4b). We noted that even though our COSYSEL tool is overall less effective, it managed to solve **117** problems (mainly UNSAT) that KISSAT-MAB could not handle. Upon taking a closer look into the classes of problems exclusively solved by COSYSEL, we further confirm that our tool is more effective at handling highly symmetrical UNSAT problems:

- UNSAT and fully symmetrical: 40/117 are several variations of the *pigeon hole problem*, 30/117 are *Tseitin formulas* [30], and 2/117 belong to the class of *n-queens problem*.
- UNSAT with more than 90% of the variables being in the symmetry generators: 17/117 from the *clique colouring* class.
- The remaining 25/117 instances are of diverse classes of problems including 6 from the SAT-*based Bitcoin mining problems (Satcoin*[4] [21] which are SAT with $\sim 0.15\%$ of variables being in the symmetry generators.

Out of those 117 instances and according to the publicly available SAT competition results, our proposal is the only tool having succeeded at solving some problems that were previously unsolved: 29 *Tseitin formulas* (28 from SAT 2016 and 1 from SAT 2019), 4 *relativised pigeonhole problems* [3,4,14] from SAT competition 2016 and 2 classic *pigeonhole problems* from SAT competition 2021.

To push our experiments even further, we compared our tool to the previously mentioned solvers (including COSYSP)[5] on fully symmetrical problems (all the variables are involved in symmetries). We collected 282 problems (Table 5) from the last SAT competitions, from [11] and from [25]. The instances represent different classes of problems like the *pigeon hole, the clique colouring and the channel routing problems*. We plotted the results in Figs. 3a and 3b for SAT and UNSAT instances respectively.

Figure 3a shows that KISSAT-MAB is more effective on SAT problems compared to the other tools, which solve more or less the same number of instances. However, Fig. 3b shows that COSYSEL stands out when it comes to UNSAT instances and even KISSAT-MAB cannot compete with the tools exploiting symmetries except for COSYSP that solves two fewer instances.

---

[4] github.com/jheusser/satcoin.

[5] We recall that COSYSP is based of MINISAT, and the comparison with the other tools is not totally fair!.

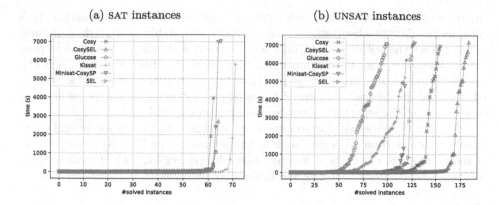

Fig. 3. Comparison of different approaches on fully symmetrical instances.

Table 5. Comparison of approaches on fully symmetrical instances using BLISS.

	GLUCOSE	COSY	SEL	COSYSEL	COSYSP	KISSAT-MAB
SAT	64	63	62	65	66	72
UNSAT	100	155	129	184	118	120
TOTAL (282)	164	218	191	249	184	192

In Table 5, we observe that COSYSP is slightly more effective than the other solvers exploiting symmetries on SAT instances. Nevertheless, on UNSAT problems, it solves fewer instances than almost all the other solvers. It is hard to identify whether this relatively low performance is due to the fact that SP is overloading the solver by keeping track of the status of all symmetries, or because it is embedded in an older CDCL solver (MINISAT). Either ways, it is clear that our implementation of COSYSEL outperforms the publicly available version of COSYSP by solving 65 more instances. This is consistent with previously made observation regarding SEL and SP (see Table 1).

5   Conclusion

We presented in this paper COSYSEL: a tool that combines ESBP, which dynamically adds symmetry breaking predicates to the formula, with SEL, which is based on learning symmetrical clauses only when they are useful.

This combination relies on the definition of *local symmetries* for a clause, which makes it possible to efficiently compute a subset of the symmetries of the formula each time symmetry breaking predicates are added to it. Our experiments investigate the effectiveness of the combined approach.

They show that COSYSEL can solve a significant number of highly symmetrical problems that state-of-the-art solvers fail to handle. We believe that COSYSEL works better with BLISS than SAUCY because BLISS detects a more significant number of symmetries, which helps ESBP cut symmetrical branches earlier. It is also more efficient when the problem is UNSAT and highly symmetrical because ESBP allows the solver to visit only few (partial) assignments per symmetrical class. Moreover, the more symmetries, the more branches ESBP cuts, and the more symmetrical clauses SEL learns. However, CosySEL seems to be less effective on SAT problems. This may be due to ESBP stopping the exploration of a sat branch of the search tree just because it is not a lex-leader.

As a future work, we plan to implement ESBP_SEL in a more recent and effective SAT solver than GLUCOSE, such as the winner of the 2021 SAT competition KISSAT-MAB [9], or the best scoring MAPLE-like solver [32], which derives most of its code-base from GLUCOSE and hence may be easier to port CosySEL to.

References

1. Aloul, F.A., Ramani, A., Markov, I.L., Sakallah, K.A.: Solving difficult sat instances in the presence of symmetry. In: Proceedings 2002 Design Automation Conference (IEEE Cat. No. 02CH37324) pp. 731–736. IEEE (2002)
2. Aloul, F.A., Sakallah, K.A., Markov, I.L.: Efficient symmetry breaking for Boolean satisfiability. IEEE Trans. Comput. **55**(5), 549–558 (2006)
3. Atserias, A., Lauria, M., Nordström, J.: Narrow proofs may be maximally long. ACM Trans. Comput. Logic (TOCL) **17**(3), 1–30 (2016)
4. Atserias, A., Müller, M., Oliva, S.: Lower bounds for DNF-refutations of a relativized weak pigeonhole principle. J. Symbolic Logic **80**(2), 450–476 (2015)
5. Audemard, G., Simon, L.: Predicting learnt clauses quality in modern sat solvers. In: Proceedings of the 21st International Joint Conference on Artificial Intelligence, IJCAI 2009, pp. 399–404 (2009)
6. Benhamou, B., Nabhani, T., Ostrowski, R., Saïdi, M.R.: Enhancing clause learning by symmetry in sat solvers. In: 2010 22nd IEEE International Conference on Tools with Artificial Intelligence, vol. 1, pp. 329–335. IEEE (2010)
7. Biere, A., Cimatti, A., Clarke, E., Zhu, Y.: Symbolic model checking without BDDs. In: Cleaveland, W.R. (ed.) TACAS 1999. LNCS, vol. 1579, pp. 193–207. Springer, Heidelberg (1999). https://doi.org/10.1007/3-540-49059-0_14
8. Biere, A., Heule, M., van Maaren, H.: Handbook of Satisfiability, vol. 185. IOS press, Amsterdam (2009)
9. Cherif, M.S., Habet, D., Terrioux, C.: Kissat MAB: combining VSIDS and CHB through multi-armed bandit. SAT Competition **2021**, 15 (2021)
10. Crawford, J., Ginsberg, M., Luks, E., Roy, A.: Symmetry-breaking predicates for search problems. KR **96**(1996), 148–159 (1996)
11. Devriendt, J., Bogaerts, B., Bruynooghe, M.: Symmetric explanation learning: effective dynamic symmetry handling for SAT. In: Gaspers, S., Walsh, T. (eds.) SAT 2017. LNCS, vol. 10491, pp. 83–100. Springer, Cham (2017). https://doi.org/10.1007/978-3-319-66263-3_6
12. Devriendt, J., Bogaerts, B., Bruynooghe, M., Denecker, M.: Improved static symmetry breaking for SAT. In: Creignou, N., Le Berre, D. (eds.) SAT 2016. LNCS, vol. 9710, pp. 104–122. Springer, Cham (2016). https://doi.org/10.1007/978-3-319-40970-2_8
13. Devriendt, J., Bogaerts, B., De Cat, B., Denecker, M., Mears, C.: Symmetry propagation: Improved dynamic symmetry breaking in SAT. In: 2012 IEEE 24th International Conference on Tools with Artificial Intelligence, vol. 1, pp. 49–56. IEEE (2012)

14. Elffers, J., Nordström, J.: Documentation of some combinatorial benchmarks. Proc. SAT Competition **2016**, 67–69 (2016)

15. Giunchiglia, F., Sebastiani, R.: Building decision procedures for modal logics from propositional decision procedures—the case study of modal K. In: McRobbie, M.A., Slaney, J.K. (eds.) CADE 1996. LNCS, vol. 1104, pp. 583–597. Springer, Heidelberg (1996). https://doi.org/10.1007/3-540-61511-3_115

16. Järvisalo, M., Le Berre, D., Roussel, O., Simon, L.: The international sat solver competitions. AI Mag. **33**(1), 89–92 (2012)

17. Junttila, T., Kaski, P.: Engineering an efficient canonical labeling tool for large and sparse graphs. In: Applegate, D., Brodal, G.S., Panario, D., Sedgewick, R. (eds.) Proceedings of the Ninth Workshop on Algorithm Engineering and Experiments and the Fourth Workshop on Analytic Algorithms and Combinatorics, pp. 135–149. SIAM (2007)

18. Katebi, H., Sakallah, K.A., Markov, I.L.: Symmetry and satisfiability: an update. In: Strichman, O., Szeider, S. (eds.) SAT 2010. LNCS, vol. 6175, pp. 113–127. Springer, Heidelberg (2010). https://doi.org/10.1007/978-3-642-14186-7_11

19. Kautz, H.A., Selman, B., et al.: Planning as satisfiability. In: ECAI, vol. 92, pp. 359–363 (1992)

20. Luks, E.M., Roy, A.: The complexity of symmetry-breaking formulas. Ann. Math. Artif. Intell. **41**(1), 19–45 (2004)

21. Manthey, N., Heusser, J.: Satcoin-bitcoin mining via SAT. SAT Competition **2018**, 67 (2018)

22. Massacci, F., Marraro, L.: Logical cryptanalysis as a sat problem. J. Autom. Reasoning **24**(1), 165–203 (2000)

23. Metin, H., Baarir, S., Colange, M., Kordon, F.: CDCLSym: introducing effective symmetry breaking in SAT solving. In: Beyer, D., Huisman, M. (eds.) TACAS 2018. LNCS, vol. 10805, pp. 99–114. Springer, Cham (2018). https://doi.org/10.1007/978-3-319-89960-2_6

24. Metin, H., Baarir, S., Kordon, F.: Composing symmetry propagation and effective symmetry breaking for SAT solving. In: Badger, J.M., Rozier, K.Y. (eds.) NFM 2019. LNCS, vol. 11460, pp. 316–332. Springer, Cham (2019). https://doi.org/10.1007/978-3-030-20652-9_21

25. Sabharwal, A.: Symchaff: exploiting symmetry in a structure-aware satisfiability solver. Constraints **14**(4), 478–505 (2009)

26. Shtrichman, O.: Tuning SAT checkers for bounded model checking. In: Emerson, E.A., Sistla, A.P. (eds.) CAV 2000. LNCS, vol. 1855, pp. 480–494. Springer, Heidelberg (2000). https://doi.org/10.1007/10722167_36

27. Shtrichman, O.: Pruning techniques for the SAT-based bounded model checking problem. In: Margaria, T., Melham, T. (eds.) CHARME 2001. LNCS, vol. 2144, pp. 58–70. Springer, Heidelberg (2001). https://doi.org/10.1007/3-540-44798-9_4

28. Tang, D., Malik, S., Gupta, A., Ip, C.N.: Symmetry reduction in SAT-based model checking. In: Etessami, K., Rajamani, S.K. (eds.) CAV 2005. LNCS, vol. 3576, pp. 125–138. Springer, Heidelberg (2005). https://doi.org/10.1007/11513988_12

29. Tchinda, R.K., Tayou Djamegni, C.: Enhancing static symmetry breaking with dynamic symmetry handling in CDCL SAT solvers. Int. J. Artif. Intell. Tools **28**(03), 1950011 (2019)

30. Tseitin, G.S.: On the complexity of derivation in propositional calculus. In: Siekmann, J.H., Wrightson, G. (eds.) Automation of reasoning, pp. 466–483. Springer, Heidelberg (1983). https://doi.org/10.1007/978-3-642-81955-1_28

31. Wang, C., Jin, H., Hachtel, G.D., Somenzi, F.: Refining the SAT decision ordering for bounded model checking. In: Proceedings of the 41st Annual Design Automation Conference, pp. 535–538 (2004)
32. Zhang, X., Cai, S., Chen, Z.: Improving CDCL via local search. SAT Competition **2021**, 42 (2021)

Sound Symbolic Execution via Abstract Interpretation and Its Application to Security

Ignacio Tiraboschi[2(✉)], Tamara Rezk[1], and Xavier Rival[2]

[1] INRIA, Université Côte d'Azur, Sophia Antipolis, France
[2] INRIA Paris, DI ENS, Ecole normale supérieure, Université PSL, CNRS, Paris, France
{ignacio.tiraboschi,xavier.rival}@inria.fr

Abstract. Symbolic execution is a program analysis technique commonly utilized to determine whether programs violate properties and, in case violations are found, to generate inputs that can trigger them. Used in the context of security properties such as noninterference, symbolic execution is precise when looking for counter-example pairs of traces when insecure information flows are found, however it is sound only up to a bound thus it does not allow to prove the correctness of programs with executions beyond the given bound. By contrast, abstract interpretation-based static analysis guarantees soundness but generally lacks the ability to provide counter-example pairs of traces.

In this paper, we propose to weave both to obtain the best of two worlds. We demonstrate this with a series of static analyses, including a static analysis called RedSoundRSE aimed at verifying noninterference. RedSoundRSE provides both semantically sound results and the ability to derive counter-example pairs of traces up to a bound. It relies on a combination of symbolic execution and abstract domains inspired by the well known notion of reduced product. We formalize RedSoundRSE and prove its soundness as well as its relative precision up to a bound. We also provide a prototype implementation of RedSoundRSE and evaluate it on a sample of challenging examples.

1 Introduction

Security properties are notoriously hard to verify. In particular, many security properties are not single-execution properties but hyperproperties [13] (also

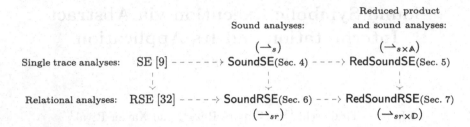

Fig. 1. Relation between different SE analyses. SE [9] is conventional symbolic execution and RSE [32,34] is its extension to relational properties. Except for RSE with invariants [23], SE and RSE are unsound in general. The rest of the analyses are sound and are our contributions: SoundSE and SoundRSE do not use abstract interpretation whereas RedSoundSE and RedSoundRSE can be combined with different abstract domains. A red dashed line represents a dependency: a relational analysis depends on a single trace analysis. A blue dashed line represents an enhancement of the analysis. (Color figure online)

referred to as relational properties), which means that refuting them sometimes requires *several* executions traces to be provided as a counter-example. In particular, noninterference [26] states that high clearance information should not impact the observation of low clearance users in any execution of the program. It has been the subject of many verification method proposals and tools (e.g. [4,7,8,23,24,33,36–38]).

Symbolic execution [9,31] (SE) is typically used to find property violations, and can be applied for policies like noninterference provided some adaptation for relational properties. SE boils down to an execution where variables initially hold symbolic values and get updated with expressions of these symbolic values whereas conditions are evaluated into symbolic path guards. The analysis involves an external tool such as an SMT solver that prunes infeasible paths and attempts to discharge verification conditions on remaining ones. SE attempts to exhaustively cover all executions paths, which is feasible only up to a bound and quickly turns out costly in presence of unbounded loops.

Conventional SE does not over-approximate executions after a fixed bound of iterations. This implies that *soundness* is lost when the program exceeds the exploration bound. Soundness ensures that, when the analysis concludes that the property of interest holds, the concrete semantics of the analyzed program is guaranteed to satisfy it. Since there is no over-approximation, when the property is violated by traces shorter than the exploration bound, tools like SMT solvers can provide instances for the symbolic values and enable the reconstruction of counter-example traces. This is of particular importance to security in order to confirm security violations. We refer to such counter-examples as **refutation models**.

The adaptation of SE to handle relational properties [32,34] requires to track several traces instead of just one. In the following, we will call this adaptation *relational symbolic execution* (or RSE). Previous work [23] has shown how to combine RSE with loop invariants, provided by the developer, in order to recover soundness at the cost of annotations and loss of precision when invariants are not strong enough.

Abstract interpretation based static analyses [14] (AI) rely on an abstraction defined as a logical approximation relation between concrete behaviors and abstract predicates and produce sound over-approximations of program semantics at the cost of completeness. However, the over-approximation entails that the analysis may fail to conclude positively even when analyzing correct programs. Moreover, most static analysis implementations lack the ability to synthesize counter-example traces.

In this paper, we formalize a combined analysis technique, which aims at bringing together advantages of both symbolic execution and abstract interpretation, in a security setup. We first show how to over-approximate SE in order to keep soundness, and call this analysis SoundSE (see Fig. 1). We use SoundSE to show the combination for conventional SE and different abstract domains, calling the resulting analyses RedSoundSE. Our analysis for relational properties is called RedSoundRSE and targets noninterference. It borrows path exploration from relational symbolic execution, parameterized by RedSoundSE, and relies on abstract interpretation based static analysis to report a sound result for all programs. Abstraction enables the early pruning of infeasible paths and the computation of sound over-approximations for program behaviors when the exploration bound is exhausted. To achieve this, RedSoundRSE automatically injects loop invariants computed by abstract domains into a relational store. Not only dependence analysis results can be used to fill security related information where the symbolic execution cannot explore paths fully but also (e.g., numerical) state abstraction information allows to improve the symbolic information extracted from the dependency analysis. Moreover, our analysis allows switching between different abstractions, and tuning specific settings, e.g., loop unroll depth (depth up to which SE is kept precise), which allows the user to change the balance between cost and precision. To summarize, we propose symbolic execution based verification methods that are sound and precise, providing refutation models up to a bound. Our contributions, illustrated in Fig. 1, are the following:

1. SoundSE and RedSoundSE: We define a sound SE analysis, and we integrate numerical abstract domains into it to prune reachable paths. As a result, we make SE [9] sound while keeping the ability of the analysis to find counter-examples.
2. SoundRSE and RedSoundRSE: We define a sound relational SE, and we combine it with dependence analysis [4] to enhance the precision of the latter while preserving soundness.
3. We prototype RedSoundRSEtogether with RedSoundSE in OCaml and show, using a series of challenging examples, that it is able to both soundly decide noninterference for secure programs and synthesize counter-examples of a size up to a given bound for insecure ones.

The structure of the paper is as follows. Section 2 defines a basic language and the noninterference notion used throughout the rest of the paper. Section 3 provides an overview on already defined analyses and highlights the main principles of RedSoundRSE. Section 4 defines SoundSE, a sound single trace symbolic execution that serves as a basis for RedSoundRSE and Sect. 5 presents RedSoundSE, a new combination of SoundSE with state abstraction. Section 6 presents SoundRSE and Sect. 7 extends it with a dependence abstraction to obtain RedSoundRSE. Section 8 evaluates our framework on small but challenging examples. Finally, Sect. 9 discusses related work and Sect. 10 concludes. The appendix contains all rules of analyses in the paper.

2   Language and Noninterference Security Notion

In this section, we introduce the language and security notion for which we formalize our analyses. We let $\mathbb{V}$ and $\mathbb{X}$ be the set of values and program variables respectively, and $\oplus$, $\ominus$ be binary operators. A boolean expression $\mathbf{b}$ is a comparison operator $\ominus$ applied to two expressions and evaluates to a boolean value $\mathbb{B} = \{\mathbf{tt}, \mathbf{ff}\}$. A statement $\mathbf{s}$ is either a skip, an assignment, a condition, or a loop. Finally, a command $\mathbf{c}$ is a finite sequence of statements. A program P is a pair $(\mathbf{c}, L)$ made of a command $\mathbf{c}$ (the body of the program), and a set of low variables $L \subseteq \mathbb{X}$, hence publicly observable (the other variables occurring in the program are high).

$$\mathbf{e} ::= v \ (v \in \mathbb{V}) \mid x \ (x \in \mathbb{X}) \mid \mathbf{e} \oplus \mathbf{e} \qquad\qquad \mathbf{b} ::= \mathbf{e} \ominus \mathbf{e}$$
$$\mathbf{s} ::= \mathtt{skip} \mid x := \mathbf{e} \mid \mathtt{if} \ \mathbf{b} \ \mathtt{then} \ \mathbf{c} \ \mathtt{else} \ \mathbf{c} \mid \mathtt{while} \ \mathbf{b} \ \mathtt{do} \ \mathbf{c} \qquad \mathbf{c} ::= \mathbf{s} \mid \mathbf{s}; \mathbf{c}$$

Semantics. Given a program $(\mathbf{c}, L)$, a *state* is a pair $(\mathbf{c}, \mu)$, where $\mathbf{c}$ is a command and μ is a function from $\mathbb{X}$ to $\mathbb{V}$, namely a *store*. In particular, a state of the form $(\mathtt{skip}, \mu)$ is final. We write $\mathbb{M}$ and $\mathbb{S}$ for the set of stores and states respectively. We use $[\mathtt{x} \mapsto x, \mathtt{y} \mapsto y, \ldots]$ to explicitly enumerate a store's contents, where $x, y, \ldots$ are concrete values. Let $(\rightarrow) \subseteq \mathbb{S} \times \mathbb{S}$ denote the small step operational semantics (which is standard) and $\rightarrow^*$ be its reflexive transitive closure.

Noninterference. Let $=_L$ be the set equality of stores restricted to low variables in L. In the rest of the paper, we focus on termination-insensitive noninterference:

Definition 1 (Termination-insensitive noninterference). *A program* $(\boldsymbol{c}, L)$ *is termination-insensitive noninterferent, written as* $\mathcal{NI}_P^{T.I}$, *if and only if, for all stores* $\mu_0, \mu_1, \mu_0', \mu_1' \in \mathbb{M}$, $\mu_0 =_L \mu_1 \wedge (\boldsymbol{c}, \mu_0) \rightarrow^* (\mathtt{skip}, \mu_0') \wedge (\boldsymbol{c}, \mu_1) \rightarrow^* (\mathtt{skip}, \mu_1') \implies \mu_0' =_L \mu_1'$.

3   Overview

In this section, we demonstrate the principle of the combination of symbolic execution and abstraction performed by RedSoundRSE so as to overcome the limitation of these two approaches taken separately. As in the rest of the paper, we focus on noninterference (NI), although the same principle would apply to other security properties as well.

```
1  if (priv > 0)        1  while (i < z) {      1  while (i > priv) {
2     y = 5;             2     i = i + 1;        2     i = i + 1;
3  else                 3     priv = priv + 5;   3     priv = priv + 2;
4     y = 5;             4  }                     4  }
   (a) Secure pro-         (b) Secure program       (c) Insecure program
   gram
```

```
1  if (priv < 0) priv = 0;
2  while (i < 10){
3     i += 1; priv += 2;
4  }
5  if (priv >= 0) y += 1;
6  else y = 0;
   (d) A secure program re-
   quiring a numerical do-
   main.
```

Fig. 2. Example programs. All variables are of type `int`, where variable is `priv` is secure.

Examples. We consider the programs displayed in Fig. 2. Essentially, programs (a) and (b) are secure with respect to the noninterference policy, where priv is high and all other variables are low, whereas (c) is not secure.

In program 2(d), variable y gets assigned 5 independently of `priv`, therefore the program is secure. For Program 2(b), let μ_0, μ_1 be two stores such that $\mu_0 =_L \mu_1$. Since μ_0 and μ_1 are low-equal executions cannot take different paths, and the loop will be executed the same amount of times. Therefore, the program is secure. Lastly, Program 2(c) is insecure, meaning that it does not satisfy noninterference. We need to provide a counter-example consisting of two executions starting from low-equal stores μ_0, μ_1 such that the corresponding output stores μ'_0, μ'_1 are not low-equal. We consider the following stores: $\mu_0 = [\text{i} \mapsto 0, \text{priv} \mapsto 0]$, and $\mu_1 = [\text{i} \mapsto 0, \text{priv} \mapsto -1]$. Finally, calculated output stores are such that $\mu'_0(\text{i}) = 0 \neq \mu'_1(\text{i}) = 1$, thus the program violates noninterference.

In the next paragraphs we study the result of verification methods for these three programs.

Verification Based on Relational Symbolic Execution. A symbolic store, referred to as ρ, maps variables to symbolic expressions of the initial values of the variables. To avoid confusion, we use an italic typewriter font for these symbolic values while program variables appear in straight typewriter font. For instance, y denotes the initial value of y. Relational symbolic execution describes pairs of executions using symbolic conditions over the initial values of variables and pairs of symbolic stores. Symbolic stores are not enough to abstract executions, since they cannot express constraints. Constraints are then provided by a *symbolic path* π that contextualizes the store. A pair (ρ, π) of a symbolic store and a symbolic path is referred to as a *symbolic precise store*.

As an example, we consider Program 2(a). Relational symbolic execution uncovers four pairs of paths depending on the sign of the initial values of `priv` in both executions. For instance, one of the diverging paths produces $\pi = (priv_0 > 0 \land priv_1 \leq 0) \implies ([\text{y}_0 \mapsto 5, \ldots], [\text{y}_1 \mapsto 5, \ldots])$, where y_0 and y_1 denote the program variable y in both executions and $priv_0, priv_1$ the initial symbolic values of `priv`. This symbolic precise store shows no information flow to y since any SMT solver can prove $\text{y}_0 = \text{y}_1$. The other three pairs of paths lead to a similar result, thus the program is proved secure.

Table 1. Analysis results compared. Symbol ✓ (resp., ✗) denotes a semantically correct (resp., incorrect) analysis outcome, with either a proof of security, a (possibly false) alarm, or a refutation model.

	Secure?	RSE	Dependence analysis	RedSoundRSE
Program 2(a)	Yes	✓ Secure	✗ False alarm	✓ Secure
Program 2(b)	Yes	✗ False alarm	✓ Secure	✓ Secure
Program 2(c)	No	✓ Refutation model	✓ Alarm	✓ Refutation model

For Program 2(b), the loop has an unbounded number of iterations, but relational symbolic execution can only cover finitely many unrollings of the loop. This prevents RSE to prove that Program 2(b) is secure.

For Program 2(c), RSE will only explore the loop up to a bound. Assuming the bound is one (any positive value would prove similar), it can determine that the program does not satisfy NI by calculating a concrete trace that violates the property. This counter-example trace is calculated by an SMT solver, for instance $i_0 = i_1 = 0$, $priv_0 = 1$ and $priv_1 = -1$ corresponds to the counter-example given previously.

Verification Based on Dependence Abstraction. Many static analyses that work for noninterference rely on some form of dependence abstraction as formalized in, e.g., [4] or [28]. We briefly summarize the abstraction of [4]. We assume an ordered set of security levels $\{\mathbb{L}, \mathbb{H}\}$ and that each value fed into a program via an input variable is given a security level. A dependency, noted as $l \rightsquigarrow \mathtt{x}$ with $l \in \{\mathbb{L}, \mathbb{H}\}$, expresses the agreement of $\mathtt{x}$ in both executions when observing from level l. This analysis, based on abstract interpretation, is *sound*.

We now discuss the analysis of some programs in Fig. 2. For Program 2(a), the analysis determines that the assignments are conditioned by the value of priv, which is initially high. Then, the dependency $\mathbb{L} \rightsquigarrow \mathtt{y}$ is dropped, indicating that y can potentially disagree between executions. In Program 2(b), the loop condition is only influenced by i and z, which are low. Then, the assignment of low variables is not affected, and i and z remain low, allowing to prove noninterference.

Lastly, Program 2(c) is not secure, and since dependence analysis is sound, the analysis discards dependency $\mathbb{L} \rightsquigarrow \mathtt{i}$ based on the illicit flow of information.

Combination of Relational Symbolic Execution and Dependence Abstraction. As observed in Table 1, relational symbolic execution fails to handle precisely program 2(b) whereas dependence abstraction fails to verify program 2(a) and provides no counter-example for program 2(c). The purpose of RedSoundRSE is to use both techniques in an alternating manner in order to increase precision and prune branches.

To achieve this, RedSoundRSE borrows from relational symbolic execution the precise analysis of assignment and condition commands, as well as the unrolled iterates of loop commands. In particular, the analysis of programs 2(a) and 2(c) is carried out as shown above. However, when the unrolling bound is reached, dependence analysis is used as a means to compute in finite time sound information about any number of further loop iterations. Indeed, when the dependence information proves that a loop induces no dependency of a given low variable on any high variable, it is possible to assume the equality of the variable in the symbolic store. This new value may not be expressed precisely in terms of the initial values, hence it may be approximated with a fresh symbol. This occurs for variable i in program 2(b).

As seen, RedSoundRSE analyzes the first three examples of Fig. 2 precisely.

Refinement of Symbolic Execution Based on State Abstraction. Program Fig. 2(d), previously not considered, cannot be proved NI by just using symbolic execution and dependence analysis. This program is secure since the assignment of i does not depend on priv, and y is conditioned by priv which is always positive after the loop.

As in Program 2(b), the loop causes the symbolic execution to stop at the unrolling bound. Dependence information allows to prove that there is no information flow to i and also that the value of y at line 8 does not depend on priv. However, the condition at line 9 depends on priv, thus dependence analysis will not prove that the assignments at lines 10 and 12 do not leak information. Symbolic execution does not succeed either as it lacks the ability to reason over the value of priv at the loop exit.

Such information may be computed using a reachability static analysis. In particular, a classical static analysis based on the abstract domains of intervals [14] computes ranges for all numeric variables and concludes in this case that priv is positive, hence only the true branch of the condition may be taken. Integrating non-relational abstract domains allows the analyzer to increase precision by automatically pruning paths.

This combination of AI and SE is referred to as RedSoundSE, defined in Sect. 5, and is later integrated into the final analysis RedSoundRSE.

4   SoundSE: Sound Symbolic Execution

We now define a type of symbolic execution, named SoundSE, as it serves as a basis for not only SoundRSE but also RedSoundSE—the product of SoundSE with abstract domains.

Symbolic Execution States. The core principle of symbolic execution is to map program variables into expressions made of *symbolic values* that denote the initial value of the program variables. We let $\overline{\mathbb{V}} = \{x, y, \ldots\}$ denote the set of symbolic values and note for clarity x the symbolic value associated to program variable x (not to be confused with concrete values). A *symbolic store* is a function ρ from program variables to *symbolic expressions* the set of which is noted $\mathbb{E}$, namely expressions defined like the programming language expressions using symbolic values instead of program variables. We write $\overline{\mathbb{M}} = \mathcal{P}(\mathbb{X} \rightarrow \mathbb{E})$ for the set of symbolic stores and write $[x \rightsquigarrow \langle x \rangle, \ldots]$ for an explicitly given symbolic store. To tie properly symbolic stores and concrete stores, we need to relate symbolic values and concrete values. To this end, we let a *valuation* be a function $\nu : \overline{V} \longrightarrow V$. Moreover, given a symbolic expression ε, we let $[\![\varepsilon]\!]$ be a partial function that maps a valuation ν to the value obtained when evaluating the expression obtained by replacing each symbolic value x in e with $\nu(x)$. We can now express the concretization of symbolic stores:

Definition 2 (Symbolic store concretization). *The* symbolic store concretization, $\gamma_{\overline{M}} : \overline{M} \longrightarrow \mathcal{P}(M \times (\overline{V} \rightarrow V))$, *maps a symbolic store to the set of pairs made of a store and a valuation that realize it, i.e.* $\gamma_{\overline{M}}(\rho) = \{(\mu, \nu) \mid \forall x \in X,\ \mu(x) = [\![\rho(x)]\!](\nu)\}$.

To precisely characterize the outcome of an execution path, a symbolic store is too abstract. Hence, SE also utilizes a symbolic expression to constrain the store, referred to as *symbolic path*, that accounts for the conditions encountered during a path. A *symbolic precise store* is a pair $\kappa = (\rho, \pi)$ where $\rho \in \overline{M}$ and π is a symbolic path. We write K for the set of symbolic precise stores. Their meaning is defined as follows:

Definition 3 (Symbolic precise store concretization). *The* symbolic precise store concretization, $\gamma_K : K \longrightarrow \mathcal{P}(M \times (\overline{V} \rightarrow V))$, *is defined by* $\gamma_K(\rho, \pi) = \{(\mu, \nu) \in \gamma_{\overline{M}}(\rho) \mid [\![\pi]\!](\nu) = tt\}$.

Example 1 (Symbolic precise store). We consider Program 2(a). Symbolic execution needs to cover two paths corresponding to each of the branches of the condition statement, i.e., depending on the sign of *priv*. Therefore, symbolic execution should produce the precise stores $(\rho_0, priv > 0)$ and $(\rho_1, priv \leq 0)$, where $\rho_0 = \rho_1 = [y \rightsquigarrow \langle 5 \rangle, \text{priv} \rightsquigarrow \langle priv \rangle]$.

Symbolic Execution Step. The main piece of the symbolic execution algorithm is the step relation, which closely follows the small step semantics of the programs. We define it by a transition relation $\rightarrow_s$ between *symbolic execution states* that are made of a program command and a symbolic precise store. Before we write down the analysis $\rightarrow_s$, we need a few definitions.

First, we define the symbolic evaluation of an expression or condition in a symbolic store, which produces a symbolic expression. We note $(e, \rho) \vdash_s \varepsilon$ the evaluation of e into symbolic expression ε in symbolic store ρ. Usually, this evaluation step boils down to the substitution of the variables in e with the symbolic expressions they are mapped to in ρ, possibly with some simplifications.

Second, we define the conservative satisfiability test of a symbolic path. This step is usually performed by an external tool such as an SMT solver, so we do not detail its internals here. We note that this test may conservatively return as a result that a symbolic path *may* be satisfiable. We note $\mathbf{may}(\pi)$ when π may be satisfiable.

$$\text{S-ASSIGN} \frac{(\mathbf{e},\rho) \vdash_s \varepsilon}{(\mathbf{x} := \mathbf{e},(\rho,\pi)) \to_s (\mathbf{skip},(\rho[\mathbf{x} \leadsto \langle \varepsilon \rangle],\pi))}$$

$$\text{S-IF-T} \frac{(\mathbf{b},\rho) \vdash_s \beta \qquad \pi' \triangleq \pi \wedge \beta \qquad \mathbf{may}(\pi')}{(\mathbf{if\ b\ then\ c_0\ else\ c_1},(\rho,\pi)) \to_s (\mathbf{c_0},(\rho,\pi))}$$

$$\text{S-LOOP-T} \frac{(\mathbf{b},\rho) \vdash_s \beta \qquad \pi' \triangleq \pi \wedge \beta \qquad \mathbf{may}(\pi')}{(\mathbf{while\ b\ do\ c},(\rho,\pi)) \to_s (\mathbf{c;\ while\ b\ do\ c},(\rho,\pi))}$$

$$\text{S-LOOP-F} \frac{(\mathbf{b},\rho) \vdash_s \beta \qquad \pi' \triangleq \pi \wedge \neg\beta \qquad \mathbf{may}(\pi')}{(\mathbf{while\ b\ do\ c},(\rho,\pi)) \to_s (\mathbf{skip},(\rho,\pi))}$$

Fig. 3. Symbolic execution step relation: a few selected rules

We now turn to the rules in Fig. 3. Rule S-ASSIGN simply updates the symbolic store with a new symbolic expression for the assigned variable. In rule S-IF-T, if the guard evaluation β is satisfiable, the true branch is accessed and β is added to the symbolic path. Finally, rules S-LOOP-T and S-LOOP-F follow similar principles as rule S-IF-T in the case of loops. We formalize the soundness of execution steps:

Theorem 1 (Soundness of a single symbolic execution step). *Let (c, μ) and $(c', \mu') \in \mathbb{S}$ be two states such that $(c, \mu) \to (c', \mu')$, $\kappa \in \mathbb{K}$ a symbolic precise store, and ν be a valuation such that $(\mu, \nu) \in \gamma_{\mathbb{K}}(\kappa)$. Then, there exists a symbolic precise store κ' such that $(\mu', \nu) \in \gamma_{\mathbb{K}}(\kappa')$ and $(c, \kappa) \to_s (c', \kappa')$.*

Sound Depth Bounded Symbolic Execution. Clearly, the exhaustive application of the symbolic execution step relation defined in Fig. 3 would not terminate. Therefore, common symbolic execution tools typically abort the exploration when they reach some sort of bound on execution lengths. This result is clearly unsound as longer executions are simply ignored. Alternatively, it is possible to over-approximate the set of precise stores that may be reachable when the bound is met. We formalize this approach here.

Essentially, symbolic states need to be augmented with two additional pieces of information, namely a boolean so-called *precision flag* which states whether symbolic execution has performed any over-approximation due to exhausting the bound, and a bound control field, called *counter*. We define set $\mathbb{W}$ as the set of counters, with a special element $w_0 \in \mathbb{W}$ that denotes the initial counter status with respect to bound control. To operate over counters, we require a function step which inputs two commands c, c', and a counter w. It produces a result of the form (b, w') where b is a boolean, and w' is the next counter. Value b is $\mathbf{tt}$ if and only if a step from c to c' can be done without exhausting the iteration bounds, and with the new counter w'. If b is $\mathbf{ff}$, the iteration bound has been reached and the state needs to be over approximated.

To perform the over approximation, a function $\mathfrak{modif}$ is required. The function inputs a symbolic store and a command, and returns a new symbolic store ρ' such that:

$$\text{S-NEXT}\ \frac{(\mathbf{c},\kappa) \rightarrow_s (\mathbf{c}',\kappa') \qquad \mathfrak{step}(\mathbf{c},\mathbf{c}',w) = (\mathbf{tt},w')}{(\mathbf{c},\kappa,w,b) \rightarrow_s (\mathbf{c}',\kappa',w',b)}$$

$$\text{S-APPROX-MANY}\ \frac{(\mathbf{c},\kappa) \rightarrow_s (\mathbf{c}',\kappa') \qquad \mathfrak{step}(\mathbf{c},\mathbf{c}',w) = (\mathbf{ff},w') \qquad \rho'' = \mathfrak{modif}(\rho,\mathbf{c})}{(\mathbf{c},(\rho,\pi),w,b) \rightarrow_s (\mathtt{skip},(\rho'',\pi),w',\mathbf{ff})}$$

Fig. 4. SoundSE: Sound bounded symbolic execution step relation

- ρ' maps each program variable that is considered to be "modified" (by a sound over approximation of the set) in $\mathbf{c}$ to a *fresh* symbolic value;
- ρ' maps all the other program variables to their image in the original store.

Example 2 (Loop iteration bounding). The most typical way to bound symbolic execution limits the number of iteration of each loop to pre-defined number k. Then, $\mathcal{W}$ consists of stacks of integers, w_0 is the empty stack, and $\mathfrak{step}$ adds a zero on top of the stack when entering a new loop and pops the value on top of the stack when exiting a loop. More importantly, it increments the value n at the top of the stack when $n \leq k$ and moving to the next iteration (rule S-LOOP-T); on the other hand, when $n > k$, it pops n and returns the $\mathbf{ff}$ precision flag.

To ensure termination, $\mathcal{W}$ and $\mathfrak{step}$ should satisfy the following *well-foundedness* property: for any infinite sequence of commands $(\mathbf{c}_i)_i$ the infinite sequence $(w_i)_i$ defined by $\mathfrak{step}(\mathbf{c}_i, \mathbf{c}_{i+1}, w_i) = (\mathbf{tt}, w_{i+1})$ should be stationary, which we assume here.

Based on these definitions, *depth bounded symbolic execution* is defined by a transition relation over 4-tuples made of a command, a symbolic state, an element of $\mathcal{W}$, and a boolean, referred to as symbolic state. We overload the notation $\rightarrow_s$ for this relation, which is defined based on the previously defined $\rightarrow_s$. The rules are provided in Fig. 4:

- Rule S-NEXT carries out an atomic step of symbolic execution that requires no over approximation; function $\mathfrak{step}$ returns the precision flag b and a new counter;
- Rule S-APPROX-MANY carries out a global approximation step; indeed, as $\mathfrak{step}$ returns $\mathbf{ff}$, the function $\mathfrak{modif}$ is applied to the symbolic state to over-approximate the effect of an arbitrary number of steps of execution of $\mathbf{c}$; alongside with the new counter state the $\mathbf{ff}$ precision is propagated forward.

Under the well-foundedness assumption, exhaustive iteration of the available symbolic execution rules from any initial symbolic state will terminate and produce finitely many symbolic states. To express the soundness of this algorithm, we need to account for the creation of symbolic values by function $\mathfrak{modif}$, which means that valuations also need to be extended. To this end, we note $\nu \preceq \nu'$ when the domain of valuation ν is included into that of ν' and when both ν and ν' agree on the intersection of their domains. We now obtain the following soundness statement:

Theorem 2 (Soundness of any sequence of single symbolic execution steps). *Let $(c, om, \mu) \in \mathbb{S}$ be a state and μ' be a store such that $(c, \mu) \rightarrow^*$ (skip, μ'). Let $\kappa \in \mathbb{K}$ be a symbolic precise store and ν be a valuation such that $(\mu, \nu) \in \gamma_\mathbb{K}(\kappa)$. Let $w \in \mathbb{W}$ be a counter. Then, there exists a symbolic precise store κ', a valuation ν', and a counter $w' \in \mathbb{W}$ such that $\nu \preceq \nu'$, $(\mu', \nu') \in \gamma_\mathbb{K}(\kappa')$, and $(c, \kappa, w, b) \rightarrow_s^* (\text{skip}, \kappa', w', b')$.*

The proof of this theorem follow from Theorem 1 (steps where step returns tt), and a global induction on the command c when rule S-APPROX-MANY applies.

Example 3 (Symbolic execution). For program 2(a), symbolic execution returns the symbolic stores shown in Example 1. We assume the bounding of Example 2 and consider program 2(b). Then, symbolic execution generates the symbolic store $[z \rightsquigarrow \langle z \rangle, i \rightsquigarrow \langle i' \rangle, \text{priv} \rightsquigarrow \langle priv' \rangle]$ with precision flag ff, and where i', $priv'$ are fresh symbolic values generated by rule S-APPROX-MANY.

Refutation up to a Bound. A very desirable feature of symbolic execution is the ability to produce counter-examples up to a bound. This feature stems from a bounded refutation result, which states that, when symbolic execution produces a final state for which the final precision flag is tt, and such that the symbolic path is satisfiable, then a matching concrete execution can be found. From the final state, the SMT solver can compute a refutation model.

Theorem 3 (Refutation up to a bound). *Let c be a command, $\kappa, \kappa' \in \mathbb{K}$ be two precise stores, $w, w' \in \mathbb{W}$, such that $(c, \kappa, w, \text{tt}) \rightarrow_s^* (\text{skip}, \kappa', w', \text{tt})$. Then, for all $(\mu', \nu') \in \gamma_\mathbb{K}(\kappa')$, it exists $(\mu, \nu) \in \gamma_\mathbb{K}(\kappa)$ such that $(c, \mu) \rightarrow^* (\text{skip}, \mu')$.*

This result follows from the fact that rule S-APPROX-MANY is never applied in the symbolic execution and from an induction on the sequence of S-NEXT steps.

Example 4 (Symbolic execution completeness up to a bound). We consider the cases discussed in Example 3. Using the bounding of Example 2, the result produced for program 2(a) is complete whereas that for program 2(b) generates some final symbolic state with precision flag ff, hence for which Theorem 3 does not apply.

5   RedSoundSE: Sound SE Combined with Abstract States

We now extend SoundSE with the ability to use the properties inferred by abstract interpretation. This combined symbolic execution is referred to as RedSoundSE, making reference to the reduced product between SoundSE and an AI based analysis.

$$\text{A-ASSIGN} \frac{a' \triangleq \mathfrak{assign}_{x,e}(a)}{(x:=e,a) \to_A (\mathfrak{skip},a')} \qquad \text{A-IF-T} \frac{a' \triangleq \mathfrak{guard}_b(a) \quad a' \neq \bot}{(\mathfrak{if} \ b \ \mathfrak{then} \ c_0 \ \mathfrak{else} \ c_1,a) \to_A (c_0,a')}$$

(a) Abstract execution step selected rules

$$\text{S-A-NEXT} \frac{\mathfrak{step}(c,c',w)=(\mathfrak{tt},w') \quad (c,a) \to_A (c',a') \quad (\kappa'',a'') \triangleq \mathfrak{reduction}(\kappa',a')}{(c,\kappa,w,b) \to_s (c',\kappa',w',b)}$$

$$\text{S-A-APPROX-MANY} \frac{(c,\kappa) \to_s (c',\kappa') \quad \mathfrak{step}(c,c',w)=(\mathfrak{ff},w')}{\kappa''=\mathfrak{modif}(\kappa,c) \quad a'=[\![c]\!]_A^\sharp(a) \quad (\kappa''',a''') \triangleq \mathfrak{reduction}(\kappa'',a')}{(c,(\kappa,a),w,b) \to_{s\times A} (\mathfrak{skip},(\kappa''',a'''),w',\mathfrak{ff})}$$

(b) Product of symbolic execution and static analysis

Fig. 5. Abstract execution step and product with symbolic execution

Abstraction of Store and Static Analysis. In the following, we assume that an *abstract domain* [14] A describing sets of stores is fixed, together with a concretization function $\gamma_A : A \longrightarrow \mathcal{P}(M)$. We assume the existence of an element $\bot \in A$ such that $\gamma_A(\bot) = \emptyset$. Additionally, we require the two following sound abstract post-condition functions for basic operations. Function $[\![\ ]\!]$ will be overloaded to replace any variable x for its mapped value in a store $\mu(x)$:

- *abstract assignment* $\mathfrak{assign}_{x,e} : A \longrightarrow A$ is parameterized by a variable x and an expression e and is such that $\forall a \in A, \{\mu[x \mapsto [\![e]\!](\mu)] \mid \mu \in \gamma_A(a)\} \subseteq \gamma_A(\mathfrak{assign}_{x,e}(a))$.
- *abstract condition* $\mathfrak{guard}_b : A \longrightarrow A$ is parameterized by a boolean expression b and is such that $\forall a \in A, \{\mu \in \gamma_A(a) \mid [\![b]\!](\mu) = \mathfrak{tt}\} \subseteq \gamma_A(\mathfrak{guard}_b(a))$.

Based on these operations, the definition of a *sound abstract execution step* relation $\to_A$ is straightforward. We show two rules in Fig. 5(a). The rules match those of $\to$ (Sect. 2) and are sound with respect to it. In the following, A is assumed to be a parameter of the analysis. It may consist of any numerical abstraction, such as the interval abstract domain [14] or the domain of convex polyhedra [16]. Moreover, the application of standard widening technique [14] allows to define a *static analysis* function $[\![c]\!]_A^\sharp : A \longrightarrow A$ that is sound in the sense that, for all command c and all abstract state a, $\{\mu' \in M \mid \exists \mu \in \gamma_A(a), (c,\mu) \to_s^* (\mathfrak{skip},\mu')\} \subseteq \gamma_A([\![c]\!]_A^\sharp(a))$

Reduced Product of Symbolic Precise Stores and Abstract States. Reduced product [15] aims at expressing precisely conjunctions of constraints expressed in distinct abstract domains. We let a precise product store be a pair $(\kappa, a) \in \mathbb{K} \times A$. In our case, the definition needs to be adapted slightly as symbolic execution and abstract domain A do not abstract exactly the same objects:

Definition 4 (Product domain). *The* product abstract domain *consists of the set* $\mathbb{K} \times \mathbb{A}$ *and the concretization function* $\gamma_{\mathbb{K} \times \mathbb{A}} : \mathbb{K} \times \mathbb{A} \longrightarrow \mathcal{P}(\mathbb{M} \times (\overline{\mathbb{V}} \to \mathbb{V}))$ *defined as follows:* $\gamma_{\mathbb{K} \times \mathbb{A}} : (\kappa, a) \longmapsto \{(\mu, \nu) \in \gamma_{\overline{\mathbb{M}}}(\kappa) \mid \mu \in \gamma_{\mathbb{A}}(a)\}$

In a precise product store (κ, a), the goal is to enhance precision by exchanging information between κ and a. This is done through a **reduction** function, which rewrites an abstract element with another of equal concretization, but that supports more precise analysis operations. This implies that $(\gamma_{\mathbb{K} \times \mathbb{A}} \circ \mathfrak{reduction})(\kappa, a) = \gamma_{\mathbb{K} \times \mathbb{A}}(\kappa, a)$. This requires the abstract domain $\mathbb{A}$ to support a function $\mathfrak{constr}$ that maps an abstract state a to a logical formula over program variables and entailed by a, namely such that, if $\mu \in \gamma_{\mathbb{A}}(a)$ then μ satisfies formula $\mathfrak{constr}(a)$. Some abstract domains—specifically intervals and abstract polyhedra—utilize an internal representation based on conjunction of constraints, in which case $\mathfrak{constr}$ is trivial. Then, $\mathfrak{reduction} : \mathbb{K} \times \mathbb{A} \longrightarrow \mathbb{K} \times \mathbb{A}$ is defined by:

$$\mathfrak{reduction}((\rho, \pi), a) \triangleq ((\rho, \pi'), a) \quad \text{where} \quad \pi' \triangleq \pi \wedge \mathfrak{constr}(a)[\mathtt{x} \mapsto \rho(\mathtt{x})]$$

Note that $[\mathtt{x} \mapsto \rho(\mathtt{x})]$ in the above definition, symbolizes the replacement of each program variable present in $\mathfrak{constr}(a)$ into its definition in ρ; this step follows from the fact that a constrains program variables whereas π constrains valuations. This general reduction function may be refined into a more precise one, where the resulting symbolic path is simplified, possibly to the **ff** formula. Furthermore, this reduction only modifies the symbolic path π, but it is possible to define a reduction operation that also rewrites the abstract state a.

Reduced Product Symbolic Execution. The product analysis, namely Red-SoundSE, takes the form of an extension of the symbolic execution function of Fig. 4. The new states are still 4-tuples, but the symbolic precise store component κ is now replaced with a precise product store (κ, a). The transition relation $\rightharpoonup_{s \times \mathbb{A}}$ between such states consists of two rules that are shown in Fig. 5(b) and that extend those in Fig. 4. In rule S-A-MANY (applied when exploration bound is met) aside from $\mathfrak{modif}$, the loop is calculated over the abstract state and then the reduction function is applied.

In both cases, the sound $\mathfrak{reduction}$ operator may be applied. In practice, for the sake of efficiency, it can be computed and applied in a lazy manner that is, only for specific steps (typically S-A-MANY and for branching commands).

Example 5 (Product analysis). For program 2(d), assuming $\mathtt{i} < 10$, and then when exiting the loop, an intervals abstract state will hold two constraints $a = \{\mathtt{i} = 10; \mathtt{priv} \geq 2\}$. Assuming a symbolic precise store $\kappa = (\rho, \pi)$ with $\rho = [\mathtt{i} \to i; \mathtt{priv} \to priv]$, the abstract constraints can be fitted to a symbolic path π' as follows: $\pi' \triangleq \pi \wedge i = 10 \wedge priv \geq 2$. A more detailed execution trace is given in Appendix A.

Soundness and Refutation Property. The RedSoundSE analysis defined in the previous paragraph satisfies the same soundness (Theorem 2) and refutation (Theorem 3) properties as standard symbolic execution, so we do not give the theorems again.

6   SoundRSE: Sound Relational Symbolic Execution

As discussed in Sect. 3, security properties like noninterference require to reason over *pairs* of execution traces thus we now set up a *sound relational symbolic execution* technique that constructs pairs of executions. This analysis will be regarded as SoundRSE.

Assumption. To keep notations lighter, we assume in this section and the next that the bounding counter step function $\mathfrak{step}$ only affects loops, namely $\mathfrak{step}(\mathbf{c}, \mathbf{c}', w) = w$ whenever $\mathbf{c}$ is not a loop command. Moreover, we do not include the product with the numerical abstract state (as in Sect. 5) in the following definitions. Since it can be added in a seamless manner, we omit it here to keep formal statements lighter.

Precise Relational Stores. We first define the notions of relational expression, relational store, and precise relational store.

Definition 5 (Relational and precise relational stores). *A* relational symbolic expression *is an element defined by the grammar:* $\tilde{\varepsilon} ::= \langle \varepsilon \rangle | \langle \varepsilon \mid \varepsilon \rangle$ *where* ε *ranges over the set* $\mathbb{E}$ *of symbolic expressions. We write* $\mathbb{E}_2$ *for the set of relational symbolic expressions. A* relational symbolic store $\tilde{\rho}$ *is a function from variables to relational symbolic expressions. We let* $\overline{\mathbb{M}}_2 = \mathbb{X} \to \mathbb{E}_2$ *stand for their set. Finally, a* precise relational store $\tilde{\kappa}$ *is a pair* $(\tilde{\rho}, \pi) \in \mathbb{K}_2$.

Before we define concretizations of $\overline{\mathbb{M}}_2$ and $\mathbb{K}_2$, we need to introduce two operations:

- The *projections* Π_0, Π_1 map relational symbolic stores into symbolic stores. They are defined in a pointwise manner, as follows: if $\tilde{\rho}(\mathbf{x}) = \langle \varepsilon \rangle$ then $\Pi_0(\tilde{\rho})(\mathbf{x}) = \Pi_1(\tilde{\rho})(\mathbf{x}) = \varepsilon$ and if $\tilde{\rho}(\mathbf{x}) = \langle \varepsilon_0 \mid \varepsilon_1 \rangle$, then $\Pi_0(\tilde{\rho})(\mathbf{x}) = \varepsilon_0$ and $\Pi_1(\tilde{\rho})(\mathbf{x}) = \varepsilon_1$. We overload the Π_0, Π_1 notation and also apply it to double symbolic expressions: $\Pi_0(\langle \varepsilon \rangle) = \Pi_1(\langle \varepsilon \rangle) = \varepsilon$ and if $\tilde{\varepsilon} = \langle \varepsilon_0 \mid \varepsilon_1 \rangle$, then $\Pi_0(\tilde{\varepsilon}) = \varepsilon_0$ and $\Pi_1(\tilde{\varepsilon}) = \varepsilon_1$.
- The *pairing* $(\rho_0 \mid \rho_1)$ of two symbolic stores ρ_0 and ρ_1 is a relational symbolic store defined such that, for all variable $\mathbf{x}$,

$$(\rho_0 \mid \rho_1)(\mathbf{x}) = \begin{cases} \langle \varepsilon \rangle & \text{if } \rho_0(\mathbf{x}) \text{ and } \rho_1(\mathbf{x}) \text{ are provably equal to } \varepsilon \in \mathbb{E} \\ \langle \rho_0(\mathbf{x}) \mid \rho_1(\mathbf{x}) \rangle & \text{otherwise} \end{cases}$$

where the notion of "provably equal" may boil down to syntactic equality of symbolic expressions or involve an external proving tool.

We can now define the concretization functions:

Definition 6 (Concretization functions). *The* concretization of relational stores $\gamma_{\overline{\mathbb{M}}_2}$ *and* concretization of precise relational stores $\gamma_{\mathbb{K}_2}$ *are defined by:*

$$\gamma_{\overline{\mathbb{M}}_2} : \overline{\mathbb{M}}_2 \longrightarrow \mathcal{P}(\mathbb{M} \times \mathbb{M} \times (\overline{\mathbb{V}} \to \mathbb{V}))$$
$$\tilde{\rho} \longmapsto \{(\mu_0, \mu_1, \nu) \mid \forall \boldsymbol{x} \in \mathbb{X}, \forall i \in \{0, 1\}, \mu_i(\boldsymbol{x}) = [\![\Pi_i(\tilde{\rho})(\boldsymbol{x})]\!](\nu)\}$$
$$\gamma_{\mathbb{K}_2} : \mathbb{K}_2 \longrightarrow \mathcal{P}(\mathbb{M} \times \mathbb{M} \times (\overline{\mathbb{V}} \to \mathbb{V}))$$
$$(\tilde{\rho}, \pi) \longmapsto \{(\mu_0, \mu_1, \nu) \in \gamma_{\overline{\mathbb{M}}_2}(\tilde{\rho}) \mid [\![\pi]\!](\nu) = \boldsymbol{tt}\}.$$

$$\text{SR-EXIT} \; \frac{}{(\texttt{skip}\bowtie\texttt{skip},(\tilde{\rho},\pi),w,b) \rightarrow_{sr} (\texttt{skip},(\tilde{\rho},\pi),w,b)}$$

$$\text{SR-COMP-R} \; \frac{(\mathbf{c}_1,(\Pi_1(\tilde{\rho}),\pi),w,b) \rightarrow_s (\mathbf{c}_1',(\rho_1',\pi'),w',b')}{(\texttt{skip}\bowtie\mathbf{c}_1,(\tilde{\rho},\pi),w,b) \rightarrow_{sr} (\texttt{skip}\bowtie\mathbf{c}_1',(\langle\!| \Pi_0(\tilde{\rho}) \,|\, \rho_1' |\!\rangle,\pi'),w',b')}$$

$$\text{SR-COMP-L} \; \frac{(\mathbf{c}_0,(\Pi_0(\tilde{\rho}),\pi),w,b) \rightarrow_s (\mathbf{c}_0',(\rho_0',\pi'),w',b')}{(\mathbf{c}_0\bowtie\mathbf{c}_1,(\tilde{\rho},\pi),w,b) \rightarrow_{sr} (\mathbf{c}_0'\bowtie\mathbf{c}_1,(\langle\!| \rho_0' \,|\, \Pi_1(\tilde{\rho}) |\!\rangle,\pi'),w',b')}$$

$$\text{SR-IF-TF} \; \frac{(\mathbf{b},\tilde{\rho})\vdash_{sr}\tilde{\beta} \quad \pi'=\pi\wedge\Pi_0(\tilde{\beta})\wedge\neg\Pi_1(\tilde{\beta}) \quad \mathbf{may}(\pi')}{(\texttt{if b then } \mathbf{c}_0 \texttt{ else } \mathbf{c}_1,(\tilde{\rho},\pi),w,b) \rightarrow_{sr} (\mathbf{c}_0\bowtie\mathbf{c}_1,(\tilde{\rho},\pi'),w,b)}$$

$$\text{SR-APPROX-MANY} \; \frac{\begin{array}{c}\mathfrak{step}(\texttt{while b do } \mathbf{c},(\mathbf{c};\texttt{ while b do } \mathbf{c}),w)=(\mathrm{ff},w') \\ \tilde{\rho}''=\mathfrak{modif}(\tilde{\rho},\mathbf{c}) \quad (\mathbf{b},\tilde{\rho}'')\vdash_{sr}\langle\beta_0,\beta_1\rangle \quad \pi'\triangleq\pi\wedge\neg\beta_0\wedge\neg\beta_1\end{array}}{(\texttt{while b do } \mathbf{c},(\tilde{\rho},\pi),w,b) \rightarrow_{sr} (\texttt{skip},(\tilde{\rho}'',\pi'),w',\mathrm{ff})}$$

Fig. 6. SoundRSE: a few selected rules of the relational symbolic execution step relation.

Example 6. We consider program 2(d) (SoundSE was discussed in Example 1). To cover pairs of executions that start with the same value for low variable y but possibly distinct values for high variable priv, relational symbolic execution should cover four pairs of paths. These four paths have the same relational symbolic store $[\texttt{priv} \rightsquigarrow \langle priv_0 \,|\, priv_1\rangle, \texttt{y} \rightsquigarrow \langle 5\rangle]$ and differ only in the symbolic path components. For instance, when the first execution takes the true branch of the condition and the second the false branch, the symbolic path is $priv_0 > 0 \wedge priv_1 \leq 0$.

Relational Symbolic Execution Algorithm. Since SoundRSE aims at describing pairs of executions, it should account for the case where the two executions follow different control flow paths. Thus, a relational symbolic state may consist of a single command when both executions follow the same path, or two commands when they diverge. We respectively note these two kinds of states $(\mathbf{c}, \tilde{\kappa}, w, b)$ and $((\mathbf{c}_0 \bowtie \mathbf{c}_1); \mathbf{c}_2, \tilde{\kappa}, w, b)$; in the latter, $\mathbf{c}_0$ (resp., $\mathbf{c}_1$) denotes the control state of the first (resp., second) execution, which they later meet in $\mathbf{c}_2$. The components w and b have the same meaning as in Sect. 4. Initial states are of the former sort.

We write $\rightarrow_{sr}$ for the relational symbolic execution step relation. A representative selection of the rules are shown in Fig. 6. Rule SR-APPROX-MANY describes a case where approximation is performed so as to ensure termination and uses the straightforward extension of $\mathfrak{modif}$ to relational symbolic states.

Soundness and Refutation Property. SoundRSE inherits similar soundness and refutation properties as SoundSE, as shown in the following theorems.

Theorem 4 (Soundness). *Let $\tilde{\kappa} \in \mathbb{K}_2$, $w \in \mathbb{W}$, and $b \in \mathbb{B}$. We let $(\mu_0, \mu_1, \nu) \in \gamma_{\mathbb{K}_2}(\tilde{\kappa})$ and assume that stores μ_0', μ_1' are such that $(\mathbf{c}, \mu_0) \rightarrow^* (\texttt{skip}, \mu_0')$ and $(\mathbf{c}, \mu_1) \rightarrow^* (\texttt{skip}, \mu_1')$. Then, there exists $\tilde{\kappa}' \in \mathbb{K}_2$, a valuation ν', and a counter state $w' \in \mathbb{W}$ such that $\nu \preceq \nu'$, $(\mu_0', \mu_1', \nu') \in \gamma_{\mathbb{K}}(\tilde{\kappa}')$, and $(\mathbf{c}, \tilde{\kappa}, w, b) \rightarrow^*_{sr} (\texttt{skip}, \tilde{\kappa}', w', b')$.*

Theorem 5 (Refutation up to a bound). *Let c be a command, $\tilde{\kappa}, \tilde{\kappa}' \in \mathbb{K}_2$ be two precise stores, $w, w' \in \mathbb{W}$, such that $(c, \kappa, w, tt) \rightarrow^*_{sr} (\text{skip}, \kappa', w', tt)$. Then, for all $(\mu'_0, \mu'_1, \nu') \in \gamma_{\mathbb{K}_2}(\kappa')$, it exists $(\mu_0, \mu_1, \nu) \in \gamma_{\mathbb{K}_2}(\kappa)$ such that $(c, \mu_0) \rightarrow^*$ (skip, μ'_0) and $(c, \mu_1) \rightarrow^* (\text{skip}, \mu'_1)$.*

SoundRSE-Based Analysis and Noninterference. We now assume a program (c, L), and show the application of SoundRSE analysis to attempt proving noninterference. The analysis proceeds according to the following steps:

1. Construction of the initial store $\tilde{\rho}_0$ such that, for all variables x present in c, $\tilde{\rho}_0(x) = \langle x \rangle$ (resp., $\tilde{\rho}_0(x) = \langle x_0 \mid x_1 \rangle$) if $x \in L$ (resp., $x \notin L$), and where x is a fresh symbolic value (resp., x_0, x_1 are fresh symbolic values).
2. Exhaustive application of semantic rules from initial state $(c, (\tilde{\rho}_0, tt), w_0, tt)$; we let $\mathcal{O}$ stand for the set of final precise relational stores with their precision flags: $\mathcal{O} \triangleq \{(\tilde{\kappa}, b) \mid \exists w \in \mathbb{W}, (c, (\tilde{\rho}_0, tt), w_0, tt) \rightarrow_{sr} (\text{skip}, \tilde{\kappa}, w, b)\}$.
3. *Attempt to prove noninterference* for each symbolic path in $\mathcal{O}$ using an external tool, such as an SMT solver; more precisely, given $((\tilde{\rho}, \pi), b) \in \mathcal{O}$,
   - if π is not satisfiable, the path is infeasible and can be ignored;
   - if it can be proved that for all variables $x \in L$, there is a unique value, i.e., $\Pi_0(\tilde{\rho})(x) = \Pi_1(\tilde{\rho})(x)$, then the program is noninterferent;
   - if a valuation ν can be found, such that $[\![\pi]\!](\nu) = tt$ (the path is satisfiable), and there exists a variable $x \in L$ such that $[\![\Pi_0(\tilde{\rho})(x)]\!](\nu) \neq [\![\Pi_1(\tilde{\rho})(x)]\!](\nu)$, and $b = tt$, then ν provides a counter-example refuting noninterference;
   - finally, if $b = ff$ and neither of the above cases occurs, no conclusive answer can be given for this path.

To summarize, the analyser either proves noninterference (when all paths are either not satisfiable or noninterferent), or it provides a valuation that refutes noninterference (when such a valuation can be found for at least one path), or it does not conclude. When a refutation is found, this refutation actually defines a real attack.

Example 7 (Noninterference). In the case of program 2(a), all paths are low-equal. The analysis of program 2(c) computes at least one interferent path if the unrolling bound is set to any strictly positive integer; in that case, a model such as the one presented in Sect. 3 can be synthesized by even basic SMT solvers. Finally, the program of Fig. 2(d) can be proved noninterferent with relational symbolic execution combined with a reduced product with a value abstract domain such as intervals (Sect. 5).

7 RedSoundRSE: Product of SoundRSE with Dependence AI

As observed in Sect. 3 some programs like that of Fig. 2(b) can be analyzed more precisely using conventional dependence analysis than by bounded symbolic execution (Sect. 4). In this section, we set up a novel form of product of abstractions, so as to benefit from this increase in precision. This notion of product is generic and does not require to fix a specific dependency abstraction. We refer to the final analysis presented in this section as RedSoundRSE.

Dependence Abstraction and Static Analysis. Although dependence abstractions may take many forms, they all characterize information flows that can be observed by comparing pairs of executions. For instance, [4] uses a lattice of security levels and abstract elements map each level to a set of variables. These variables are left unmodified when the input value of variables of higher levels change. Other works use relational abstract domains, where relational means that relations are maintained *across pairs of executions*. Therefore, we can characterize such analyses with an abstraction of pairs of stores:

Definition 7 (Dependence abstraction and analysis). *A dependence abstraction is defined by an abstract lattice* $\mathbb{D}$ *from security levels to variables and a concretization function*

$$\gamma_{\mathbb{D}} : \mathbb{D} \longrightarrow \mathcal{P}(\mathbb{M} \times \mathbb{M})$$
$$d \longmapsto \{(\mu_0, \mu_1) \in \mathbb{M} \times \mathbb{M} \mid \mu_0 =_{d(\mathbb{L})} \mu_1\}$$

A sound dependency analysis is defined by a function $[\![c]\!]_{\mathbb{D}}^{\sharp} : \mathbb{D} \to \mathbb{D}$ *such that, for all* $d \in \mathbb{D}$, $(\mu_0, \mu_1) \in \gamma_{\mathbb{D}}(d)$, $\{(\mu_0', \mu_1') \in \mathbb{M} \times \mathbb{M} \mid \forall i \in \{0, 1\}, (c, \mu_i) \to (\text{skip}, \mu_i')\} \subseteq \gamma_{\mathbb{D}} \circ [\![c]\!]_{\mathbb{D}}^{\sharp}(d)$.

Example 8 (Standard dependence based abstraction [4]). The abstraction of [4] is an instance of Definition 7. Let $\{\mathbb{L}, \mathbb{H}\}$ be the set of security levels. Assume an initial abstract state d that captures pairs of concrete stores that are low equal for some program (c, L). By applying the dependence analysis, if the final dependence state has a low dependency for each initially low variable, the program is noninterferent.

In practice such information is computed by forward abstract interpretation, using syntactic dependencies for expressions and conditions, and conservatively assuming conditions may generate (implicit) flows to any operation that they guard.

We note that Definition 7 accounts not only for dependence abstractions such as that of [4]. In particular, [22] proposes a semantic patch analysis which can also be applied to security properties by using a relational abstract domain to relate pairs of executions; such analyses use an abstraction that also writes as in Definition 7. In the following, we assume a sound dependence analysis is fixed.

Product of Symbolic Execution and Dependence Analysis. We now combine dependence analysis and symbolic execution. For most statements, SoundRSE rules defined in Fig. 6 introduce no imprecision. The notable exception is the case where the execution bound is reached as in rule SR-APPROX-MANY. Therefore, the principle of the combined analysis is to replace this imprecise rule with another that uses dependence analysis results to strengthen relational stores.

$$\text{step}(\texttt{while b do c},(\texttt{c; while b do c}),w) = (\texttt{ff},w')$$
$$d = [\![\texttt{while b do c}]\!]_{\mathbb{D}}^{\sharp}(\tau_{s\rightarrow\mathbb{D}}(\tilde{\rho}))   \tilde{\rho}'' = \mathfrak{modif}_{\mathbb{D}}(\tilde{\rho},\texttt{c},\lambda_{\mathbb{D}\rightarrow\mathbb{L}}(d))$$
$$\text{SR-APPROX-MANY-DEP} \frac{(\mathbf{b},\tilde{\rho}'')\vdash_{\text{sr}}\langle\beta_0,\beta_1\rangle   \pi' \triangleq \pi \wedge \neg\beta_0 \wedge \neg\beta_1}{(\texttt{while b do c},(\tilde{\rho},\pi),w,b) \rightharpoonup_{sr\times\mathbb{D}}(\texttt{skip},(\tilde{\rho}'',\pi'),w',\mathbf{ff})}$$

Fig. 7. RedSoundRSE: Symbolic execution approximation and product with dependence information.

First, we introduce two operations to transport information in a sound manner into and from the dependence abstract domain:

Definition 8 (Information translation and dependence abstraction). *The* translation from symbolic to dependence *is a function* $\tau_{s\rightarrow\mathbb{D}} : \overline{\mathbb{M}}_2 \rightarrow \mathbb{D}$ *that is sound in the following sense:* $\forall\tilde{\rho} \in \overline{\mathbb{M}}_2$, $\forall(\mu_0,\mu_1,\nu) \in \gamma_{\overline{\mathbb{M}}_2}(\tilde{\rho})$, $(\mu_0,\mu_1) \in \gamma_{\mathbb{D}} \circ \tau_{s\rightarrow\mathbb{D}}(\tilde{\rho})$. *The* extraction of dependence information *is a function* $\lambda_{\mathbb{D}\rightarrow\mathbb{L}} : \mathbb{D} \rightarrow \mathcal{P}(\mathbb{X})$ *that is sound in the following sense:* $\forall d \in \mathbb{D}$, $\forall(\mu_0,\mu_1) \in \gamma_{\mathbb{D}}(d)$, $\mu_0 =_{\lambda_{\mathbb{D}\rightarrow\mathbb{L}}(d)} \mu_1$

Intuitively, $\tau_{s\rightarrow\mathbb{D}}$ should compute a dependence abstract domain element that expresses a property implied by the relational symbolic store it is applied to. In the set-up of Example 8, a straightforward way to achieve that is to map $\tilde{\rho}$ to an element d that maps $\mathbb{L}$ to the set: $\{\mathtt{x} \in \mathbb{X} \mid \mathbf{may}(\Pi_0(\tilde{\rho})(\mathtt{x}) = \Pi_1(\tilde{\rho})(\mathtt{x}))\}$

When $\tilde{\rho}(\mathtt{x}) = \langle\varepsilon\rangle$, this equality is clearly satisfied; when $\tilde{\rho}(\mathtt{x}) = \langle\varepsilon_0 \mid \varepsilon_1\rangle$, the equality $\varepsilon_0 = \varepsilon_1$ needs to be discharged by an external tool such as an SMT solver. Similarly, the function $\lambda_{\mathbb{D}\rightarrow\mathbb{L}}$ extracts a set of variables which are proved to remain low by the its argument. In the setup of Example 8, this boils down to returning $d(\mathbb{L})$.

We now present the combined analysis. The symbolic execution step SR-APPROX-MANY-DEP is shown in Fig. 7 and replaces rule SR-APPROX-MANY (Fig. 6). When the execution bound is reached for a loop statement, it performs the dependence analysis of the whole loop from the dependence state derived by applying $\tau_{s\rightarrow\mathbb{D}}$ to the relational symbolic store. Then, it applies $\lambda_{\mathbb{D}\rightarrow\mathbb{L}}$ to derive the set of variables that are proved to be low by the dependence analysis. Finally, it computes a new relational symbolic store by modifying the variables according to the set of variables determined low:

- if variable $\mathtt{x}$ is low based on the $\lambda_{\mathbb{D}\rightarrow\mathbb{L}}$ output, $\mathfrak{modif}_{\mathbb{D}}$ synthesizes one fresh symbolic value x_{new} and maps it to $\langle x_{\text{new}}\rangle$;
- if variable $\mathtt{x}$ cannot be proved low, $\mathfrak{modif}_{\mathbb{D}}$ synthesizes two fresh symbolic values x_{new0}, x_{new1} and maps $\mathtt{x}$ to $\langle x_{\text{new0}} \mid x_{\text{new1}}\rangle$.

Remark 1 (Reduced product property). We stress the fact that the rule SR-APPROX-MANY may be applied multiple times during the analysis, essentially whenever a loop statement is analyzed, which is generally many times more than the number of loop commands in the program due to abstract iterations. Therefore, our analysis *cannot* be viewed as a fixed sequence of analyses. Such a decomposition (e.g., where dependence analysis is ran first and SE second) would be strictly less precise than our reduced product based approach.

```
1  if (priv > 0)        1  i = 0; w = 2;       1  i = 0;
2     i = 0;            2  x = 100;            2  while (i < 3) {      1  i = 0;
3  else                 3  while(i < x) {      3     y0 = y1;           2  while (i < 100) {
4     i = 0;            4     if (x <= 0)      4     y1 = y2;           3     if (priv > 0)
5  while (i < 10) {     5        w = priv;     5     y2 = priv;         4        y = 5;
6     i += 1;           6     i += 2;          6     i += 1;            5     i += 1;
7     priv += 5;        7     x += 1;          7  }                    6  }
8  }                    8  }                   8  y1 = 0; y2 = 0;
       (a) Secure              (b) Secure          (c) Insecure              (d) Insecure
```

Fig. 8. Programs illustrating different properties of the analyzer. Variable `priv` is high.

Soundness and Refutation Properties. Under the assumption that the dependence analysis and translation operations are sound, so is the combined symbolic execution, thus Theorem 4 still holds. Moreover, the refutation property of Theorem 5 also holds.

Example 9 (Combined analysis). We consider program 2(b). As discussed in Sect. 3, the loop statement may execute unboundedly many times, thus relational symbolic execution applies rule SR-APPROX-MANY-DEP. The initial dependence abstract element computed for the loop by $\tau_{s \rightarrow D}$ maps $\mathbb{L}$ to $\{i, z\}$ and $\mathbb{H}$ to all variables. The dependence analysis of the loop returns the same element. Thus, the set of low variables returned by $\lambda_{D \rightarrow L}$ is $\{i, z\}$, which allows to compute a precise relational symbolic store and to successfully verify the program is noninterferent.

8 Comparison

In this section we compare our analyses among them as well as with the dependency analysis of Assaf et al. [4]. To do so, we implemented prototypes of all the analyses. Our goal is not to evaluate the analyses in large code bases but to assess their differences based on programs that are small but challenging for typical noninteference analysers.

Implementation. We prototype the analyses proposed in this work as well as the dependency analysis, intervals and convex polyhedra analysis. The prototype is implemented in around 4k lines of OCaml code, using the Apron library [29] for the numerical domains and the Z3 SMT solver [21]. By defining a shared interface for SoundSE and RedSoundSE, the implementation of RedSoundRSE is parameterized by these. An artifact of the implementation has been provided.

Evaluation. We compare the 3 different relational techniques using different single-trace analyses by evaluating them on a set of challenging examples. Our results are shown in Table 2. In the following, we split NI programs from non NI ones. For the latter we look at the refutation capabilities of the analysis.

Table 2. Evaluation and comparison of analyses combination. $\mathbb{D}$ denotes the dependency analysis of [4]. Symbol $\checkmark$ (resp., $\times$) denotes a semantically correct (resp., incorrect) analysis outcome, with either a proof of security, a (possibly false) alarm, or a refutation model. For RedSoundSE columns, when the analyses succeed to prove NI, we mark the result with **I** (resp. **P**) to indicate that the intervals (resp. polyhedra) domain is being used.

Relational Analysis	$\mathbb{D}$	SoundRSE		RedSoundRSE ($\mathbb{D}$)	
relational analysis input:	None	SoundSE	RedSoundSE	SoundSE	RedSoundSE
Program	**Secure?**				
Fig. 2(a)	Yes	✗ False alarm	✓ Secure	✓ Secure (**I,P**)	✓ Secure	✓ Secure (**I,P**)
Fig. 2(b)	Yes	✓ Secure	✗ False alarm	✓ Secure (**P**)	✓ Secure	✓ Secure (**I,P**)
Fig. 2(d)	Yes	✗ False alarm	✗ False alarm	✓ Secure (**I,P**)	✗ False alarm	✓ Secure (**I,P**)
Fig. 8(a)	Yes	✗ False alarm	✗ False alarm	✗ False alarm	✓ Secure	✓ Secure (**I,P**)
Fig. 8(b)	Yes	✗ False alarm	✗ False alarm	✗ False alarm	✗ False alarm	✓ Secure (**I,P**)
Fig. 8(c)	No	✓ Alarm	✓ Refutation model	✓ Refutation model	✓ Refutation model	✓ Refutation model
Fig. 8(c)	No	✓ Alarm	✓ Refutation model	✓ Refutation model	✓ Refutation model	✓ Refutation model
Fig. 8(d)	No	✓ Alarm	✓ Alarm	✓ Alarm	✓ Alarm	✓ Alarm

Comparison of the Verification Capabilities of Different Relational Analyses. Programs of Fig. 2 were already explained in Sect. 3 and our prototype confirmed these results, which are summarized in Table 2.

In Program 8(a), the first condition renders dependence analysis useless as it will consider variable i high. This program will also fail to be verified by SoundRSE if the iteration bound is lower than 10: in this case, i will be assigned a fresh symbolic value and hence be deemed high. In contrast, RedSoundRSE can determine that the value of i in the loop does not depend on priv.

Program 8(b) is more convoluted. The analysis requires both numerical and dependence abstractions in order to prove its NI. The analysis will determine (conservatively) that three variables are modified in the loop: x, i and w. Dependence analysis can determine that variable i and x are low even if both are modified. However, since w depends on x, and the exact value of x is unknown, it is not possible to determine that w is low. By adding a numerical domain, it is easy to track that the value of x is always positive, which implies that the if statement can never be executed.

Comparison of the Refutation Capabilities of Different Relational Analyses. Since SoundRSE and RedSoundRSE unroll loops a bounded number of times, there are insecure programs for which a refutation model can be found, and programs where this is not possible. Notice that, to refute a program with a model, it is required that the symbolic execution did not perform any over approximation, i.e. that the precision flag is set to false when the analysis finds the violation. Therefore, the results for insecure programs of SoundRSE are similar to those of the different combinations that rely on symbolic execution, as reflected on Fig. 2. For Program 2(c), a valuation can be found by doing one iteration: $\nu(i_0) = \nu(i_1) = 1$ and $\nu(priv_0) = 0$, $\nu(priv_1) = 1$. For Program 8(c), a model can be found if the bound of iterations is set to 4 or higher. The valuation ν just needs to map variable priv to two different values: $\nu(priv_0) \neq \nu(priv_1)$.

In Program 8(d), for any user-set bound lower than 100 the execution will have to overapproximate, losing refutation capabilities.

Conclusion of the Evaluation. We have evaluated and compared our analyses among them and with the state-of-the-art on dependency analyses [4] on a set of 8 challenging examples. Our results show that, in contrast to dependencies [4], analyses inherit the capacity of providing a refutation model up to a bound from symbolic execution. Moreover, RedSoundRSE instantiated with RedSoundSE is capable of soundly verifying all the examples, in contrast to all the other compared analyses, as summarized in Table 2.

Limitations. As RedSoundSE is sound and automatic, it necessarily fails to achieve completeness (by Rice's Theorem [3,30]). In return, we provide completeness up to a bound. Another more subtle limitation is that the numerical abstraction are applied at the level of the single symbolic execution (RedSoundSE). This means that these abstractions cannot track down relations between executions, but just local constraints.

9 Related Work

Hyperproperties. Noninterference was first defined by Goguen and Meseguer [26], and also generalized to more powerful attacker models under the property name of declassification. We refer the reader to a survey on declassification policies [37] up to 2005. As discussed in the introduction, noninterference is not a safety property but a safety hyperproperty [13], a.k.a. hypersafety. Several works in the literature have shown that hypersafety verification can be reduced to verification of safety properties [7,13,20,39], however this reduction is not always efficient in practice [39]. In our work, we do not reduce noninterference to verification of safety but rather apply relational analyses. We only show our results using noninterference but the methodology can be easily generalized to more relaxed declassification properties, provided sound abstract domains exist.

Symbolic Execution. SE is a static analysis technique that was born in the 70s [9, 31] and that is now deployed in several popular testing tools, such as KLEE [11] and NASA's Symbolic PathFinder [35], to name a few. A primary goal and strength of SE is to find paths leading to counter-examples to generate concrete input values exercising that path. This is of particular importance to security in order to debug and confirm the feasibility of an attack when a vulnerability is detected.

Alatawi et al. [2] use AI to enhance the precision of a dynamic symbolic execution aimed at path coverage. Their approach consists of first doing an analysis of the program with AI to capture indirect dependences in order to enhance path predicates. Furthermore, their analysis does not maintain soundness (nor completeness). Meanwhile, our approach continuously alternates between abstract domains and symbolic execution, keeping soundness and completeness up to a bound. Lastly, Alatawi et al. [2] do not analyze relational properties such as noninterference but just safety properties.

We focus the rest of the related work on static analysis techniques for relational security properties: for a broader discussion on symbolic execution we refer the interested reader to a survey [10] up to 2011 and an illuminating discussion on SE challenges in practice up to 2013 [12].

Relational Symbolic Execution. In order to apply SE to security properties such as noninterference, Milushev et al. [32] propose a form of relational symbolic execution (RSE) to use KLEE to analyze noninterference by means of a technique called self-composition [7,20,39] to reduce a relational property of a program p to a safety property of a transformation of p. More recently, Daniel et al. have optimized RSE to be applicable to binary code to analyze relational properties such as constant time [17] and speculative constant time [18,19] and discovered violations of these properties in real-world cryptographic libraries. All these approaches are based on pure (relational) SE static techniques and, as such, they are not capable of recovering soundness beyond a fixed bound as in our case. The closest work to RedSoundRSE is RelSym [23] which supports interactive refutation, as well as soundness. In order to recover soundness, Chong et al. [23] propose to use RelSym on manually annotated programs with loop invariants. Precision of refutation is guaranteed only if the invariants are strong enough, which cannot be determined by the tool itself. Precision is not guaranteed in any other cases. In contrast, our invariants are automatically generated via AI and precision of refutation is always guaranteed up to a bound, which is automatically computed by our tool.

Sound Static Analyses for Hyperproperties. As discussed in the introduction, many sound verification methods have been proposed for relational security properties. We refer the reader to an excellent survey on this topic [36] up to 2003. After 2003, several sound (semi-) static verification methods of noninterference-like properties have been proposed by means of type systems (e.g. [6,24]), hybrid types, (e.g. [38]), relational logics (e.g. [1]), model checking (e.g. [5,27]), and pure AI [4]. We expand on the ones based on AI since they are the closest to our work. Giacobazzi and Mastroeni [25] define abstractions for attacker's views of program secrets and design sound automatic program analyses based on AI for sets of executions (in contrast to relational executions). Assaf et al. [4] are the first to express hyperproperties entirely within the framework of AI by defining a Galois connection that directly approximates the hyperproperty of interest. We utilize the abstract domain of Assaf et al. [4] combined with SE to obtain RedSoundSE. Notice that because the framework of Assaf et al. [4] relies on incomplete abstraction, their analysis is not capable of precise refutation nor provide refutations models. To the best of our knowledge, no previous work has combined abstract domains and SE to achieve soundness.

10 Conclusion

In this work, we propose a series of analyses, summarized in Fig. 1, combining SE and AI. Our analyses are sound, precise, and able to synthesize counter-examples up to a given bound. We prototype these analyses as well as several AI domains and a dependency analysis to verify noninterference. Our results, summarized in Table 2, show that on a set of challenging examples for noninterference, our analysis performs better than the dependency analysis and is able to precisely-blank and soundly conclude on whether programs are noninterferent or not and provide refutation models up to a bound. Given these encouraging results, we plan to generalize the target security property and make the analyses scale to other languages as future work.

Acknowledgements. The authors thank the anonymous reviewers for their comments, helpful for improving the paper. This project was funded by INRIA Challenge SPAI and by the VeriAMOS ANR Project. This research was partially supported by the ANR17- CE25-0014-01 CISC project We would also like to thank Josselin Giet and Adam Khayam for their observations.

A Trace of Program 2(d) with **RedSoundSE** Using Intervals

This section aims to show the execution of one symbolic trace of program 2(d). Initial precise store κ will capture the initial low-equality of variables i and y. The abstract state is a. Changes to the product store are marked in red.

$$\kappa = \begin{cases} \rho = [\mathtt{i} \rightarrow \langle i_0 \rangle, \mathtt{y} \rightarrow \langle y_0 \rangle, \mathtt{priv} \rightarrow \langle priv_0 \rangle] \\ \pi = \mathtt{tt} \\ a_l = [\,] \end{cases}$$

In line 3, since priv is unconstrained, the semantics can choose either path. Let us assume that our trace follows rule S-IF-T. Then, by line 5 the state is as follows.

$$\kappa = \begin{cases} \rho = [\mathtt{i} \rightarrow \langle i_0 \rangle, \mathtt{y} \rightarrow \langle y_0 \rangle, \mathtt{priv} \rightarrow \langle 0 \rangle] \\ \pi = priv_0 < 0 \\ a = [\mathtt{priv} = 0] \end{cases}$$

Since this loop has an unbounded amount of iterations, we know that an over approximation will happen. Let us assume that the iteration bound is 1 (meaning that the semantics will execute the loop once at most before over approximating), and that i < 10. By executing one full iteration the following symbolic state is reached.

$$\kappa = \begin{cases} \rho = [\mathtt{i} \rightarrow \langle i_0 + 1 \rangle, \mathtt{y} \rightarrow \langle y_0 \rangle, \mathtt{priv} \rightarrow \langle 2 \rangle] \\ \pi = i_0 < 10 \wedge priv_0 < 0 \\ a = [\mathtt{priv} = 0; \mathtt{i} < 11] \end{cases}$$

Since now the limit of iterations is reached, next step is over approximating the loop. For the example we will next show the state just before the reduction. Notice that the new constraints in π are the result of negating the guard.

$$\kappa = \begin{cases} \rho = [\mathtt{i} \to \langle i_1 \rangle, \mathtt{y} \to \langle y_0 \rangle, \mathtt{priv} \to \langle priv_1 \rangle] \\ \pi = i_1 \geq 10 \wedge i_0 < 10 \wedge priv_0 < 0 \\ a = [\mathtt{priv} \geq 2; \mathtt{i} = 10] \end{cases}$$

Because variables $\mathtt{i}$ and $\mathtt{priv}$ were modified, new symbolic values are assigned. This generates a big inaccuracy, but abstract states can compensate. By reducing we add the constraints of a to π.

$$\kappa = \begin{cases} \rho = [\mathtt{i} \to \langle i_1 \rangle, \mathtt{y} \to \langle y_0 \rangle, \mathtt{priv} \to \langle priv_1 \rangle] \\ \pi = priv_1 \geq 2 \wedge i_1 = 10 \wedge i_1 \geq 10 \wedge i_0 < 10 \wedge priv_0 < 0 \\ a = [\mathtt{priv} \geq 2; \mathtt{i} = 10] \end{cases}$$

Thanks to the reduction, we get information allowing for the low equality of $\mathtt{i}$ but also we get information about $\mathtt{priv}$ being positive. Finally, the last $\mathtt{if}$ statement will not be executed.

B   SE Step Relation

This section shows the full set of rules of SE, the standard not-sound symbolic execution.

$$\text{S-ASSIGN} \quad \frac{(\mathbf{e}, \rho) \vdash_s \varepsilon}{(\mathtt{x} := \mathbf{e}, (\rho, \pi)) \to_s (\mathtt{skip}, (\rho[\mathtt{x} \rightsquigarrow \langle \varepsilon \rangle], \pi))}$$

$$\text{S-SEQ-EXIT} \quad \frac{}{(\mathtt{skip};\ \mathbf{c}_1, \kappa) \to_s (\mathbf{c}_1, \kappa)} \qquad \text{S-SEQ} \quad \frac{(\mathbf{c}_0, \kappa) \to_s (\mathbf{c}_0', \kappa')}{(\mathbf{c}_0;\ \mathbf{c}_1, \kappa) \to_s (\mathbf{c}_0';\ \mathbf{c}_1, \kappa')}$$

$$\text{S-IF-T} \quad \frac{(\mathbf{b}, \rho) \vdash_s \beta \qquad \pi' \triangleq \pi \wedge \beta \qquad \mathbf{may}(\pi')}{(\mathtt{if}\ \mathbf{b}\ \mathtt{then}\ \mathbf{c}_0\ \mathtt{else}\ \mathbf{c}_1, (\rho, \pi)) \to_s (\mathbf{c}_0, (\rho, \pi))}$$

$$\text{S-IF-F} \quad \frac{(\mathbf{b}, \rho) \vdash_s \beta \qquad \pi' \triangleq \pi \wedge \neg\beta \qquad \mathbf{may}(\pi')}{(\mathtt{if}\ \mathbf{b}\ \mathtt{then}\ \mathbf{c}_0\ \mathtt{else}\ \mathbf{c}_1, (\rho, \pi)) \to_s (\mathbf{c}_1, (\rho, \pi))}$$

$$\text{S-LOOP-T} \quad \frac{(\mathbf{b}, \rho) \vdash_s \beta \qquad \pi' \triangleq \pi \wedge \beta \qquad \mathbf{may}(\pi')}{(\mathtt{while}\ \mathbf{b}\ \mathtt{do}\ \mathbf{c}, (\rho, \pi)) \to_s (\mathbf{c};\ \mathtt{while}\ \mathbf{b}\ \mathtt{do}\ \mathbf{c}, (\rho, \pi))}$$

$$\text{S-LOOP-F} \quad \frac{(\mathbf{b}, \rho) \vdash_s \beta \qquad \pi' \triangleq \pi \wedge \neg\beta \qquad \mathbf{may}(\pi')}{(\mathtt{while}\ \mathbf{b}\ \mathtt{do}\ \mathbf{c}, (\rho, \pi)) \to_s (\mathtt{skip}, (\rho, \pi))}$$

C   SoundSE Step Relation

This section shows the full set of rules of SoundSE by using SE, in Appendix B.

$$\text{S-NEXT} \quad \frac{(\mathbf{c}, \kappa) \to_s (\mathbf{c}', \kappa') \qquad \mathtt{step}(\mathbf{c}, \mathbf{c}', w) = (\mathbf{tt}, w')}{(\mathbf{c}, \kappa, w, b) \to_s (\mathbf{c}', \kappa', w', b)}$$

$$\text{S-APPROX-MANY} \quad \frac{(\mathbf{c}, \kappa) \to_s (\mathbf{c}', \kappa') \qquad \mathtt{step}(\mathbf{c}, \mathbf{c}', w) = (\mathbf{ff}, w') \qquad \rho'' = \mathfrak{modif}(\rho, \mathbf{c})}{(\mathbf{c}, (\rho, \pi), w, b) \to_s (\mathtt{skip}, (\rho'', \pi), w', \mathbf{ff})}$$

D Abstract Step Relation

This section shows the full set of rules of the abstract analysis used in Red-SoundSE.

$$\text{A-ASSIGN} \frac{a' \triangleq \mathfrak{assign}_{x,e}(a)}{(x := e, a) \rightarrow_A (\text{skip}, a')} \qquad \text{A-IF-T} \frac{a' \triangleq \mathfrak{guard}_b(a) \qquad a' \neq \bot}{(\text{if b then } c_0 \text{ else } c_1, a) \rightarrow_A (c_0, a')}$$

$$\text{A-IF-F} \frac{a' \triangleq \mathfrak{guard}_{\neg b}(a) \qquad a' \neq \bot}{(\text{if b then } c_0 \text{ else } c_1, a) \rightarrow_A (c_0, a')}$$

$$\text{A-SEQ-EXIT} \frac{}{(\text{skip}; c_1, a) \rightarrow_A (c_1, a)} \qquad \text{A-SEQ} \frac{(c_0, a) \rightarrow_A (c_0', a')}{(c_0; c_1, a) \rightarrow_A (c_0'; c_1, a')}$$

$$\text{A-LOOP-F} \frac{a' \triangleq \mathfrak{guard}_{\neg b}(a) \qquad a' \neq \bot}{(\text{while b do } c_0, a) \rightarrow_A (\text{skip}, a')}$$

$$\text{A-LOOP-T} \frac{a' \triangleq \mathfrak{guard}_b(a) \qquad a' \neq \bot}{(\text{while b do } c_0, a) \rightarrow_A (c; \text{ while b do } c_0, a')}$$

E RedSoundSE Step Relation

RedSoundSE is defined by rules of Appendix B, Appendix C and Appendix D.

$$\text{S-A-NEXT} \frac{\mathfrak{step}(c, c', w) = (\mathsf{tt}, w') \quad (c, a) \rightarrow_A (c', a') \quad (\kappa'', a'') \triangleq \mathfrak{reduction}(\kappa', a')}{(c, \kappa, a, w, b) \rightarrow_{s \times A} (c', \kappa'', a'', w', b)}$$

$$\text{S-A-APPROX-MANY} \frac{(c, \kappa) \rightarrow_s (c', \kappa') \quad \mathfrak{step}(c, c', w) = (\mathsf{ff}, w') \quad \kappa'' = \mathfrak{modif}(\kappa, c)}{(c, (\kappa, a), w, b) \rightarrow_{s \times A} (\text{skip}, (\kappa''', a'''), w', \mathsf{ff})}$$

F RSE and SoundRSE Step Relations

This section shows the full set of rules for SoundRSE. RSE is a subset of SoundRSE, by removing rule SR-APPROX-MANY, and removing the counter and boolean flag.

$$\text{SR-ASSIGN} \frac{(e, \tilde{\rho}) \vdash_{sr} \tilde{\varepsilon}}{(x := e, (\tilde{\rho}, \pi), w, b) \rightarrow_{sr} (\texttt{skip}, (\tilde{\rho}[x \rightsquigarrow \langle \tilde{\varepsilon} \rangle], \pi), w, b)}$$

$$\text{SR-SEQ-EXIT} \frac{}{(\texttt{skip};\ c_1, \tilde{\kappa}) \rightarrow_{sr} (c_1, \tilde{\kappa})} \qquad \text{SR-SEQ} \frac{(c_0, \tilde{\kappa}) \rightarrow_{sr} (c_0', \tilde{\kappa}')}{(c_0;\ c_1, \tilde{\kappa}) \rightarrow_{sr} (c_0';\ c_1, \tilde{\kappa}')}$$

$$\text{SR-IF-TT} \frac{(b, \tilde{\rho}) \vdash_{sr} \tilde{\beta} \qquad \pi' = \pi \wedge \Pi_0(\tilde{\beta}) \wedge \Pi_1(\tilde{\beta}) \qquad \textbf{may}(\pi')}{(\texttt{if b then } c_0 \texttt{ else } c_1, (\tilde{\rho}, \pi), w, b) \rightarrow_{sr} (c_0, (\tilde{\rho}, \pi'), w, b)}$$

$$\text{SR-IF-TF} \frac{(b, \tilde{\rho}) \vdash_{sr} \tilde{\beta} \qquad \pi' = \pi \wedge \Pi_0(\tilde{\beta}) \wedge \neg\Pi_1(\tilde{\beta}) \qquad \textbf{may}(\pi')}{(\texttt{if b then } c_0 \texttt{ else } c_1, (\tilde{\rho}, \pi), w, b) \rightarrow_{sr} (c_0 \bowtie c_1, (\tilde{\rho}, \pi'), w, b)}$$

$$\text{SR-IF-FT} \frac{(b, \tilde{\rho}) \vdash_{sr} \tilde{\beta} \qquad \pi' = \pi \wedge \neg\Pi_0(\tilde{\beta}) \wedge \Pi_1(\tilde{\beta}) \qquad \textbf{may}(\pi')}{(\texttt{if b then } c_0 \texttt{ else } c_1, (\tilde{\rho}, \pi), w, b) \rightarrow_{sr} (c_1 \bowtie c_0, (\tilde{\rho}, \pi'), w, b)}$$

$$\text{SR-IF-FF} \frac{(b, \tilde{\rho}) \vdash_{sr} \tilde{\beta} \qquad \pi' = \pi \wedge \neg\Pi_0(\tilde{\beta}) \wedge \neg\Pi_1(\tilde{\beta}) \qquad \textbf{may}(\pi')}{(\texttt{if b then } c_0 \texttt{ else } c_1, (\tilde{\rho}, \pi), w, b) \rightarrow_{sr} (c_0, (\tilde{\rho}, \pi'), w, b)}$$

$$\text{SR-LOOP-TT} \frac{\texttt{step}(c, c', w) = (\textbf{tt}, w') \qquad (b, \tilde{\rho}) \vdash_{sr} \tilde{\beta} \qquad \pi' = \pi \wedge \Pi_0(\tilde{\beta}) \wedge \Pi_1(\tilde{\beta}) \qquad \textbf{may}(\pi')}{(\texttt{while b do } c_0, (\tilde{\rho}, \pi), w, b) \rightarrow_{sr} (c_0;\ \texttt{while b do } c_0, (\tilde{\rho}, \pi), w', b)}$$

$$\text{SR-LOOP-TF} \frac{\texttt{step}(c, c', w) = (\textbf{tt}, w') \qquad (b, \tilde{\rho}) \vdash_{sr} \tilde{\beta} \qquad \pi' = \pi \wedge \Pi_0(\tilde{\beta}) \wedge \neg\Pi_1(\tilde{\beta}) \qquad \textbf{may}(\pi')}{(\texttt{while b do } c_0, (\tilde{\rho}, \pi), w, b) \rightarrow_{sr} ((c_0;\ \texttt{while b do } c_0) \bowtie \texttt{skip}, (\tilde{\rho}, \pi'), w', b)}$$

$$\text{SR-LOOP-FT} \frac{\texttt{step}(c, c', w) = (\textbf{tt}, w') \qquad (b, \tilde{\rho}) \vdash_{sr} \tilde{\beta} \qquad \pi' = \pi \wedge \neg\Pi_0(\tilde{\beta}) \wedge \Pi_1(\tilde{\beta}) \qquad \textbf{may}(\pi')}{(\texttt{while b do } c_0, (\tilde{\rho}, \pi), w, b) \rightarrow_{sr} (\texttt{skip} \bowtie (c_0;\ \texttt{while b do } c_0), (\tilde{\rho}, \pi'), w', b)}$$

$$\text{SR-LOOP-FF} \frac{\texttt{step}(c, c', w) = (\textbf{tt}, w') \qquad (b, \tilde{\rho}) \vdash_{sr} \tilde{\beta} \qquad \pi' = \pi \wedge \neg\Pi_0(\tilde{\beta}) \wedge \neg\Pi_1(\tilde{\beta}) \qquad \textbf{may}(\pi')}{(\texttt{while b do } c_0, (\tilde{\rho}, \pi), w, b) \rightarrow_{sr} (\texttt{skip}, (\tilde{\rho}, \pi'), w', b)}$$

$$\text{SR-EXIT} \frac{}{(\texttt{skip} \bowtie \texttt{skip}, (\tilde{\rho}, \pi), w, b) \rightarrow_{sr} (\texttt{skip}, (\tilde{\rho}, \pi), w, b)}$$

$$\text{SR-COMP-R} \frac{(c_1, (\Pi_1(\tilde{\rho}), \pi), w, b) \rightarrow_s (c_1', (\rho_1', \pi'), w', b')}{(\texttt{skip} \bowtie c_1, (\tilde{\rho}, \pi), w, b) \rightarrow_{sr} (\texttt{skip} \bowtie c_1', (\langle\!|\Pi_0(\tilde{\rho}) \mid \rho_1'|\!\rangle, \pi'), w', b')}$$

$$\text{SR-COMP-L} \frac{(c_0, (\Pi_0(\tilde{\rho}), \pi), w, b) \rightarrow_s (c_0', (\rho_0', \pi'), w', b')}{(c_0 \bowtie c_1, (\tilde{\rho}, \pi), w, b) \rightarrow_{sr} (c_0' \bowtie c_1, (\langle\!|\rho_0' \mid \Pi_1(\tilde{\rho})|\!\rangle, \pi'), w', b')}$$

$$\text{SR-APPROX-MANY} \frac{\begin{array}{c}\texttt{step}(\texttt{while b do } c, (c;\ \texttt{while b do } c), w) = (\textbf{ff}, w') \\ \tilde{\rho}'' = \mathfrak{modif}(\tilde{\rho}, c) \qquad (b, \tilde{\rho}'') \vdash_{sr} \langle \beta_0, \beta_1 \rangle \qquad \pi' \triangleq \pi \wedge \neg\beta_0 \wedge \neg\beta_1\end{array}}{(\texttt{while b do } c, (\tilde{\rho}, \pi), w, b) \rightarrow_{sr} (\texttt{skip}, (\tilde{\rho}'', \pi'), w', \textbf{ff})}$$

G   RedSoundRSE Step Relation

RedSoundRSE is defined by rules of Appendix F plus rule SR-APPROX-MANY-DEP.

$$\text{SR-APPROX-MANY-DEP} \frac{\begin{array}{c}\texttt{step}(\texttt{while b do } c, (c;\ \texttt{while b do } c), w) = (\textbf{ff}, w') \\ d = [\![\texttt{while b do } c]\!]_{\mathbb{D}}^{\sharp}(\tau_{s \rightarrow \mathbb{D}}(\tilde{\rho})) \\ \tilde{\rho}'' = \mathfrak{modif}_{\mathbb{D}}(\tilde{\rho}, c, \lambda_{\mathbb{D} \rightarrow \mathbb{L}}(d)) \\ (b, \tilde{\rho}'') \vdash_{sr} \langle \beta_0, \beta_1 \rangle \qquad \pi' \triangleq \pi \wedge \neg\beta_0 \wedge \neg\beta_1\end{array}}{(\texttt{while b do } c, (\tilde{\rho}, \pi), w, b) \rightarrow_{sr \times \mathbb{D}} (\texttt{skip}, (\tilde{\rho}'', \pi'), w', \textbf{ff})}$$

References

1. Aguirre, A., Barthe, G., Gaboardi, M., Garg, D., Strub, P.Y. : A relational logic for higher-order programs. Proc. ACM Program. Lang. 1(ICFP), 1–29 (2017)
2. Alatawi, E., Søndergaard, H., Miller, T.: Leveraging abstract interpretation for efficient dynamic symbolic execution. In: Rosu, G., Penta, M.D., Nguyen, T.N., (eds.), Proceedings of the 32nd IEEE/ACM International Conference on Automated Software Engineering, ASE 2017, Urbana, IL, USA, 30 October - 03 November 2017, pp. 619–624. IEEE Computer Society (2017)
3. Asperti, A., Armentano, C.: A page in number theory. J. Formaliz. Reason. 1(1), 1–23 (2008)
4. Assaf, M., Naumann, D. A., Signoles, J., Totel, É., Tronel, F.: Hypercollecting semantics and its application to static analysis of information flow. In: Symposium on Principles of Programming Languages (POPL), pp. 874–887. ACM (2017)
5. Backes, M., Köpf, B., Rybalchenko, A.: Automatic discovery and quantification of information leaks. In: 30th IEEE Symposium on Security and Privacy (S&P 2009), 17–20 May 2009, Oakland, California, USA, pp. 141–153 (2009)
6. Banerjee, A., Naumann, D.A., Rosenberg, S.: Expressive declassification policies and modular static enforcement. In: IEEE Symposium on Security and Privacy (S&P 2008), 18–21 May 2008, Oakland, California, USA. IEEE Computer Society (2008)
7. Barthe, G., D'Argenio, P.R., Rezk, T.: Secure information flow by self-composition. In: Proceedings of the IEEE Computer Security Foundations Workshop (CSF) vol. 17, pp. 100–114 (2004)
8. Bielova, N., Rezk, T.: A taxonomy of information flow monitors. In: Piessens, F., Viganò, L. (eds.) POST 2016. LNCS, vol. 9635, pp. 46–67. Springer, Heidelberg (2016). https://doi.org/10.1007/978-3-662-49635-0_3
9. Boyer, R.S., Elspas, B., Levitt, K.N.: SELECT - a formal system for testing and debugging programs by symbolic execution. In: Proceedings of the International Conference on Reliable Software 1975, Los Angeles, California, USA, 21–23 April 1975, pp. 234–245. ACM (1975)
10. Cadar, C., Godefroid, P., Khurshid, S., Pasareanu, C.S., Sen, K., Tillmann, N., Visser, W.: Symbolic execution for software testing in practice: preliminary assessment. In: Proceedings of the 33rd International Conference on Software Engineering, ICSE 2011, Waikiki, Honolulu, HI, USA, 21–28 May 2011, pp. 1066–1071. ACM (2011)
11. Cadar, C., Nowack, M.: KLEE symbolic execution engine in 2019. Int. J. Softw. Tools Technol. Transf. (2021)
12. Cadar, C., Sen, K.: Symbolic execution for software testing: three decades later. Commun. ACM 56(2), 82–90 (2013)
13. Clarkson, M.R., Schneider, F.B.: Hyperproperties. In: Proceedings of the IEEE Computer Security Foundations Symposium (CSF), pp. 51–65. IEEE (2008)
14. Cousot, P., Cousot, R.: Abstract interpretation: a unified lattice model for static analysis of programs by construction or approximation of fixpoints. In: Symposium on Principles of Programming Languages (POPL), pp. 238–252. ACM (1977)
15. Cousot, P., Cousot, R.: Systematic design of program analysis frameworks. In: Symposium on Principles of Programming Languages (POPL). ACM (1979)
16. Cousot, P., Halbwachs, N.: Automatic discovery of linear restraints among variables of a program. In: Symposium on Principles of Programming Languages (POPL), pp. 84–97. ACM (1978)

17. Daniel, L.A., Bardin, S., Rezk, T.: Binsec/rel: efficient relational symbolic execution for constant-time at binary-level. In: 2020 IEEE Symposium on Security and Privacy, SP 2020, San Francisco, CA, USA, 18–21 May 2020, pp. 1021–1038 (2020)

18. Daniel, L., Bardin, S., Rezk, T.: Hunting the haunter - efficient relational symbolic execution for spectre with haunted relse. In: 28th Annual Network and Distributed System Security Symposium, NDSS 2021, virtually, 21–25 February 2021. The Internet Society (2021)

19. Daniel, L., Bardin, S., Rezk, T.: Reflections on the experimental evaluation of a binary-level symbolic analyzer for spectre. In: Post-proceedings of the LASER@NDSS 2021. The Internet Society (2022)

20. Darvas, Á., Hähnle, R., Sands, D.: A theorem proving approach to analysis of secure information flow. In: Hutter, D., Ullmann, M. (eds.) SPC 2005. LNCS, vol. 3450, pp. 193–209. Springer, Heidelberg (2005). https://doi.org/10.1007/978-3-540-32004-3_20

21. de Moura, L., Bjørner, N.: Z3: an efficient SMT solver. In: Ramakrishnan, C.R., Rehof, J. (eds.) TACAS 2008. LNCS, vol. 4963, pp. 337–340. Springer, Heidelberg (2008). https://doi.org/10.1007/978-3-540-78800-3_24

22. Delmas, D., Miné, A.: Analysis of software patches using numerical abstract interpretation. In: Chang, B.-Y.E. (ed.) SAS 2019. LNCS, vol. 11822, pp. 225–246. Springer, Cham (2019). https://doi.org/10.1007/978-3-030-32304-2_12

23. Farina, G.P., Chong, S., Gaboardi, M.: Relational symbolic execution. In: Komendantskaya, E., editor, Proceedings of the 21st International Symposium on Principles and Practice of Programming Languages, PPDP 2019, Porto, Portugal, 7–9 October 2019, pp. 10:1–10:14. ACM (2019)

24. Fournet, C., Planul, J., Rezk, T.: Information-flow types for homomorphic encryptions. In: Chen, Y., Danezis, G., Shmatikov, V., (eds.), Proceedings of the 18th ACM Conference on Computer and Communications Security, CCS 2011, Chicago, Illinois, USA, 17–21 October 2011, pp. 351–360 (2011)

25. Giacobazzi, R., Mastroeni, I.: Abstract non-interference: parameterizing non-interference by abstract interpretation. In: Proceedings of the 31st ACM SIGPLAN-SIGACT Symposium on Principles of Programming Languages, POPL 2004, Venice, Italy, 14–16 January 2004, pp. 186–197. ACM (2004)

26. Goguen, J.A., Meseguer, J.: Security policies and security models. In: IEEE Symposium on Security and Privacy, Oakland, pp. 11–20. IEEE Computer Society (1982)

27. Huisman, M., Worah, P., Sunesen, K.: A temporal logic characterisation of observational determinism. In: 19th IEEE Computer Security Foundations Workshop (CSFW 2006) (2006)

28. Hunt, S., Sands, D.: On flow-sensitive security types. In: Symposium on Principles of Programming Languages (POPL), pp. 79–90. ACM (2006)

29. Jeannet, B., Miné, A.: APRON: a library of numerical abstract domains for static analysis. In: Bouajjani, A., Maler, O. (eds.) CAV 2009. LNCS, vol. 5643, pp. 661–667. Springer, Heidelberg (2009). https://doi.org/10.1007/978-3-642-02658-4_52

30. Rogers, H., Jr.: Theory of Recursive Functions and Effective Computability (Reprint from 1967). MIT Press, Cambridge (1987)

31. King, J.C.: Symbolic execution and program testing. Commun. ACM **19**(7), 385–394 (1976)

32. Milushev, D., Beck, W., Clarke, D.: Noninterference via symbolic execution. In: Giese, H., Rosu, G. (eds.) FMOODS/FORTE -2012. LNCS, vol. 7273, pp. 152–168. Springer, Heidelberg (2012). https://doi.org/10.1007/978-3-642-30793-5_10

33. Ngo, M., Bielova, N., Flanagan, C., Rezk, T., Russo, A., Schmitz, T.: A better facet of dynamic information flow control. In: Champin, P., Gandon, F., Lalmas, M., Ipeirotis, P.G., (eds.), Companion of the The Web Conference 2018 on The Web Conference 2018, WWW 2018, Lyon, France, 23–27 April 2018, pp. 731–739 (2018)

34. Palikareva, H., Kuchta, T., Cadar, C.: Shadow of a doubt: testing for divergences between software versions. In: Proceedings of the 38th International Conference on Software Engineering, ICSE 2016, Austin, TX, USA, 14–22 May 2016, pp. 1181–1192. ACM (2016)

35. Pasareanu, C.S., Mehlitz, P.C., Bushnell, D.H., Gundy-Burlet, K., Lowry, M.R., Person, S., Pape., M.: Combining unit-level symbolic execution and system-level concrete execution for testing NASA software. In: Ryder, B.G., Zeller, A., (eds.), Proceedings of the ACM/SIGSOFT International Symposium on Software Testing and Analysis, ISSTA 2008, Seattle, WA, USA, 20–24 July 2008, pp. 15–26. ACM (2008)

36. Sabelfeld, A., Myers, A.C.: Language-based information-flow security. IEEE J. Sel. Areas Commun. **21**(1), 5–19 (2003)

37. Sabelfeld, A., Sands, D.: Dimensions and principles of declassification. In: 18th IEEE Computer Security Foundations Workshop, (CSFW-18 2005), 20–22 June 2005, Aix-en-Provence, France, pp. 255–269. IEEE Computer Society (2005)

38. Fragoso Santos, J., Jensen, T., Rezk, T., Schmitt, A.: Hybrid typing of secure information flow in a Javascript-like language. In: Ganty, P., Loreti, M. (eds.) TGC 2015. LNCS, vol. 9533, pp. 63–78. Springer, Cham (2016). https://doi.org/10.1007/978-3-319-28766-9_5

39. Terauchi, T., Aiken, A.: Secure information flow as a safety problem. In: Hankin, C., Siveroni, I. (eds.) SAS 2005. LNCS, vol. 3672, pp. 352–367. Springer, Heidelberg (2005). https://doi.org/10.1007/11547662_24

Result Invalidation for Incremental Modular Analyses

Jens Van der Plas$^{(\boxtimes)}$ ⓘ, Quentin Stiévenart ⓘ, and Coen De Roover ⓘ

Software Languages Lab, Vrije Universiteit Brussel, Brussels, Belgium
{jens.van.der.plas,quentin.stievenart,coen.de.roover}@vub.be

Abstract. To reduce the running time of static analysis tools upon program changes, incremental static analyses reuse and update pre-existing results. Such analyses must efficiently detect and remove outdated results. We introduce three novel, complementary result invalidation strategies for incremental modular analyses. The core idea of our work is to alternate invalidation with computation. We apply our strategies to a recent, state-of-the-art incremental modular analysis that suffers from imprecision, and evaluate them on soundness, precision, and performance. Our strategies lead to precision improvements compared to an incremental analysis without invalidation, though the precision of a full reanalysis is not yet matched. On most benchmarks, our incremental analysis performs well. However, on some benchmarks our analysis performs poorly as the changes drastically change program behaviour, for which the changes are difficult for an incremental analysis to handle.

Keywords: Static program analysis · Incremental program analysis · Modular program analysis

1 Introduction

Static analysis is an approach to computing properties of programs without running them. It is the foundation of code smell, bug, and vulnerability detection tools (e.g., [14,15,21,28,35]) used in modern software engineering processes such as continuous integration pipelines [31]. An analysis that is fast in the presence of small and frequent code changes can even be incorporated into a development environment. To meet these demands, incremental static analyses have been proposed [2,8,13,24,27]. Given the results of an *initial analysis*, an *incremental analysis* updates the results given the code changes. The goal of an incremental analysis is to produce results faster than a *full reanalysis* by reusing and updating previous results.

Recently, Van der Plas et al. [18] introduced a general approach to rendering any modular static analysis incremental. Modular analyses divide a program into parts which are (re-)analysed separately but whose analyses may be interdependent. The authors posit that modularity facilitates bounding the impact of changes. While the evaluation shows that incremental updates are often faster

C. Dragoi et al. (Eds.): VMCAI 2023, LNCS 13881, pp. 296–319, 2023.
https://doi.org/10.1007/978-3-031-24950-1_14

than a full reanalysis, incremental updates may be less precise than a full reanalysis as the presented analysis cannot delete outdated results. In this paper, we improve upon the approach by Van der Plas et al. [18] as follows:

- We introduce three complementary strategies to regain lost precision. The idea is to *interleave* invalidation with recomputation, to maximise reuse of previously computed results. Our strategies can be applied to modular static analyses that employ global-store widening and infer dependencies amongst components.
- We implemented these strategies and evaluate their impact on the precision and performance of the incremental analysis, when used alone or in combination.

2 Background

We now introduce modular static analysis, following a recent formulation by Nicolay et al. [16]. We obtain an incremental version of this formulation by applying the incrementalisation approach by Van der Plas et al. [18].

2.1 Modular Static Analysis

A modular static analysis [5] divides a program into static parts, e.g., function definitions, referred to as *modules*. A module may have multiple runtime instantiations, e.g., function calls, which the analysis might discern as well. We refer to their reification in the analysis as *components*. A component consists of a module and a context used to discern the different instantiations. Depending on the definition of contexts, more instantiations may be discerned, increasing the analysis precision (and complexity). A modular analysis analyses its components in isolation. The analysis of one component may however trigger the (re-)analysis of another. The remainder of this paper focuses on function-modular analyses. All examples use a lattice representing each value as a set of its possible types, and empty contexts, i.e., every module will correspond to at most one component.

Effect-Driven Modular Static Analysis. Recently, ModF, an effect-driven formulation of function-modular analysis has been introduced [16]. ModF is a control-flow analysis also computing value information. It *reifies* the *computational dependencies* between components and uses these to drive a fixed-point computation, alternating between an *inter-component analysis*, scheduling components for analysis, and an *intra-component analysis*, analysing individual components. The inter-component analysis, referred to as INTER and shown in Algorithm 1, uses a worklist of components to be analysed. Initially, this worklist contains a MAIN component[1], representing the program's entry point (line 1). Every analysis step removes a component from the worklist (lines 6–7) and analyses it (line 8); the analysis terminates when the worklist is empty.

---

[1] In formalisms, lowercase Greek letters denote components (e.g., α and β). Otherwise, we denote them by their corresponding name in small caps (e.g., MAIN and FIB).

Algorithm 1: The inter-component analysis (INTER) of ModF.

```
 1  WL := {Main}; // The work list, initially containing the MAIN component.
 2  V := ∅; // The visited set.
 3  D := λr.∅; // Set of dependencies (read effects).
 4  σ := λa.⊥; // Global value store, initially all addresses map to bottom.
 5  while WL ≠ ∅ do
 6  │   α ∈ WL;
 7  │   WL := WL \ {α};
 8  │   (C,' R', W', σ') = intra(α, σ); // Intra-component analysis.
 9  │   σ := σ';
10  │   V := V ∪ {α};
11  │   WL := WL ∪ (C' \ V);
12  │   foreach r ∈ R' do D := D[r ↦ D(r) ∪ {α}];
13  │   foreach w ∈ W' do WL := WL ∪ D(w);
14  end
15  return (σ, V, D);
```

The *store*, mapping abstract addresses to abstract values, abstractly represents the heap. ModF uses global-store widening [30], i.e., there is a single global value store σ within the analysis [16]. For every component, σ contains an abstract return value. Upon a function call, ModF does not step into the function, but retrieves the stored return value (or ⊥ if no value had been stored).

A component's analysis returns a set of *effects* reifying its computational dependencies, together with an updated store (line 8). Dependencies are function calls (generating *call effects*) and reads/writes in the store (generating resp. *read/write effects* – the latter is only generated when σ actually changes).[2] These effects are used to determine the component(s) to be added to the worklist, causing components depending on updated information to be reanalysed.

A ModF analysis results in (line 15): (1) the store σ, (2) the set of components created, and (3) the set of dependencies (read effects). We consider all parts of the result equally relevant, though in practice one might only be interested in σ.

Example. We illustrate how ModF computes the control-flow and value information of the Scheme[3] program in Listing 1. ModF analyses it as follows (omitting some effects for brevity):

1. The analysis starts with MAIN. Binding x generates a write effect for this variable. Then, a call effect to fun is generated, and the corresponding component, FUN, is added to the worklist. As no return value had been computed for FUN, ⊥ is read from the store; a read effect on this return value is registered.
2. FUN is analysed, producing a call effect for inc and read effects for x and for the return value of INC. The new component INC is added to the worklist,

---

[2] For brevity, in pseudocode, the set C represents the set of all components corresponding to the emitted call effects, and the sets R and W represent the addresses corresponding to the emitted read and write effects respectively.

[3] In this work, we use Scheme, a dynamically-typed dialect of Lisp with support for higher-order functions. Its dynamic nature makes it difficult to analyse as control and data flow are intertwined, precluding the computation of a call graph ahead of time. Scheme is representative for a whole class of languages such as JavaScript.

```
1   (define x 0)                            ; Definition of a variable x.
2   (define (fun) (inc) x)                  ; Function that reads x.
3   (define (inc) (set! x (+ x 1)) #t)      ; Function that reads and writes x.
4   (fun)
```

Listing 1. Example Scheme program of two functions.

as is MAIN because FUN's return value is updated to `Int`, generating a write effect.

3. Either MAIN or INC can now be analysed. Assuming INC is analysed (the order does not affect the result [16]), INC reads x, generating a read effect, and also writes to this variable. As the value in the store is not updated, no write effect is generated. As the return value of INC is updated to `Bool`, a write effect is generated and FUN is added to the worklist again.

4. The analysis continues until the worklist is empty.

The principle of effect-driven flow analysis is applicable to different module granularities, e.g., thread-modular analyses [22], and can be used with any abstract domain without infinite ascending chains and with any context-sensitivity.

The Component Graph. The analysis of a component generates call effects, each corresponding to a component discovered by the analysis. After the analysis of a component α, INTER collects the set of components *called* by α, denoted C_α. This gives rise to a cyclic directed graph, the *component graph*, representing how components are created: for every component $\beta \in C_\alpha$ there is an edge from α to β. Figure 1 depicts the component graph from previous example.

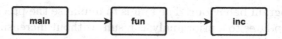

Fig. 1. The component graph corresponding to the analysis of the program in Listing 1: inc is called from fun, which is called from the program's entry point.

2.2   Incremental Modular Static Analysis

Van der Plas et al. [18] present an approach to rendering an effect-driven modular static analysis incremental. It requires the analysed program to be annotated with *change expressions*, which are akin to the patch annotations of Palikareva et al. [17]. A change expression specifies how a given expression is updated. Its first argument represents the original expression; its second argument represents the expression that replaces the original. Change expressions can be added manually, or be inserted by a change distiller (e.g., [6,7]) or change logger (e.g., [10,12,32]). In the following function, the predicate is updated from (= n 0) to (< n 2):

```
1   (define (factorial n)
2       (if (<change> (= n 0) (< n 2))
3           n
4           (* n (factorial (- n 1)))))
```

For a given set of change expressions, Van der Plas et al. [18] compute the affected analysis results and update them accordingly. Their analysis tracks which change expressions within the source code of a module were encountered during the analysis of the corresponding components. Every component whose analysis encountered a change expression is considered to be *directly affected*. If an expression in a module changes, only the components that encountered this expression during their analysis are affected. All directly affected components are added to the worklist and the fixed-point computation is restarted. The modular analysis design ensures that indirectly affected components are reanalysed too.

Sources of Imprecision. Table 1 shows the three parts of the result of a ModF analysis. The approach by Van der Plas et al. [18] only updates prior results monotonically: no outdated information can be removed; the result of the analysis over-approximates the behaviour of both the updated and original program. All parts of the result may suffer from imprecision, as shown in Table 1. This means that components and dependencies no longer representing the program's behaviour cannot be removed. In σ, values cannot become more precise. Imprecision in one part of the result may cause imprecision in other parts. E.g., when a value in σ is imprecise, the analysis may explore more paths and thus infer more components and dependencies, which may in turn degrade the store's precision.

3   Strategies for Precision Recovery

We now introduce three complementary strategies that improve the precision of an incremental analysis result by invalidating the information that corresponds to outdated program behaviour. The aim is to minimise the precision loss caused by monotonic updates to a prior analysis result, without increasing analysis time.

3.1   Invalidation Principle

The presented strategies treat the intra-component analysis as a black box and do not put any restrictions on the lattice nor on the context-sensitivity used by the analysis. The intra-component analysis must only compute a set of effects.

The aim is to invalidate as few valid results as possible, so that results not impacted by a change need not be needlessly recomputed. Related work [2,13] often consist of an *invalidation phase*, which *over-approximates* and clears outdated results, and a *recomputation phase*, which updates the analysis results. To avoid over-approximating outdated results, we *interleave* invalidation with recomputation, maximising reuse. After an intra-component analysis, INTER computes which parts of the results have become obsolete and removes them; information is only removed when it is no longer computed by an intra-component analysis. Mapping this onto Algorithm 1, invalidation happens after line 8. Our approach leads to a recompute-and-invalidate cycle: the analysis of a component may lead to a result invalidation, which in turn can lead to more analyses of components.

Table 1. Overview of the parts of the analysis result, of the sources of imprecision for each part, and of the corresponding strategies to invalidate outdated results.

COMPONENTS	
Explanation	Set of components created during the analysis, each abstractly representing an aspect of the runtime behaviour of the program, e.g., a function call
Imprecision	Components no longer representing the program's behaviour cannot be removed
Solution	COMPONENT INVALIDATION (CI): remove components that are no longer created

DEPENDENCIES	
Explanation	Set of inter-component dependencies (read effects) computed during the analysis, each marking a link between a component and an address in the global value store σ. Using these dependencies, the analysis of one component takes into account information computed by the analysis of other components
Imprecision	Dependencies that are no longer valid cannot be removed
Solutions	DEPENDENCY INVALIDATION (DI): remove dependencies that are no longer computed by the reanalysis of an impacted component
	CI: removing a component clears its dependencies

VALUE STORE σ	
Explanation	Over-approximates the heap. Mapping of abstract addresses to abstract values
Imprecision	Values in σ are updated monotonically, since they are joined upon updates
Solutions	WRITE INVALIDATION (WI): improve the precision of values in the store σ by removing values that are no longer written
	CI: when WI is enabled, the removal of a component may allow σ to be refined.

Table 1 outlines the developed strategies, one for each part of the analysis result: component invalidation, dependency invalidation, and write invalidation. Though, invalidations in one part of the result may impact the other parts.

3.2   Component Invalidation (CI)

Component invalidation (CI) removes components from the analysis result that are no longer created by any other component, plus the dependencies related to these components. Consider e.g., the program in Listing 2. The initial analysis creates four components, shown by the component graph on top of Fig. 2. The change expression replaces the call to `fac-loop` by a call to `fac`; `fac-loop` (and transitively `loop`) are no longer called. The reanalysis of MAIN now finds that FAC-LOOP is no longer called: FAC-LOOP and LOOP can both be removed.

```
1   (define (fac n)
2     (if (< n 2)
3         n
4         (* n (fac (- n 1)))))
5   (define (fac-loop n) ; Executes the 'fac' function in a loop.
6     (define (loop i)
7       (if (< i n)
8           (begin
9             (display (fac i))
10            (display " ")
11            (loop (+ i 1)))))
12    (loop 0))
13  (<change> (fac-loop 10) (fac 10)) ; Updated to call 'fac' directly.
```

Listing 2. A change causing components to be removed.

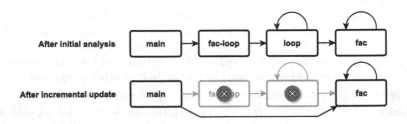

Fig. 2. ModF components for the program in Listing 2. On top, the components after the initial analysis of the program; at the bottom, the components after the incremental update. Arrows depict generated call effects.

CI uses the component graph to detect outdated components: all components no longer transitively reachable from MAIN, i.e., the entry point of the program, can be removed. Algorithm 2 extends INTER with CI. For every component α, INTER caches C_α, the set of components called by α's last analysis, using a cache $\mathbb{C}$. The set of dependencies R_α, cached in $\mathbb{R}$, allows the efficient removal of the dependencies of deleted components ($\mathbb{R}$ holds the same information as D but in the reverse order, avoiding a full traversal of D). After the analysis of a component α, the set of components called by the analysis of this component, C'_α, is returned. INTER then retrieves C_α, the set of components called during the *previous* analysis of α, and updates the cache $\mathbb{C}$ (lines 12–13). It then computes the set containing all components that are no longer called by α. If this set is non-empty, one or more edges were removed from the component graph and some components may have become outdated (line 14). In this case, the transitive closure of $\mathbb{C}$ is computed, starting from MAIN; all components that are not part of it are removed (lines 15–16). All dependencies of these components are removed too, avoiding the existence of dependencies to non-existent components. The transitive closure is needed because a component can only be removed if it is no longer created by any other component. Finally, $\mathbb{R}$ is updated (line 18). Note that lines 15 and 16 will never be executed during the initial analysis of the program. To avoid the needless but possibly expensive computation of set differences in the condition, we first check whether an incremental update is taking place (line 14). For similar reasons, we do the same for DI and WI.

Algorithm 2: INTER extended with component invalidation (in blue) and dependency invalidation (in purple).

```
    // Assumes the existence of a cache for the sets C, ℂ, initialised as ℂ := λα.∅ before
      the initial analysis, and the existence of a cache for the sets R, ℝ, initialised
      as ℝ := λα.∅ before the initial analysis.
 1  Function deleteComponent(β) is
 2  |    foreach r ∈ ℝ(β) do D := D[r ↦ D(r) \ {β}]; // Delete dependencies.
    |    // Remove β from all data structures.
 3  |    V := V \ {β}; WL := WL \ {β}; ℝ := ℝ \ {β}; ℂ := ℂ \ {β};
 4  end
 5  while WL ≠ ∅ do
 6  |    ... // Ditto Alg. 1.
 7  |    foreach w ∈ W' do WL := WL ∪ D(w);
 8  |    if incremental update then
 9  |    |    R := ℝ(α);
10  |    |    foreach r ∈ (R \ R') do  D := D[r ↦ D(r) \ {α}];
11  |    end
12  |    C := ℂ(α);
13  |    ℂ := ℂ[α ↦ C']; // Update ℂ immediately to use the updated C'.
14  |    if incremental update and C \ C' ≠ ∅ then
15  |    |    reachable := ℂ(MAIN) ∪ {β|γ ∈ reachable ∧ β ∈ ℂ(γ)};
16  |    |    foreach β ∈ (V \ reachable) do  deleteComponent(β);
17  |    end
18  |    ℝ := ℝ[α ↦ R']; // Both for component invalidation and dependency invalidation.
19  end
20  return (σ, V, D);
```

3.3   Dependency Invalidation (DI)

The second strategy, *dependency invalidation* (DI), removes outdated dependencies. This ensures that components are not spuriously reanalysed. Consider, e.g., the program in Listing 3. Initially, READ has a dependency on a_x. During the incremental update, the analysis of READ will find a new dependency on a_y, whilst the dependency on a_x can be removed.

Algorithm 2 also extends INTER with DI. The set of dependencies computed during the last analysis of every component α, R_α, is cached using the cache $\mathbb{R}$ (also used by CI). After the (re-)analysis of a component α, INTER collects the computed dependencies, R'_α. It then fetches the dependencies computed during the *previous* analysis of α from $\mathbb{R}$ and computes the set of outdated dependencies which are then removed (lines 9–10). Finally, as for CI, $\mathbb{R}$ is updated (line 18).

```
1  (define x 1)
2  (define y 2)
3  (define (write) (<change> (set! x 7) (set! y 7)))
4  (define (read) (<change> x y))
5  (read)
6  (write)
```

Listing 3. Example program with changing dependencies. Initially READ has a dependency on the address of variable x, a_x. In the new version of the program, READ solely has a dependency on a_y, the address of variable y.

```
1  (define (fromBool b)
2    (if b
3        (<change> 'aSymbol        "aString")
4        (<change> 'anotherSymbol "anotherString")))
5  (define x (fromBool (some-complicated-predicate)))
6  (display x)
```

Listing 4. Example program. Initially, x only holds a symbol, whereas after the update it can only contain a string.

3.4 Write Invalidation (WI)

Write invalidation (WI) aims to increase the precision of abstract values in the store. It is motivated by Listing 4. Variable x is changed from storing symbols to strings. A strong update would *overwrite* the abstract value Symbol by String in σ. A monotonic update instead joins the values together, resulting into the less precise value {Symbol, String}. Clearly, a strong update is desired.

The values in σ are part of an abstract domain, forming a complete lattice. The *higher* a value resides in the lattice, the less precise information it represents. WI aims to *lower* all values as much as possible by monitoring the values computed for every address in σ, and by lowering values that no longer correspond to the program's behaviour. We first describe the required monitoring.

Provenance Tracking. Values in σ result from one or more writes, each monotonically updating the value. In this process, the analysis loses information w.r.t. the *constituents* and origins of the values. E.g., when α writes 1 to a and β writes -1 to a, $\sigma(a)$ contains {Int}, without information about the values written by α and β, nor about which components wrote these values. We introduce *provenance tracking* to regain this information. For every component and address in $dom(\sigma)$, the analysis maintains the *contribution* of the component to the address, i.e., the join of all values written to the address during the analysis of the component. This requires intercepting to *write* operations to the store.

Consider the case in Fig. 3: components α and β read and write two variables, x and y: both write y, α reads x, and β reads y. When α writes Int to y and β writes Boolean to y, σ holds join of these values, {Int, Boolean}, for y.

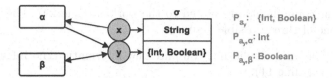

Fig. 3. Interaction of intra-component analyses with variables and their values in σ illustrated. On the right, the provenance and contributions of a_y are shown.

During the analysis of a component α, we track, for each written address a, the join of all values written to that address. We call this joined value the *contribution* of α to a, denoted $P_{a,\alpha}$. For every address, the contributions of all components are cached. We call this cache the *provenance* of the address, P_a. We define the *provenance value* of an address a as the join of all values in its provenance. Figure 3 depicts this information on the right in grey.

Non-monotonic Store Updates. The intra-component analyses perform all updates monotonically. INTER thus has to restore precision after it has been lost. Provenance tracking enables WI to perform non-monotonic updates to σ, improving its precision. This is possible when a previously-written address is no longer written by a component, and when the contribution of a component to an address changes in a non-monotonic way.[4] The code for WI is shown in Algorithm 3.

Outdated Writes. The analysis of a component tracks all addresses written to. For every component α, INTER caches this set, W_α, using a cache $\mathbb{W}$. After the analysis of a component α, INTER collects the set of written addresses, W'_α, and computes the set containing all addresses previously written by the component that are no longer written (line 26). Finally, the cache $\mathbb{W}$ is updated (line 28). When the contribution $P_{a,\alpha}$ of α to an address a is removed, its provenance value, no longer influenced by $P_{a,\alpha}$, is used as the new value for the address (lines 2–3). If the provenance value equals the value at $\sigma(a)$, deletion is completed. Else, the provenance value replaces the value σ. All dependent components are scheduled for reanalysis (line 5), allowing the new value to be taken into account during their reanalysis, possibly leading to further refinements of the result. When an address is no longer written by any component, all information in the analysis' data structures related to this address can be removed (line 6).

More Precise Writes. After every intra-component analysis, INTER compares the contribution of the component for every written address, to the corresponding contribution computed by the component's previous analysis. Based on this comparison, the value at the given address in σ may be updated, in which case

---

[4] Conceptually, the first case corresponds to the second case for which the contribution of the component to an address has become $\bot$. We treat it separately since no write to the address is performed any more.

all dependent components are added to the worklist (line 29). The comparison may yield one of three possible results:

$P_{a,\alpha} = P'_{a,\alpha}$ The analysis did not compute new information, no information can be discarded (line 11).

$P_{a,\alpha} \sqsubseteq P'_{a,\alpha}$ The update is monotonic, no information can hence be discarded. The updated contribution is stored (line 12).

$P_{a,\alpha} \not\sqsubseteq P'_{a,\alpha}$ The contribution changes non-monotonically. The value for a can be replaced by the new provenance value (computed on line 14), now taking into account the updated contribution $P'_{a,\alpha}$ (stored in $\mathbb{P}$ on line 12).

The second and third case may not lead to an update of σ as the value computed on line 14 can be the same as the value already in σ. Only when the new value is different, dependent components need to be scheduled for reanalysis.

Reinforcing Component Invalidation. Section 3.2 introduced CI. However, CI does not allow for the removal of information from σ: values written by removed components cannot be deleted, a limitation that can be remedied by combining CI with WI. When a component α is removed, all addresses in the set $\mathbb{W}(\alpha)$ are treated as outdated writes, described in Sect. 3.4. This allows σ to become more precise, which may in turn invoke the analysis of other components. The updated code for component deletion is shown in Algorithm 4.

4   Evaluation

We evaluated the presented strategies to answer the following research questions:

RQ1 How well do the three invalidation strategies improve the precision of the analysis, both when applied individually and when applied in combination?

RQ2 What is the impact of the invalidation strategies on the time needed to perform an incremental update?

RQ3 How much does the incremental analysis reduce the analysis time compared to a full reanalysis of the program?

We tested soundness of the initial analysis and the incremental update experimentally (1) by ensuring that the analysis over-approximates multiple runs of a concrete interpreter [1,29], and (2) by comparing the incremental analysis results to the results of a non-incremental analysis. We performed these tests for a thread-modular analysis for a concurrent Scheme, for a function-modular analysis, for all possible combinations of the invalidation strategies, and for a constant propagation and a type abstract domain; no unsound results were encountered.

Algorithm 3: INTER extended with write invalidation (in teal).

```
    // Assumes the existence of a cache for the sets W, 𝕎, initialised as 𝕎 := λα.∅
        before the initial analysis, and the existence of a cache ℙ, the provenance,
        initialised to ℙ := λa.(λα.⊥) before the initial analysis.
 1  Function deleteContribution(α, a) is
 2  │    ℙ := ℙ[a ↦ (ℙ(a) \ {α})];
 3  │    v := ⊔_{β∈dom(ℙ(a))} ℙ(a)(β);
 4  │    if v ≠ σ(a) then
 5  │    │    WL := WL ∪ D(a);
         │    │    // If an address is no longer written by any component, it is deleted.
         │    │        Otherwise, the store is updated.
 6  │    │    if ℙ(a) = ∅ then  σ := σ \ {a}; ℙ := ℙ \ {a}; D := D \ {a};  else σ := σ[a ↦ v];
 7  │    end
 8  end
    // updateAddressIncremental compares the new contribution v' of α to a to the previous
        contribution v, and improves the store if possible.
 9  Function updateAddressIncremental(α, a, v') is
10  │    v := ℙ(a)(α); // Previous contribution of α to a, P_{a,α}.
11  │    if v = v' then return false; // Identical contribution: no precision gain.
12  │    ℙ := ℙ[a ↦ (ℙ(a)[α ↦ v'])];
13  │    old := σ(a);
14  │    new := if v ⊑ v' then old ⊔ v' else ⊔_{β∈dom(ℙ(a))} ℙ(a)(β);
15  │    if old = new then return false;
16  │    σ := σ[a ↦ new]; // Update the store.
17  │    return true;
18  end
19  while WL ≠ ∅ do
20  │    ... // Ditto Alg. 1.
21  │    σ := σ'; // This line can now be omitted.
22  │    ... // Ditto Alg. 1.
23  │    foreach w ∈ W' do WL := WL ∪ D(w); // This line can now be omitted.
24  │    if incremental update then
25  │    │    W := 𝕎(α);
26  │    │    foreach w ∈ (W \ W') do deleteContribution(α, w);
27  │    end
28  │    𝕎 := 𝕎[α ↦ W'];
         │    // P computed during the intra-component analysis. P maps every written address
         │        to the join of all values written to it during the component's analysis.
29  │    foreach (a, v) ∈ P do  if updateAddressIncremental(α, a, v) then WL := WL ∪ D(α);
30  end
31  return (σ, V, D);
```

4.1   Experimental Design

Our evaluation uses a context-insensitive ModF analysis for Scheme, with a
LIFO-ordered worklist and a product lattice[5]. We implemented our contributions
in the open-source MAF framework[6] [29]. Our evaluation is run on a 2015 Dell

---

[5] The lattice represents primitive values by their possible types, except booleans which
are represented as their respective value when possible. Pointers are represented as
sets of addresses (in $dom(\sigma)$); closures and primitives are represented using sets as
well. A join of two values is the pointwise join of the corresponding elements of the
product, where the join of two sets is their union.

[6] A repository containing our implementation can be found online: https://github.
com/softwarelanguageslab/maf (branch **incremental-experiments**).

Algorithm 4: deleteComponent (in blue) reinforced with WI (in teal).

```
1  Function deleteComponent(β) is
2      foreach r ∈ ℝ(β) do D := D[r ↦ D(r) \ {β}]; // Delete dependencies.
       // Remove β from all data structures.
3      V := V \ {β}; WL := WL \ {β}; ℝ := ℝ \ {β}; ℂ := ℂ \ {β};
4      W := 𝕎(β);
5      forall w ∈ W do deleteContribution(β, w);
6      𝕎 := 𝕎 \ {β};
7  end
```

PowerEdge R730[7] running OpenJDK 1.8.0_312 and Scala 3.1.0. The JVM was given a maximum of 32GB RAM, and all analyses used a timeout of 30 min.

To evaluate the precision of the incremental update (RQ1), we inspect the store σ at the end of the analysis. For each address, we measure the precision of the incremental update by comparing its value to its counterpart in the store of a full reanalysis. The proportion of addresses in the final store that contain values equally or less precise than the values obtained by a full reanalysis shows us how much precision can still be improved. We also compare to the store resulting from an incremental analysis without result invalidation. Here, the proportion of addresses in the final store that contain values equally or more precise than the values obtained by an incremental update without invalidation shows us how many addresses have an improved precision thanks to our strategies. We perform these comparisons for all possible combinations of the invalidation strategies.

To evaluate the performance of our strategies (RQ2 & RQ3), we measure the time needed to (1) analyse the initial program, (2) fully analyse the updated program, and (3) perform the incremental update given a set of enabled strategies. For (1) and (2), no strategy is enabled; the analysis will not maintain the caches required by any strategy. For (3), the initial analysis initialises all caches used by the strategies. Each measurement is repeated 15 times preceded by a warm-up of 3 repetitions or of maximally 30 min. Garbage collection is forced prior to each analysis.

Comparing the precision and performance of an incremental update using all strategies to (1) an (imprecise but fast) update without invalidation, (2) an update using only one or two strategies, and (3) a (precise but slow) full reanalysis, allows us to investigate a trade-off between precision and performance.

Benchmarking Suites. Our evaluation uses two benchmarking suites.[8] Each benchmark program is a Scheme program containing real-world code, annotated with change expressions. As such, a benchmark corresponds to program changes.

Curated Benchmarks. We curated a suite of 32 programs to which we manually added changes resembling possible developer edits, shown in Table 2. The

---

[7] The computer has 2 Intel Xeon 2637 processors and 256 GB of RAM.
[8] In our online repository, the curated benchmarks can be found in the folders /test/changes/scheme and /test/changes/scheme/reinforcingcycles. The generated benchmarks can be found in the folder /test/changes/scheme/generated.

programs originate from different sources, e.g., a university course with programming exercises in Scheme, together with the solutions for solving particular exercises, and benchmarking suites used by other researchers. Example edits include changing representations of data structures (e.g., replacing lists by vectors in nbody-processed), or updating a meta-interpreter (e.g., adding the ability to make variables immutable in freeze or making procedures dynamically scoped in mceval-dynamic). In programs like slip-0-to-1, slip-1-to-2, and slip-2-to-3, edits convert the program to a later version. A new abstraction is introduced and used throughout peval. Some edits were constructed to be tricky for an incremental update to process accurately, as they trigger *cyclic reinforcement of lattice values* [23,24] (see Sect. 4.2). Also, certain programs contain the same changes but use a different granularity of change expressions; this is e.g., the case for multiple-dwelling (coarse) and multiple-dwelling (fine), and for satFine, satMiddle, and satCoarse. The runtimes of the initial analyses of programs the curated suite vary from 0 s to 117 s.

Table 2. The curated suite, retrieved from various sources. For every benchmark, we list the lines of code as counted with cloc and the number of change expressions.

Benchmark	LOC	#Chg	Benchmark	LOC	#Chg
baseline	6	1	primtest	43	11
browse	164	1	cycleCreation	3	1
collatz	18	1	higher-order-paths1	4	2
fact	5	1	higher-order-paths2	4	1
fib-loop	15	1	implicit-paths	3	1
fib	5	2	ring-rotate	32	2
freeze	327	11	sat	16	4
gcipd	9	2	satCoarse	17	1
leval	379	11	satFine	13	3
matrix	617	3	satMiddle	16	3
mceval-dynamic	246	4	satRem	20	2
multiple-dwelling (coarse)	434	1	slip-0-to-1	123	6
multiple-dwelling (fine)	404	3	slip-1-to-2	117	3
nbody-processed	1252	10	slip-2-to-3	397	9
nboyer	636	2	tab-inc	317	3
peval	507	38	tab	307	3

Generated Benchmarks. We automatically generated 5 mutations for each of 190 programs, originating from various sources, obtaining 950 programs. We use a set of edit patterns of one or more change expressions that are inserted randomly, with a certain probability and at an arbitrary depth in the program. We consider the following patterns: expression deletion (7.5%), inserting a random sub-expression (5%), swapping expressions (10%), wrapping an expression with a call to the identity function (7.5%), negating the predicate of an if (7.5%), and swapping the branches of an if (7.5%). A valid mutation has at least one edit, is unique, and does not lead to an error after running it with a Scheme interpreter for one minute. The runtimes of the initial analyses of programs the generated suite vary from 0 s to 148 s, most programs complete in under 10 s.

4.2 Precision Evaluation (RQ1)

We evaluate the precision improvement caused by our invalidation strategies as follows. On every benchmark program, and for all possible configurations, we count the percentage of addresses in σ that is less precise than a full reanalysis. Figure 4 depicts the results of our precision evaluation. These allow us to (1) evaluate the precision improvement caused by the application of the presented strategies, and (2) to see whether additional opportunities for precision improvement are possible. As a precision improvement of σ can only be expected when WI is enabled, we only show results for an incremental update without result invalidation, with WI, and with all strategies enabled (where CI reinforces WI).

Precision Improvements over Naive Incremental Analysis. For the curated suite, in some cases such as higher-order-paths1, we observe a big precision improvement. On other programs, the improvement remains minor. fib-loop shows that reinforcing CI can lead to additional precision improvements. On benchmarks such as browse and nbody-processed, the benefit is smaller, though browse now reaches full precision. Unexpectedly, and only on slip-0-to-1, reinforcement decreases precision (this is not visible on the figure). The reason for this seems to be that, although sound, the obtained fixed-point depends on the analysis order of the components. On the generated suite, the number of imprecise values in the store is reduced by 15%–20% on average (geometric mean over all generated benchmarks): there is an improvement of about 10% with WI and an additional improvement of about 10% using all strategies.

Table 3 shows the quartiles of the distribution of the store's precision among all benchmarks in the generated suite for the same configurations. Without invalidation, more than 50% of all benchmark programs do not achieve full precision. However, using all strategies, the analysis reaches full precision on most benchmarks. The table shows the added benefit of reinforcing CI.

Remaining Imprecision in the Analysis Result. Figure 4 also shows remaining possibilities for precision improvement. On 13 curated benchmarks for which the incremental update without invalidation did not achieve full precision, the update with all strategies now does (indicated by a bar reaching 100%). However, on other benchmarks, more improvements remain possible.

The precision of σ influences the control flow explored by the analysis, and so the number of components and dependencies: precision gains due to WI can lead to the invalidation of components and dependencies when all strategies are

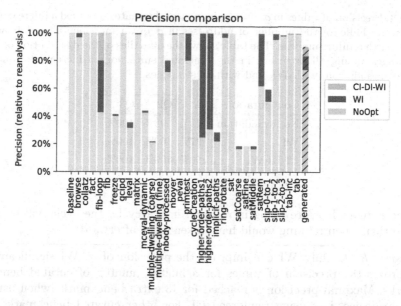

Fig. 4. Precision of values in σ after an incremental update compared to a full reanalysis. Bars represent the percentage of addresses in σ of an incremental update whose values match a full reanalysis. In grey, precision of an incremental update without invalidation is shown. In dark green, the additional percentage of matching addresses due to WI is shown. In light green, the further additional percentage of matching addresses using all strategies is shown. The rightmost bar shows the geometric mean of all benchmarks in the generated suite. (Color figure online)

enabled. Of course, CI and DI can also be beneficial in without WI, though only WI can propagate precision gains to other components.

The imprecision in σ is worsened by our change representation: change expressions always require an old and new expression. For example, to introduce a new variable in a program, a placeholder value for the old program needs to be used, e.g., #f (false): (define x (<change> #f 10)). As this value will reside in σ and cannot be removed by the incremental update when WI is not enabled, some values in σ may be artificially imprecise. However, imprecision still remains for some benchmarks when WI is enabled. One reason we found is *cyclic reinforcement of lattice values* [23,24], which arises when, due to the abstractions in the analysis, the computation of a value at an address is influenced by the value at that address itself, thereby influencing its own provenance.[9] WI cannot restore the precision of values in such a cycle. We also believe that this phenomenon causes the result to depend on the exploration order, e.g., when

---

[9] Some programs in our curated suite, such as cycleCreation and implicit-paths, are explicitly created to contain this behaviour.

Table 3. Precision of values in σ after an incremental update compared a full reanalysis. Percentages indicate the number of addresses in σ of an incremental update whose values match a full reanalysis. The table shows the quartiles of the distribution of these percentages among all programs in the generated suite, for an incremental analysis without invalidation, with WI, and with all strategies.

Configuration	Q1	Q2	Q3
No invalidation	73%	98%	100%
WI	97%	100%	100%
CI-DI-WI	100%	100%	100%

a value is refined before being introduced into a cycle, the cycle will be more precise than when refining would have taken place afterwards.

> *Answer RQ1.* Only WI can improve the precision of σ. WI significantly improves the precision of values for a limited number of curated benchmarks. Maximal precision is reached for 13 extra benchmarks when using all strategies, i.e., using reinforced CI. For other curated benchmarks, a large percentage of addresses remains less precise. We also observe a big improvement on the generated suite, though several addresses still remain imprecise. Once again, the combination of CI and WI leads to a substantial additional precision improvement.

4.3   Performance w.r.t. No Invalidation (RQ2)

Figure 5 shows the results of the performance evaluation for RQ2. Times are shown relative to an incremental update without invalidation. CI and DI do not cause a significant slowdown of the incremental analysis. A slowdown appears when using WI, but, overall, the incremental update remains faster than a full reanalysis (see Sect. 4.4). This slowdown can be explained as follows. As WI refines σ, updates may trigger the reanalysis components, leading to further reanalyses and impacting performance. On the curated benchmarks, this increase in running time is more moderate for the combination of CI and WI.

CI and WI combined reduce, in some cases, the analysis time as outdated components are not analysed anymore. Also, WI may create more opportunities for CI: when values become more refined, this may lead to more outdated components, which may in turn lead to an improvement of values in σ.

> *Answer RQ2.* CI and DI have no substantial negative impact on the running time of an incremental update. Only WI causes a slowdown: as WI regains precision, changes to σ may cause components to be scheduled for reanalysis.

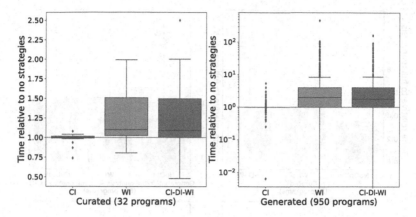

Fig. 5. Analysis time of the incremental update relative to an incremental update without invalidation. Benchmarks for which the incremental update completed in 0 ms are omitted in the graphs, because a relative time cannot be computed.

4.4 Performance w.r.t. Full Reanalysis (RQ3)

Figure 6 shows the results of our performance evaluation for RQ3. Times are shown relative to the time needed by a full reanalysis.

For the curated suite, overall, the incremental update is faster than a full reanalysis. The medians are consistently under 0.2, meaning that on more than half of the benchmark programs, the incremental update is more than 5 times faster. When both CI and WI are used, we see one outlier which corresponds to the `primtest` benchmark for which the running times are very low, meaning that there is no opportunity for the incremental analysis to gain time.

The results of the generated suite are grouped based on the time taken by the initial analysis and the full reanalysis. The slowdown caused by WI is most outspoken for short-running generated benchmark programs, where the overhead of the strategies may be relatively high. When both the initial analysis and full reanalysis complete in under a second, and when both analyses run a second or longer, overall, the incremental update remains faster than a full reanalysis. Although WI may cause minor slowdowns, the incremental update remains more than 10× faster compared to a full reanalysis. On programs that have an initial analysis taking a second or more but a shorter full reanalysis, the incremental update is slower: for almost all configurations, the incremental analysis takes at least as long as a full reanalysis for most benchmarks, with median slowdowns of up to 100 and outliers showing slowdowns larger than 1000. It is difficult to pinpoint the exact root cause for each performance difference. We list several possible reasons that may explain this behaviour:

- The change representation may cause less result reuse. In our implementation, change expressions cannot be placed at all program points. Some changes must be represented with coarse-grained change expressions. E.g., to rename a function parameter, the change expression must wrap around the entire

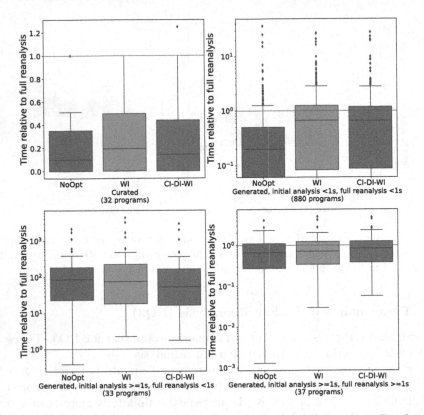

Fig. 6. Analysis time of the incremental update relative to a full reanalysis. Benchmarks for which the full reanalysis completed in 0 ms are counted but omitted in the graphs because a relative time cannot be computed.

function definition, thus components corresponding to the function cannot be reused.

- The generated programs may contain too many changes, leading to many impacted components: 25 programs have over 30 changes and 79 programs have over 20 changes. On almost half of the programs, more than 20% of the components is directly affected. As many components are affected, the incremental analysis may not benefit from its modularity to bound the impact of the changes.
- Changes may significantly alter program behaviour. 33 benchmarks had a long-running initial analysis and short-running full reanalysis. In these cases, the incremental analysis performs very poorly. It is possible that the randomly inserted changes prune away a lot of program functionality, leading to a very fast reanalysis, whereas an incremental update needs to propagate information deletion. Although we haven't verified the behaviour of all benchmarks individually, the reduced running time of the full analysis indicates that in

these cases, an incremental update is inadequate due to the nature of the program changes.
- No dedicated worklist algorithm is used. Components may be scheduled for analysis due to newly inferred information or due to invalidation, but neither is prioritised. By intertwining recomputation by invalidation, information may be added or removed in an unspecified order; information may be removed that is later readded, or vice versa. We assume that the analysis of components in an unordered way may negatively impact the analysis performance.

To improve performance, future work should consider imposing an order on the worklist. It may also be useful to investigate heuristics to determine which changes would better be processed by a full reanalysis, e.g., when a program update leads to a big removal of program functionality.

> *Answer RQ3.* On the curated suite, on the short-running generated benchmarks, and on the generated benchmarks with a long-running initial analysis and reanalysis, overall, the incremental update is faster. Yet, on the generated benchmarks with a long-running initial analysis but with a short-running full reanalysis, almost all incremental updates are slower. The nature of the changes may be to blame for this: a very fast reanalysis may indicate a serious reduction in program behaviour, in which case the incremental update has to invalidate many results, causing high relative runtimes.

5 Related Work

Nichols et al. [13] introduce *fixpoint reuse* to incrementally analyse JavaScript programs. They map program points to corresponding program points in the new program, allowing reuse of analysis results for mapped points. The mapping function plays a key role: more mapped points lead to more reuse and a faster analysis, but incorrect matches can cause the analysis to lose precision.

IncA [24–27] is a Datalog-based analysis framework that produces the same results as a full reanalysis. It uses an incremental Datalog with a semi-naïve, stratified evaluation strategy [25]. For every tuple, a *support count* indicating the number different derivations of the tuple is maintained and used to invalidate tuples after program updates. Contrary to IncA, our approach does not require programs and analyses to be converted into a Datalog-like representation.

Andromeda [28] is an incremental, demand-driven taint analysis. Its relies on a *support graph* to find taint facts that are outdated. In contrast, our analysis is not tailored to a specific client analysis. Saha and Ramakrishnan [20] also use support graphs in their framework for implementing incremental, demand-driven analyses. They require analyses and programs under analysis to be specified as Horn clauses and represent changes by means of the addition or deletion of facts.

Reviser [2] is an incremental, inter-procedural data-flow analysis for analyses expressible in the IDE or IFDS frameworks. Its results match a full program analysis but it requires a static call-graph; dynamic languages are unsupported.

Our approach is not limited to specific analyses and does not require a static call graph. Other incremental approaches relying on static call graphs comprise a.o. alias analyses [33], interval analyses [3], dataflow analyses [4,19,34], and analyses tailored to specific client tools, such as race detection [35]. Liu et al. [11] present an incremental points-to analysis not requiring a prebuilt call graph. It preserves precision but is limited to flow-insensitive analyses, unlike ours.

Garcia-Contreras et al. [8] present a context-sensitive incremental modular analysis which achieves incrementality at the inter-modular and intra-modular level. The analysis requires an encoding of the program in constrained Horn clauses. Contrary to ours, the analysis does not divide the program into modules itself and does not use components but a programmer-defined lexical module partitioning is used, but it is claimed that any partitioning is possible. Thus, their analysis can, e.g., not be used with thread-modular analyses, in contrast to ours. Later work [9] presents an updated approach, also capable of handling external modules, together with a formal description and a further evaluation.

6   Conclusion

We presented three complementary invalidation strategies to improve the precision and performance of the incremental modular analysis approach presented by Van der Plas et al. [18]. Our approach interleaves reanalysis of components with invalidation. Component invalidation removes outdated components and their dependencies, and, when combined with write invalidation, can also improve the precision of the values in the store σ. Dependency invalidation removes outdated dependencies. Write invalidation uses provenance tracking to retract and replace outdated contributions from components to σ, enabling non-monotonic updates.

We tested our strategies for unsoundness and evaluated their precision and performance empirically on real-world programs using a small suite of 32 programs with possible developer edits and a large corpus 950 of programs with generated edits. Our strategies allow the incremental analysis to reach the same result as a full reanalysis on 13 more programs in the curated suite in comparison to when none of the proposed strategies is used. On other programs, the precision loss is reduced, yet the results did not match the precision of a full reanalysis. For the generated suite, using all strategies, on average, the number of less precise addresses in σ is reduced from 30% to about 10%. The best improvements were realised by the combination of write invalidation with component invalidation.

Performance-wise, overall, the incremental analysis scores well. We did find some benchmarks with particular program changes for which the incremental update proved to be slower than a full reanalysis, e.g., in 33 of the 950 programs in the generated suite where the changes removed a big part of a program's functionality. Future work includes handling cyclic reinforcement of lattice values, stratifying the worklist of the analyses, and investigating heuristics for triggering a full reanalysis rather than an incremental update.

Acknowledgements. This work was partially supported by the Research Foundation – Flanders (FWO) (grant number 11F4822N) and by the *Cybersecurity Initiative Flanders.*

References

1. Andreasen, E.S., Møller, A., Nielsen, B.B.: Systematic approaches for increasing soundness and precision of static analyzer. In: Proceedings of the 6th ACM SIG-PLAN International Workshop on State of the Art in Program Analysis, SOAP 2017, pp. 31–36. Association for Computing Machinery, New York (2017). https://doi.org/10.1145/3088515.3088521

2. Arzt, S., Bodden, E.: Reviser: efficiently updating IDE-/IFDS-based data-flow analyses in response to incremental program changes. In: Jalote, P., Briand, L.C., van der Hoek, A. (eds.) Proceedings of the 36th International Conference on Software Engineering, ICSE 2014, Hyderabad, India, 31 May–07 June 2014, pp. 288–298. ACM Press, New York (2014). https://doi.org/10.1145/2568225.2568243

3. Burke, M.G.: An interval-based approach to exhaustive and incremental interprocedural data-flow analysis. ACM Trans. Program. Lang. Syst. **12**(3), 341–395 (1990). https://doi.org/10.1145/78969.78963

4. Carroll, M.D., Ryder, B.G.: Incremental data flow analysis via dominator and attribute updates. In: Ferrante, J., Mager, P. (eds.) Conference Record of the Fifteenth Annual ACM Symposium on Principles of Programming Languages, San Diego, California, USA, 10–13 January 1988, pp. 274–284. ACM Press (1988). https://doi.org/10.1145/73560.73584

5. Cousot, P., Cousot, R.: Modular static program analysis. In: Horspool, R.N. (ed.) CC 2002. LNCS, vol. 2304, pp. 159–179. Springer, Heidelberg (2002). https://doi.org/10.1007/3-540-45937-5_13

6. Falleri, J., Morandat, F., Blanc, X., Martinez, M., Monperrus, M.: Fine-grained and accurate source code differencing. In: Crnkovic, I., Chechik, M., Grünbacher, P. (eds.) ACM/IEEE International Conference on Automated Software Engineering, ASE 2014, Vasteras, Sweden, 15–19 September 2014, pp. 313–324. ACM, New York (2014). https://doi.org/10.1145/2642937.2642982

7. Gall, H.C., Fluri, B., Pinzger, M.: Change analysis with evolizer and changedistiller. IEEE Softw. **26**(1), 26–33 (2009). https://doi.org/10.1109/MS.2009.6

8. Garcia-Contreras, I., Caballero, J.F.M., Hermenegildo, M.V.: An Approach to Incremental and Modular Context-Sensitive Analysis (2018). http://oa.upm.es/53067/

9. Garcia-Contreras, I., Morales, J.F., Hermenegildo, M.V.: Incremental and modular context-sensitive analysis. Theory Pract. Logic Program. **21**(2), 196–243 (2021). https://doi.org/10.1017/S1471068420000496

10. Hattori, L., Lanza, M.: Syde: a tool for collaborative software development. In: Proceedings of the 32nd ACM/IEEE International Conference on Software Engineering, ICSE 2010, vol. 2, p. 235–238. Association for Computing Machinery, New York (2010). https://doi.org/10.1145/1810295.1810339

11. Liu, B., Huang, J., Rauchwerger, L.: Rethinking incremental and parallel pointer analysis. ACM Trans. Program. Lang. Syst. **41**(1), 6:1–6:31 (2019)

12. Negara, S., Vakilian, M., Chen, N., Johnson, R.E., Dig, D.: Is it dangerous to use version control histories to study source code evolution? In: Noble, J. (ed.) ECOOP 2012. LNCS, vol. 7313, pp. 79–103. Springer, Heidelberg (2012). https://doi.org/10.1007/978-3-642-31057-7_5

13. Nichols, L., Emre, M., Hardekopf, B.: Fixpoint reuse for incremental JavaScript analysis. In: Grech, N., Lavoie, T. (eds.) Proceedings of the 8th ACM SIGPLAN International Workshop on State of the Art in Program Analysis, SOAP@PLDI 2019, Phoenix, AZ, USA, 22 June 2019, pp. 2–7. ACM (2019). https://doi.org/10.1145/3315568.3329964
14. Nicolay, J., Noguera, C., De Roover, C., De Meuter, W.: Determining dynamic coupling in JavaScript using object type inference. In: 2013 IEEE 13th International Working Conference on Source Code Analysis and Manipulation (SCAM), pp. 126–135. IEEE (2013)
15. Nicolay, J., Stiévenart, Q., De Meuter, W., De Roover, C.: Purity analysis for JavaScript through abstract interpretation. J. Softw.: Evol. Process **29**(12), e1889 (2017)
16. Nicolay, J., Stiévenart, Q., De Meuter, W., De Roover, C.: Effect-driven flow analysis. In: Enea, C., Piskac, R. (eds.) VMCAI 2019. LNCS, vol. 11388, pp. 247–274. Springer, Cham (2019). https://doi.org/10.1007/978-3-030-11245-5_12
17. Palikareva, H., Kuchta, T., Cadar, C.: Shadow of a doubt: testing for divergences between software versions. In: Dillon, L.K., Visser, W., Williams, L. (eds.) Proceedings of the 38th International Conference on Software Engineering, ICSE 2016, Austin, TX, USA, 14–22 May 2016, pp. 1181–1192. ACM, New York (2016). https://doi.org/10.1145/2884781.2884845
18. Van der Plas, J., Stiévenart, Q., Van Es, N., De Roover, C.: Incremental flow analysis through computational dependency reification. In: 20th IEEE International Working Conference on Source Code Analysis and Manipulation, SCAM 2020, 27–28 September 2020, pp. 25–36. IEEE Computer Society (2020). https://doi.org/10.1109/SCAM51674.2020.00008
19. Pollock, L.L., Soffa, M.L.: An incremental version of iterative data flow analysis. IEEE Trans. Softw. Eng. **15**(12), 1537–1549 (1989). https://doi.org/10.1109/32.58766
20. Saha, D., Ramakrishnan, C.R.: Incremental and demand-driven points-to analysis using logic programming. In: Barahona, P., Felty, A.P. (eds.) Proceedings of the 7th International ACM SIGPLAN Conference on Principles and Practice of Declarative Programming, Lisbon, Portugal, 11–13 July 2005, pp. 117–128. ACM (2005). https://doi.org/10.1145/1069774.1069785
21. Stievenart, Q., Nicolay, J., De Meuter, W., De Roover, C.: Detecting concurrency bugs in higher-order programs through abstract interpretation. In: Proceedings of the 17th International Symposium on Principles and Practice of Declarative Programming, pp. 232–243 (2015)
22. Stiévenart, Q., Nicolay, J., De Meuter, W., De Roover, C.: A general method for rendering static analyses for diverse concurrency models modular. J. Syst. Softw. **147**, 17–45 (2019). https://doi.org/10.1016/j.jss.2018.10.001
23. Szabó, T.: Incrementalizing static analyses in datalog. Doctoral dissertation, Johannes Gutenberg-Universität Mainz, Mainz, Germany (2021). http://doi.org/10.25358/openscience-5613
24. Szabó, T., Bergmann, G., Erdweg, S., Voelter, M.: Incrementalizing lattice-based program analyses in Datalog. Proc. ACM Program. Lang. **2**(OOPSLA), 1–29 (2018). https://doi.org/10.1145/3276509
25. Szabó, T., Erdweg, S., Bergmann, G.: Incremental whole-program analysis in datalog with lattices. In: Freund, S.N., Yahav, E. (eds.) Proceedings of the 42nd ACM SIGPLAN International Conference on Programming Language Design and Implementation, PLDI 2021, pp. 1–15. ACM, New York (2021). https://doi.org/10.1145/3453483.3454026

26. Szabó, T., Bergmann, G., Erdweg, S.: Incrementalizing inter-procedural program analyses with recursive aggregation in Datalog, p. 3 (2019). Presented at the Second Workshop on Incremental Computing, IC 2019, Athens, Greece, 21 October 2019
27. Szabó, T., Erdweg, S., Voelter, M.: IncA: a DSL for the definition of incremental program analyses. In: Lo, D., Apel, S., Khurshid, S. (eds.) Proceedings of the 31st IEEE/ACM International Conference on Automated Software Engineering, ASE 2016, pp. 320–331. ACM, New York (2016). https://doi.org/10.1145/2970276.2970298
28. Tripp, O., Pistoia, M., Cousot, P., Cousot, R., Guarnieri, S.: ANDROMEDA: accurate and scalable security analysis of web applications. In: Cortellessa, V., Varró, D. (eds.) FASE 2013. LNCS, vol. 7793, pp. 210–225. Springer, Heidelberg (2013). https://doi.org/10.1007/978-3-642-37057-1_15
29. Van Es, N., Van der Plas, J., Stiévenart, Q., De Roover, C.: MAF: a framework for modular static analysis of higher-order languages. In: 20th IEEE International Working Conference on Source Code Analysis and Manipulation, SCAM 2020, Adelaide, Australia, 27–28 September 2020. IEEE Computer Society (2020)
30. Van Horn, D., Might, M.: Abstracting abstract machines. In: Hudak, P., Weirich, S. (eds.) Proceedings of the 15th ACM SIGPLAN International Conference on Functional Programming, ICFP 2010, Baltimore, MD, USA, 27–29 September 2010, pp. 51–62. ACM, New York (2010). https://doi.org/10.1145/1863543.1863553
31. Vassallo, C., Panichella, S., Palomba, F., Proksch, S., Gall, H.C., Zaidman, A.: How developers engage with static analysis tools in different contexts. Empir. Softw. Eng. 25(2), 1419–1457 (2019). https://doi.org/10.1007/s10664-019-09750-5
32. Yoon, Y., Myers, B.A.: Capturing and analyzing low-level events from the code editor. In: Proceedings of the 3rd ACM SIGPLAN Workshop on Evaluation and Usability of Programming Languages and Tools, PLATEAU 2011, pp. 25–30. Association for Computing Machinery, New York (2011). https://doi.org/10.1145/2089155.2089163
33. Yur, J., Ryder, B.G., Landi, W.: An incremental flow- and context-sensitive pointer aliasing analysis. In: Boehm, B.W., Garlan, D., Kramer, J. (eds.) Proceedings of the 1999 International Conference on Software Engineering, ICSE 1999, Los Angeles, CA, USA, 16–22 May 1999, pp. 442–451. ACM (1999). https://doi.org/10.1145/302405.302676
34. Zadeck, F.K.: Incremental data flow analysis in a structured program editor. In: Deusen, M.S.V., Graham, S.L. (eds.) Proceedings of the 1984 SIGPLAN Symposium on Compiler Construction, Montreal, Canada, 17–22 June 1984, pp. 132–143. ACM (1984). https://doi.org/10.1145/502874.502888
35. Zhan, S., Huang, J.: ECHO: instantaneous in situ race detection in the IDE. In: Proceedings of the 24th ACM SIGSOFT International Symposium on Foundations of Software Engineering, FSE 2016, Seattle, WA, USA, 13–18 November 2016, pp. 775–786 (2016). https://doi.org/10.1145/2950290.2950332

Synthesizing History and Prophecy Variables for Symbolic Model Checking

Cole Vick and Kenneth L. McMillan[✉]

UT Austin, Austin, USA
cvick@cs.utexas.edu

Abstract. Introduction of history and prophecy variables can allow a proof to be expressed in a weaker logic or a more localized form. This fact has been used, for example, to allow purely propositional, quantifier-free, invariant generators to produce proofs for parameterized systems requiring universal quantification in the inductive invariant. However, automatic synthesis of history and prophecy variables remains an open problem. We introduce counterexample-guided heuristics for this purpose based on property-driven refutation of counterexamples and Craig interpolation. The approach is evaluated on a set of benchmarks based on array manipulating programs with multiple loops.

1 Introduction

The addition of auxiliary variables is a common tactic in program verification. These can be *history variables* that record some information about past program state, or *prophecy variables* that predict some aspect of future program state. In some cases auxiliary variables may be necessary for (relative) completeness of a proof system. For example, in the Owicki/Gries system, history variables are necessary [27], while prophecy variables are needed to prove program refinement using refinement maps [1]. In other cases, auxiliary variables are used to *simplify* a proof, allowing it to be constructed in a less expressive language. For example, history and prophecy variables are used in [25] in a scheme that reduces the proofs of parameterized protocols to a propositional invariant generation problem. More subtly, we can think of automated compositional proof using grammatical inference [5] as inference of a history variable (the state of an automaton) that reduces the inductive invariant to a conjunction of local invariants.

C. Dragoi et al. (Eds.): VMCAI 2023, LNCS 13881, pp. 320–340, 2023.
https://doi.org/10.1007/978-3-031-24950-1_15

(a) Original program

```
for (int i = 0; i < N; i++) {
    a[i] = i;
}
for (int i = 0; i < N; i++) {
    x = f(i);
    b[i] = a[x];
}
assert ∀ 0 ≤ j < N. b[j] ≥ 0;
```

(b) Instrumented program

```
for (int i = 0; i < N; i++) {
    a[i] = i;
}
for (int i = 0; i < N; i++) {
    x = f(i);
    b[i] = a[x];
    η₂ = (i = π₁) ? x : η₂;
}
assert 0 ≤ π₁ < N ∧ π₂ = η₂ → b[π₁] ≥ 0;
```

Fig. 1. Array scattering program, original and instrumented. The function $f(i)$ returns a non-deterministically chosen index of a.

Eliminating Quantifiers with Auxiliary Variables. A particularly important application of auxiliary variables is to eliminate the need for quantifiers in an inductive invariant (or to eliminate quantifier alternation). This can allow us to apply a model checker to the invariant generation problem, even if the model checker cannot handle quantifiers. The key question is how to introduce auxiliary variables in an *automated* way. This might provide an important advantage, as the direct synthesis of quantified inductive invariants has proved challenging, even for problems of modest size.

As an example, consider the program of Fig. 1(a), a fragment in a notional C-like language. In this program, the array a is first filled with non-negative values. Then the array b is filled with values from a chosen by an unknown function f. We then assert that all elements of b are non-negative. In the proof of this program, we need universal quantifiers over array indices. For example, a suitable invariant between the loops would be $\forall\, 0 \leq j < N.\ a[j] \geq 0$. We can't express the invariant in the array theory without quantifiers, because a quantifier-free invariant can only reference a bounded number of elements of the arrays, while the property depends on all N elements and N is arbitrarily large.

On the other hand, by adding auxiliary variables we can reduce the dependence of the property to only a single element of a and b. Figure 1(b) shows the program instrumented in this way. In this program π_1 and π_2 are prophecy variables and η_2 is a history variable. Variable π_1 predicts an index of b for which the assertion fails. It replaces the quantified variable j in the assertion (using a process called Herbrandization). Prophecy variable π_2 predicts the index of a that is assigned to this element of b. The history variable η_2 records the index of this element of a so the prophecy can be validated when the assertion fails. A suitable quantifier-free inductive invariant of the second loop in this program is $0 \leq i \leq N$ and $0 \leq \pi_2 < N \to a[\pi_2] \geq 0$ and $\pi_2 = \eta_2 \wedge \pi_1 < i \to b[\pi_1] \geq 0$. In effect, the prophecy variables have replaced the quantifiers, allowing us to write the invariant in a way that refers to only one element each of the arrays a and b.

History Variables from Counterexamples. Apart from [5], little attention has been paid to the problem of *automated* generation of auxiliary variables. Recently, Mann, *et al.*, have considered automated generation of auxiliary variables for array-manipulating programs such as Fig. 1(a) [20]. This method allows the programs to be proved by an invariant generator that does not handle quantifiers or even implement the theory of arrays.

The method works by abstracting away the array theory, effectively replacing the array operations by uninterpreted functions. It then uses a CEGAR approach to refine the abstraction by adding ground instances of the array theory axioms to the program's transition relation. A false counterexample is one that violates an array axiom instance. If this violation spans only a single transition in the counterexample, the counterexample is easily eliminated by adding a single instance of the axiom to the transition relation. The trick is to handle axiom instances that span multiple transitions. In this case, we replace symbols in the violated axiom with prophecy variables until it spans only a single transition. As an example, this instance of an array theory axiom results in prophecy variable π_2:

$$i^2 = i^0 \rightarrow \texttt{select}(\texttt{store}(a^0, i^0, i^0), i^2) = i^0.$$

Here the superscripts represent states in a counterexample where each loop is executed once, i.e. $N = 1$. That is, if in the second loop (at state 2) we read the same index of a that was written in the first loop (at state 0) then we obtain the value written. To make this instance span a single transition, we replace i^2 by a prophecy variable π_2 that predicts its value. Thus we obtain:

$$\pi_2 = i^0 \rightarrow \texttt{select}(\texttt{store}(a^0, i^0, i^0), \pi_2) = i^0.$$

The method then synthesizes a history variable η_2 that captures the value of i^2. We then implement the prophecy variable π_2 by conditioning the assertion on $\pi_2 = \eta_2$.

Unfortunately, the method of [20] does not synthesize the *conditional* history variable shown in Fig. 1(b). It generates instead a variable that stores the value of i unconditionally, effectively delaying it by one iteration. This eliminates counterexamples with $N = 1$, resulting in a new counterexample with $N = 2$, giving a new history variable with a delay of two steps, and so on to infinity. In this paper, we introduce a method of synthesizing a history variable that stores a value *conditionally*, as shown in the figure. This allows the method to converge in cases of unbounded loops. Our method is based on searching for a *property-directed refutation* of the counterexample, using the array axioms. From this proof, we extract both the relevant axiom instance and the capture condition of the history variable ($i = \pi_1$ in the Fig. 1(b)).

Capture Conditions from Invariants. Although this approach is effective in some instances, we encounter many problems for which the appropriate capture condition involves reasoning beyond the array theory. Consider the program in Fig. 2(a) that sums up an array of non-negative integers, asserting that the

(a) Original program

```
for (int i = 0; i < N; i++) {
    a[i] = i;
}
j = 0;
for (int i = 0; i < N; i++) {
    j = j + a[i];
}
assert j ≥ 0;
```

(b) Instrumented program

```
for (int i = 0; i < N; i++) {
    a[i] = i;
}
j = 0;
for (int i = 0; i < N; i++) {
    j = j + a[i];
    η₁ = j ≥ 0 ? i : η₁;
}
assert π₁ = η₁ → j ≥ 0;
```

Fig. 2. Array summing program, original and instrumented.

sum is non-negative. The second loop maintains an invariant that the sum j is non-negative. To prove this program with a quantifier-free invariant, we need to capture the loop index i at a point when the invariant goes from true to false, which implies that $a[i]$ is negative. One suitable capture condition is shown in the instrumented program of Fig. 2(b). The history variable captures i at the last moment when the invariant is true. We can prove the instrumented program using the following inductive invariant for the second loop: $0 \leq i \leq N$ and $a[\pi_1] \geq 0$ and $\pi_1 = \eta_1 \wedge \pi_1 < i- > j \geq 0$.

To discover this capture condition, we need a way to synthesize relevant invariants of the program (though not necessarily an inductive invariant). We will present a way to do this using sequence interpolants [14].

Contributions. The primary contributions of this paper are *(1)* a method of inferring *conditional history variables* from counterexamples, based on *property-directed refutations* of abstract counterexamples, *(2)* a method of inferring capture conditions from sequence interpolants and *(3)* a benchmark evaluation, showing that this approach is substantially more effective than the unconditional approach of [20] and that it out-performs state-of-the-art CHC solvers that produce quantified invariants on small-scale benchmarks.

Limitations. We abstract only the array theory. Our implementation handles only prenex-universal assertions. We do not consider recursive programs. We do not consider the question of scalability to large programs.

2   Related Work

We can divide the related work into two categories. Methods in the first category differ from the current approach in that they construct and verify a quantified inductive invariant [10,11,15,16,19,28]. Some of these restrict the verification conditions to decidable fragments. These include the Invisible Invariants method [28] which relies on a small model theorem, and UPDR [15] which uses the decidable EPR fragment. Other methods (*e.g*, [19]) rely on incomplete heuristic quantifier instantiation.

In the second category, we have methods that, like the current method, transform the problem in some way, allowing verification without the use of quantifiers and reusing an existing invariant generator or CHC solver. The most common approach is to transform the problem into a *non-linear* CHC satisfiability problem [2,12,26]. These methods differ from the present method in two ways: they require a non-linear CHC solver and they are pre-processing techniques (eager abstraction) while our method is a CEGAR (lazy abstraction) approach.

Many works use manually-introduced auxiliary variables to eliminate the need for quantifiers [3,21–23,25]. We are aware of only one approach, however, that automates this process, that of Mann, *et al.* [20], which is the starting point for this work. That approach is strongly limited by the restriction to unconditional history variables, a limitation that we address here.

3   Preliminaries

Logic. Let $FO_=(\mathbb{S}, \mathbb{V})$ be standard sorted first-order logic with equality, where $\mathbb{S}$ is a collection of first-order sorts and $\mathbb{V}$ is a vocabulary of sorted non-logical symbols. We assume a special sort $\mathbb{B} \in \mathbb{S}$ that is the sort of propositions. Each symbol $f\mathord{:}S \in \mathbb{V}$ has an associated sort S of the form $D_1 \times \cdots \times D_n \to R$, where $D_i, R \in \mathbb{S}$ and $n \geq 0$ is the *arity* of the symbol. If $n = 0$, we say $f\mathord{:}S$ is a *constant*, and if $R = \mathbb{B}$ it is a *relation*. We write vocab(t) for the set of non-logical symbols occurring in term t.

Given a set of sorts $\mathbb{S}$, a *universe* U maps each sort in $\mathbb{S}$ to a non-empty set (with $U(\mathbb{B}) = \{\mathbf{tt}, \mathbf{ff}\}$). An *interpretation* of a vocabulary $\Sigma \subseteq \mathbb{V}$ over universe U maps each symbol $f\mathord{:}D_1 \times \cdots \times D_n \to R$ in Σ to a function in $U(D_1) \times \cdots \times U(D_n) \to U(R)$. A Σ-structure is a pair $\mathcal{M} = \langle U, \mathcal{I} \rangle$ where U is a universe and $\mathcal{I}$ is an interpretation of Σ over U. The structure is a *model* of a proposition ϕ in $FO_=(\mathbb{S}, \mathbb{V})$ if ϕ evaluates to $\mathbf{tt}$ under $\mathcal{I}$ according to the standard semantics of first-order logic. In this case, we write $\mathcal{M} \models \phi$. An *extension* of structure $\mathcal{M}$ to vocabulary $\hat{\Sigma} \supseteq \Sigma$ is a $\hat{\Sigma}$-structure $\langle U, \hat{\mathcal{I}} \rangle$ such that $\hat{\mathcal{I}}(x) = \mathcal{I}(x)$ for all symbols $x \in \Sigma$. Given an interpretation $\mathcal{J}$ with domain disjoint from $\mathcal{I}$, we write $\mathcal{M}, \mathcal{J}$ to abbreviate the structure $\langle U, \mathcal{I} \cup \mathcal{J} \rangle$.

Vocabularies. We divide $\mathbb{V}$ into several disjoint classes. The *background symbols* $\mathbb{V}_B$ are used to represent the signature of a background theory such as linear integer arithmetic or the theory of arrays. A *background theory* (theory in the sequel) is a collection of formulas over the background symbols $\mathbb{V}_B$. The *state symbols* $\mathbb{V}_S$ are used to represent the state of a system. A *state formula* is a formula over $\mathbb{V}_B \cup \mathbb{V}_S$. The *primed symbols* $\mathbb{V}'_S$ contain, for each state symbol s, a distinct symbol s'. For any term t in the logic, we denote by t' the result of replacing every state symbol occurring in t with the corresponding primed symbol s'. A *transition formula* is a formula over $\mathbb{V}_B \cup \mathbb{V}_S \cup \mathbb{V}'_S$. We write unprime$(t)$ for the result of replacing every primed symbol s' in t with s. We also distinguish a sequence of disjoint vocabularies V_S^i, for $i = 0, 1, \ldots$, such that V_S^i contains a distinct symbol denoted s^i for every state symbol s. For term t, we write t^i for the result of replacing every state symbol s with s^i and s' with s^{i+1} in t. We write $\text{unindex}_i(t)$ for the result of replacing s^i with s and s^{i+1} with s' in t, for every state symbol s. For any $\Sigma \subseteq V$, we write Σ_B for $\Sigma \cap V_B$, Σ_S for $\Sigma \cap V_S$, Σ_S' for $\Sigma \cap V_S'$ and Σ_S^i for $\Sigma \cap V_S^i$.

Transition Systems. A Σ-trace, for vocabulary Σ, is a pair $\langle n, \mathcal{M} \rangle$ where $n \geq 0$ and $\mathcal{M}$ is a structure over $\Sigma_B \cup \Sigma^0 \cup \cdots \cup \Sigma^n$. An *extension* of τ is a $\hat{\Sigma}$-trace $\hat{\tau} = \langle n, \hat{\mathcal{M}} \rangle$, for some vocabulary $\hat{\Sigma} \supseteq \Sigma$, such that $\hat{\mathcal{M}}$ is an an extension of $\mathcal{M}$.

A *transition system* is a triple $M = \langle \Sigma, I, T \rangle$ where $\Sigma \subset V$ is a vocabulary, I is a state formula over Σ and T is a transition formula over Σ. For theory $\mathcal{T}$, a $\mathcal{T}$-*trace* of M is a Σ-trace $\langle n, \mathcal{M} \rangle$ such that:

- $\mathcal{M} \models \mathcal{T}$ and
- $\mathcal{M} \models I^0$, and
- for $0 \leq i < n$, $\mathcal{M} \models T^i$.

A *safety problem* (problem in the sequel) $\Pi = \langle \Sigma, I, T, \phi \rangle$ is a system $M = \langle \Sigma, I, T \rangle$ equipped with a *safety condition* ϕ over Σ_S. A $\mathcal{T}$-counterexample to Π is a $\mathcal{T}$-trace $\langle n, \mathcal{M} \rangle$ of M such that $\mathcal{M} \not\models \phi^n$. A problem is $\mathcal{T}$-*valid* if it has no $\mathcal{T}$-counterexamples.

The *bounded model checking unfolding* $\text{BMC}(\Pi, n)$ is the sequence of formulas $I^0, T^0, \ldots, T^{n-1}, \neg\phi^n$. We note that the $\mathcal{T}$-counterexamples of Π are exactly the $\mathcal{T}$-models of $\text{BMC}(\Pi)$.

In the sequel, we assume without loss of generality that the safety conditions of all problems are prenex-existential. This can be achieved by Herbrandization, the dual process of Skolemization. In particular, this replaces leading universal quantifiers with fresh background constants (as π_1 replaces $\forall j$ in Fig. 1).

4   Theory Abstraction and Refinement

As in [20], our procedure begins by abstracting away the array theory, effectively treating the array operators as uninterpreted functions. In the sequel, we fix a background theory $\mathcal{T}$, an abstract theory $\mathcal{T}_A$ and a refinement theory $\mathcal{T}_R$ such that $\mathcal{T} = \mathcal{T}_A \cup \mathcal{T}_R$. In practice, $\mathcal{T}_A$ is EUFLIA (uninterpreted functions with equality and linear integer arithmetic) while $\mathcal{T}_R$ is the array theory. The following theorem states that model checking with $\mathcal{T}_A$ is sound:

Theorem 1. *If problem Π is $\mathcal{T}_A$-valid, then Π is $\mathcal{T}$-valid.*

Proof Sketch. A $\mathcal{T}$-counterexample of Π is also a $\mathcal{T}_A$-counterexample of Π since $\mathcal{T}$ implies $\mathcal{T}_A$. $\qquad\square$

4.1   Refinement

After theory abstraction we may obtain false counterexamples. A refinement preserves the set of concrete counterexamples, and hence is sound. However, it may eliminate abstract counterexamples. We employ three classes of refinements: theory refinements, prophecy refinements and history refinements.

Formally, a *refinement* is a partial function $\mathcal{R}$ from problems to problems such that, if problem Π is in $\mathrm{dom}(\mathcal{R})$, then for every $\mathcal{T}$-counterexample τ of Π, there exists a $\mathcal{T}$-counterexample $\hat{\tau}$ of $\mathcal{R}(\Pi)$ that is an extension of τ.

The *composition* of two partial functions $f \circ g$ is the partial function such that $(f \circ g)(x) = f(g(x))$ when $x \in \mathrm{dom}(g)$ and $g(x) \in \mathrm{dom}(f)$. We say $f \sqsubseteq g$ if for all $x \in \mathrm{dom}(f)$ we have $x \in \mathrm{dom}(g)$ and $g(x) = f(x)$. We write $\mathbf{1}$ for the identify function over any domain.

Lemma 1. *Refinements are closed under composition.*

A refinement $\mathcal{R}$ is said to *kill* $\mathcal{T}_A$-counterexample τ of problem Π if no extension of τ is a $\mathcal{T}_A$-counterexample of $\mathcal{R}(\Pi)$.

Theory Refinement. A theory refinement adds some validity of $\mathcal{T}$ to the initial condition or the transition relation. As these formulas are equivalent to tt modulo $\mathcal{T}$, this leaves the concrete traces unchanged.

Formally, a *theory refinement* is a problem transformer we denote $\mathcal{R} = \mathrm{THEORYREF}(\psi_I, \psi_T)$ where ψ_I is a ground state formula and ψ_T is a ground transition formula such that $\mathcal{T} \models \psi_I, \psi_T$. If problem $\Pi = \langle \Sigma, I, T, \phi \rangle$ is such that $\mathrm{vocab}(\psi_I, \psi_T) \subseteq \Sigma$, then $\mathcal{R}(\Pi) = \langle \Sigma, I \wedge \psi_I, T \wedge \psi_T, \phi \rangle$.

Theorem 2. $\mathrm{THEORYREF}(\phi_I, \phi_T)$ *is a refinement.*

Prophecy Refinement. A prophecy refinement introduces a fresh background constant that predicts the value of some expression at the end of a trace. Since it is a background symbol, it is invariant over time.

A *prophecy refinement* is a problem transformer we denote $\mathcal{R} = \mathrm{PROPHREF}(x, t)$ where x is a background constant and t is a state term. If $\Pi = \langle \Sigma, I, T, \phi \rangle$ is a problem such that $x \notin \Sigma$ and $\mathrm{vocab}(t) \subseteq \Sigma$, then $\mathcal{R}(\Pi) = \langle \Sigma \cup \{x\}, I, T, x = t \rightarrow \phi \rangle$.

Theorem 3. $\mathrm{PROPHREF}(x, t)$ *is a refinement.*

Proof Sketch. Let τ be a $\mathcal{T}$-counterexample of Π and let n be the length of τ. Extend τ such that $x = t^n$. This is a $\mathcal{T}$-counterexample of $\mathcal{R}(\Pi)$. $\square$

History Refinement. A *history refinement* is a problem transformer we denote $\mathrm{HISTREF}(x, \psi, t)$ where x is a state constant, t is a state term and ψ is a transition formula. If $\Pi = \langle \Sigma, I, T, \phi \rangle$ is a problem such that $x \notin \Sigma$ and $\mathrm{vocab}(t, \psi) \subseteq \Sigma$, then $\mathrm{HISTREF}(x, \psi, t)(\Pi) = \langle \Sigma \cup \{x\}, I, T \wedge x' = \mathrm{ite}(\psi, t, x), \phi \rangle$.

N.B. Our class of refinements differs from that of [20] in that the history variable x stores the value of t conditionally, while in the prior work t is delayed unconditionally.

Theorem 4. $\text{HISTREF}(x, \psi, t)$ *is a refinement.*

Proof Sketch. Let τ be a $\mathcal{T}$-counterexample of Π. Extend τ with a fresh variable x such that $x^{i+1} = \text{ite}(\psi, t, x)$ for $i = 1 \ldots n$. This is a $\mathcal{T}$-counterexample of $\mathcal{R}(\Pi)$. □

We use history refinements to store the value of term t at some given time j until the end of the trace, so that $x^n = t^j$. The storage time j is the last time such that the capture condition ψ holds. With respect to a trace $\tau = \langle n, \mathcal{M} \rangle$, we say that refinement $\text{HISTREF}(x, \psi, t)$ *captures* t^j if $\mathcal{M} \models \psi^j$ and for all $j < i < n$, $\mathcal{M} \not\models \psi^i$.

Lemma 2. *If τ is a trace of problem Π and refinement $\mathcal{R} = \text{HISTREF}(x, \psi, t)$ captures t^j in τ, then for every extension $\hat{\tau}$ of τ in $\mathcal{R}(\Pi)$, $\hat{\tau} \models x^n = t^j$.*

Proof Sketch. By the definition of history refinement, in $\hat{\tau}$ we have $x^{j+1} = t^j$ and for all $j < k < n$, $x^{k+1} = x^k$. By induction on k we have $x^n = t^j$. □

5   Counterexample-Guided Refinement

The algorithms we present are non-deterministic, with non-deterministic choice indicated by the "choose" keyword. Choice could be implemented by backtracking, but in practice we use heuristics that are detailed in Sect. 8.

We introduce refinements lazily with counterexample-guided abstraction refinement (CEGAR). The general refinement loop is shown in Fig. 3. If model checking determines that the property is true modulo the abstract theory, we return true. Else, we obtain an abstract counterexample τ. If τ is a concrete counterexample, we return false. Else, we call REFINEMENTS, which generates a sequence of refinements that kill τ. We choose a refinement among those found (or abort if none are found). Then we apply CEGAR to the refined problem.

Theorem 5 (Soundness). *Assume* REFINEMENTS*(Π, τ) is a set of refinements for all problems Π and counterexamples τ to Π. If* CEGAR*(Π) terminates, the result is true iff Π is valid.*

Proof Sketch. Refinements preserve concrete counterexamples. Thus, if we return true, by Theorem 1 there are none. Moreover, if we return false, τ is a $\mathcal{T}$-counterexample of the original Π. □

If the refinements returned by REFINEMENTS always kill τ, and if it always returns a refinement, then we say that CEGAR makes *refinement progress* in the sense that each refinement eliminates at least one abstract counterexample. This is a heuristically useful property, but it does not guarantee termination.

> **Function** CEGAR(Π)
> **Input** problem Π
> **Returns** true if Π is $\mathcal{T}$-valid
>
> **if** Π is $\mathcal{T}_A$-valid:
>   **return** true
> **let** τ be a $\mathcal{T}_A$-counterexample to Π
> **if** τ is a $\mathcal{T}$-counterexample to Π:
>   **return** false
> **choose** $\mathcal{R}$ in REFINEMENTS(Π,τ)
> **return** CEGAR($\mathcal{R}(\Pi)$)

Fig. 3. Counterexample-guided refinement

5.1   Refinement with Local Axiom Instances

Figure 4 shows our general algorithm for refinement. It is similar to [20], except in that the history variables are conditional. We are given a ground instance ψ of an axiom in $\mathcal{T}_R$ that is false in abstract counterexample τ and a set Γ of potential history variable conditions. If the counterexample is of length $n = 0$, we kill it by conjoining $\text{unindex}_0(\psi)$ to the initial condition. If the vocabulary of the axiom instance ψ spans a single transition from time i to time $i + 1$, we kill the counterexample by conjoining $\text{unindex}_i(\psi)$ to the transition relation. Otherwise, we *localize* ψ to a single transition from time i to time $i + 1$ by replacing each state symbol not indexed by either i or $i+1$ with a prophecy variable. Formally, we replace x^j for $j \neq i, i+1$ by a fresh prophecy variable that captures x^j in the trace. We capture the value of x^j by introducing a history variable η that uses a Boolean combination of predicates in Γ as the storage condition. We then introduce a prophecy variable π that predicts η at time n, i.e. the time when the property is checked.

We now argue correctness of this procedure. Say that formula ϕ is *initial* if $\text{vocab}(\phi) \subseteq \mathbb{V}_B \cup \mathbb{V}_S^0$ and *i-local* for $i \geq 0$ if $\text{vocab}(\phi) \subseteq \mathbb{V}_B \cup \mathbb{V}_S^i \cup \mathbb{V}_S^{i+1}$.

Lemma 3. *If τ is a $\mathcal{T}_A$-counterexample for problem Π and ψ is an initial formula such that $\mathcal{T} \models \psi$ and $\tau \not\models \psi$, then* THEORYREF($\text{unindex}_0(\psi)$, $\textbf{tt}$) *kills τ.*

Proof Sketch. Since $\tau \not\models \text{unindex}_0(\psi)^0$ it follows that no extension of τ can be a $\mathcal{T}_A$-trace of the refinement. $\qquad\square$

Lemma 4. *If τ is a $\mathcal{T}_A$-counterexample for problem Π and ψ is an i-local formula such that $\mathcal{T} \models \psi$ and $\tau \not\models \psi$, then* THEORYREF($\textbf{tt}$, $\text{unindex}_i(\psi)$) *kills τ.*

Proof sketch Since $\tau \not\models \text{unindex}_i(\psi)^i$ it follows that no extension of τ can be a $\mathcal{T}_A$-trace of the refinement. $\qquad\square$

Theorem 6. *If τ is a $\mathcal{T}_A$-counterexample to problem Π, and $\mathcal{T} \models \psi$ and $\tau \not\models \psi$ then every refinement in $AxiomRefine(\tau, \psi, \Gamma)$ kills τ.*

Function AXIOMREFINE($\tau = \langle n, \mathcal{M} \rangle, \psi, \Gamma$)
    Input counterexample τ, axiom instance ψ false in $\mathcal{M}$,
      condition set Γ
    Yields sequence of refinements that kill τ

    if $n = 0$:
      yield THEORYREF(unindex$_0$(ψ), **tt**)
    else:
      $\mathcal{R} \leftarrow \mathbf{1}$ (* the identity refinement *)
      choose $0 \leq i < n$
      for each x^j in ψ such that $j \neq i, i+1$:
        choose Boolean combination ρ over Γ such that:
          (1) ρ is j-local and
          (2) unindex$_j$(ρ) captures x^j in τ
        let π, η be fresh constants in $\mathbb{V}_B, \mathbb{V}_S$
        $\psi \leftarrow \psi[\pi/x^j]$
        $\mathcal{R} \leftarrow$ PROPHREF(π, η) $\circ$ HISTREF(η, unindex$_j$(ρ), x) $\circ \mathcal{R}$
      yield THEORYREF(**tt**, unindex$_i$(ψ)) $\circ \mathcal{R}$

Fig. 4. Refinement with an axiom instance

Proof Sketch. Let n be the length of τ. If $n = 0$ we apply Lemma 3. Else, construct a trace $\hat{\tau}$ by extending τ at each iteration of the loop in *AxiomRefine* such that $\eta^k = x^j$ for all $j < k \leq n$ and $\pi = x^j$. The loop maintains the invariant that $\hat{\tau} \not\models \psi$ and, by Theorems 3 and 4, that $\hat{\tau}$ is a $\mathcal{T}_A$-counterexample to $\mathcal{R}(\Pi)$. Moreover, on termination of the loop, ψ is i-local, thus by Lemma 4, the returned refinement kills $\hat{\tau}$ and hence τ.

6   Proof-Based Prophecy Heuristic

To use the above algorithm to eliminate a counterexample, we must find a suitable axiom instance violation and predicates from which to construct the capture condition. Heuristically, we wish to find ground instances of $\mathcal{T}_R$ axioms that are violated by the abstract counterexample and are *causally related* to the property failure. That is, we seek axiom violations that do not depend on accidental aspects of the counterexample that are unrelated to the property. In this way, we hope to produce refinements that kill a large space of abstract counterexamples.

We achieve this by constructing a property-driven refutation of the counterexample. We start with a *goal term* whose value in the counterexample we wish to contradict. We then use trigger-based quantifier instantiation, as introduced in the Simplify theorem prover [8] to match the goal term against a *trigger pattern* in one of the axioms. This gives us a ground axiom instance in which the trigger term is equal to the goal term. If this axiom instance is false in the counterexample, we use it to generate a refinement. Otherwise, we extract from the axiom instance a new goal term and try to contradict the value of this term.

We also extract relevant predicates to use in history variable conditions from the axiom instances in the refutation.

A *trigger-form* axiom is of the form $\forall V.\ \psi \rightarrow t_t = t_g$ where V is a set of variables, quantifier-free formula ψ is the *precondition*, term t_t is the *trigger* and term t_g is the *goal*. Each variable in V must occur in the trigger t_t exactly once. In the sequel, we assume the refinement theory $\mathcal{T}_R$ is a set of trigger-form axioms.

Here are the array theory axioms in trigger form:

$$\forall A, X, Y, V.\ (X = Y) \rightarrow \mathtt{select}(\mathtt{store}(A, X, V), Y) = V \tag{1}$$

$$\forall A, X, Y, V.\ (X \neq Y) \rightarrow \mathtt{select}(\mathtt{store}(A, X, V), Y) = \mathtt{select}(A, Y) \tag{2}$$

$$\forall X, V.\ \mathtt{tt} \rightarrow \mathtt{select}(\mathtt{constArr}(V), X) = V \tag{3}$$

Example of Property-Driven Counterexample Refutation. Suppose the following sub-formulas appear in the BMC unfolding:

$$a^1 = \mathtt{store}(a^0, i^0, x^0) \tag{4}$$

$$a^2 = a^1 \tag{5}$$

$$b^3 = \mathtt{store}(b^2, j^2, \mathtt{select}(a^2, k^2)) \tag{6}$$

$$\neg p(\mathtt{select}(b^3, l^3)) \tag{7}$$

The last of these represents the failure of the safety property. Let us take $\mathtt{select}(b^3, l^3)$ as our goal term. By substitution with (6), this term is equal to $\mathtt{select}(\mathtt{store}(b^2, j^2, \mathtt{select}(a^2, k^2)), l^3)$. Thus, modulo equality, we can match it against the trigger term $\mathtt{select}(\mathtt{store}(A, X, V), Y)$ of axiom (1) with the assignment $A = b^2$, $X = j^2$, $V = \mathtt{select}(a^2, k^2)$, $Y = l^3$. This gives us the following axiom instance:

$$(j^2 = l^3) \rightarrow \mathtt{select}(\mathtt{store}(b^2, j^2, \mathtt{select}(a^2, k^2)), j^2) = \mathtt{select}(a^2, k^2)$$

Suppose that the precondition $j^2 = l^3$ of this instance is true. In this case we consider the axiom instance to be relevant to the goal. If the instance is false, we use it as a refinement. It is 2-local (referring only to symbols at times 2 and 3) therefore we can add it directly to the transition relation to kill the counterexample. On the other hand, suppose the instance is true. This implies our goal term is equal to $\mathtt{select}(a^2, k^2)$, the goal term of the axiom. We therefore take this as our new goal, attempting to contradict its value in the counterexample. Moreover, we consider the precondition $j^2 = l^3$ as a potential history variable condition. Intuitively, this is the condition under which the new goal term influences the property.

The new goal $\mathtt{select}(a^2, k^2)$ again matches axiom (1) yielding this axiom instance:

$$(i^0 = k^2) \rightarrow \mathtt{select}(\mathtt{store}(a^0, i^0, x^0), k^2) = x^0$$

Again supposing that the precondition is true but the axiom instance is false, we can use this instance to refine. This instance, however, is not local because it contains k^2. We can localize it by capturing the value of k^2 with a history variable η under the condition $j^2 = l^3$ derived from our first inference. This adds $\eta' = \text{ite}(j = l', k, \eta)$ to the transition relation and rewrites the safety property to $\pi = \eta \rightarrow p(\text{select}(b, l))$. Substituting the non-local term k^2 by π we add the following axiom instance to the transition relation, killing the counterexample:

$$(i = \pi) \rightarrow \text{select}(\text{store}(a, i, x), \pi) = x$$

E-Graphs. We now describe our algorithm for refinement based on property-driven counterexample refutations.

An E-graph [8] is a structure that maps a set of terms to equality classes. If we are given a model of a formula, the corresponding E-graph is a partial interpretation of the symbols over the universe that is sufficient to evaluate all the sub-terms of the formula. In our case, the formula is a BMC unfolding and the model is a counterexample to the safety property. As in Simplify, we match triggers to goal terms by substituting the free variables with terms occurring in the E-graph. Heuristically, by using existing terms in the BMC formula we hope to obtain axiom instances that are generally useful and not specific to one counterexample.

A *partial Σ-interpretation* over a universe U maps each symbol $f^{D_1 \times \cdots \times D_n \rightarrow R}$ in Σ to a *partial* function $U(D_1) \times \cdots \times U(D_n) \rightarrow U(R)$. For partial Σ-interpretations $\mathcal{I}, \hat{\mathcal{I}}$, we say $\mathcal{I} \sqsubseteq \hat{\mathcal{I}}$ if $f[\mathcal{I}] \sqsubseteq f[\hat{\mathcal{I}}]$ for all $f \in \Sigma$. An *E-graph* over symbols Σ is a pair $\mathcal{E} = \langle U, \mathcal{I} \rangle$ where U is a universe and $\mathcal{I}$ is a partial Σ-interpretation over U. For E-graphs $\mathcal{E} = \langle U, \mathcal{I} \rangle$ and $\hat{\mathcal{E}} = \langle U, \hat{\mathcal{I}} \rangle$ over Σ, we say $\mathcal{E} \sqsubseteq \hat{\mathcal{E}}$ if $\mathcal{I} \sqsubseteq \hat{\mathcal{I}}$.

We assume a special value $\bot$ not present in any universe. The interpretation $t[\mathcal{E}]$ of a term t in an E-graph $\mathcal{E} = \langle U, \mathcal{I} \rangle$ is defined as follows:

- $x[\mathcal{E}] = \mathcal{I}(x)$ if x is a constant in $\text{dom}(\mathcal{I})$ else $\bot$,
- $f(t_1, \ldots, t_n)[\mathcal{E}] = \mathcal{I}(f)(t_1[\mathcal{I}], \ldots, t_n[\mathcal{I}])$ if $f \in \text{dom}(\mathcal{I})$ and $(t_1[\mathcal{I}], \ldots, t_n[\mathcal{I}]) \in \text{dom}(\mathcal{I}(f))$ else $\bot$.

Given a Σ-structure $\mathcal{M}$ and a set of terms $\mathcal{L}$ over Σ, let $\text{EGRAPH}(\mathcal{M}, \mathcal{L})$ denote the least $\mathcal{E} \sqsubseteq \mathcal{M}$ such that $t[\mathcal{E}] \neq \bot$ for all terms $t \in \mathcal{L}$.

Matching Modulo Equality. Given an E-graph $\mathcal{E}$, a term t_t with free variables V and a ground term t_m, a *match modulo equality* $t_t \rightarrow t_m$ is an assignment σ of ground terms to variables in V such that $\mathcal{E} \models t_t[\sigma] = t_m$.

Given a Σ-structure τ, an E-graph $\mathcal{E}$, a term t_m and a set of trigger-form axioms $\mathcal{T}$, a *trigger match* for t_m is $\phi[\sigma]$ for any $\phi \in \mathcal{T}$ of the form $\forall V. \psi \rightarrow t_t = t_g$ such that σ is a match $t_t \rightarrow t_m$ modulo equality and $\tau \models \psi[\sigma]$.

The algorithm shown in Fig. 5 generates a stream of trigger matches given τ, $\mathcal{E}$ and t_m. It is described as a generator (as in [8]) in which each "yield" statement appends an element to the stream. We rely on a procedure EMATCH that yields a stream of matches $t_t \rightarrow t_m$ in a given E-graph. This is implemented in the same way as in [8].

Generator TRIGGERMATCHES(τ, $\mathcal{E}$, t_m)
Input trace τ, E-graph $\mathcal{E}$, term t_m
Yields sequence of trigger matches for t_m

> **for** $\forall V.\psi \to t_t = t_g$ **in** $\mathcal{T}_R$:
>   **for** σ **in** EMATCH($\mathcal{E}$, t_t, t_m):
>   **if** $\tau \models \psi[\sigma]$:
>   **yield** $(\psi \to t_t = t_g)[\sigma]$

Fig. 5. Trigger matching algorithm

Generator VIOLATIONS(τ, $\mathcal{E}$, t_m, Γ)
Input trace τ, E-graph $\mathcal{E}$, term t_m, conditions Γ
Yields sequence of violations $\langle \phi, \Gamma \rangle$

> **for** $\psi \to t_t = t_g$ **in** TRIGGERMATCHES(τ, $\mathcal{E}$, t_m):
>   **if** $\tau \models \psi \to t_t = t_g$:
>   **for** v **in** VIOLATIONS(τ, $\mathcal{E}$, t_g, $\Gamma \cup \{\psi\}$):
>   **yield** v
>   **else**:
>   **yield** $\langle \psi \to t_t = t_g, \Gamma \rangle$

Fig. 6. Property-guided search for counterexample refutations

Figure 6 shows a procedure that searches for a refutation of a counterexample. A *violation* is a pair consisting of an axiom instance that is false in the counterexample, and a set of preconditions for the instance to be relevant to the goal. The procedure searches for trigger matches against the goal term. For each match found, if the axiom instance is false, it returns a violation. If the axiom instance is true, it recurs on the instance's goal term, adding its precondition to the list of preconditions.

Finally, Fig. 7 shows our procedure for generating refinements from a counterexample. We begin by building the E-graph $\mathcal{E}$ for the terms in the BMC formula. We call a procedure MINECONDITIONS to collect a set of predicates from the problem that are heuristically likely to be useful history variable conditions. Similarly MINETERMS returns a stream of terms that are likely to be causally related to the property, ordered from most to least relevant. See Sect. 8 for details of these functions in our implementation. For each term, we call VIOLATIONS to search for relevant axiom violations, and for each of these, we call AXIOMREFINE to generate refinements.

Theorem 7. *If τ is an abstract $\mathcal{T}_A$-counterexample for problem Π then every refinement in $Refinements(\Pi, \tau)$ kills τ.*

Generator REFINEMENTS($\Pi = \langle \Sigma, I, T, \phi \rangle$, $\tau = \langle n, \mathcal{M} \rangle$)
    Input problem Π, counterexample τ
    Yields sequence of refinements that kill τ

$\mathcal{E} \leftarrow$ EGRAPH(τ, terms(BMC(Π, n)))
$\Gamma \leftarrow$ MINECONDITIONS(Π, τ)
for t_m in MINETERMS(Π, τ)
    for $\langle \psi, \Gamma \rangle$ **in** VIOLATIONS(τ, $\mathcal{E}$, t_m, Γ):
        for $\mathcal{R}$ **in** AXIOMREFINE(τ, ψ, Γ):
            yield $\mathcal{R}$

Fig. 7. Procedure for generating refinements

Proof Sketch. Every ϕ generated by VIOLATIONS is an axiom instance false in τ. Thus by Theorem 6 every generated refinement kills τ.   □

7   Capture Conditions from Interpolants

Consider the program of Fig. 2. In this program, non-negative values from array a are added to variable j, initially zero. The safety property requires that j is non-negative on termination. Suppose that, as before, we capture the last index of array a that flows to j with prophecy variable π_1. This refinement can kill a counterexample in which the *last* write to j causes it to become negative. However, it does not kill a counterexample in which j is already negative at the last time it is written. In this case, we need a more nuanced notion of causality. That is, the second loop maintains the invariant that $j \geq 0$. The cause of the property failure is actually the last write that turned the invariant from true to false. A suitable condition for capturing the index x would thus be $j \geq 0$, or perhaps $j \geq 0 \wedge \neg(j' \geq 0)$.

The question is how to guess the invariant that the loop maintains. In simple cases such as this, it is just the safety condition. However it is easy to construct examples where this is not the case (for example, with multiple loops). A common approach to guessing an invariant of a loop is to construct a sequence interpolant for the BMC unfolding.

If $\beta = \beta_0, \ldots, \beta_n$, for $n \geq 2$, is a formula sequence, formula sequence $\mathcal{I} = \mathcal{I}_0, \ldots, \mathcal{I}_{n-1}$ is a *sequence interpolant modulo* $\mathcal{T}$ if:

- $\mathcal{T} \models \beta_0 \rightarrow \mathcal{I}_0$, and
- for $i = 1 \ldots n - 1$, $\mathcal{T} \models \beta_i \wedge \mathcal{I}_{i-1} \rightarrow \mathcal{I}_i$, and
- $\mathcal{T} \models \mathcal{I}_{n-1} \wedge \beta_n \rightarrow \mathbf{ff}$, and
- for $0 \leq i < n$, vocab($\mathcal{I}_i$) $\subseteq$ (vocab($\beta_0, \ldots, \beta_i$) $\cap$ vocab($\beta_{i+1}, \ldots, \beta_n$)) $\cup \mathbb{V}_B$.

A sequence interpolant can be constructed from an unsatisfiable sequence of formulas using an interpolating theorem prover [24] such as SMTInterpol [4]. Figure 8 shows a modified version of MINECONDITIONS that adds predicates derived from an interpolant for the BMC unfolding of the problem. This makes

Function MINECONDITIONSITP($\Pi = \langle \Sigma, I, T, \phi \rangle, \tau = \langle n, \mathcal{M} \rangle$)
Input problem Π, counterexample τ to Π
Yields set of conditions

$\Gamma \leftarrow$ MINECONDITIONS(Π, τ)
if $\mathcal{T} \models \neg$BMC(Π, n):
    let $\mathcal{I}$ be a sequence interpolant for BMC(Π, n) modulo $\mathcal{T}$
    for ψ in $\mathcal{I}$:
        for atomic p occurring in ψ:
            $\Gamma \leftarrow \Gamma \cup \{p, \neg p\}$
return Γ

Fig. 8. Capture conditions from interpolants

it possible to handle problems such as Fig. 2 in which an array holds aggregate values from another array. Note that we extract from the interpolant only atomic predicates and their negations. In principle, we could use a larger class of Boolean combinations, at the expense of a more expensive search for a refinement.

8   Evaluation

In our evaluation, we consider looping programs using arrays that require quantifiers in the inductive invariant to prove safety properties. We address two research questions: *(1)* Does our approach to synthesizing conditional history variables effectively allow a quantifier-free model checker to verify the programs, and *(2)* are conditional history variables more effective than unconditional history variables. We use two baselines: the algorithm of [20] and an ablation that uses our property-driven refutation method to find axiom violations, but uses unconditional history variables.

8.1   Implementations

We implemented our algorithm in the Python programming language. We will call this implementation CONDHIST[1]. We use the tool IC3ia [6] for model checking in CEGAR, Fig. 3. As in [20], we chose this tool because it supports uninterpreted function symbols, which we need for the abstract theory. For counterexample generation, we use Z3 [7] to solve the BMC unfolding at the depth of the IC3ia counterexample. This is needed to obtain the values of the uninterpreted function symbols. For satisfiability and interpolation in algorithm MINECONDITIONSITP we use SMTInterpol [4].

---

[1] The code and instructions for reproducing our results is available at https://github.com/cvick32/ConditionalHistory. Including our tool code and the benchmark transformation code there are roughly 3800 SLOC in Python.

Input and Preprocessing. We take input in the form of linear CHC solving problems in the SMTLIB2 format using the Z3 fixpoint convention or the HORN logic convention. These problems can have multiple uninterpreted predicates corresponding to program control locations. We translate this form to a simple transition system by adding an integer control location symbol pc. We also Herbrandize the safety condition by replacing leading universals with background symbols. As in [20], we replace large numeric constants with free symbols, which helps to prevent IC3ia from diverging. If we obtain a false counterexample, we run again without this abstraction. Additionally, we replace any reference to a primed variable in a transition with an equivalent unprimed expression if possible. These preprocessing tactics are held constant across the ablation experiments.

Heuristics. There are a few nondeterministic choices in the algorithm. Here, we detail the heuristics we use to make each of these choices. These heuristics are primitive, and we expect that they can be significantly improved.

In CEGAR we choose the first refinement obtained. In algorithm AXIOM-REFINE we chose the time i in which to localize the axiom instance as the least time index occurring in the instance. Intuitively, this is because our refutations chain backward in time from `select` operations to corresponding `store` operations. The heuristic places the axiom instance in the transition with the `store`, with the result that the address of the `select` becomes a prophecy variable. The Boolean combination ρ is the conjunction of all of the predicates in Γ that are j-local and true in the counterexample. Intuitively, making the history condition as strong as possible makes it more likely that the desired term x^j will be captured. We try this first without using interpolant-derived predicates and then add them if this fails to achieve capture of x^j. This is done to avoid the overhead of the interpolant computation if possible. We could perhaps improve this by greedily removing predicates as long as capture is maintained.

In MINECONDITIONS we produce a set of predicates of the form $pc = i$ where pc is a special symbol representing the program control location and i is an integer. We also produce p and $\neg p$ for every program branch condition p. We found this control flow information to be useful in constructing capture conditions. In MINETERMS we list all of the terms in the BMC unfolding in reverse time order. In this way, we prioritize refutations of counterexamples that are causally connected to the safety property.

Baseline Implementations. The implementation of the algorithm of [20] by its authors is called PROPHIC3. Unfortunately, we found that this tool produced incorrect instrumented programs in some cases. Because of this, we implemented their algorithm as described in [20], using the same underlying tools and preprocessing as CONDHIST. We will call this implementation UNCONDHIST1. It differs from CONDHIST in several ways. First, it captures a value x^j at time n using a sequence of $n - j$ unconditional history variables each delaying the value by one step. Second, it produces axiom violations by enumerating and evaluating all of the axiom instances that can be constructed using index terms, `select` terms and `constArr` terms in the BMC formula. This process continues until all BMC counterexamples at depth n are eliminated, at which point the UNSAT core is used to select a sufficient set of instances. Additionally, for each BMC counterexample, we select just the i-local violations if there are any, to avoid generating unnecessary prophecy. We implemented only one version of the algorithm, the so-called "strong abstraction", which abstracts only the array theory and not the "weak abstraction" which also abstracts the equality axioms for equality between arrays. This version gives a more direct comparison to CONDHIST.

We also implemented a version of our algorithm UNCONDHIST2 that is identical to CONDHIST except that it generates unconditional history variables in the same way as UNCONDHIST1.

8.2 Experiments

For evaluation, we use as a benchmark a set of 193 problems from the distribution of the FreqHorn tool [9]. These represent properties of small array manipulating programs that typically have from one to three loops. Single loop benchmarks are grouped in the Single category, totaling 118 benchmarks, while all other benchmarks are grouped in the Multi category, totaling 75 benchmarks. All require quantifiers in the inductive invariant. This tests the ability of the algorithm to avoid divergence by introducing suitable history and prophecy variables, though it does not address the question of scalability.

On this benchmark set, we compare against the performance of the algorithm of [20]. This tests the hypothesis that conditional history variables are more effective in preventing divergence than unconditional history variables. We should note, however, that even in cases where the generated history and prophecy variables *allow* an unquantified inductive invariant, the underlying model checker (IC3ia) may still diverge. Improvements in the model checker could improve the performance of all of the algorithms that we compare.

We applied the tools on a AWS EC2 instance, with 8 GB of physical memory, using a timeout of 120 s and no limit on memory usage.

Results. Table 1 shows the benchmark results. For each tool we show the number of solved problems and the number of timeouts. In both "Solved" columns the number to the left represents the total number of benchmarks solved and the number to the right, the number of those solutions which required auxiliary variables. We observe that the new algorithm solves a substantially larger number of problems than either baseline. This supports a positive answer to research question 2. We noted that, of the 30 problems solved by UNCONDHIST1 using history and prophecy variables, all 30 were solved by UNCONDHIST2 with *no* history and prophecy variables (not counting prophecy variables introduced by Herbrandization). A possible explanation of this is that history variables in the method of [20] are serving only to compensate for poor choices of axiom violations. We note that all the single-loop probems that we solved can be handled by Herbrandization alone.

Table 1. Comparison of tools on benchmark problems. Here, "single" refers to single-loop benchmarks and "multiple" to multiple-loop benchmarks. The notation $x \mid y$ means that x problems were solved, and among those, y problems used history and prophecy variables.

Tool	Single solved	Single timeouts	Multi solved	Multi timeouts
CONDHIST	95 | 0	23	66 | 29	9
Freqhorn	81	37	60	15
Quic3	81	37	36	39
GSpacer	59	59	30	45
UNCONDHIST1	59 | 23	59	28 | 7	47
UNCONDHIST2	95 | 0	23	28 | 0	47

The table also shows results for two existing tools that generate quantified invariants without auxiliary variables. These are Freqhorn [9], Quic3 [13] and GSpacer [18], two versions of Spacer [17] extended with universally quantified predicates. These tools have both performed well recently in the arrays category of CHC-COMP [29], the CHC solving competition. We observe that COND-HIST outperforms state-of-the-art CHC solvers, supporting a positive answer to research question 1.

On the 29 problems solved by CONDHIST using at least one auxilary variable, there were on average 3.89 refinements, with 1.24 history and prophecy variables. On the 30 benchmarks solved by UNCONDHIST1 using at least one auxiliary variable, there were on average 6.7 refinements, using 5.5 history and prophecy variables[2]. This suggests that the property-driven refutation heuristic produces more relevant refinements than the UNSAT core-based heuristic of PROPHIC3 [20].

9 Conclusion and Future Work

We introduced an approach to the introduction of auxiliary variables that can allow array-manipulating programs to be verified using quantifier-free invariants. This makes it possible to apply existing model checkers and CHC solvers that cannot generate quantified invariants. The two key aspects of this approach are conditional history variables and property-directed refutations of counterexamples.

Experimentally, we observed that the prior approach using unconditional history variables (*i.e.*, fixed delays) is not effective for verifying programs with arrays when prophecy is actually required. This is not surprising, since most loops in such programs are unbounded. This means that for any finite delay,

---

[2] For a benchmark-by-benchmark comparison of all the tools that were evaluated, see. https://github.com/cvick32/ConditionalHistory/tree/main/paper-results.

we can construct a long enough loop execution that the finite-delay history variable becomes irrelevant. In our observation, with more relevant choices of axiom violations, unconditional history variables were almost never helpful. On the other hand, the conditional history variable approach can effectively use prophecy to solve problems like the examples in Sect. 1 that the unconditional approach cannot solve.

There are several limitations of the current approach to be addressed in future work. The method applies to theories that can be axiomatized in what we called "trigger form". This applies to the array theory but other theories will likely require some generalization of the method. Also, for properties with quantifier alternation, we will have to instantiate quantifiers in the property. The current method does not handle this. Moreover, it is easy to construct problems for which history and prophecy are insufficient to eliminate quantifiers from the invariant. For these cases, auxiliary variables may still be helpful, but we cannot make do with a quantifier-free invariant generator and the current approach will not work with a quantified invariant generator. Despite this, it is interesting that the method outperforms theoretically more capable quantified invariant generators on small programs. More engineering work is needed to test the scalability of the approach on problems of realistic size.

Finally, of course, conditionally storing the value of an existing term in the program is common in manual proofs, but still represents a small class of possible history variables. Synthesizing history variables from a richer template may present difficult heuristic challenges.

From a broader perspective, the synthesis of new terms in a proof is a key strategy (perhaps *the* key strategy) in decomposing proofs into simpler lemmas and is known to reduce proof complexity for propositional logic. However, it has proven very challenging to automate. Despite limitations in the form of history variables, we consider the general approach of PROPHIC3 [20] to be significant and promising. We believe that history variable synthesis in general is an important topic for future research.

References

1. Abadi, M., Lamport, L.: The existence of refinement mappings. Theoret. Comput. Sci. **82**(2), 253–284 (1991)
2. Bjørner, N., McMillan, K., Rybalchenko, A.: On solving universally quantified horn clauses. In: Logozzo, F., Fähndrich, M. (eds.) SAS 2013. LNCS, vol. 7935, pp. 105–125. Springer, Heidelberg (2013). https://doi.org/10.1007/978-3-642-38856-9_8
3. Chou, C.-T., Mannava, P.K., Park, S.: A simple method for parameterized verification of cache coherence protocols. In: Hu, A.J., Martin, A.K. (eds.) FMCAD 2004. LNCS, vol. 3312, pp. 382–398. Springer, Heidelberg (2004). https://doi.org/10.1007/978-3-540-30494-4_27
4. Christ, J., Hoenicke, J., Nutz, A.: SMTInterpol: an interpolating SMT solver. In: Donaldson, A., Parker, D. (eds.) SPIN 2012. LNCS, vol. 7385, pp. 248–254. Springer, Heidelberg (2012). https://doi.org/10.1007/978-3-642-31759-0_19

5. Cobleigh, J.M., Giannakopoulou, D., PǎsǍreanu, C.S.: Learning assumptions for compositional verification. In: Garavel, H., Hatcliff, J. (eds.) TACAS 2003. LNCS, vol. 2619, pp. 331–346. Springer, Heidelberg (2003). https://doi.org/10.1007/3-540-36577-X_24

6. Daniel, J., Cimatti, A., Griggio, A., Tonetta, S., Mover, S.: Infinite-state liveness-to-safety via implicit abstraction and well-founded relations. In: Chaudhuri, S., Farzan, A. (eds.) CAV 2016. LNCS, vol. 9779, pp. 271–291. Springer, Cham (2016). https://doi.org/10.1007/978-3-319-41528-4_15

7. de Moura, L., Bjørner, N.: Z3: an efficient SMT solver. In: Ramakrishnan, C.R., Rehof, J. (eds.) TACAS 2008. LNCS, vol. 4963, pp. 337–340. Springer, Heidelberg (2008). https://doi.org/10.1007/978-3-540-78800-3_24

8. Detlefs, D., Nelson, G., Saxe, J.B.: Simplify: a theorem prover for program checking. J. ACM **52**(3), 365–473 (2005)

9. Fedyukovich, G., Prabhu, S., Madhukar, K., Gupta, A.: Quantified invariants via syntax-guided synthesis. In: Dillig, I., Tasiran, S. (eds.) CAV 2019. LNCS, vol. 11561, pp. 259–277. Springer, Cham (2019). https://doi.org/10.1007/978-3-030-25540-4_14

10. Ghilardi, S., Ranise, S.: Backward reachability of array-based systems by SMT solving: Termination and invariant synthesis. Log. Methods Comput. Sci. **6**(4) (2010)

11. Goel, A., Sakallah, K.: On symmetry and quantification: a new approach to verify distributed protocols. In: Dutle, A., Moscato, M.M., Titolo, L., Muñoz, C.A., Perez, I. (eds.) NFM 2021. LNCS, vol. 12673, pp. 131–150. Springer, Cham (2021). https://doi.org/10.1007/978-3-030-76384-8_9

12. Gurfinkel, A., Shoham, S., Meshman, Y.: SMT-based verification of parameterized systems. In: Zimmermann, T., Cleland-Huang, J., Su, Z. (eds.) Proceedings of the 24th ACM SIGSOFT International Symposium on Foundations of Software Engineering, FSE 2016, Seattle, WA, USA, 13–18 November 2016, pp. 338–348. ACM (2016)

13. Gurfinkel, A., Shoham, S., Vizel, Y.: Quantifiers on demand. CoRR, abs/2106.00664 (2021)

14. Henzinger, T.A., Jhala, R., Majumdar, R., McMillan, K.L.: Abstractions from proofs. In: Jones, N.D., Leroy, X. (eds.) Proceedings of the 31st ACM SIGPLAN-SIGACT Symposium on Principles of Programming Languages, POPL 2004, Venice, Italy, 14–16 January 2004, pp. 232–244. ACM (2004)

15. Karbyshev, A., Bjørner, N.S., Itzhaky, S., Rinetzky, N., Shoham, S.: Property-directed inference of universal invariants or proving their absence. J. ACM **64**(1), 7:1–7:33 (2017)

16. Koenig, J.R., Padon, O., Immerman, N., Aiken, A.: First-order quantified separators. In: Proceedings of the 41st ACM SIGPLAN Conference on Programming Language Design and Implementation, PLDI 2020, New York, NY, USA, pp. 703–717. Association for Computing Machinery (2020)

17. Komuravelli, A., Gurfinkel, A., Chaki, S.: SMT-based model checking for recursive programs (2014)

18. Krishnan, H.G.V., Gurfinkel, A.: CHC-COMP 2020 submission (2020)

19. Lahiri, S.K., Bryant, R.E.: Constructing quantified invariants via predicate abstraction. In: Steffen, B., Levi, G. (eds.) VMCAI 2004. LNCS, vol. 2937, pp. 267–281. Springer, Heidelberg (2004). https://doi.org/10.1007/978-3-540-24622-0_22

20. Mann, M., Irfan, A., Griggio, A., Padon, O., Barrett, C.: Counterexample-guided prophecy for model checking modulo the theory of arrays. In: TACAS 2021. LNCS,

vol. 12651, pp. 113–132. Springer, Cham (2021). https://doi.org/10.1007/978-3-030-72016-2_7

21. McMillan, K.L.: Circular compositional reasoning about liveness. In: Pierre, L., Kropf, T. (eds.) CHARME 1999. LNCS, vol. 1703, pp. 342–346. Springer, Heidelberg (1999). https://doi.org/10.1007/3-540-48153-2_30

22. McMillan, K.L.: Verification of infinite state systems by compositional model checking. In: Pierre, L., Kropf, T. (eds.) CHARME 1999. LNCS, vol. 1703, pp. 219–237. Springer, Heidelberg (1999). https://doi.org/10.1007/3-540-48153-2_17

23. McMillan, K.L.: Parameterized verification of the FLASH cache coherence protocol by compositional model checking. In: Margaria, T., Melham, T. (eds.) CHARME 2001. LNCS, vol. 2144, pp. 179–195. Springer, Heidelberg (2001). https://doi.org/10.1007/3-540-44798-9_17

24. McMillan, K.L.: An interpolating theorem prover. Theor. Comput. Sci. **345**(1), 101–121 (2005)

25. McMillan, K.L.: Eager abstraction for symbolic model checking. In: Chockler, H., Weissenbacher, G. (eds.) CAV 2018. LNCS, vol. 10981, pp. 191–208. Springer, Cham (2018). https://doi.org/10.1007/978-3-319-96145-3_11

26. Monniaux, D., Gonnord, L.: Cell morphing: from array programs to array-free horn clauses. In: Rival, X. (ed.) SAS 2016. LNCS, vol. 9837, pp. 361–382. Springer, Heidelberg (2016). https://doi.org/10.1007/978-3-662-53413-7_18

27. Owicki, S.S., Gries, D.: Verifying properties of parallel programs: An axiomatic approach. Commun. ACM **19**(5), 279–285 (1976)

28. Pnueli, A., Ruah, S., Zuck, L.: Automatic deductive verification with invisible invariants. In: Margaria, T., Yi, W. (eds.) TACAS 2001. LNCS, vol. 2031, pp. 82–97. Springer, Heidelberg (2001). https://doi.org/10.1007/3-540-45319-9_7

29. Rümmer, P.: Competition report: CHC-COMP-20. Electron. Proc. Theor. Comput. Sci. **320**, 197–219 (2020)

Solving Constrained Horn Clauses over Algebraic Data Types

Lucas Zavalía[iD], Lidiia Chernigovskaia[iD], and Grigory Fedyukovich[✉][iD]

Florida State University, Tallahassee, FL, USA
lrzavalia@fsu.edu, lidiya.chernigovskaya@gmail.com, grigory@cs.fsu.edu

Abstract. Safety verification problems are often reduced to solving the satisfiability of Constrained Horn Clauses (CHCs), a set of constraints in first-order logic involving uninterpreted predicates. Synthesis of interpretations for the predicates, also known as *inductive invariants synthesis*, is challenging in the presence of Algebraic Data Types (ADTs). Defined inductively, ADTs describe possibly unbounded chunks of data, thus they often require synthesizing recursive invariants. We present a novel approach to this problem based on *functional synthesis*: it attempts to extract recursive functions from constraints that capture the semantics of unbounded computation over the chunks of data encoded in CHCs. Recursive function calls are beneficial since they allow rewriting the constraints and introducing equalities that further can be simplified away. This largely simplifies the problem of generating invariants and lets them have simple interpretations that are recursion-free at the highest level and have function calls. We have implemented the approach in a new CHC solver called ADTCHC. Our algorithm relies on an external automated theorem prover to conduct proofs by structural induction, as opposed to a black-box constrained solver. With two alternative solvers of choice, ADTIND and VAMPIRE, the new toolset has been evaluated on a range of public benchmarks, and it exhibited its strengths against state-of-the-art CHC solvers on particular benchmarks that require recursive invariants.

1 Introduction

The trend in programming languages to organize data recursively originates from the fundamentals of logics, and it was first proposed as an alternative to pointers by Hoare [21] almost half a century ago. Since then, algebraic data types (ADTs) found their use as a modeling language in various software verification problems and enjoy tailored decision procedures [3, 45–47, 55]. With recursive functions over ADTs, verification conditions have a compact and elegant structure and can be handled by structural induction. However, induction-based methods often require adding helper lemmas that themselves require proofs [48, 59]. Recent approaches to lemma synthesis are based on Satisfiability Modulo Theories (SMT) and suggest using proof-failure generalization and Syntax Guided

---

The second author is currently employed at JetBrains N.V., The Netherlands.

ⓒ The Author(s), under exclusive license to Springer Nature Switzerland AG 2023
C. Dragoi et al. (Eds.): VMCAI 2023, LNCS 13881, pp. 341–365, 2023.
https://doi.org/10.1007/978-3-031-24950-1_16

Synthesis (SyGuS) [1]. It still needs improvements both in terms of scalability and expressiveness of the supported formulas.

Constrained Horn clauses (CHCs) over ADTs serve as a model for recursive computation and enable formulating safety verification tasks. CHCs make use of uninterpreted predicate symbols and a set of first-order logic implications that can use these predicates either in their left-hand side, right-hand side, or in both. Cyclic logic relations, formulated this way, correspond to loops and recursive functions. Interpretations to predicate symbols that satisfy all implications, can be treated as inductive invariants. In fact, the applicability of CHCs goes far beyond deductive verification conditions over traditionally defined recursive data structures. In software model checking [25], ADTs might encode strings, in synthesis problems [32] – the unrealizability, in relational verification [16] – simulation relations. That is, a CHC solver gradually becomes a *push-the-button technique* applicable in many domains, and thus it exempts the user from doing any specific preparation of the code and providing insights to the solver. A richer arsenal of low-level approaches that a CHC solver might employ is therefore required, e.g., new approaches to *functional synthesis* that are capable of extracting a function definition from a declarative specification.

Although there are many CHC solvers [2,5,8,9,13,19,28,29,34,38,40,41,44, 51,57,61] available for various SMT theories (e.g., integer/real arithmetic, bitvectors, and/or arrays), only a few solvers, e.g., [11,24,52], can actually support ADTs. In fact, there is a big challenge while solving CHCs for ADTs. Because ADTs are defined inductively, all the functions that process them need to be recursive too. To capture the behavior of these recursive functions over ADTs, invariants often need to describe properties over all elements of these ADTs. Specifically, this often requires the invariant to be recursive itself, thus allowing one to express properties over potentially unbounded data structures. However, when validating such invariants, a (set of) universally quantified formula(s) over ADTs needs to be constructed, and an automated proof checker should conduct the validity proofs by structural induction.

Our contribution lies in the approach to generate recursive invariants over inductively defined data structures that capture the semantics of recursive functions precisely. In particular, our approach seeks to extract a functional representation from the CHC constraints and exploit an automated theorem prover to validate this functional representation with respect to the given safety property. Our solver called ADTCHC builds on top of recent advances of automated theorem proving [36,59] that are capable of validating the interpretations constructed by the invariant synthesizer on the fly. Provers split a goal into a base case and an inductive step, prove each of them separately, generate and exploit inductive hypotheses. Whenever needed, provers can also generate a set of helper lemmas to be used for future subgoals.

Our secondary contribution in this paper is in the amendments to the ADTIND [59] prover that is the primary backend solver of ADTCHC. We present two new features of ADTIND that help in its proving process: generation of helper lemmas from common subterms and filtering possibly invalid lemmas. These features are needed when a current subgoal requires an additional induction, which could be expensive. We thus synthesize a candidate lemma and attempt to prove it by induction, such that if it is successful then the lemma helps to prove the ultimate goal. However, during the synthesis, there are often many invalid lemma candidates. Our approach thus relies on a filtering procedure to remove some lemma candidates quickly.

ADTCHC and ADTIND are built on top of the Z3 SMT solver [12]. In addition to ADTs, they support constraints in linear arithmetic and uninterpreted functions. We have evaluated ADTCHC on a range of public benchmarks originated from the safety verification tasks written in functional programming languages. We have compared ADTCHC to the top of CHC solvers presented in the CHC-COMP [50], namely ELDARICA [24], HOICE [8], PCSAT [52] and RACER [27] (implemented on top of GSPACER [37]). The experiments show that our tool is able to solve more benchmarks than competitors.

2   Preliminaries

A many-sorted first-order theory is defined as a tuple $\langle S, \mathcal{F}, \mathcal{P} \rangle$, where S is a set of sorts, $\mathcal{F}$ is a set of function symbols, and $\mathcal{P}$ is a set of predicate symbols, including equality. A formula φ is called satisfiable if there exists a model where φ evaluates to $true$. If every model of φ is also a model of ψ, then we write $\varphi \implies \psi$. A formula φ is called valid if $true \implies \varphi$.

Definition 1 (ADT). An ADT is a tuple $\langle s, C \rangle$, where s is a sort and C is a set of uninterpreted functions (called *constructors*), such that each $c \in C$ has some type $A \to s$ for some A. If for some s, A is s-free, we say that c is a *base* constructor denoted bc_s (otherwise, an *inductive* constructor denoted ic_s).

In this paper, we assume that all ADTs are *well-defined* in the sense that for each a of sort s, if a is constructed using some $c_i \in C$, i.e., $\exists b . a = c_i(b)$ is true, then for all other constructors $c_j \in C \setminus \{c_i\}$, $\forall b . a \neq c_j(b)$ is true. Well-definedness allows for pattern matching, which is the key vehicle for defining recursive functions over the ADT.

Example 1. A single-linked list $\mathbb{L}$ over elements of sort $\mathbb{Z}$ is defined as *nil* (i.e., a base constructor) or *cons* (i.e., an inductive constructor that takes as input an integer, called the head, and another list, called the tail). Examples of recursive functions over lists include the length, append, and reverse. We use a mnemonic notation to represent lists as sequences of elements, i.e., $\langle 1, 2 \rangle$ stands for a list constructed by $cons(1, cons(2, nil))$.

For proving the validity of a formula $\forall x . \varphi(x)$, where variable x has sort s, we follow the well-known principle of *structural induction*. That is, we prove independently the base case (i.e., that $\varphi(bc_s)$ holds), then generate inductive hypotheses (i.e., formulas of form $\varphi(x_i)$ for fresh variables x_i, which correspond to sort s) and prove the inductive step, (i.e., that all $\varphi(x_1), \ldots, \varphi(x_n)$ imply $\varphi(ic_s(y_1, \ldots, y_k, x_1, \ldots, x_n, y_{k+1}, \ldots, y_m)))$, where y_i has sort s_i.

Throughout the paper, we are interested in determining the validity of formulas of the form $\forall x.\varphi(x)$, where φ may have nested universal quantifiers:

$$\forall x.\Big(\forall y.\psi(y)\Big) \wedge \ldots \wedge \Big(\forall z.\gamma(z)\Big) \implies \theta(x) \tag{1}$$

Formulas $\psi, \ldots, \gamma$ on the left side of the implication (1) are called *assumptions*. If an assumption is not implied by any combination of other assumptions, it is called an *axiom* (otherwise, a *lemma*). The formula θ on the right side of the implication is called a *goal*.

Definition 2 (CHC). Assume that X is a countable set of variables associated with a sort S. A first-order language A of quantifier-free formulas over F, P, and X is called a *constraint language*. A formula $\varphi \in A$ is called a *constraint*. A *constrained Horn clause* (*CHC*) is a formula in first-order logic of the form:

$$\varphi \wedge r_1(x_1, \ldots, x_n) \wedge \ldots \wedge r_p(y_1, \ldots, y_k) \implies H$$

where we consider a fixed set R of *uninterpreted relation symbols*, such that $R \cap (F \cup P) = \varnothing$. Expression H, called the *head* of the clause, is either an application $r_0(z_1, \ldots, z_m)$ or constant $\bot$. Each r_i is an uninterpreted relation symbol ($r_i \in R$) and φ is a constraint. Each x_i, y_j, and z_k is a variable from X.

The left side of a CHC C is called the *body*. If there are no symbols from R in the body of C, then C is called a *fact*. If the head of C is $\bot$, then it is called a *query*. Otherwise, C is called *inductive*. If an inductive CHC contains one application of an uninterpreted relation symbol in the body, it is called *linear*, and *non-linear*, if more than one.

Definition 3. Given a set S of CHCs and $r \in R$, the *definitive rules* (denoted just *rules*) of r is a subset of S, such that:

$$rules(S, r) \overset{\text{def}}{=} \{C \in S \mid head(C) = r(\cdot)\}$$

Definition 4. A system of CHCs S over R is called satisfiable if there exists an *interpretation* I for each uninterpreted relation symbol from R in A, that make all CHCs from S valid. It defines a *solution* of the system, also referred to as *inductive invariant*.

Technically Definition 4 needs substitutions defined as follows. Let φ be a formula, and I be an interpretation for R. Then a substitution is the formula $\varphi[I/R]$ obtained from φ by replacing each occurrence of a formula of the form $r(x_1, \ldots, x_n)$ by $I(r)(x_1, \ldots, x_n)$, where $r \in R$. We naturally generalize this notation to sets of formulas (e.g., CHCs).

Systems of CHCs serve as compact representations of symbolic program encodings (i.e., for any number of loop iterations and any recursive depth). Automated verification is then reduced to determining the satisfiability of the corresponding systems of CHCs, and their solutions represent safe inductive invariants, i.e., formulas that over-approximate the sets of reachable states, but precise enough to prove unreachability of the error state. In this paper, we focus on finding recursive invariants for CHC systems, having no assumptions about the programming language used for writing the original program (thus, the approach can be used at the backends of verification tools such as [25,39]).

3 Recursive Functional Synthesis

The problem of *functional synthesis* (FS) is intuitively formulated as extracting a function implementation from its declarative specification. More formally, the problem is concerned with determining the satisfiability of a second-order formula $\exists f.\forall \vec{x}.p(f, \vec{x})$ over the uninterpreted function symbol f.

3.1 From CHC to FS

When it comes to representing declarative specification over ADTs, it is convenient to rely on CHCs over auxiliary uninterpreted predicates to represent unbounded computation over the structure of these ADTs. At the same time, if we assume that interpretations of predicates can only involve equalities and uninterpreted function symbols, then conjunctions of (universally quantified) CHCs can be rewritten to FS tasks that, if solved, provide solutions to the initial CHC tasks. In general, answering FS queries is difficult (undecidable), but if the structure of formulas is known, some successful heuristics can apply. In the rest of the section, we formulate a syntactic fragment and a tailored heuristic for solving FS problems over ADTs.

Definition 5. Given a set of CHCs S over a single relational symbol r, we say that a set of (universally quantified) first-order formulas $S_{f,\vec{x},y}$ is *CHC-inspired* if $S_{f,\vec{x},y} \overset{\text{def}}{=} \{c \mid c[r/\lambda \vec{x}, y \,.\, y = f(\vec{x})] \in S\}$.

That is, a CHC-inspired set of formulas can be constructed from CHCs after replacing all uninterpreted predicates by equalities involving uninterpreted functions. In the following, we assume we are given a set of CHCs S over some r, all the CHCs are *definitive* (recall Definition 3), and the are implicit syntax assumptions about the shape of CHCs (which we overcome in Sect. 3.2). Then, we show how we can mechanically construct a CHC-inspired set after some analysis and transformation of the CHCs.

Definition 6. Let,

- r be a relation with arity n,
- i be a natural number such that $i < n$,
- C be a set of constructors such that $T = \langle s, C \rangle$ is an ADT, and
- $D = \{B \implies r(a_1, ..., a_n)\}$ be a set of definitive CHCs.

We define C_i to be the set of i-th arguments of heads of CHCs in D. That is,

$$C_i = \{a_i \mid (B \implies r(a_1, ..., a_N)) \in D\}.$$

We say that the set D is $\langle T, r, i \rangle - complete$ if:

1. $|C_i| = |C|$,
2. For each constructor $c \in C$ there is an element $c(\cdot) \in C_i$.

Note: this means there is a bijection between the set C_i and the set of constructors C for the ADT, $T = \langle s, C \rangle$.

We call such argument position i an *inductive input argument* position. Intuitively, a set of implications should have a representative for each constructor of some ADT among arguments of the head, and the argument position should be the same for all the implications. It is useful in the next phase of our functional synthesis procedure: since targeting the construction of recursive definitions, we need to separately construct the base and inductive cases to satisfy all constructors of the inductive input argument.

Example 2. The CHCs below over the relational symbol p represent the computation of the length and the sum of a linked list using a single traversal of the data structure.[1]

$$\ell = 0 \land s = 0 \implies p(nil, \ell, s)$$
$$p(xs', \ell', s') \land \ell = \ell' + 1 \land s = s' + x \implies p(cons(x, xs'), \ell, s)$$

This set is a set of definitive CHCs since the head of each implication has a relation symbol. Now we define the set $C_1 = \{nil, cons(x, xs')\}$. Since C is defined as $C = \{nil, cons\}$ it is clear that $|C_1| = |C|$. Similarly it is clear to for all elements of C_0 there is a corresponding element in C. Thus we say that this set of CHCs is $\langle \mathbb{L}, p, 1 \rangle$-complete.

The corresponding CHC-inspired set $S_{f,xs,s}$ allows us to *embed* a recursive function for computing a sum into the relation p:

$$\ell = 0 \land s = 0 \implies s = f(nil)$$
$$s' = f(xs') \land \ell = \ell' + 1 \land s = s' + x \implies s = f(cons(x, xs')) \tag{2}$$

Similarly, we can define another CHC-inspired set $S_{g,xs,\ell}$ for a g function for computing the length of the list. Functions f and g can then be discovered separately and they do not contradict each other in a sense that the conjunction $s = f(xs) \land \ell = g(xs)$ is an invariant for the initial CHC system. Solutions for the FS problems $\exists f . \bigwedge_i S_{f,xs,s}$ and $\exists g . \bigwedge_i S_{g,xs,\ell}$, respectively, are as follows:

$$f = \lambda xs. \begin{cases} 0 & \text{if } xs = nil \\ f(xs') + x & \text{if } xs = cons(x, xs') \end{cases}$$

$$g = \lambda xs. \begin{cases} 0 & \text{if } xs = nil \\ g(xs') + 1 & \text{if } xs = cons(x, xs') \end{cases} \tag{3}$$

---

[1] Although it is conventional in practice to compute the length and the sum in different traversals, it is not necessarily more efficient to do it this way. Also, combining traversals might be needed in verification purposes, see e.g. [42].

3.2 The Eq-Prop Transformation

Our approach to recursive functional synthesis is driven by a transformation
of the formulas originated from the given CHCs. The key idea is to ultimately
rewrite the head by as many equalities from the body, aiming to produce a
recurrence relation. We introduce an EQ-PROP transformation whose purpose
is twofold: by moving constraints from left to right, we 1) aim at constructing
an equality having two (or more) applications of the function symbol f (but
with different arguments), and 2) normalize the given CHCs with respect to the
Definition 6, thus facilitating the inductive input arguments detection.

More formally, if the head H has an instance of term b, and the body has
equality $a = b$, then the transformation replaces b by a in H and removes $a = b$
from the body:

$$\frac{(a = b) \wedge C \implies H(b, \cdot)}{C \implies H(a, \cdot)} \text{ [EQ-PROP]}$$

Example 3. Applying EQ-PROP once to the formulas below would replace xs
in the heads of both implications and yield (2):

$$xs = nil \wedge \ell = 0 \wedge s = 0 \implies s = f(xs)$$
$$s' = f(xs') \wedge xs = cons(x, xs') \wedge \ell = \ell' + 1 \wedge s = s' + x \implies s = f(xs)$$

Further, applying EQ-PROP two more times sequentially replaces s and then s'
in the heads, yielding:

$$\ell = 0 \implies 0 = f(nil)$$
$$\ell = \ell' + 1 \implies f(xs') + x = f(cons(x, xs'))$$

The remaining formulas in the bodies are then removed by quantifier elimination,
and the resulting recurrence relation for f (also, an interpretation for the function
symbol) can be used to extract function interpretation (3).

Theorem 1. *Given a $\langle T, r, i \rangle$-complete set of CHCs S for some r, let $S_{f,xs,s}$ be
its CHC-inspired set of formulas. If applying EQ-PROP (possibly, multiple times)
to $S_{f,xs,s}$ yields a recurrence relation for f, then a solution for the functional
synthesis problem $\exists f \,.\, \bigwedge_i S_{f,xs,s}$ can be extracted from the recurrence relation.*

The proof of the theorem follows from definitions and the soundness of EQ-
PROP: by propagating equalities and rewriting the formula at the right, we essentially perform quantifier elimination, thus preserving the satisfiability of the formula. When checking the validity of the functional synthesis solution on the
initial CHCs, the bodies of CHC will compensate for the equalities that were
eliminated during EQ-PROP application. Lastly, Definition 6 guarantees that the
constructed recurrence relation is well-formed and its branches do not contradict
each other.

4   Recursive Invariants

Solving arbitrary CHCs over ADTs is challenging. Because of the unbounded nature of data structures, the creation, modification, or folding of them requires the introduction of multiple recursive functions, as well as proving inductive properties about them. In particular, CHC may not only represent function definitions, but also assumptions/assertions about data and reachability information.

Example 4. A CHC system below gives a number of constraints over the theory of $\mathbb{L}$:

$$xs = nil \implies app(xs, ys, ys)$$
$$xs = cons(x, xs') \wedge zs = cons(x, zs') \wedge app(xs', ys, zs') \implies app(xs, ys, zs)$$
$$app(xs, ys, rs) \wedge app(ys, zs, ts) \wedge app(xs, ts, us) \implies app(rs, zs, us)$$
$$xs = nil \implies rev(xs, xs)$$
$$xs = cons(x, xs') \wedge rev(xs', ys') \wedge app(ys', cons(x, nil), ys) \implies rev(xs, ys)$$
$$rev(xs, xs') \wedge rev(ys, ys') \wedge app(xs', ys', zs') \wedge$$
$$app(ys, xs, rs) \wedge rev(rs, rs') \wedge \neg(zs' = rs') \implies \bot$$

The first two CHCs define the append of one list xs to another list ys. The first CHC gives the base case of app, i.e., if xs is empty then the result equals ys. (Technically, predicate app has arity three, and the first two arguments of each $app(\cdot, \cdot, \cdot)$ represent inputs, and the last one represents the output. Note that the last two arguments in the head of the first CHC are the same, indicating that appending nil to any ys does not change ys.) The second CHC gives the inductive case of app, i.e., to append some $cons(x, xs')$ to some ys, we first need to append xs' to ys and then to cons x to the result.

The third CHC gives an additional associativity-like constraint over app. Note that this can be derived from the previous CHCs and does not affect the satisfiability of the CHC system. We give it mainly for the following reasons: 1) the provided CHC system is syntactically a feasible input to our algorithm, and it still needs to be handled, and 2) our algorithm is capable of separating such CHCs from the remaining definitive CHCs for app and further solving the FS problem.

The next two CHCs describe the process of reversing a list using app. Again, the base rule applies to an empty list, and the inductive rule applies to some $cons(x, xs')$, i.e., xs' needs to be reversed first and then placed in the resulting list before x. Lastly, the query CHC gives a constraint on both app and rev: given lists xs and ys, and their reverses xs' and ys', then appending xs' to ys' yields the same result as reversing the append of ys and xs.

We target the invariant generation via recursive functional synthesis. Note that this is different from a direct way of generating recursive invariants, i.e., where an interpretation of the predicate has applications of the same predicate (see e.g., [16]). Finding such interpretations, however, could be tricky in the cases of nonlinear or nested recursion. We propose to extract function definitions from CHCs and extend the syntax of the underlying constraint language for each CHC problem by these functions, thus allowing us to formulate invariants using the recursive functions.

Example 5. Recall Example 4. The CHC system is satisfied by the following invariants:

$$\boldsymbol{app} \mapsto \lambda xs, ys, zs \,.\, zs = f_{\boldsymbol{app}}(xs, ys)$$
$$\boldsymbol{rev} \mapsto \lambda xs, ys \,.\, ys = f_{\boldsymbol{rev}}(xs)$$

where:

$$f_{\boldsymbol{app}} = \lambda xs, ys. \begin{cases} ys & \text{if } xs = nil \\ cons(x, f_{\boldsymbol{app}}(xs', ys)) & \text{if } xs = cons(x, xs') \end{cases}$$

$$f_{\boldsymbol{rev}} = \lambda xs. \begin{cases} nil & \text{if } xs = nil \\ f_{\boldsymbol{app}}(f_{\boldsymbol{rev}}(xs'), cons(x, nil)) & \text{if } xs = cons(x, xs') \end{cases}$$

5   Solving CHCs over ADTs

In this section we introduce our main contribution: an algorithm to solve CHCs over ADT using functional synthesis.

5.1   Challenges of Recursive Functional Synthesis When Dealing with Arbitrary CHCs

When dealing only with a subset of definitive rules, Recursive Functional Synthesis (RFS) is straightforward. In general, CHC solving may provide additional challenges due to 1) presence of multiple relations, 2) additionally provided lemmas (syntactically, in the form of definitive rules, but outside of any $\langle T, r, i \rangle$-complete sets) that need to be validated, and 3) queries that need to be validated.

However, the first obstacle for applying any RFS reasoning is the possible uncertainty when deciding which argument of an uninterpreted relation symbol should be picked as a return argument. We can however apply EQ-PROP multiple times and eliminate as many equalities from the body of a CHC as possible, and then proceed to guessing an equality such that EQ-PROP could be applied again. This is achieved by introducing a fresh uninterpreted *function symbol* and using it to replace an uninterpreted *relation symbol*, thus posing an RFS query. Rule NEW-FUN formulates precisely how this transformation is applied to a CHC with applications of predicate r in the body and in the head. Additionally, to guarantee that EQ-PROP can be applied afterwards, our algorithm picks a common subterm a_i (if it exists) among arguments of r:

$$\frac{[r(a_1, .., a_i, .., a_n) \wedge]B \implies r(b_1, .., b_i, .., b_n)}{[a_i = f_{r,i}(a_1, .., a_{i-1}, a_{i+1}, .., a_n) \wedge]B \implies b_i = f_{r,i}(b_1, .., b_{i-1}, b_{i+1}, .., b_n)} \text{ [NEW-FUN]}$$

Here, $n = arity(r)$ and $f_{r,i}$ is a fresh symbol from $\mathcal{F}$. Note that this transformation does not guarantee success, i.e., it may make the system of constraints imposed by the CHC system unsatisfiable. However, the opposite claim is more optimistic: if after applying this transformation, all implications in the CHC system are valid, then the synthesized predicate interpretation:

$$\lambda x_1 \ldots x_n . x_i = f_{r,i}(x_1 \ldots x_{i-1}, x_{i+1}, \ldots x_n)$$

for some interpretation of $f_{r,i}$ can be used in the rest of the CHC solving process. In other words, after applying NEW-FUN, we update the current solution of the CHC system to map $I(r) = \lambda x_1 \ldots x_n . x_i = f_{r,i}(x_1 \ldots x_{i-1}, x_{i+1}, \ldots x_n)$, then apply EQ-PROP again and substitute the interpretations from I to predicates in the whole CHC system.

5.2 Core Algorithm

Our algorithm takes a system of CHCs S and a set of uninterpreted relation symbols $\mathcal{R}$ as input and determines the satisfiability of S. The main idea behind creating a recursive interpretation for an uninterpreted predicate is to replace it with an uninterpreted function, and then derive the *definition* for these functions. We require identifying the *inductive input argument* and the *return argument* among the arguments of the uninterpreted predicate. To well-define a recursive function, the inductive input argument should have the ADT sort, and we should select enough CHCs to cover functionality for the base and recursive cases.

The first step of the algorithm (line 1) is the ordering of uninterpreted predicates $\mathcal{R}$ so that if r_i will be processed before r_j, then r_i must not depend on r_j. This partial ordering enables us to use the already discovered interpretations of dependent predicate.

Definition 7 (Predicate Dependency Ordering). Let $rules(S,r)$ be as defined in Definition 3, given $r_i, r_j \in \mathcal{R}$, we say that r_i depends on r_j (written $r_i \prec r_j$) if:

- $r_i \neq r_j$, and
- $r_j \in rules(S, r_i)$, or there exists another $r_k \in \mathcal{R}$, such that $r_i \prec r_k$ and $r_k \prec r_j$, and
- $r_i \notin rules(S, r_j)$, and $\nexists r_k \in \mathcal{R}$ such that $r_j \prec r_k$ and $r_k \prec r_i$.

Example 6. The system of CHCs in Example 4 has two uninterpreted relation symbols, **app** and **rev**, and the inductive CHC for **rev** applies **app** in the body, making **rev** $\prec$ **app**. The algorithm thus finds an interpretation for **app** first and then proceeds to **rev**.

If all predicates in the CHC system can be ordered, the algorithm then proceeds to synthesizing an implementation for every relation symbol, beginning with the ones that do not have any dependencies (line 2). At each iteration of this loop, the algorithm aims at synthesizing an interpretation for a single relation $r \in \mathcal{R}$. It maintains an invariant (line 4) that all relation symbols except

---

Algorithm 1: ADT-CHC: Solving CHCs over ADTs.

Input: CHC system S, uninterpreted predicates symbols $\mathcal{R}$

Output: $res \in \{\text{SAT}, \text{UNKNOWN}\}$, interpretations I for $\mathcal{R}$

1    $order \leftarrow \text{ORDERPREDICATES}(\mathcal{R})$;

2   **for** $(r \leftarrow \text{TOP}(order); r \in order; r \leftarrow \text{NEXT}(order))$ **do**

3       $rules \leftarrow \{C \in S \mid head(C) = r(\cdot)\}$;

4   **assert** $\forall r' \in \mathcal{R} \setminus \{r\}$. if r' is used in $rules$ then $I(r')$ is defined;

5       $rules \leftarrow rules[I/\mathcal{R}]$;

6       $\big(rules \leftarrow \text{EQ-PROP}(rules)\big)^{*}$;

7   **if** $i \in \varnothing$ **then**

8   **return** UNKNOWN;

9   let $rules_T$ be a $\langle T, r, i\rangle$-complete subset of $rules$ for some T;

10   **for** $j \in [1, i) \cup (i, arity(r)]$ **do**

11   let C' be the result of applying $\text{EQ-PROP} \circ \text{NEW-FUN}_j$

                            to some $C \in rules_T$ such that $C' \neq C$;

12   **if** $C' \in \varnothing$ **then continue**;

13        $sol \leftarrow \lambda x_1 \ldots x_n . x_j = f_{r,j}(x_1 \ldots x_{j-1}, x_{j+1}, \ldots x_n)$;

14   **if** $I(r)$ is not defined **then** $I(r) \leftarrow sol$;

15   **else** $I(r) \leftarrow \lambda x_1 \ldots x_n . I(r)(x_1 \ldots x_n) \wedge sol$;

16       $rules' \leftarrow rules[I(r)/r]$;

17   let $rules'_T$ be a $\langle T, r, i\rangle$-complete subset of $rules'$ for some T;

18       $D \leftarrow \bigwedge\limits_{C \in rules'_T} \forall vars(C) . C$;

19   **if** $\text{ISVALID}(Lemmas \wedge D \implies rules')$ **then**

20        $Lemmas \leftarrow Lemmas \wedge D$;

21   **else**

22   **return** UNKNOWN;

23 **assert** $\forall r \in \mathcal{R} \implies I(r)$ is defined;

24 **if** $\text{ISVALID}\Big(Lemmas \implies \text{APPLY}\big(\{C \in S \mid head(C) = \bot\}, I\big)\Big)$ **then**

25   **return** SAT;

26 **return** UNKNOWN;

---

r that occur in the definitive $rules$ of r are already mapped to their interpretations, i.e., that all previous iterations of the loop succeeded. This enables us to use all interpretations (line 5): we simply replace all predicate symbols in all CHCs with the corresponding interpretations.

For the further processing of $rules$ of r, we require the rules to deterministically identify the branches of the recursive function, denoted $f_{r,i}$, that corresponds to r. To precisely determine that, the algorithm first identifies the inductive input argument of $f_{r,i}$. It applies rule EQ-PROP (line 6 which uses the Kleene star notation to reflect the continuous nature of the rule application until a fixedpoint is reached) for every rule in $rules$. If no inductive input argument is found (line 7), the algorithm cannot proceed. Otherwise, it attempts to find a return argument.

The nested loop in lines 10–15 approaches various possible return arguments of r (i.e., excluding the inductive input arguments). It searches for an implication C if $rules_T$ where the composition of NEW-FUN and EQ-PROP successfully applies, i.e., the body permits a replacement of some relation symbol by a new equality. This gives a new interpretation of r as a conjunction of equalities over new function symbols $f_{r,i}$ (line 15) that is recorded in I. A definition of $f_{r,i}$ is created by rewriting this interpretation in all $rules_T$ (line 18) and universally quantifying all free variables. To check the correctness of the constructed definitions and interpretations for r, the algorithm uses all the remaining $rules$ after the substitution and check their actual validity using a theorem prover (line 19).

Example 7. In order to confirm that the third CHC in $rules(S, \boldsymbol{app})$ is valid after the substitution of the interpretation that uses the definition of f_{app} in Example 5, we prove the validity of the following formula (which succeeds by induction on xs):

$$\forall ys \,.\, f_{app}(nil, ys) = ys \wedge$$
$$\forall xs, ys, x \,.\, f_{app}(cons(x, xs), ys) = cons(x, f_{app}(xs, ys)) \implies$$
$$\forall xs, ys, rs, zs, ts, us, f_{app}(xs, ys) = rs \wedge f_{app}(ys, zs) = ts \wedge$$
$$f_{app}(xs, ts) = us \implies f_{app}(rs, zs) = us)$$

Interestingly, this query can be recycled in the remainder of the algorithm to accelerate the solving process of the query.

A successful ending of the algorithm is when the theorem prover returns VALID for all the queries.

Theorem 2. *If the algorithm terminates with the* SAT *result (line 25), the input CHC system is satisfiable.*

The theorem can be proved by observing that the algorithm succeeds when all interpretations are found and a recursive function is synthesized for each predicate in $\mathcal{R}$. The soundness of interpretations with respect to intermediate goals is captured in the nested loop (line 19), and if the theorem prover does not succeed for some goal, the next possible return argument is considered. If no suitable return argument is found, then the algorithm does not find an interpretation: it either terminates with an UNKNOWN, or violates either of assertions in lines 4 or 23 (and thus, terminates with an UNKNOWN too).

Lastly, note that the backtracking in our pseudocode is simplified away for demonstration reasons. In fact, it could be the case that for a couple of relations $r_k \prec r_j$, there are two valid return arguments (or inductive input arguments) for an interpretation of r_j, but only one of them works for r_k. In this case, the algorithm needs to backtrack from processing r_k to r_j and re-synthesize the interpretation and the recursive function w.r.t. another argument(s). In our pseudocode, this can be simulated by running the algorithm again and making different decisions in lines 19, and/or 11, or being more selective in the loop in line 10. Evidently, in practice, it can be implemented in a more efficient way.

6   Automated Induction with AdtInd

In this section, we give an overview of our ADTIND prover that handles quantified formulas over ADT. It is specifically applied to prove the validity of formulas that arise at different stages of Algorithm 1, in this paper. However, it can also be used as a standalone tool and attempt user-given inputs.

6.1   Overview

The prover is a partial reimplementation of the work initially published in [59] and extended with new features. It follows the structural induction principle: it picks one quantified ADT-variable at a time and generates the base-case subgoal, inductive hypotheses, and the inductive-step subgoal. It then either uses an SMT solver to derive the subgoals directly from the assumptions (i.e., inductive hypotheses, recursive function definitions, or automatically generated lemmas), rewrites subgoals using the assumptions, or splits subgoals into a series of smaller subgoals to be solved recursively.

We refer the reader to the high-level presentation in [59] for a precise pseudocode. To simplify the presentation, we demonstrate the flow of ADTIND on a particular example of validating the synthesized interpretation on the query from Example 4.

Example 8. ADTIND begins with posing a quantified query and then simplifies it:

$$\forall ys \cdot f_{app}(nil, ys) = ys \wedge$$
$$\forall xs, ys, x \cdot f_{app}(cons(x, xs), ys) = cons(x, f_{app}(xs, ys)) \wedge$$
$$f_{rev}(nil) = nil \wedge$$
$$\forall xs, x \cdot f_{rev}(cons(x, xs)) = f_{app}(f_{rev}(xs), cons(x, nil)) \implies$$
$$\forall xs, ys \cdot f_{rev}(f_{app}(xs, ys)) = f_{app}(f_{rev}(ys), f_{rev}(xs))$$

Structurally, the formula above is a logical implication, and the conjunction on its left consists of recursive definitions of f_{app} and f_{rev}. These are the universally quantified formulas that initially form the set of assumptions. Further, on the right of the formula, there are two quantified ADT variables xs and ys, and ADTIND initiates a proof by induction over one of them, xs. The base case is just:

$$\forall ys \cdot f_{rev}(f_{app}(nil, ys)) = f_{app}(f_{rev}(ys), f_{rev}(nil))$$

After rewriting the base cases of the definitions of f_{app} and f_{rev}, the goal becomes:

$$\forall ys \cdot f_{rev}(ys) = f_{app}(f_{rev}(ys), nil)$$

The description of the steps to prove it is deferred to Example 9. Then, ADTIND generates an inductive hypothesis for a fixed xs that is added to the assumptions:

$$\forall ys \cdot f_{rev}(f_{app}(xs, ys)) = f_{app}(f_{rev}(ys), f_{rev}(xs))$$

and formulates a new subgoal over the same xs which is further proved valid:

$$\forall ys, x \,.\, f_{rev}(f_{app}(cons(x, xs), ys)) = f_{app}(f_{rev}(ys), f_{rev}(cons(x, xs))).$$

Two important features that ADTIND relies on are *lemma generation* (see Sect. 6.2) and *filtering lemma candidates* (Sect. 6.3). Both of them are designed for situations when a current subgoal does not immediately follow from the current assumptions (e.g., it may require a proof by induction). Our strategy is to synthesize a lemma, the validity of which is substantially easier to be proved than the validity of the goal. However, during the synthesis, we may end up with invalid lemma candidates. In this case, our approach leverages a filtering procedure that helps to remove some lemma candidates quickly.

6.2   Extracting Common Subterms for Helper Lemmas

Our approach generates auxiliary lemmas by replacing common subterms in the subgoal by fresh variables. An important condition for soundness of this method is that such newly introduced lemmas should themselves follow from the given assumptions. In particular, ADTIND separates the failure formula from the context, universally quantifies the variables, picks a subset of assumptions, and initializes the new solving process. If succeeded, this newly discovered assumption is added to the set of existing assumptions, and the solving process of the initial formula resumes.

Recall our motivating example. We demonstrate how the query can be proved using the recursive definitions of f_{app} and f_{rev}.

Example 9. Recall Example 8 and the following subgoal:

$$\forall ys \,.\, f_{rev}(ys) = f_{app}(f_{rev}(ys), nil)$$

At this point, ADTIND needs a helper lemma that can be discovered by proving the current goal by induction. However, the presence of f_{rev} unnecessarily complicates the process. In this case, ADTIND finds a common subterm, $f_{rev}(ys)$, replaces it by a fresh quantified variable and gets a new goal which is easily provable by induction:

$$\forall zs \,.\, zs = f_{app}(zs, nil).$$

ADTIND then adds this quantified formula as a new assumption, and the restarted proof process immediately concludes that this assumption implies the base case.

ADTIND has a systematic way for finding helper lemmas, demonstrated in the example above. Specifically, at the point when no assumption is applicable, our solver explores the parse tree of the goal and finds common patterns. Of particular interest are the applications of the same functions to the same tuples of arguments, as well as arithmetic and Boolean constraints. ADTIND then ranks them (more common patterns are considered first) and attempts to prove them one-by-one until either something is proved, or everything tried. In the latter case, the solver performs backtracking.

6.3   Filtering Procedure

As shown in the previous subsection, ADTIND implements a procedure to expand its own search space by generating helper lemmas from a set of candidate expressions. Each of these candidate formulas should be verified by an instance of ADTIND before it can be used. However, ADTIND is not designed to deal with invalid formulas, i.e., the ones for which no proving strategy could succeed, and thus ADTIND diverges. To resolve this problem, we present a filtering procedure that we call DISPROOF which quickly tries to filter *potentially invalid* lemmas.

The filtering procedure begins by enumerating ADT literals up to a certain depth. Then, the filtering procedure substitutes each quantified variable in the current goal for ADT literals to create a set of quantifier-free formulas. Finally, the filtering procedure rewrites the quantifier-free formulas to eliminate functions and sends the negations of the resulting formulas to an SMT solver. The DISPROOF procedure returns FILTER when the SMT solver finds at least one satisfiable negation, and thus the candidate formula is not considered in the process any longer. Otherwise DISPROOF returns UNKNOWN, and ADTIND attempts to prove it. Note that the procedure may filter a valid candidate formula that could be potentially useful. However since we use this procedure only to accelerate the search of lemmas (i.e., whenever a candidate lemma is filtered, ADTIND quickly jumps to another candidate), the soundness of the entire procedure is not compromised.

Algorithm 2 gives a pseudocode of this procedure. It receives a formula $\forall x_1, \ldots, x_n \cdot G(x_1, \ldots, x_n)$ over n universally quantified ADT variables. The algorithm begins with generating n sets of ADT literals (line 4) for each of the ADT variables $x_1, \ldots, x_n$, where each literal has depth at most k. Intuitively, this procedure is recursive:

- At level 0, ADTLITGEN$(x_i, 0)$ returns a singleton set T_0 consisting of an application of the base constructor of variable x_i.
- At level k, it assumes a set T_{k-1} is generated for level $k - 1$. Then, for the inductive constructor ic for the sort of x_i with arity m that uses p arguments of the sort of x_i, and each subset of p literals $\ell_1, \ldots, \ell_p \in T_{k-1}$, ADTLITGEN$(x_i, k)$ generates $m - p$ fresh variables $v_1, \ldots, v_{m-p}$ and applies ic to $\ell_1, \ldots, \ell_p, v_1, \ldots, v_{m-p}$. The resulting literal is added to T_k.

Importantly, ADTLITGEN does not generate concrete literals, except of the one at level 0. We let the literals to use fresh variables and use an SMT solver to evaluate them, if possible, such that the resulting concrete literals violate the goal.

The algorithm further substitutes each combination of the generated ADT literals for all $x_1, \ldots, x_n$ in G (line 6) and proceeds to rewriting the resulting formula using the given assumptions. For each substitution, the algorithm performs rewriting using assumptions until no more rewrites are possible (for more information, see [59]). Finally, if a rewritten term does not have occurrences of any functions or predicates defined in the assumptions, then its negation can be checked for the satisfiability with an SMT solver (line 9). If it is satisfiable, then

---

Algorithm 2: DISPROOF: Fitering candidate expressions.

Input: candidate expression of form $\forall x_1, \ldots, x_n . G(x_1, \ldots x_n)$,
set of assumptions A, exploration depth: k
Output: $res \in \{\text{FILTER}, \text{UNKNOWN}\}$

1 **for** $i \in [1, n]$ **do**
2      $lits_i \leftarrow \varnothing$;
3   **for** $j \in [0, k]$ **do**
4          $lits_i \leftarrow lits_i \cup \text{ADTLITGEN}(x_i, j)$;
5 **for** $\langle \ell_1, \ldots, \ell_n \rangle \in lits_1 \times \ldots \times lits_n$ **do**
6      $t \leftarrow G[\ell_1/x_1, \ldots \ell_n/x_n]$;
7      $\left(t \leftarrow \text{REWRITE}(t, A) \right)^*$;
8   **if** t has occurrences only of constructors and equality **then**
9   **if** ISSAT($\neg t$) **then**
10   **return** FILTER;
11 **return** UNKNOWN

---

the violation of goal G is found (and a concrete ADT literal is extracted from the model generated by the solver for the variables we introduced).

Example 10. Suppose ADTIND generates the following helper lemma with f_{rev} defined as in previous examples:

$$\forall x. f_{rev}(x) = x$$

The filtering procedure then instantiates variable x with three ADT literals, enumerated up to depth 2, where v_0 and v_1 are fresh variables: nil, $cons(v_0, nil)$, and $cons(v_1, cons(v_0, nil))$, resulting in the following list of formulas:

$$f_{rev}(nil) = nil$$
$$f_{rev}(cons(v_0, nil)) = cons(v_0, nil)$$
$$f_{rev}(cons(v_1, cons(v_0, nil))) = cons(v_1, cons(v_0, nil))$$

Next it unrolls each of these formulas by applying the definition of f_{rev}, resulting in:

$$nil = nil$$
$$cons(v_0, nil) = cons(v_0, nil)$$
$$cons(v_0, cons(v_1, nil)) = cons(v_1, cons(v_0, nil))$$

Finally, our procedure tests the negation of each of these terms using an SMT solver and determines that for $v_0 \mapsto 0$ and $v_1 \mapsto 1$, the negation of the last equality is true. The procedure then returns FILTER, and ADTIND jumps to another candidate.

Note that when doing filtering the procedure does not send any assumptions to the solver, and the constructors are treated as uninterpreted functions. Without any extra axiomatization, some of the solver's SAT results might be spurious.

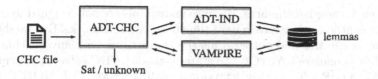

Fig. 1. CHC solving process with ADTCHC with ADTIND or VAMPIRE at the backend.

In principle, our procedure can be extended to become a sound refutation procedure, if a sufficient number of constraints about constructors are supplied. In our application, we trade precision for the speed of lemma generation, so even if we miss some potentially useful lemmas, the procedure continues with the next candidates and still has a chance to prove the main goal valid.

7 Implementation and Evaluation

In this section we present the overview of the implemented CHC solver and provide its evaluation compared to state-of-the-art.

7.1 Framework

Figure 1 gives an overview of the flow of the solving process. The tool takes a CHC file (in the conventional SMT-LIBv2 format) as input. In addition to ADTs, the inputs may have constraints over Linear Integer Arithmetic (LIA). During the solving process, ADTCHC tightly communicates with its backend solvers, ADTIND (as was described in Sect. 5.2) and VAMPIRE [36]. While posing ADT queries and receiving the confirmations of their validity, ADTCHC converts a subset of CHCs to recursive functions and makes their definitions available for future use.

While solving for the validity, ADTIND relies on the recursive definitions of functions over ADT produced by ADTCHC and successfully proved queries, and ADTIND automatically generates some helper lemmas on the fly. Lemmas are then shared among ADTCHC and backend solver and can be observed by the user. On the lower level, ADTIND reduces the reasoning over ADTs to equality and uninterpreted functions (EUF), and uses the Z3 SMT solver [12] to discharge auxiliary formulas. VAMPIRE, to the best of our knowledge, has its own satisfiability and theory solvers and it uses a portfolio approach for solving formulas.

The source code of ADTCHC and its benchmarks are available at https://github.com/grigoryfedyukovich/aeval/tree/adt-chc.

7.2 Experiments

We have considered publicly available CHC benchmarks that encode well-known verification problems. Our ultimate goal is to find invariants to prove safety properties in these benchmarks. Because many tools are not designed to recursive invariant synthesis, they have a hard time to solve benchmarks. But we show that ADTCHC is effective and outperforms the competitors on many benchmarks.

We have compared ADTCHC to state-of-the-art CHC solvers participated in the CHC-COMP [50], namely ELDARICA[2] [24], PCSAT[3] [52], HOICE [8], and RACER[4] [27] – the extension of GSPACER [37]. We considered three sets of benchmarks in the CHC format complying with the CHC-COMP rules converted to the CHC format from benchmarks used by various theorem provers: the first set, contains 28 problems, is derived from benchmarks for ADTIND [59], the second one with 17 problems comes from CLAM [26], and the last with 26 problems is taken from LEON [56]. The safety verification properties are concerned about the correctness of various operations on lists, amortized queues and binary trees.

We configured ADTCHC to run in four different modes (that correspond to four first columns of Table 1 and Table 2): the first two use ADTIND as the backend solver, and the last two use VAMPIRE as the backend solver. In the first configuration of ADTIND (denoted **w/A(1)**), the backend solver performs exactly as described in Sect. 6 but without the candidate filtering method, so it tries to prove valid *all candidate lemmas* that are generated. In the second configuration (denoted **w/A(2)**), we added the candidate filtering method, and thus, the solver finds *potentially invalid* lemmas first and skips them (if the filtering is unsuccessful, then the solver tries to prove the candidate). It allows the solver to save time and proceed to discovery of new lemmas. For VAMPIRE, in its first configuration (denoted **w/V(1)**) we use the default setting, and in the second configuration of VAMPIRE (denoted **w/V(2)**), we force it conduct the proofs by structural induction.

We used a timeout of 300 s CPU time for each tool and configuration. Overall, there are 71 benchmarks, and 41 of them were solved by either the configuration of ADTCHC +ADTIND. 42 benchmarks were solved by either the configuration of ADTCHC +VAMPIRE. More importantly, in total by either of four configurations of ADTCHC, our tool solved 52 benchmarks. Among them, 31 benchmarks were not solved by any other other competing tool. ELDARICA solved 24 benchmarks, RACER solved 16 and PCSAT solved 9, and HOICE solved 19.

Comparing backends of ADTCHC, it is apparent that ADTIND is on average significantly faster than VAMPIRE. It could be attributed to the fact that the latter uses the portfolio mode. However, both tools have they strengths since there are benchmarks solely solved only with ADTIND and only with VAMPIRE.

For benchmarking, we used a workstation equipped with 2.8 GHz Intel Core i7 4-Core (11th generation) and 12 GB of DDR4 RAM running Ubuntu 21.04.

---

[2] version 2.0.6.

[3] https://github.com/hiroshi-unno/coar.

[4] https://github.com/hgvk94/z3/tree/racer.

Table 1. Results (sec); "—" stands for "unknown".

Benchmark	AdtChc w/A(1)	w/A(2)	w/V(1)	w/V(2)	Eldarica	Racer	PCSat	HOICE
ADTIND/heap_size	0.65	0.62	84.12	80.62	1.31	0.01	0.89	0.09
ADTIND/list_append_ass	0.48	0.4	119.93	0.06	—	—	—	—
ADTIND/list_append_len	1.1	1.03	120.5	0.1	24.1	—	—	—
ADTIND/list_append_min	1.97	2.3	120.47	120.45	—	—	—	—
ADTIND/list_append_min2	0.55	0.5	60.31	60.29	—	—	—	—
ADTIND/list_append_nil	0.71	0.67	60.28	0.06	1.38	0.02	—	0.1
ADTIND/list_append_sum	0.64	0.58	74.63	0.11	—	—	—	0.13
ADTIND/list_interleave	—	—	60.22	60.22	—	—	—	0.22
ADTIND/list_len_butlast	0.57	0.51	60.3	60.29	4.81	—	—	—
ADTIND/list_len_stren	4.09	5.53	—	—	1.36	0.01	1.21	1.12
ADTIND/list_len	0.35	0.31	66.32	0.08	1.11	0.02	0.96	0.08
ADTIND/list_min_max	0.76	0.72	60.31	15.91	1.66	0.02	1.53	—
ADTIND/list_min_sum_len	3.22	3.28	66.65	15.71	1.52	0.02	1.57	—
ADTIND/list_min_sum	1.72	1.75	74.74	73.89	—	—	—	—
ADTIND/list_rev_append	6.36	—	284.61	282.75	—	—	—	0.01
ADTIND/list_rev_len	1.87	1.69	126.6	66.12	—	—	—	—
ADTIND/list_rev	—	—	—	—	—	—	—	—
ADTIND/list_rev2_append	0.95	0.86	—	120.44	—	—	—	—
ADTIND/list_rev2_len	—	—	120.7	120.75	—	—	—	—
ADTIND/queue_amort	137.05	—	—	—	1.15	0.02	1.3	0.15
ADTIND/queue_len	—	—	—	—	—	—	—	—
ADTIND/queue_popback	—	—	—	—	49.11	—	—	—
ADTIND/queue_push_to_list	—	—	—	—	—	—	—	—
ADTIND/queue_push	—	—	—	—	—	—	—	—
ADTIND/tree_insert_all_size	3.22	3.18	217.8	213.72	—	—	—	—
ADTIND/tree_insert_size	0.33	0.25	—	—	—	—	—	—
ADTIND/tree_insert_sum	0.34	0.26	—	—	—	—	—	—
ADTIND/tree_size	0.64	0.58	77.69	0.1	1	0.01	0.78	0.07
CLAM/goal10	—	—	238.76	239.38	—	—	—	—
CLAM/goal11	—	—	240.36	239.98	—	—	—	—
CLAM/goal12	—	5.06	120.64	119.92	—	—	—	—
CLAM/goal17	—	—	238.32	239.82	—	—	—	—
CLAM/goal18	—	—	247.09	243.21	—	—	—	—
CLAM/goal19	—	—	238.4	239.56	—	—	—	—
CLAM/goal2	74.49	9.45	127.41	1.19	2.04	—	—	—
CLAM/goal21	—	—	—	—	—	—	—	—
CLAM/goal27	—	—	241.2	240.97	—	—	—	—
CLAM/goal3	1.29	1.02	120.52	0.12	—	—	—	—
CLAM/goal4	—	—	124.06	123.77	8.67	—	—	—
CLAM/goal5	—	5.2	126.74	126.58	—	—	—	—
CLAM/goal6	—	—	241.15	241.06	—	—	—	—
CLAM/goal7	2.76	2.7	120.78	120.75	—	—	—	—
CLAM/goal72	0.49	0.41	60.3	0.06	—	—	—	—
CLAM/goal8	—	31.15	206.28	205.47	—	—	—	—
CLAM/goal9	—	31.32	205.37	204.55	—	—	—	—
LEON/amortize-queue-goal1	0.6	0.54	66.28	0.11	—	—	—	—
LEON/amortize-queue-goal10	0.93	0.81	186.23	185.46	—	—	—	—
LEON/amortize-queue-goal11	0.47	0.41	120.38	60.2	—	—	—	—
LEON/amortize-queue-goal12	—	—	—	—	—	—	—	—
LEON/amortize-queue-goal13	—	—	—	—	—	—	—	—
LEON/amortize-queue-goal14	—	—	—	—	—	—	—	—
LEON/amortize-queue-goal15	—	—	—	—	—	—	—	—
LEON/amortize-queue-goal3	—	—	—	—	—	—	—	—
LEON/amortize-queue-goal4	—	—	—	—	2.33	—	—	0
LEON/amortize-queue-goal5	—	—	—	—	4.14	—	—	—
LEON/amortize-queue-goal6	—	—	—	—	—	—	—	0
LEON/amortize-queue-goal8	0.8	0.71	120.43	60.18	1.49	0.01	—	0.08
LEON/amortize-queue-goal9	—	—	—	—	—	—	—	—

Table 2. Results (cont).

Benchmark	ADTCHC				ELDARICA	RACER	PCSAT	HOICE
	w/A(1)	w/A(2)	w/V(1)	w/V(2)				
LEON/bsearch-tree-goal1	198.64	—	—	—	—	—	—	—
LEON/bsearch-tree-goal10	—	—	—	—	2.81	0.05	—	0.29
LEON/bsearch-tree-goal11	67.1	—	—	—	0.84	0.01	—	0.52
LEON/bsearch-tree-goal12	17.92	15.56	—	—	1.01	0.01	—	0.29
LEON/bsearch-tree-goal13	30.73	34.32	—	180.62	—	—	—	—
LEON/bsearch-tree-goal14	—	—	—	—	0.56	0	0.91	0.09
LEON/bsearch-tree-goal2	—	127.23	271.82	263.31	—	—	—	—
LEON/bsearch-tree-goal3	—	—	278.39	281.68	—	—	—	—
LEON/bsearch-tree-goal4	—	—	—	—	—	—	—	—
LEON/bsearch-tree-goal5	—	—	—	—	—	—	—	—
LEON/bsearch-tree-goal6	—	273.95	—	—	0.94	0.01	—	0.4
LEON/bsearch-tree-goal8	89.45	—	—	—	0.63	0	0.79	0.07
LEON/bsearch-tree-goal9	—	280.87	—	—	0.66	0.01	—	0.23

8   Related Work

Existing CHC solvers [2,5,8,9,13,19,24,27–29,34,38,40,41,44,51,57,61] are utilized by the software model checkers for imperative languages [20,22], object-oriented languages [30,31], dataflow languages [17,18], and functional programming languages [7,15,33,43,58]. Algorithmically, solvers are based on Counterexample-Guided Abstraction Refinement (CEGAR) [10], Counterexample-Guided Inductive Synthesis (CEGIS) [52,54], Property Directed Reachability (PDR) [6,14], Machine Learning [53], but currently, there is no clear witness that any of these approaches are, in general, better than others. Furthermore, with the exception of [24], solvers are limited to relatively lightweight SMT theories and not ADT.

There is a plethora of proposed quantifier elimination algorithms and decision procedures for the first-order ADT fragment [3,45–47,55] and for an extension of ADT with constraints on term sizes [60]. As often useful for solving, the Craig interpolation procedure for ADT constraints has been proposed by [23]. Such techniques are being incorporated by various SMT solvers, like Z3 [12], CVC4 [4], and PRINCESS [49]. Our SMT-based approach to handling ADTs uses a new functional synthesis approach: it works by rewriting CHCs and obtaining new definition from declarative CHC constraints. Lastly, there are approaches for CHC-based relational verification over ADTs [11,42] that effectively reduce reasoning to CHCs over lightweight SMT theories. These approaches do not generate inductive invariants over ADTs, while our approach does.

There is an approach where inductive invariants are represented by finite tree automata implemented in RINGGEN [35]. A system of CHCs over ADTs is rewritten into a formula over uninterpreted function symbols by eliminating all disequalities, testers, and selectors from the clause bodies. Then they reduce the satisfiability modulo theory of ADTs to satisfiability modulo EUF and apply off-the-shelf finite model finder to build a finite model of the reduced verification conditions. The automaton representing the safe inductive invariant are derived using the correspondence between finite models and tree automata. Unfortunately, RINGEN works only on pure ADT (i.e., it defines natural numbers inductively as zero and +1, but we make use of Presburger arithmetic).

9   Conclusion and Future Work

We have presented a new approach to solve CHC problems over ADT using recursive function synthesis. Instead of generating recursive predicates, the approach generates recursive functions by applying semantics-preserving transformations to a subset of given CHCs determined on the fly. The remaining CHCs are used to validate the solutions and the approach reduces this problem to an off-the-shelf theorem prover that is expected to prove the validity of each universally quantified formula following the principle of structural induction. Our implementation called ADTCHC exploits the Z3 SMT solver to process a number of quantifier-free queries over arithmetic, uninterpreted functions, and arrays. While ADTCHC outputs a number of recursive definitions of functions that are used in interpretations of predicates, theorem provers ADTIND and VAMPIRE automatically discharge the validity checks, often generating a number of useful lemmas that can be exchanged among queries and accelerate the solving process. We also presented two new features of ADTIND that help in its proving process: generation of helper lemmas from common subterms and filtering potentially useless candidate lemmas. We experimentally compared our tools with state-of-the-art, and it shows promising results. In the future, we plan to extend the set of features of the tools, and in particular support solving queries with nested (and possibly, alternating) quantifiers.

Acknowledgments. The work is supported in parts by a gift from Amazon Web Services.

References

1. Alur, R., et al.: Syntax-guided synthesis. In: FMCAD, pp. 1–17. IEEE (2013)
2. Bakhirkin, A., Monniaux, D.: Combining forward and backward abstract interpretation of horn clauses. In: Ranzato, F. (ed.) SAS 2017. LNCS, vol. 10422, pp. 23–45. Springer, Cham (2017). https://doi.org/10.1007/978-3-319-66706-5_2
3. Barrett, C., Shikanian, I., Tinelli, C.: An abstract decision procedure for a theory of inductive data types. J. Satisfiability, Boolean Model. Comput. **3**, 21–46 (2007)
4. Barrett, C., et al.: CVC4. In: Gopalakrishnan, G., Qadeer, S. (eds.) CAV 2011. LNCS, vol. 6806, pp. 171–177. Springer, Heidelberg (2011). https://doi.org/10.1007/978-3-642-22110-1_14
5. Beyene, T.A., Popeea, C., Rybalchenko, A.: Solving existentially quantified horn clauses. In: Sharygina, N., Veith, H. (eds.) CAV 2013. LNCS, vol. 8044, pp. 869–882. Springer, Heidelberg (2013). https://doi.org/10.1007/978-3-642-39799-8_61
6. Bradley, A.R.: SAT-based model checking without unrolling. In: Jhala, R., Schmidt, D. (eds.) VMCAI 2011. LNCS, vol. 6538, pp. 70–87. Springer, Heidelberg (2011). https://doi.org/10.1007/978-3-642-18275-4_7

7. Champion, A., Chiba, T., Kobayashi, N., Sato, R.: ICE-based refinement type discovery for higher-order functional programs. In: Beyer, D., Huisman, M. (eds.) TACAS 2018. LNCS, vol. 10805, pp. 365–384. Springer, Cham (2018). https://doi.org/10.1007/978-3-319-89960-2_20

8. Champion, A., Kobayashi, N., Sato, R.: HoIce: an ICE-based non-linear horn clause solver. In: Ryu, S. (ed.) APLAS 2018. LNCS, vol. 11275, pp. 146–156. Springer, Cham (2018). https://doi.org/10.1007/978-3-030-02768-1_8

9. Chen, Y.-F., Hsieh, C., Tsai, M.-H., Wang, B.-Y., Wang, F.: Verifying recursive programs using intraprocedural analyzers. In: Müller-Olm, M., Seidl, H. (eds.) SAS 2014. LNCS, vol. 8723, pp. 118–133. Springer, Cham (2014). https://doi.org/10.1007/978-3-319-10936-7_8

10. Clarke, E., Grumberg, O., Jha, S., Lu, Y., Veith, H.: Counterexample-guided abstraction refinement. In: Emerson, E.A., Sistla, A.P. (eds.) CAV 2000. LNCS, vol. 1855, pp. 154–169. Springer, Heidelberg (2000). https://doi.org/10.1007/10722167_15

11. De Angelis, E., Fioravanti, F., Pettorossi, A., Proietti, M.: Solving horn clauses on inductive data types without induction. TPLP 18(3–4), 452–469 (2018)

12. de Moura, L., Bjørner, N.: Z3: an efficient SMT solver. In: Ramakrishnan, C.R., Rehof, J. (eds.) TACAS 2008. LNCS, vol. 4963, pp. 337–340. Springer, Heidelberg (2008). https://doi.org/10.1007/978-3-540-78800-3_24

13. Dietsch, D., Heizmann, M., Hoenicke, J., Nutz, A., Podelski, A.: Ultimate TreeAutomizer. In: HCVS/PERR, vol. 296 of EPTCS, pp. 42–47 (2019)

14. Eén, N., Mishchenko, A., Brayton, R.K.: Efficient implementation of property directed reachability. In: FMCAD, pp. 125–134. IEEE (2011)

15. Fedyukovich, G., Ahmad, M.B.S., Bodík, R.: Gradual synthesis for static parallelization of single-pass array-processing programs. In: PLDI, pp. 572–585. ACM (2017)

16. Fedyukovich, G., Ernst, G.: Bridging arrays and ADTs in recursive proofs. In: TACAS 2021. LNCS, vol. 12652, pp. 24–42. Springer, Cham (2021). https://doi.org/10.1007/978-3-030-72013-1_2

17. Garoche, P., Gurfinkel, A., Kahsai, T.: Synthesizing modular invariants for synchronous code. In: HCVS, vol. 169 of EPTCS, pp. 19–30 (2014)

18. Garoche, P. Kahsai, T., Thirioux, X.: Hierarchical state machines as modular horn clauses. In: HCVS, vol. 219 of EPTCS, pp. 15–28 (2016)

19. Grebenshchikov, S., Lopes, N.P., Popeea, C., Rybalchenko, A.: Synthesizing software verifiers from proof rules. In: PLDI, pp. 405–416. ACM (2012)

20. Gurfinkel, A., Kahsai, T., Komuravelli, A., Navas, J.A.: The seahorn verification framework. In: Kroening, D., Păsăreanu, C.S. (eds.) CAV 2015. LNCS, vol. 9206, pp. 343–361. Springer, Cham (2015). https://doi.org/10.1007/978-3-319-21690-4_20

21. Hoare, C.A.R.: Recursive data structures. Int. J. Parallel Program. 4(2), 105–132 (1975)

22. Hojjat, H., Konečný, F., Garnier, F., Iosif, R., Kuncak, V., Rümmer, P.: A verification toolkit for numerical transition systems. In: Giannakopoulou, D., Méry, D. (eds.) FM 2012. LNCS, vol. 7436, pp. 247–251. Springer, Heidelberg (2012). https://doi.org/10.1007/978-3-642-32759-9_21

23. Hojjat, H., Rümmer, P.: Deciding and interpolating algebraic data types by reduction. In: SYNASC, pp. 145–152. IEEE (2017)

24. Hojjat, H., Rümmer, P.: The ELDARICA horn solver. In: FMCAD, pp. 158–164. IEEE (2018)

25. Hojjat, H., Rümmer, P., Shamakhi, A.: On strings in software model checking. In: Lin, A.W. (ed.) APLAS 2019. LNCS, vol. 11893, pp. 19–30. Springer, Cham (2019). https://doi.org/10.1007/978-3-030-34175-6_2

26. Ireland, A., Bundy, A.: Productive use of failure in inductive proof. In: Zhang, H. (ed.) Automated Mathematical Induction, pp. 79–111. Springer, Cham (1996). https://doi.org/10.1007/978-94-009-1675-3_3

27. Hari Govind, V.K., Shoham, S., Gurfinkel, A.: Solving constrained horn clauses modulo algebraic data types and recursive functions. Proc. ACM Program. Lang. 6(POPL), 1–29 (2022)

28. Kafle, B., Gallagher, J.P., Ganty, P.: Solving non-linear Horn clauses using a linear Horn clause solver. In: HCVS, vol. 219 of EPTCS, pp. 33–48 (2016)

29. Kafle, B., Gallagher, J.P., Morales, J.F.: RAHFT: a tool for verifying horn clauses using abstract interpretation and finite tree automata. In: Chaudhuri, S., Farzan, A. (eds.) CAV 2016. LNCS, vol. 9779, pp. 261–268. Springer, Cham (2016). https://doi.org/10.1007/978-3-319-41528-4_14

30. Kahsai, T., Kersten, R., Rümmer, P., Schäf, M.: Quantified heap invariants for object-oriented programs. In: LPAR, vol. 46 of EPiC Series in Computing, pp. 368–384. EasyChair (2017)

31. Kahsai, T., Rümmer, P., Sanchez, H., Schäf, M.: JayHorn: a framework for verifying java programs. In: Chaudhuri, S., Farzan, A. (eds.) CAV 2016. LNCS, vol. 9779, pp. 352–358. Springer, Cham (2016). https://doi.org/10.1007/978-3-319-41528-4_19

32. Kim, J., Hu, Q., D'Antoni, L., Reps, T.: Semantics-guided synthesis. Proc. ACM on Program. Lang. 5(POPL), 1–32 (2021)

33. Kobayashi, N., Sato, R., Unno, H.: Predicate abstraction and CEGAR for higher-order model checking. In: ACM, pp. 222–233. ACM (2011)

34. Komuravelli, A., Gurfinkel, A., Chaki, S.: SMT-based model checking for recursive programs. In: Biere, A., Bloem, R. (eds.) CAV 2014. LNCS, vol. 8559, pp. 17–34. Springer, Cham (2014). https://doi.org/10.1007/978-3-319-08867-9_2

35. Kostyukov, Y., Mordvinov, D., Fedyukovich, G.: Beyond the elementary representations of program invariants over algebraic data types. In: PLDI, pp. 451–465 (2021)

36. Kovács, L., Voronkov, A.: First-order theorem proving and VAMPIRE. In: Sharygina, N., Veith, H. (eds.) CAV 2013. LNCS, vol. 8044, pp. 1–35. Springer, Heidelberg (2013). https://doi.org/10.1007/978-3-642-39799-8_1

37. Vediramana Krishnan, H.G., Chen, Y.T., Shoham, S., Gurfinkel, A.: Global guidance for local generalization in model checking. In: Lahiri, S.K., Wang, C. (eds.) CAV 2020. vol. 12225, pp. 101–125. Springer, Cham (2020). https://doi.org/10.1007/978-3-030-53291-8_7

38. Krishnan, H.G.V., Fedyukovich, G., Gurfinkel, A.: Word level property directed reachability. In: ICCAD, pp. 1–9. IEEE (2020)

39. Matsushita, Y., Tsukada, T., Kobayashi, N.: RustHorn: CHC-based verification for rust programs. In: ESOP 2020. LNCS, vol. 12075, pp. 484–514. Springer, Cham (2020). https://doi.org/10.1007/978-3-030-44914-8_18

40. McMillan, K.L.: Lazy annotation revisited. In: Biere, A., Bloem, R. (eds.) CAV 2014. LNCS, vol. 8559, pp. 243–259. Springer, Cham (2014). https://doi.org/10.1007/978-3-319-08867-9_16

41. McMillan, K.L., Rybalchenko, A.: Solving constrained Horn clauses using interpolation. In Technical report MSR-TR-2013-6 (2013)

42. Mordvinov, D., Fedyukovich, G.: Synchronizing constrained horn clauses. In: LPAR, vol. 46 of EPiC Series in Computing, pp. 338–355. EasyChair (2017)

43. Mordvinov, D., Fedyukovich, G.: Verifying safety of functional programs with rosette/unbound. CoRR, abs/1704.04558 (2017). https://github.com/dvvrd/rosette
44. Mordvinov, D., Fedyukovich, G.: Property directed inference of relational invariants. In: FMCAD, pp. 152–160. IEEE (2019)
45. Oppen, D.C.: Reasoning about recursively defined data structures. J. ACM (JACM) **27**(3), 403–411 (1980)
46. Pham, T., Gacek, A., Whalen, M.W.: Reasoning about algebraic data types with abstractions. J. Autom. Reason. **57**(4), 281–318 (2016)
47. Reynolds, A., Blanchette, J.C.: A decision procedure for (co) datatypes in SMT solvers. J. Autom. Reason. **58**(3), 341–362 (2017)
48. Reynolds, A., Kuncak, V.: Induction for SMT solvers. In: D'Souza, D., Lal, A., Larsen, K.G. (eds.) VMCAI 2015. LNCS, vol. 8931, pp. 80–98. Springer, Heidelberg (2015). https://doi.org/10.1007/978-3-662-46081-8_5
49. Rümmer, P.: A constraint sequent calculus for first-order logic with linear integer arithmetic. In: Cervesato, I., Veith, H., Voronkov, A. (eds.) LPAR 2008. LNCS (LNAI), vol. 5330, pp. 274–289. Springer, Heidelberg (2008). https://doi.org/10.1007/978-3-540-89439-1_20
50. Fedyukovich, G., Rümmer, P.: Competition report: CHC-COMP-21. In: Hojjat, H., Kafle, B. (eds.) Proceedings 8th Workshop on Horn Clauses for Verification and Synthesis, HCVS@ETAPS 2021, Virtual, 28th March 2021. EPTCS, vol. 344, pp. 91–108 (2021). https://doi.org/10.4204/EPTCS.344.7
51. Rümmer, P., Hojjat, H., Kuncak, V.: Disjunctive interpolants for horn-clause verification. In: Sharygina, N., Veith, H. (eds.) CAV 2013. LNCS, vol. 8044, pp. 347–363. Springer, Heidelberg (2013). https://doi.org/10.1007/978-3-642-39799-8_24
52. Satake, Y., Unno, H., Yanagi, H.: Probabilistic inference for predicate constraint satisfaction. In: AAAI, pp. 1644–1651. AAAI Press (2020)
53. Sharma, R., Aiken, A.: From invariant checking to invariant inference using randomized search. In: Biere, A., Bloem, R. (eds.) CAV 2014. LNCS, vol. 8559, pp. 88–105. Springer, Cham (2014). https://doi.org/10.1007/978-3-319-08867-9_6
54. Solar-Lezama, A., Tancau, L., Bodík, R., Seshia, S.A., Saraswat, V.A.: Combinatorial sketching for finite programs. In: ASPLOS, pp. 404–415. ACM (2006)
55. Suter, P., Dotta, M., Kuncak, V.: Decision procedures for algebraic data types with abstractions. ACM Sigplan Not. **45**(1), 199–210 (2010)
56. Suter, P., Köksal, A.S., Kuncak, V.: Satisfiability modulo recursive programs. In: Yahav, E. (ed.) SAS 2011. LNCS, vol. 6887, pp. 298–315. Springer, Heidelberg (2011). https://doi.org/10.1007/978-3-642-23702-7_23
57. Unno, H., Terauchi, T.: Inferring simple solutions to recursion-free horn clauses via sampling. In: Baier, C., Tinelli, C. (eds.) TACAS 2015. LNCS, vol. 9035, pp. 149–163. Springer, Heidelberg (2015). https://doi.org/10.1007/978-3-662-46681-0_10
58. Unno, H., Torii, S., Sakamoto, H.: Automating induction for solving horn clauses. In: Majumdar, R., Kunčak, V. (eds.) CAV 2017. LNCS, vol. 10427, pp. 571–591. Springer, Cham (2017). https://doi.org/10.1007/978-3-319-63390-9_30
59. Yang, W., Fedyukovich, G., Gupta, A.: Lemma synthesis for automating induction over algebraic data types. In: Schiex, T., de Givry, S. (eds.) CP 2019. LNCS, vol. 11802, pp. 600–617. Springer, Cham (2019). https://doi.org/10.1007/978-3-030-30048-7_35

60. Zhang, T., Sipma, H.B., Manna, Z.: Decision procedures for recursive data structures with integer constraints. In: Basin, D., Rusinowitch, M. (eds.) IJCAR 2004. LNCS (LNAI), vol. 3097, pp. 152–167. Springer, Heidelberg (2004). https://doi.org/10.1007/978-3-540-25984-8_9
61. Zhu, H., Magill, S., Jagannathan, S.: A data-driven CHC solver. In: PLDI, pp. 707–721. ACM (2018)

ARENA: Enhancing Abstract Refinement for Neural Network Verification

Yuyi Zhong[✉], Quang-Trung Ta, and Siau-Cheng Khoo

School of Computing, National University of Singapore, Singapore, Singapore
{yuyizhong,taqt,khoosc}@comp.nus.edu.sg

Abstract. As neural networks have taken on a critical role in real-world applications, formal verification is earnestly needed to guarantee the safety properties of the networks. However, it remains challenging to balance the trade-off between precision and efficiency in abstract interpretation based verification methods. In this paper, we propose an abstract refinement process that leverages the convex hull techniques to improve the analysis efficiency. Specifically, we introduce the double description method in the convex polytope domain to detect and eliminate multiple *spurious* adversarial labels simultaneously. We also combine the new activation relaxation technique with the iterative abstract refinement method to compensate for the precision loss during abstract interpretation. We have implemented our proposal into a verification framework named ARENA, and assessed its effectiveness by conducting a series of experiments. These experiments show that ARENA yields significantly better verification precision compared to the existing abstract-refinement-based tool DeepSRGR. It also identifies falsification by detecting adversarial examples, with reasonable execution efficiency. Lastly, it verifies more images than the state-of-the-art verifier PRIMA.

Keywords: Abstract refinement · Double description method · Neural network verification

1 Introduction

As neural networks have been proverbially applied to safety-critical systems, formal guarantee about the safety properties of the networks, such as robustness, fairness, etc., is earnestly needed. For example, researchers have been working on robustness verification of neural networks, to ascertain that the network classification result can remain the same when the input image is perturbed subtly and imperceptibly during adversarial attacks [1,2].

There exists sound and complete verification techniques where the robustness property can be ascertained but regrettably at high complexity and execution cost [3,4]. For better scalability, several incomplete verifiers have been proposed to analyze larger networks with abstract interpretation technique while bearing

---

The original version of the chapter has been revised. The acknowledgment section have been corrected. A correction to this chapter can be found at
https://doi.org/10.1007/978-3-031-24950-1_18

C. Dragoi et al. (Eds.): VMCAI 2023, LNCS 13881, pp. 366–388, 2023.
https://doi.org/10.1007/978-3-031-24949-5_17

exactness sacrifices [5–7]. To mitigate this shortcoming, there have been investigations into better convex relaxation [8,9] or iterative abstract refinement [10] to make up with the precision loss in abstract interpretation techniques.

This work is inspired by the counterexample guided abstraction refinement (CEGAR) method [11] in program analysis, aiming to improve the precision of abstract interpretation results by identifying *spurious counterexamples*: these are examples which appear to have violated desired analysis outcome – due to over-approximated calculation inherent in the abstract interpretation computation – but can be shown to be fake by the refinement method. Proof of the existence of spurious counterexamples can diminish the range of inconclusive results produced by abstract interpretation. In the context of neural network verification, such spurious counterexamples can be conceptualized as adversarial *regions* that are perceived to have lent support (spuriously) on certain adversarial labels; ie., labels which differ from the designated label in the robustness test.

An existing work that has successfully employed abstract refinement technique to improve the precision of the abstract-interpretation based verification tool is DeepSRGR [10]. That work repetitively selects an adversarial label and attempts to eliminate the corresponding spurious region progressively through iteration of refinements. Technically, it encodes a spurious region as a linear inequality, adds it to the constraint encoding of the network, and employs linear programming with the objective set to optimize the concrete bounds of selected ReLU neurons in the network. This process is repeated until either the spurious region is found to be inconsistent with the encoded network, or time out.

In this paper, we enhance the existing effectiveness of DeepSRGR by introducing *convex hull* techniques (ie., techniques that observe and conform to convex property) to abstract refinement computation. Together, these techniques facilitate *simultaneous* elimination of multiple spurious regions, and capitalize the *dependencies* among ReLU neurons. Specifically, we tighten the looseness of ReLU approximation during abstract refinement process through a *mutli-ReLU convex abstraction technique* (cf. [9]) that captures dependencies within a set of ReLU neurons. Moreover, we leverage a double-description method (cf. [12]) used in convex polytope computation to eliminate multiple spurious regions simultaneously; this circumvents the challenges faced with the application of linear programming technique to optimize disjunction of linear inequalities.

We have implemented our proposed techniques in a CPU-based prototypical analyzer named ARENA (Abstract Refinement Enhancer for Neural network verificAtion). In addition to verifying the robustness property of a network with respect to an image, ARENA is also capable of detecting adversarial examples that ascertain the falsification of the network property. We conducted experiments to assess the effectiveness of ARENA against the state-of-the-art tools, including the CPU-based verifiers DeepSRGR [10] and PRIMA [9], and the GPU-based verifier α, β-CROWN [13]. The results show conclusively that ARENA returns an average of 15.8% more conclusive images compared with DeepSRGR while terminates in comparable amount of time; and it also outperforms PRIMA by returning 16.6% more conclusive images. Furthermore, ARENA can verify or falsify 79.3% images of that of the state-of-the-art complete tool α, β-CROWN on average for selected networks.

We summarize our contributions below:

⋄ We adapt the double description method proposed in the convex polytope domain [12] to solve disjuncts of constraints in Linear Programming (LP) encoding, allowing us to prune multiple adversarial labels together to increase overall efficiency.
⋄ We leverage the multi-ReLU convex abstraction in PRIMA [9] to further refine the abstraction in the analysis process to increase verification precision.
⋄ We utilize the solutions returned by the LP solver to detect adversarial examples and assert property violation when counter-examples are discovered.
⋄ We conducted experiments comparing our prototypical analyzer ARENA against state-of-the-art verification tools, and demonstrate high effectiveness in our verification framework. To the best of our knowledge, ARENA outperforms the current state-of-the-art approximated methods that run on CPU.

In the remaining part of the paper, we give an illustrative example showing the overall process of our method in Sect. 2, followed by a formal description of our methodologies in Sect. 3. We demonstrate our evaluation process and experimental results in Sect. 4. Section 5 discusses the current limitation, plan for future work and the generalization of our work. We give a literature review in Sect. 6, which contains closely related works with respect to our research scope. Finally, we summarize our work and conclude in Sect. 7.

2 Overview

In this section, we first describe the abstract refinement technique implemented in DeepSRGR [10]. Then, we discuss its limitations and introduce our approach to overcome them. Table 1 displays the notations we use throughout this section.

Table 1. Notations and descriptions of Sect. 2

Π	The network constraint set/encoding
Υ	A potential adversarial region
P	The over-approximate convex hull

2.1 Spurious Region Guided Refinement

DeepSRGR is a sound but incomplete verification method that relies on the polyhedral abstract domain in DeepPoly [7], where the abstract value of each network neuron x_i is designed to contain four elements (l_i, u_i, l_i^s, u_i^s). The *concrete* lower bound l_i and upper bound u_i pair forms a closed interval $[l_i, u_i]$ that over-approximates all the values that neuron x_i could take. The *symbolic* constraints l_i^s, u_i^s are linear expressions of x_i defined over preceding neurons with the requirement that $l_i^s \leq x_i \leq u_i^s$.

In the following, we illustrate the verification process where the abstract domain is used to verify the robustness property of a fully-connected network with ReLU activation (Fig. 1) w.r.t the input space $I = [-1, 1] \times [-1, 1]$ of 2 input neurons x_1, x_2. This network has 3 output neurons y_1, y_2, y_3, corresponding to the three labels L_1, L_2, L_3 that an input in I can be classified as. Here, the robustness property which we aim to verify is that the neural network can always classify the entire input space I as label L_1, which corresponds to the output neuron y_1. More specifically, the verifier should be able to prove that the conditions $y_1 - y_2 > 0$ and $y_1 - y_3 > 0$ always hold for the entire input space I.

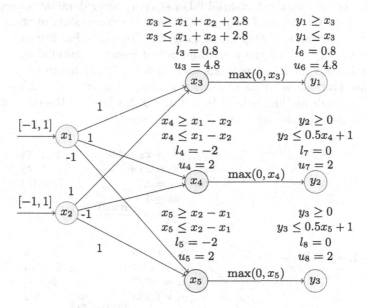

Fig. 1. The example network to perform DeepPoly abstract interpretation

Through the abstract interpretation technique, as deployed by DeepPoly, we can compute the abstract values for each neuron; these are displayed near the corresponding nodes. Specifically, the computed value for the lower bound of $y_1 - y_2$ and $y_1 - y_3$ are both -0.2 (the process of the lower bound computation is provided in Appendix A in our technical report), which fails to assert that $y_1 - y_2 > 0$ and $y_1 - y_3 > 0$. In other word, DeepPoly cannot ascertain the robustness of the network for the given initial input space I. Given the over-approximation nature of abstract interpretation technique, it is not clear if the robustness property can be verified.

In order to further improve the robustness verification of the considered neural network, DeepSRGR conducts a *spurious region guided refinement* process that includes the following steps:

1. Obtain the conjunction of all linear inequities that encode the network, including the input constraint $x_1, x_2 \in [-1, 1]$ and the constraints within the abstract values of all neurons (*i.e.*, the constraints in Fig. 1). We denote this network encoding constraint set as Π.

2. Take the conjunction of the current network encoding and the negation of the property to solve a potential *spurious* region. For example, the feasible region of $\Pi \wedge (y_1 - y_2 \leq 0)$ refers to a *potential* adversarial region (denote as Υ) that may contain a counterexample with adversarial label L_2 (corresponding to output neural y_2); whereas the region *outside* of Υ is already a safe region that will not be wrongly classified as L_2. However, the region Υ may exist only due to the over-approximate abstraction but does not contain any true counterexample. Therefore, this region is spuriously constructed and could be eliminated. If we successfully eliminate Υ, then we can conclude that label L_2 will not be a valid adversarial label since y_2 never dominates over y_1.

3. To eliminate the region Υ, DeepSRGR uses the constraints of the region to refine the abstraction using linear programming (LP). For instance, we take Π and $y_1 - y_2 \leq 0$ as the constraint set of linear programming. To obtain tighter bounds for input neurons and unstable ReLU neurons[1], we set the objective function of LP as $\min(x_i)$ and $\max(x_i)$ where $i \in [1, 2, 4]$. The new solved intervals are highlighted in red in Fig. 2, where all the current neuron intervals now specify the region Υ.

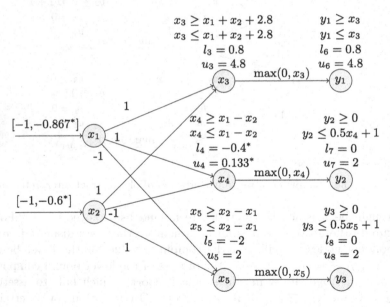

Fig. 2. The effect of applying LP-based interval refinement (in red marked by *) (Color figure online)

4. DeepSRGR leverages those tighter bounds to guide the abstract interpretation of the region Υ in the next iteration. It performs a second run on DeepPoly and makes sure that this second run compulsorily follows the new

---

[1] Unstable ReLU neuron refers to a ReLU neuron whose input range can be both negative and positive (like y_2, y_3).

bounds computed in the previous step. As shown in Fig. 3, the blue colored part refers to the updated abstract values during the second execution of DeepPoly, where the abstraction of all neurons are refined due to the tighter bounds (red colored part) returned by LP solving. Now the lower bound of $y_1 - y_2$ is 0.7, making $y_1 - y_2 \leq 0$ actually infeasible within the region Υ. Therefore, we conclude that Υ is a spurious region that does not contain any true counterexample, and we can eliminate adversarial label L_2.

5. If we fail to detect $y_1 - y_2 \leq 0$ to be infeasible, DeepSRGR iterates the process from step 2–4 where it calls LP solving and re-executes DeepPoly on the new bounds until it achieves one of the termination conditions: (i) It reaches the maximum number of iterations (DeepSRGR sets it to be 5 by default); or (ii) it detects infeasibility for the spurious region.

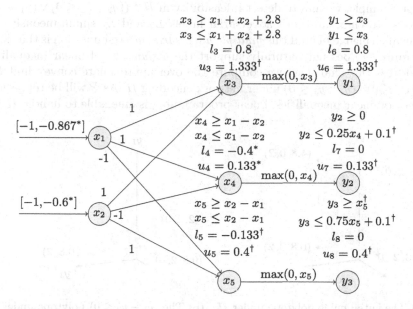

Fig. 3. Results of the second run of DeepPoly (in blue marked by †) (Color figure online)

Similarly, after eliminating the adversarial label L_2, DeepSRGR will apply the same process to eliminate the spurious region defined by $\Pi \wedge (y_1 - y_3 \leq 0)$, which corresponds to the output neural y_3 and the adversarial label L_3.

In summary, DeepSRGR uses iterative LP solving and DeepPoly execution to attempt to eliminate spurious regions which do not contain counterexamples. Assuming the ground-truth label to be L_c, DeepSRGR runs this refinement process for each region $\Pi \wedge (y_c - y_t \leq 0)$ where $t \neq c$. If DeepSRGR is able to eliminate all adversarial labels related to output neurons y_t where $t \neq c$, then it successfully ascertains the robustness property of the image. If DeepSRGR fails to eliminate one of the adversarial labels within the iteration boundary, the robustness result remains inconclusive.

2.2   Scaling up with Multiple Adversarial Label Elimination

We mention three contributions in Sect. 1 including efficiency improvement, precision improvement and adversarial example detection. In this section, we only give an overview of our multiple adversarial label elimination method which aims to improve the analysis efficiency; we defer the discussion of the remaining part of our system to Sect. 3.

As mentioned in Sect. 2.1, DeepSRGR invokes the refinement process to sequentially eliminate each spurious region $\Pi \wedge (y_c - y_t \leq 0)$, which corresponds to the adversarial label L_t ($t \neq c$). For an n-label network, it requires $n-1$ refinement invocations in the worst case, with each invocation taking possibly several iterations. To speed up the analysis, we eliminate multiple spurious regions at the same time in one refinement process.

For example, we aim to detect infeasibility in $\Pi \wedge ((y_1 - y_2 \leq 0) \vee (y_1 - y_3 \leq 0))$ so as to eliminate both adversarial labels L_2 and L_3 simultaneously. The technical challenge behind this *multiple adversarial label* elimination is that linear programming does not naturally support the *disjunction* of linear inequalities. To address this challenge, we compute the over-approximate convex hull P of $(y_1 - y_2 \leq 0) \vee (y_1 - y_3 \leq 0)$ under network encoding Π. As P will be represented as a set of linear inequalities, linear programming is amenable to handle $\Pi \wedge P$.

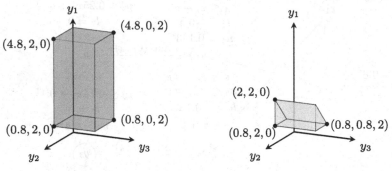

(a) The initial cubic polytope under Π (b) The $(y_1 - y_2 \leq 0)$ polytope under Π

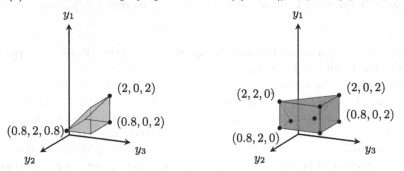

(c) The $(y_1 - y_3 \leq 0)$ polytope under Π (d) The convex hull of union of (b),(c)

Fig. 4. The convex polytopes under network encoding, with respect to (y_1, y_2, y_3).

In detail, the initial convex polytope associated with y_1, y_2, y_3 is a 3-D cube pictured in Fig. 4a, where $y_1 \in [0.8, 4.8], y_2 \in [0, 2], y_3 \in [0, 2]$ after we perform DeepPoly as shown in Fig. 1. The convex polytope for the constraint $y_1 - y_2 \leq 0$ under the network encoding Π corresponds to the shape in Fig. 4b where $y_1 - y_2 \leq 0$ is a cutting-plane imposed on the initial cube in Fig. 4a. Similarly, the projection of $y_1 - y_3 \leq 0$ to a convex polytope can be visualized in Fig. 4c. We further compute the over-approximate convex hull P of the union of the two polytopes as in Fig. 4d.

We can observe that P is defined by 8 vertices (annotated as eight black extreme points). It is worth-noting that these 8 vertices actually come from either vertices in Fig. 4b or vertices in Fig. 4c. We will provide the explanation and the theory on how to compute the convex hull of the union of two polytopes in Sect. 3.2. Explicitly, P can also be represented by the following constraint set (1), which correspond to the 7 red-colored surfaces in Fig. 4d:

$$-y_1 + y_2 + y_3 \geq 0 \qquad y_2 \geq 0 \qquad y_3 \geq 0 \qquad -1 + 1.25y_1 \geq 0$$
$$2 - y_1 \geq 0 \qquad 2 - y_2 \geq 0 \qquad 2 - y_3 \geq 0 \tag{1}$$

We take the network encoding Π and constraint set of P as the input to the LP solver, and conduct interval solving as in Sect. 2.1. We annotate the new bounds obtained through LP solving as red color, and the updated abstract values after the second abstract interpretation as blue color in Fig. 5. The lower bounds of both $y_1 - y_2$ and $y_1 - y_3$ now become 0.2, making it infeasible to achieve $y_1 - y_2 \leq 0$ or $y_1 - y_3 \leq 0$. Therefore, we successfully do the verification with just one refinement process invocation.

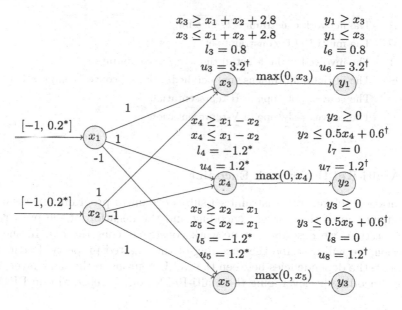

Fig. 5. The new intervals (in red with *) with two-adversarial labels encoding and new abstraction introduced by the second run of DeepPoly (in blue with †) (Color figure online)

3 Methodologies

As described in Sect. 2, we identify the feasible region of the network encoding and the negation of a property (i.e. $\Pi \wedge (y_c - y_t \leq 0) : t \neq c$) as a potential spurious region and leverage the refinement process to ascertain and possibly eliminate such spurious regions. To further improve the precision and efficiency, we propose three techniques as we have summarized in Sect. 1:

1. We update the negation of the property encoding to capture multiple spurious regions at the same time, as we demonstrate the example on $(y_1 - y_2 \leq 0) \vee (y_1 - y_3 \leq 0)$ in Sect. 2.2. This method allows us to reuse the linear programming part among several spurious regions and improve efficiency.
2. We leverage the multi-ReLU convex abstraction proposed in PRIMA [9] to obtain a more precise network encoding Π, which helps to increase the verification precision.
3. We detect adversarial examples to falsify robustness property. In particular, as the LP solver finds the conjunction of the network encoding and the negation of the property to be feasible, its optimization solution could actually ascertain a property violation and help us conclude with falsification.

We will discuss the three methodologies in separate subsections, and conclude this section with an overall description of our verification framework ARENA. Table 2 shows the notations we use in the main text of this section.

Table 2. Notations and descriptions of Sect. 3

Π	The network constraint set/encoding
Ω	The multi-ReLU constraint set
Λ	The involved variable set during convex computation
Θ	The initial multidimensional octahedra during convex computation
P_i^H	The convex polytope in H-representation
P_i^V	The convex polytope in V-representation

3.1 Multi-ReLU Network Encoding

As mentioned before, the constraint set subject to linear programming resolution is a conjunction of the network encoding and the negation of the property. In this subsection, we describe our network constraint construction. In the next subsection, we will describe the encoding of the negated property. Particularly, we capture the dependencies between the ReLU neurons in the same layer in our network encoding by leveraging the multi-ReLU convex relaxation in PRIMA.

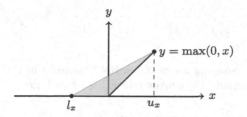

Fig. 6. The triangle approximation of a single ReLU neuron

As depicted in Fig. 6, DeepSRGR uses a triangular shape to encode each ReLU neuron independently, where the ReLU node $y = \max(0, x)$ with $x \in [l_x, u_x]$. The triangular shape is defined by three linear constraints:
$$y \geq x \qquad\qquad y \geq 0 \qquad\qquad y \leq \frac{u_x}{u_x - l_x}(x - l_x)$$
This looseness of ReLU encoding can inhibit precision improvement in Deep-SRGR. As a matter of fact, it has been reported that, when they increase the maximum number of iterations from 5 to 20, only two more properties can be verified additionally, and no more properties can be verified when they further increase from 20 to 50 [10].

To break this precision barrier, we deploy the technique of *multi-ReLU relaxation* in PRIMA [9] where they compute the convex abstraction of k-ReLU neurons via novel convex hull approximation algorithms. For instance, if $k = 2$ and the ReLU neurons in the same layer are denoted by y_1, y_2, and the inputs to these two ReLU neurons are x_1, x_2 respectively, PRIMA will compute a convex hull in (y_1, y_2, x_1, x_2) space to capture the relationship between the two ReLU neurons and their inputs. An example of the convex hull is defined as:
$$\Omega = \{\, x_1 + x_2 - 2y_1 - 2y_2 \geq -2, \qquad 0.375x_2 - y_2 \geq -0.75,$$
$$-x_1 + y_1 \geq 0, \qquad -x_2 + y_2 \geq 0, \qquad y_1 \geq 0, \qquad y_2 \geq 0 \,\}$$
As we can see, Ω contains the constraint $x_1 + x_2 - 2y_1 - 2y_2 \geq -2$ that correlates (y_1, y_2, x_1, x_2) all together, which is beyond the single ReLU encoding. In general, PRIMA splits the input region into multiple sub-regions and then computes the convex hull of multiple ReLU neurons. For example, splitting the input region along $x_1 = 0$ results in two sub-regions where $y_1 = x_1$ (y_1 is activated) and $y_1 = 0$ (y_1 is deactivated). In each sub-region, the behavior of y_1 is determinate and this yields a tighter or even exact convex approximation. Finally, PRIMA computes a joint convex over-approximation (as in Ω) of the convex polytopes computed for each sub-region.

For deployment, we consider 3-ReLU neurons in our paper. We filter out the unstable ReLU neurons in each ReLU layer, and divide them into a set of 3-ReLU groups with one overlapping neuron between two adjacent groups as shown in Fig. 7, where a dashed box identifies a 3-ReLU group. We then leverage PRIMA to compute the constraints for each 3-ReLU group, and add those additional constraints into the original network encoding in order to obtain a more precise network abstraction and better verification precision.

Fig. 7. The 3-ReLU grouping for unstable ReLU neurons in the same layer i, where we use PRIMA to compute the convex relaxation for each group.

3.2   Multiple Adversarial Label Elimination

We now explain how we encode the negated property, especially when we take multiple spurious regions into consideration. As demonstrated in Sect. 2.2, to make it amenable for LP encoding, we need to compute the over-approximate convex hull of the union of multiple convex polytopes like in Fig. 4d. To explain the theory behind, we first introduce the required knowledge with respect to convex polytope representation. The convex polytope in this paper refers to a bounded convex polytope that is also a convex region contained in the $n-$dimensional Euclidean space R^n. There are two essential definitions of a convex polytope: as the intersection of half-space constraints (H-representation) and as the convex hull of a set of extremal vertices (V-representation) [14].

H-Representation. A convex polytope can be defined as the intersection of a finite number of closed half-spaces. A closed half-space in an n-dimensional space can be expressed by a linear inequality:

$$a_1x_1 + a_2x_2 + \cdots + a_nx_n \leq b \tag{2}$$

A closed convex polytope can be taken as the set of solutions to a linear constraint set, just like the constraint set (1) shown in Sect. 2.2.

V-Representation. A closed convex polytope can also be defined as the convex hull with a finite number of points where this finite set must contain the set of extreme points of the polytope (i.e. the black-colored dots in Fig. 4d).

Double Description. The Double description method [12] aims to maintain both V-representation and H-representation during computation. This "duplication" is beneficial because to compute the intersection of two polytopes in H-representation is trivial since we only need to take the union set of the half-space constraints. On the other hand, to compute the convex hull of the union of two polytopes is trivial in V-representation as we take the union set of the vertices. The program `cddlib`[2] is an efficient implementation of the double description method, which provides functionalities that enable transformation from V-representation to H-representation (named as convex hull problem); and vice versa (named as vertex enumeration problem).

We leverage this V-H transformation in cddlib to compute the convex hull of the union of multiple convex polytopes. We set up a *batch size* δ to be in $[2, 5]^3$, which defines the number of spurious regions to be considered simultaneously.

---

[2] https://github.com/cddlib/cddlib.

[3] We explain our parameter range setting in Sect. 5 and also provide the batch size study experiments in Sect. 4.4.

Assume that the ground truth label is L_c, the related adversarial labels are $L_1, \cdots, L_\delta$, the convex-hull computation of $(y_c - y_1 \leq 0) \vee \cdots \vee (y_c - y_\delta \leq 0)$ is conducted as follows:

Polytope Computation for Each Spurious Region. We compute the H-representation of the polytope for each spurious region in the $(\delta+1)$-dimensional space with respect to the variable set $\Lambda = y_c, y_1, \cdots, y_\delta$. Intuitively, we obtain the H-representation of polytope $(y_c - y_i \leq 0)$ by taking the interval constraints of Λ (which is a multidimensional cube) conjunct with $y_c - y_i \leq 0$, as our example in Fig. 4. But this encoding is coarse as we neglect the dependencies between Λ that are in the same layer. For a more precise encoding, we follow the idea of [8] and compute the multidimensional octahedra Θ of $y_c, y_1, \cdots, y_\delta$, which yields $3^{\delta+1} - 1$ constraints defined over Λ. Therefore, the H-representation of polytope $(y_c - y_i \leq 0)$ will be the constraint set Θ and $y_c - y_i \leq 0$.

Union of Convex Polytopes. We obtain the H-representation of the δ polytopes in the previous step and denote them by $P_1^H, \cdots, P_\delta^H$ respectively. Since the union of polytopes is trivial in V-representation – as mentioned earlier, we use the H-V transformation in cddlib to generate these V-representations of the δ polytopes (referred to as $P_1^V, \cdots, P_\delta^V$). As illustrated in Fig. 8, we then produce the union set P_u^V of these vertices sets and transform it to its H-representation P_u^H, which is the convex hull of the union of δ polytopes. As P_u^H is represented by a set of linear inequalities, we conjunct it with the network encoding Π and submit the constraints for LP solving.

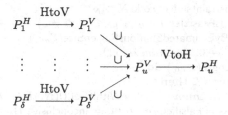

Fig. 8. The convex hull computation of the union of δ convex polytopes

3.3 Adversarial Example Detection

As mentioned previously, we take the conjunction of the network encoding and the negation of the property as the input constraint set to the LP solver and aim to eliminate spurious region(s) when detecting infeasibility. On the other hand, when the constraint set is feasible, we can set the input neurons and the unstable ReLU neurons as objective function and try to resolve for tighter intervals. In fact, a feasible constraint set indicates the possibility of a property violation. The LP solver not only returns the optimized value of the objective function, it also returns a solution that leads to the optimization, which could be a potential counter-example of robustness. Therefore, we include a supplementary procedure that takes each optimal solution obtained from the LP solver and checks if it constitutes an adversarial example.

This process brings forth two benefits: (1) it detects counter-examples and asserts the violation of robustness; (2) it enables the process to *terminate early with falsification* (once a counter-example is discovered) instead of exhausting all the iterations.

3.4   The Verification Framework ARENA

We now present an overview of our verification framework ARENA. In addition to the implementation of the three main technical points covered earlier, our framework includes the following optimizations as well.

Algorithm 1: Overall analysis procedure in ARENA

Input:
- N: input neural network with input layer γ_{in}, and output neurons $y_1, ..., y_n$
- y_c: the output neuron corresponding to the ground truth label L_c ($1 \leq c \leq n$)
- δ is the refinement batch size (the number of adversarial labels to be eliminated by batch in each iteration).

Output: Verification result (Verified for robustness verified, Falsified for robustness violated, Inconclusive for inconclusive result).

```
 1: (res, A_N) ← VerifyByDeepPoly(N)              // Result and network abstraction
 2: if res = Verified then
 3:     return Verified
 4: else
 5:     Π ← GetConstraintsInNetwork(N, A_N)
 6:     L_adv ← {L_i |IsFeasible(Π ∧ y_c − y_i ≤ 0), ∀i ≠ c}    // All adversarial labels
 7:     L_adv ← SortByEstimatedSpuriousRegionSize(L_adv)       // Sort decreasingly
 8:     L_elim ← ∅; i ← 0; iter_num ← 999                      // Initialization
 9:     while i < Length(L_adv) do
10:         if iter_num > 2 then
11:             status, iter_num ← RefineWithKReLU(N, Π, L_c, L_adv[i], L_elim)
12:             if status = Falsified or status = Inconclusive then
13:                 return status
14:             else                                           // status = Verified
15:                 L_elim ← L_elim ∪ {L_adv[i]}
16:                 i ← i + 1
17:         else
18:             L' ← GetNextAdversarialLabelBatch(L_adv, δ)
19:             status ← EliminateAdversarialLabels(N, Π, L_c, L', L_elim)
20:             if status = Falsified or status = Inconclusive then
21:                 return status
22:             else                                           // status = Verified
23:                 L_elim ← L_elim ∪ {L'}
24:                 i ← i + SizeOf(L')
25:     if L_elim = L_adv then
26:         return Verified
27:     else
28:         return Inconclusive
```

Optimization 1: *Prioritising elimination of larger spurious regions.* We choose to order the sequence of the spurious regions according to the descending order of the respective regions' sizes, by eliminating the "toughest" spurious region first. Since robustness only holds when all spurious regions are eliminated, we terminate the refinement process early if we fail to prune a larger spurious region. As it is difficult to compute the actual size of the spurious region, we deploy the metric in DeepSRGR where they take the lower bound of expression $y_c - y_i$ given by DeepPoly as the estimation of the region size, i.e. the smaller this value is, the larger the region is likely to be and thus it would be tougher for us to eliminate the region.

Optimization 2: *Cascading refinement.* Our system is designed to apply increasingly more scalable and less precise refinement methods. We hereby define process RefineWithKReLU as the refinement method with multi-ReLU encoding for the network and multi-adversarial label pruning feature *disabled*. Similarly, process EliminateAdversarialLabels(δ) is the refinement method with multi-ReLU encoding, and *taking into consideration δ spurious regions simultaneously*. With additional over-approximation error potentially being introduced by computing the union of polytopes, EliminateAdversarialLabels(δ) is less precise method compared to RefineWithKReLU but more scalable as it eliminates δ spurious regions simultaneously. We first use RefineWithKReLU to eliminate the larger spurious regions and record the number of iterations ς required to prune the current spurious region. If $\varsigma \leq 2$, this indicates that it is rather amiable to prune the current spurious region, and affordable to call upon EliminateAdversarialLabels(δ) to eliminate the remaining smaller spurious regions.

We present the overall analysis procedure in Algorithm 1. To begin with, we only apply the refinement process to images that fail to be verified by DeepPoly (lines 4). For refinement, we first obtain all the network constraints generated during abstract interpretation (line 5) and all potential adversarial labels (line 6). Then we call upon the processes RefineWithKReLU (line 11) or EliminateAdversarialLabels(δ) (line 19) to eliminate one or multiple spurious regions as stated in optimization 2 mechanism. The analyzer returns "Falsified" value if it detects an adversarial example (lines 13, 21); or it returns "Inconclusive" value if it fails to eliminate one of the adversarial labels and fails to find a counter-example (lines 13, 21, 28). We declare verification to be successful if and only if we can eliminate all the adversarial labels (lines 25–26).

Details of the two refinement processes are presented in Algorithms 2 and 3 (in Appendix B in our technical report). These two algorithms only differ in the property encoding (line 3–4 in Algorithm 3 vs lines 3–4 in Algorithm 2). Reading Algorithm 2 more closely: it first computes the convex hull of the union of the spurious regions according to Sect. 3.2 (line 3). Next, it conjuncts the convex hull with the network encoding in line 4, and add constraint $y_c - y_t > 0$ for each of the previously eliminated adversarial label L_t (lines 5–6); this helps to reduce the solution space further. If the combined constraint set Σ is found to be infeasible, the process returns "verified" since violation of the property cannot be attained (lines 7–8). But if Σ is feasible, the process leverages it to further tighten the bounds for input and unstable ReLU neurons and updates the network (lines 9, 14). Moreover, the process checks if each LP solution is a valid counter-example; if so, it returns "Falsified" (lines 10–11, 15–16). With the newly solved bounds, the process then re-runs DeepPoly to obtain a tighter network encoding (lines 17–18) that is more amenable to encounter infeasibility in the latter iterations. Finally, the process returns "inconclusive" if it fails to conclude within the maximum number of iterations (line 20).

Algorithm 2: The refinement procedure EliminateAdversarialLabels

Function Name: EliminateAdversarialLabels($N, \Pi, L_c, \mathcal{L}', \mathcal{L}_{elim}$)

Input:
- N: input neural network with input layer γ_{in}, and output neurons $y_1, ..., y_n$
- Π: the constraint set of N
- y_c: the output neuron corresponding to the ground truth label L_c ($1 \leq c \leq n$)
- $\mathcal{L}', \mathcal{L}_{elim}$: the batch of adversarial labels to be refined, and the list of previously eliminated labels

Output: the refinement status

```
 1: counter = 0
 2: while counter < τ do                           // τ is an iteration threshold
 3:    P_u^H ← ComputeConvexHull(N, L')   // P_u^H is the convex hull of polytopes
 4:    Σ ← Π ∧ P_u^H                               // Initialize constraint set
 5:    for all L_t ∈ L_elim do
 6:        Σ ← Σ ∧ (y_c − y_t > 0)
 7:    if IsInfeasible(Σ) then
 8:        return Verified
 9:    N ← LPSolveInputInterval(Σ, γ_in)   // Update network with new bounds
10:    if ExistsAnAdversarialExample(N) then
11:        return Falsified
12:    for all ReLU layer γ'_k in N do
13:        γ_k ← GetPrecedingInputAffineLayer(γ'_k)
14:        N ← LPSolveUnstableReLUs(Σ, γ_k)               // Update new bounds
15:        if ExistsAnAdversarialExample(N) then
16:            return Falsified
17:    A ← RecomputeNetworkAbstractionByDeepPoly(N)
18:    Π ← GetConstraintsInNetwork(N, A)
19:    counter = counter + 1
20: return Inconclusive
```

---

4   Experiments

We implemented our method in a prototypical verifier called ARENA in both C and C++ programming languages (C++ is used for the k-ReLU computation feature, while the rest of the system is implemented in C). Our verifier is built on top of DeepPoly in [15]: it utilizes DeepPoly as the back-end abstract interpreter for neural networks. Moreover, it uses Gurobi[4] version 9.5 as the LP solver for constraints generated during abstract refinement.

We evaluate the performance of ARENA with state-of-the-art CPU-based incomplete verifiers including DeepSRGR [10], PRIMA [9], and DeepPoly [7]. Furthermore, we compare with a complete verifier α, β-CROWN [13], which is GPU-based and the winning tool of VNN-COMP 2022 [16]. The evaluation machine is equipped with two 2.40 GHz Intel(R) Xeon(R) Silver 4210R CPUs with 384 GB of main memory and a NVIDIA RTX A5000 GPU. The implementation is 64-bit based.

Note that DeepSRGR [10] was purely implemented in Python, while the main analysis in ARENA, PRIMA and DeepPoly were implemented in C/C++. Furthermore, this original version of DeepSRGR does not support convolutional networks nor the ONNX network format in our benchmark. Therefore, to avoid any runtime discrepancy introduced by different languages and to support our tested networks, we re-implemented the refinement technique of DeepSRGR in C, and conducted the experiment on the re-implemented DeepSRGR, where we release our re-implementation of DeepSRGR at this link: https://github.com/arena-verifier/DeepSRGR. The source code of our verifier ARENA is available online at: https://github.com/arena-verifier/ARENA.

4.1   Experiment Setup

Evaluation Datasets and Testing Networks. We chose the commonly used MNIST [17] and CIFAR10 [18] datasets. MNIST is an image dataset with handwritten digits, containing gray-scale images with 28×28 pixels. CIFAR10 includes RGB three-channel images with size 32×32. Our testing image set consists of the first 100 images of the test set of each dataset, which is accessible from [15].

We selected fully-connected (abbreviated as FC) and convolutional (abbreviated as Conv) networks from [15] as displayed in Table 3, with up to around 50k neurons. We explicitly list the number of hidden neurons, the number of activation layers, trained defense[5], and the number of candidate images for each network. Here, the candidate images refer to those testing images that can be correctly classified by the network and we only apply robustness verification on the candidate images.

Robustness Analysis. We conducted robustness analysis against L_∞ norm attack [19] with a perturbation parameter ϵ. Assuming that each pixel in the test image originally takes an intensity value p_i, it now takes an intensity interval $[p_i - \epsilon, p_i + \epsilon]$ after applying L_∞ norm attack with a specified constant ϵ.

This naturally forms an input space defined by $\times_{i=1}^n [p_i - \epsilon, p_i + \epsilon]$, and all the tools attempt to verify if all the "perturbed" images within the input space will be classified the same as the original image by the tested network. If so, we

---

[4] https://www.gurobi.com/.

[5] A trained defense refers to a defense method against adversarial samples, with the purpose of improving the robustness property of the network.

claim that the robustness property is verified. On the contrary, if we detect a counter-example with a different classification label, we assert the falsification of the robustness property. Finally, if we fail to conclude with verification or falsification, we return *unknown* to the user, meaning that the analysis result is inconclusive. We set up a challenging perturbation ϵ for each network and show in Table 3.

Table 3. Experimental fully connected and convolutional networks

Network	Dataset	Type	ϵ	#Layer	#Neurons	Defense	Candidates
M_3_100	MNIST	FC	0.028	3	210	None	98
M_5_100	MNIST	FC	0.08	6	510	DiffAI	98
M_6_100	MNIST	FC	0.025	6	510	None	99
M_9_100	MNIST	FC	0.023	9	810	None	97
M_6_200	MNIST	FC	0.016	6	1,010	None	99
M_9_200	MNIST	FC	0.015	9	1,610	None	97
M_convSmall	MNIST	Conv	0.11	3	3,604	None	100
M_convMed	MNIST	Conv	0.1	3	5,704	None	100
M_convBig	MNIST	Conv	0.306	6	48,064	DiffAI [20]	95
C_6_500	CIFAR10	FC	0.0032	6	3,000	None	56
C_convMed	CIFAR10	Conv	0.006	3	7,144	None	67

4.2   Comparison with the CPU-Based Verifiers

We present the robustness analysis results for ten networks in Table 3 and we describe the parameter configurations of our tool ARENA in Appendix C in our technical report. To execute PRIMA, we use the "refinepoly" domain in ERAN [15]. We report the experiment results drawn from different tools for MNIST and CIFAR10 networks in Table 4. For ARENA, we report the number of verified images, the number of falsified images and the average execution time for each testing image. DeepSRGR does not detect adversarial examples, neither does it attempt to assert violation of the property. PRIMA, on the other hand, returns two unsafe image for one MNIST network only (the detailed results and parameter setting are given in Appendix E in our report). Thus we omit the falsification column from the report for these two methods. Due to time limitation, for networks M_9_200, C_6_500 and C_convMed, we set a 2 h timeout for each image. If the refinement process fails to terminate before timeout, we consider the verification as inconclusive.

We observe from Table 4 that ARENA returns significantly more conclusive images (including both verified and falsified images) for all the networks than DeepSRGR, with comparable or even less execution time than that of Deep-SRGR. ARENA also returns more conclusive images for all the networks than PRIMA, except for the subject MNIST_3_100, where ARENA returns less verified images than PRIMA. Our in-depth investigation reveals that it is because

Table 4. The number of verified/falsified images and average execution time (in seconds) per image for MNIST and CIFAR10 network experiments

Neural Net	ARENA			DeepSRGR		PRIMA		DeepPoly	
	Verify	Falsify	Time	Verify	Time	Verify	Time	Verify	Time
M_3_100	63	5	87.65	54	68.76	69	123.73	24	0.105
M_5_100	77	7	250.39	67	153.75	53	19.15	25	0.522
M_6_100	45	6	650.10	38	324.14	38	173.03	23	0.280
M_9_100	44	10	1527.2	34	1004.4	34	191.60	30	0.587
M_6_200	48	3	1514.2	35	1312.3	34	222.45	25	0.313
M_9_200	43	6	3857.8	35	3536.7	29	238.63	29	0.536
M_convSmall	69	7	176.93	66	251.27	70	84.23	31	0.605
M_convMed	66	5	2054.9	60	2826.6	59	125.88	24	1.646
C_6_500	31	9	2703.3	24	3985.2	20	269.96	16	12.22
C_convMed	31	7	3417.1	30	4385.4	30	230.74	21	3.87

two out of the three hidden layers have their ReLU neurons being encoded *exactly* with MILP in PRIMA.

These analysis results are better visualized in Fig. 9 and 10 in Appendix D in our technical report. As can be seen in Appendix D, ARENA generally returns more conclusive images than the rest of the tools. On average, ARENA returns 15.8% more conclusive images than DeepSRGR and 16.6% more conclusive images than PRIMA respectively for the testing networks. In summary, to the best of our knowledge, ARENA outperforms the current state-of-the-art approximated methods that run on CPU.

4.3 Comparison with the GPU-Based Verifier α, β-CROWN

Furthermore, we compare with the state-of-the-art tool α, β-CROWN (alpha-beta-CROWN) [13]. Note that this is a *complete* verification tool in the sense that it will produce a conclusive answer given sufficient amount of time.

We started our experiments with the version of α, β-CROWN available in August 2022. We report in detail here the average execution time for each image, the number of verified images and falsified images in Table 5 with five networks. We rerun our experiments with the availability of the November 2022 version of α, β-CROWN for all tested networks and present the results in Table 6. In terms of execution speed, we observe that α, β-CROWN is much superior to ARENA, mainly due to the deployment of GPU acceleration. In terms of the numbers of verified and falsified images, we note that ARENA can verify or falsify 79.3% of that of α, β-CROWN on average, for the upper seven subject tests. For the last four subject tests, α, β-CROWN introduces MIP encoding in their solution to capture *exact* ReLU functionality, and thus further enhancing the number of verified images. We are currently investigating techniques for implementing ARENA on GPU, with the goal to improve the number of verified/falsified images with reasonable time bound.

Table 5. The number of verified/falsified images and average execution time for ARENA and α, β-CROWN (version dated Aug 2022), time is presented in *seconds*

Neural Net	ARENA			α, β-CROWN		
	Verify	Falsify	Average time	Verify	Falsify	Average time
M_3_100	63	5	87.65	54	11	25.15
M_5_100	77	7	250.3	53	10	48.67
M_convSmall	69	7	176.9	39	16	3.06
M_convMed	66	5	1625.9	30	18	2.90
M_convBig	53	30	589.52	49	24	4.21

Table 6. The number of verified/falsified images and average execution time of all tested networks for ARENA and α, β-CROWN (version dated Nov 2022), time is presented in *seconds*

Neural Net	ARENA			α, β-CROWN		
	Verify	Falsify	Average time	Verify	Falsify	Average time
M_3_100	63	5	87.65	81	13	40.81
M_5_100	77	7	250.3	87	12	30.89
C_6_500	31	9	2703.3	21	20	600.2
C_convMed	31	7	3417.1	36	22	283.7
M_convSmall	69	7	176.9	83	16	9.10
M_convMed	66	5	1625.9	82	18	7.24
M_convBig	53	30	589.52	60	29	109.17
M_6_100	45	6	650.1	82	8	182.44
M_6_200	48	3	1514.2	87	4	313.37
M_9_100	44	10	1527.2	77	13	278.45
M_9_200	43	6	3857.8	79	9	455.53

4.4   Multi-adversarial Label Parameter Study

In this experiment, we selected three networks from Table 3 to assess how the batch size parameter δ may impact the verification precision and execution time.

Theoretically, a larger value of δ may lead to a more efficient analysis process as it allows more adversarial regions to be eliminated at the same time. However, setting the parameter to be δ requires the computation of the union of δ convex polytopes. This in turn may introduce more over-approximation error and may jeopardize the analysis precision.

As δ aims to speed up the refinement process, we present the number of images that are verified through iterative refinement process and the average verification time for *those refined images*. As we only apply the refinement process to those testing images that DeepPoly fails to verify, the *refined images*

Table 7. The number of verified images through the refinement process (VTR) and average verification time per refined image for different δ setting.

Network	ARENA									
	(disabled)		($\delta = 2$)		($\delta = 3$)		($\delta = 4$)		($\delta = 5$)	
	VTR	Time(s)	VTR	Time(s)	VTR	Time(s)	VTR	Time(s)	VTR	Time(s)
M_3_100	41	142.91	39	144.67	39	135.62	39	124.99	38	127.77
M_6_100	22	1414.4	22	1312.6	22	1250.5	22	1202.4	22	1047.9
M_6_200	26	4809.7	23	2552.2	23	1828.2	23	1663.2	23	1297.0

refer to those images that are successfully verified through our refinement process, NOT through the original DeepPoly process. The experimental results are shown in Table 7 where we compare among parameters $\delta = 2, 3, 4, 5$ and with multi-adversarial label feature being disabled (the same as setting $\delta = 1$).

The experiment results show that the choice of a larger δ still allows us to achieve closely comparable precision while requires less execution time. Since an appropriate set-up of parameters leads to a better combination of precision and efficiency, we describe our configuration of each tested network in Appendix C in our technical report.

5 Discussion

We now discuss the limitation of our work. As described in Sect. 3.2, our batch size parameter δ is bounded to 5 at maximum for both precision and time-efficiency concern. In consideration for precision solely, as we compute the over-approximate convex hull of the union of multiple convex polytopes, the process will inevitably introduce additional over-approximate error into the LP encoding, yielding coarser neuron intervals. Thus we bound the value of δ to mitigate the degree of precision sacrifice. For time-efficiency issue, the transformation between V-representation and H-representation (refer to Sect. 3.2) – in either direction – is generally NP-hard, thus incurring exponential overhead with larger dimensions. As the parameter δ yields a $(\delta + 1)$-dimensional space, it is advisable to keep δ-value small so that the convex hull computation process will not become an execution bottleneck. For future work, we will explore the possibility of assigning δ dynamically for different networks to strike a better trade-off between speed and precision.

Our proposed refinement process could be applied to other verification techniques for improved precision, as long as they use linear constraints to approximate the underlying network [6,8,9].

6 Related Work

Network verification methods can be generally categorized as complete or incomplete methods. Complete methods conduct exact analysis over the network, especially for ReLU-activated networks. Given adequate time and resources, the complete methods return deterministic verification or violation of the robustness property to the user. Typical existing works are usually SMT (satisfiability modulo theory) based, MILP (mixed integer liner program) based or branch and bound (BaB) based [3,4,21,22]. For instance, β-CROWN [22] is a GPU-based verifier which uses branch and bound method to enable exact reasoning over the ReLU activation function. Furthermore, β-CROWN could also perform as an incomplete verifier with early termination.

On the other hand, the incomplete methods choose to over-approximate the non-linearity of the network using abstract interpretation or bound propagation etc. [6,7,23,24]. They are faced with precision loss due to the over-approximation of network behaviour. Consequently, the analysis result becomes inconclusive when the incomplete verifiers fail to verify the property. To rectify this deficiency, researchers have proposed various techniques like [8–10]. The work in [8] presents a new convex relaxation method that considers multiple ReLUs jointly in order to capture the correlation between ReLU neurons in the same layer. This idea has been further developed in PRIMA [9] which reduces the complexity of ReLU convex abstraction computation via a novel convex hull approximation algorithm. In comparison, DeepSRGR [10] elects to refine the abstraction in an iterative manner, where it repeatedly uses the spurious regions to stabilize the ReLU neurons until the abstraction is precise enough to eliminate the adversarial label linked to that specified spurious region. In our work, we combine both these refinement methods [9,10] to break the precision barrier and also leverages the double-description method to retain efficiency as well.

7   Conclusion

We leverage the double description method in convex polytope area to compute the convex hull of the union of multiple polytopes, making it amenable for eliminating multiple adversarial labels simultaneously and boosting the analysis efficiency. Furthermore, we combine the convex relaxation technique with the iterative abstract refinement method to improve the precision in abstract interpretation based verification system. We implemented our prototypical analyzer ARENA to conduct both robustness verification and falsification. Experiment results show affirmatively that ARENA enhances abstract refinement techniques by attaining better verification precision compared to DeepSRGR, with reasonable execution time; it also competes favourably in comparison with PRIMA. Finally, it is also capable of detecting adversarial examples.

We believe that our proposed method can positively boost the effectiveness of sound but incomplete analyses and be applied to other methods that use linear constraints to approximate the network for effective precision enhancement.

Acknowledgement. This research is supported by a Singapore Ministry of Education Academic Research Fund Tier 1 T1-251RES2103 and the National Research Foundation, Singapore under its Emerging Areas Research Projects (EARP) Funding Initiative. Any opinions, findings and conclusions or recommendations expressed in this material are those of the author(s) and do not reflect the views of National Research Foundation, Singapore. We are grateful to Julian Rüth saraedum and Komei Fukuda for their prompt answer to our queries on cddlib. And we appreciate Mark Niklas Müller's assistance to our queries on PRIMA.

References

1. Ren, K., Zheng, T., Qin, Z., Liu, X.: Adversarial attacks and defenses in deep learning. Engineering **6**(3), 346–360 (2020)
2. Yuan, X., He, P., Zhu, Q., Li, X.: Adversarial examples: attacks and defenses for deep learning. IEEE Trans. Neural Netw. Learn. Syst. **30**(9), 2805–2824 (2019)
3. Tjeng, V., Xiao, K., Tedrake, R.: Evaluating robustness of neural networks with mixed integer programming. In: International Conference on Learning Representations (ICLR). OpenReview.net (2019)
4. Katz, G., Barrett, C., Dill, D.L., Julian, K., Kochenderfer, M.J.: Reluplex: an efficient SMT solver for verifying deep neural networks. In: Majumdar, R., Kunčak, V. (eds.) CAV 2017. LNCS, vol. 10426, pp. 97–117. Springer, Cham (2017). https://doi.org/10.1007/978-3-319-63387-9_5
5. Pulina, L., Tacchella, A.: An abstraction-refinement approach to verification of artificial neural networks. In: Touili, T., Cook, B., Jackson, P. (eds.) CAV 2010. LNCS, vol. 6174, pp. 243–257. Springer, Heidelberg (2010). https://doi.org/10.1007/978-3-642-14295-6_24
6. Gehr, T., Mirman, M., Drachsler-Cohen, D., Tsankov, P., Chaudhuri, S., Vechev, M.: AI2: safety and robustness certification of neural networks with abstract interpretation. In: IEEE Symposium on Security and Privacy (SP), pp. 3–18. IEEE Computer Society (2018)
7. Singh, G., Gehr, T., Püschel, M., Vechev, M.T.: An abstract domain for certifying neural networks. Proc. ACM Program. Lang. **3**(POPL), 41:1–41:30 (2019)
8. Singh, G., Ganvir, R., Püschel, M., Vechev, M.T.: Beyond the single neuron convex barrier for neural network certification. In: Wallach, H.M., Larochelle, H., Beygelzimer, A., d'Alché-Buc, F., Fox, E.B., Garnett, R., (eds.) Advances in Neural Information Processing Systems, vol. 32: Annual Conference on Neural Information Processing Systems 2019, NeurIPS 2019, December 8–14, 2019, Vancouver, BC, Canada, pp. 15072–15083 (2019)
9. Müller, M.N., Makarchuk, G., Singh, G., Püschel, M., Vechev, M.T.: PRIMA: general and precise neural network certification via scalable convex hull approximations. Proc. ACM Program. Lang. **6**(POPL), 1–33 (2022)
10. Yang, P.: Improving neural network verification through spurious region guided refinement. In: Groote, J.F., Larsen, K.G. (eds.) TACAS 2021. LNCS, vol. 12651, pp. 389–408. Springer, Cham (2021). https://doi.org/10.1007/978-3-030-72016-2_21
11. Clarke, E., Grumberg, O., Jha, S., Lu, Y., Veith, H.: Counterexample-guided abstraction refinement. In: Emerson, E.A., Sistla, A.P. (eds.) CAV 2000. LNCS, vol. 1855, pp. 154–169. Springer, Heidelberg (2000). https://doi.org/10.1007/10722167_15
12. Fukuda, K., Prodon, A.: Double description method revisited. In: Deza, M., Euler, R., Manoussakis, I. (eds.) CCS 1995. LNCS, vol. 1120, pp. 91–111. Springer, Heidelberg (1996). https://doi.org/10.1007/3-540-61576-8_77

13. CMU. Alpha-Beta-CROWN: a fast and scalable neural network verifier with efficient bound propagation (2022). https://github.com/huanzhang12/alpha-beta-CROWN. Accessed 11 Aug 2022
14. McMullen, P.: Convex polytopes, by Branko Grunbaum, second edition (first edition (1967) written with the cooperation of V. L. Klee, M. Perles and G. C. Shephard. Comb. Probab. Comput. **14**(4), 623–626 (2005)
15. ETH. ETH Robustness Analyzer for Neural Networks (ERAN) (2022). https://github.com/eth-sri/eran. Accessed 11 Aug 2022
16. 3rd International Verification of Neural Networks Competition (VNN-COMP'22) (2022). https://sites.google.com/view/vnn2022. Accessed 11 Aug 2022
17. LeCun, Y., Cortes, C.: MNIST handwritten digit database (2010)
18. Krizhevsky, A., Nair, V., Hinton, G.: Cifar-10 (Canadian institute for advanced research)
19. Carlini, N., Wagner, D.: Towards evaluating the robustness of neural networks. In: IEEE Symposium on Security and Privacy (SP), pp. 39–57 (2017)
20. Mirman, M., Gehr, T., Vechev, M.T.: Differentiable abstract interpretation for provably robust neural networks. In: International Conference on Machine Learning (ICML), pp. 3575–3583 (2018)
21. Botoeva, E., Kouvaros, P., Kronqvist, J., Lomuscio, A., Misener, R.: Efficient verification of relu-based neural networks via dependency analysis. In: The Thirty-Fourth AAAI Conference on Artificial Intelligence, AAAI 2020, The Thirty-Second Innovative Applications of Artificial Intelligence Conference, IAAI 2020, The Tenth AAAI Symposium on Educational Advances in Artificial Intelligence, EAAI 2020, New York, NY, USA, 7–12 February 2020, pp. 3291–3299. AAAI Press (2020)
22. Wang, S.: Beta-crown: efficient bound propagation with per-neuron split constraints for complete and incomplete neural network verification. CoRR, abs/2103.06624 (2021)
23. Zhang, H., Weng, T.W., Chen, P.Y., Hsieh, C.J., Daniel, L.: Efficient neural network robustness certification with general activation functions. In: Bengio, S., Wallach, H.M., Larochelle, H., Grauman, K., Cesa-Bianchi, N., Garnett, R., (eds.) Advances in Neural Information Processing Systems Annual Conference on Neural Information Processing Systems 2018, NeurIPS 2018, 3–8 December 2018, Montréal, Canada, vol. 31, pp. 4944–4953 (2018)
24. Wang, S., Pei, K., Whitehouse, J., Yang, J., Jana, S.: Formal security analysis of neural networks using symbolic intervals. In: Enck, W., Felt, A.P., (eds.) 27th USENIX Security Symposium, USENIX Security 2018, Baltimore, MD, USA, 15–17 August 2018, pp. 1599–1614. USENIX Association (2018)

Correction to: ARENA: Enhancing Abstract Refinement for Neural Network Verification

Yuyi Zhong, Quang-Trung Ta, and Siau-Cheng Khoo

Correction to:
Chapter 17 in: C. Dragoi et al. (Eds.): *Verification,*
***Model Checking, and Abstract Interpretation*, LNCS 13881,**
https://doi.org/10.1007/978-3-031-24950-1_17

In the originally published version of chapter 17, the acknowledgment section had been rendered incorrectly. This has been corrected.

The updated version of this chapter can be found at
https://doi.org/10.1007/978-3-031-24950-1_17

Author Index

Printed in the United States
by Baker & Taylor Publisher Services